Introduction to Lorentz Geometry

Introduction to Lorentz Geometry

Curves and Surfaces

Ivo Terek Couto
Alexandre Lymberopoulos

CRC Press
Taylor & Francis Group
Boca Raton London New York

CRC Press is an imprint of the
Taylor & Francis Group, an **informa** business
A CHAPMAN & HALL BOOK

First edition published 2021
by CRC Press
6000 Broken Sound Parkway NW, Suite 300, Boca Raton, FL 33487-2742

and by CRC Press
2 Park Square, Milton Park, Abingdon, Oxon, OX14 4RN

CRC Press is an imprint of Taylor & Francis Group, LLC

Library of Congress Cataloging-in-Publication Data

Names: Couto, Ivo Terek, author. | Lymberopoulos, Alexandre, author.
Title: Introduction to Lorentz geometry : curves and surfaces / Ivo Terek Couto, Alexandre Lymberopoulos.
Other titles: Introdução à geometria Lorentziana. English.
Description: First edition. | Boca Raton : C&H/CRC Press, 2020. | Translation of: Introdução à Geometria Lorentziana. | Includes bibliographical references and index.
Identifiers: LCCN 2020028007 (print) | LCCN 2020028008 (ebook) | ISBN 9780367468644 (hardback) | ISBN 9781003031574 (ebook)
Subjects: LCSH: Geometry, Differential. | Lorentz transformations. | Curves. | Surfaces. | Mathematical physics.
Classification: LCC QA641 .C65 2020 (print) | LCC QA641 (ebook) | DDC 516.3/6--dc23
LC record available at https://lccn.loc.gov/2020028007
LC ebook record available at https://lccn.loc.gov/2020028008

ISBN: 978-0-367-46864-4 (hbk)
ISBN: 978-0-367-62411-8 (pbk)
ISBN: 978-1-003-03157-4 (ebk)

Contents

Preface of the Portuguese Version vii

Preface ix

CHAPTER 1 ▪ Welcome to Lorentz-Minkowski Space 1

1.1 PSEUDO–EUCLIDEAN SPACES 2
1.1.1 Defining $\mathbb{R}^n_\nu$ 2
1.1.2 The causal character of a vector in $\mathbb{R}^n_\nu$ 3
1.2 SUBSPACES OF $\mathbb{R}^n_\nu$ 4
1.3 CONTEXTUALIZATION IN SPECIAL RELATIVITY 19
1.4 ISOMETRIES IN $\mathbb{R}^n_\nu$ 29
1.5 INVESTIGATING $O_1(2,\mathbb{R})$ AND $O_1(3,\mathbb{R})$ 43
1.5.1 The group $O_1(2,\mathbb{R})$ in detail 43
1.5.2 The group $O_1(3,\mathbb{R})$ in (a little less) detail 44
1.5.3 Rotations and boosts 47
1.6 CROSS PRODUCT IN $\mathbb{R}^n_\nu$ 53
1.6.1 Completing the toolbox 57

CHAPTER 2 ▪ Local Theory of Curves 63

2.1 PARAMETRIZED CURVES IN $\mathbb{R}^n_\nu$ 64
2.2 CURVES IN THE PLANE 77
2.3 CURVES IN SPACE 97
2.3.1 The Frenet-Serret trihedron 98
2.3.2 Geometric effects of curvature and torsion 106
2.3.3 Curves with degenerate osculating plane 118

CHAPTER 3 ▪ Surfaces in Space 129

3.1 BASIC TOPOLOGY OF SURFACES 130
3.2 CAUSAL TYPE OF SURFACES, FIRST FUNDAMENTAL FORM 155
3.2.1 Isometries between surfaces 169
3.3 SECOND FUNDAMENTAL FORM AND CURVATURES 178

3.4 THE DIAGONALIZATION PROBLEM 190
3.4.1 Interpretations for curvatures 194
3.5 CURVES IN A SURFACE 207
3.6 GEODESICS, VARIATIONAL METHODS AND ENERGY 219
3.6.1 Darboux-Ribaucour frame 222
3.6.2 Christoffel symbols 228
3.6.3 Critical points of the energy functional 232
3.7 THE FUNDAMENTAL THEOREM OF SURFACES 250
3.7.1 The compatibility equations 250

CHAPTER 4 ■ Abstract Surfaces and Further Topics 259

4.1 PSEUDO-RIEMANNIAN METRICS 260
4.2 RIEMANN'S CLASSIFICATION THEOREM 280
4.3 SPLIT-COMPLEX NUMBERS AND CRITICAL SURFACES 286
4.3.1 A brief introduction to split-complex numbers 286
4.3.2 Bonnet rotations 298
4.3.3 Enneper-Weierstrass representation formulas 304
4.4 DIGRESSION: COMPLETENESS AND CAUSALITY 314

APPENDIX ■ Some Results from Differential Calculus 325

Bibliography 331

Index 335

Preface of the Portuguese Version

This book has the goal of simultaneously approaching the Differential Geometry of curves and surfaces in two distinct ambient spaces: Euclidean and Lorentzian. The essential difference between them is the the way of measuring lengths, with the former being the one that we are already used to. What is new is the study of such objects in the Lorentzian space, making systematic comparisons between the two cases.

Differential Geometry is an area of Mathematics which consists of the study of geometric properties of certain objects, by using techniques from Differential Calculus and Linear Algebra, occasionally employing more sophisticated tools.

Lorentzian Differential Geometry, in turn, is interesting not only for Mathematics itself, but it reaches Physics as well, being the mathematical language of General Relativity. The main motivation for preparing the present text was to present, in a sufficiently self-contained (and, we hope, simple) way, part of the vast bibliography on the subject, which is scattered in scientific papers and articles, often with non-uniform language and notation.

To begin our journey, we need to understand the "static" geometry of the Lorentzian space, studying the Linear Algebra of this ambient space. This is done in Chapter 1, which also presents results about the transformations that preserve this geometry. Such results are stated in parallel with results from Euclidean Geometry. Some interactions of the Lorentz-Minkowski space with the theory of Special Relativity also appear here.

In Chapter 2, we study in detail the geometry of curves. To do so, we'll combine what was covered in the previous chapter with results from single-variable Differential Calculus. The basic theory is the same, no matter the dimension of the ambient space, and we'll develop this in the first section, leaving the particular cases of the plane and three-dimensional space for the next sections.

To complete this introduction to Lorentzian Differential Geometry, we'll do a systematic study of surfaces in three-dimensional spaces, developing simultaneously the theory in both Euclidean and Lorentzian spaces, emphasizing the differences between the two cases, whenever those occur. Here, we will need results from the previous chapters, and tools from Multivariable Calculus. In the end of the book, we present an appendix with the main results used.

The fourth chapter has essentially three goals: showing the direction to which the previously seen results can be generalized and the modern way to approach Differential Geometry; establishing relations between certain surfaces and complex variables (or an equivalent, there defined); and, finally, briefly discussing the interpretation of General Relativity in this more abstract mathematical setting. Naturally, for this chapter, we expect more mathematical maturity, and some familiarity with the theory of complex variables and point-set topology.

All the sections in this book are accompanied by many exercises, and there are over

300 of them. Those that are more important for the development of the theory are indicated with a †, and many others are mentioned along the text. We do not expect the reader to find the time to work through all of the exercises, but we do expect all the statements to be at least read and understood.

Throughout the text, we will follow the usual notations from mathematical literature, but to make the reading easier, we will adopt bold font for vectors in the ambient space, while Greek letters are reserved for curves, and Roman letters for fixed vectors or parametrized surfaces. When the parameter of a curve has any special meaning, some alteration in font will be made to emphasize such change.

Lastly, several colleagues have contributed suggestions and constructive criticism to improve the presentation and content of the book. In particular, we would like to thank Prof. Antonio de Padua Franco Filho for the careful reading of the manuscript. We would also like to acknowledge FAPESP and CNPq for the partial support. And, to all, our sincere thanks.

São Paulo, June of 2018

Ivo Terek Couto
Alexandre Lymberopoulos

Preface

This is a translation of the Portuguese version originally published by the Brazilian Mathematical Society in 2018. The organization of the text in chapters and sections remains the same as in the original version. So, what has changed? During the translation, several small corrections were made here and there, but especially in statements of exercises which were not precise enough, or missing important details. And this time, there is a tentative solution manual available for instructors, at the following address:

`www.routledge.com/9780367468644`

Still on the topic of exercises, we cannot miss the opportunity to emphasize that while the main goal of this book is indeed to introduce Differential Geometry of Curves and Surfaces to students, but with a (Lorentzian) twist, this text can nevertheless be used for teaching a "vanilla" course (as the reader should note by setting $\nu = 0$ and $\epsilon_\alpha = \eta_\alpha = \epsilon_M = 1$ in everything to appear). If this path is chosen, most of Chapter 1 may be omitted, since a standard pre-requisite Linear Algebra class should have already covered most of the important results regarding positive-definite inner products. Section 1.4, however, is crucial (as it deals with the notion of isometry). Proceeding to Chapters 2 and 3, Subsection 2.3.3 and the first half of Section 3.4 are *purely* Lorentzian, and thus could be skipped. Topics in Chapter 4 could serve as ideas for student presentations in small classes, according to the instructor's discretion.

Lastly, we would like to thank Prof. Jan Lang, not only for finally convincing us to try and publish this English version of the text, but also for actually suggesting to work with CRC Press, whose outstanding support and professionalism were essential throughout the whole translation process.

Columbus, OH — Ivo Terek Couto
São Paulo, SP — Alexandre Lymberopoulos

CHAPTER 1

Welcome to Lorentz-Minkowski Space

INTRODUCTION

In this chapter we define the pseudo-Euclidean spaces $\mathbb{R}^n_\nu$ and, in particular, the Lorentz-Minkowski space $\mathbb{L}^n = \mathbb{R}^n_1$, the ambient where the geometric objects studied throughout this book are contained. Beyond definition and terminology, we provide the main results from Linear Algebra derived from this new way to measure angles and lengths.

In Section 1.1 we present the space $\mathbb{L}^n$ and the causal character of its elements, as well as a geometric interpretation of the position of vectors and their causal character when $n = 2$ or $n = 3$.

In Section 1.2 the concept of orthogonality is introduced in this context and we extend the definition of causal character, previously made for vectors, to subspaces of $\mathbb{L}^n$. We use a Sylvester-like criterion to determine the causal character of a subspace, and we establish a relation between the causal characters of a subspace and its orthogonal "complement" (the reason for the quotation marks used here will also be explained). We finish this section with a modified Gram-Schmidt process and establish conditions for the existence of orthonormal bases for subspaces of $\mathbb{R}^n_\nu$.

Continuing into Section 1.3, we focus on a physical feature of $\mathbb{L}^4$: it is the model of a spacetime free of gravity. We provide a physical interpretation of the causal character of vectors and discuss, in brief, causal and temporal cones, as well as the concepts of causal and chronological precedence. The new versions of the Cauchy-Schwarz and triangle inequalities are used to explain the Twin Paradox, a highlight in Special Relativity. Surprising results are revealed here, such as the fact that two lightlike vectors are Lorentz-orthogonal if and only if they are parallel. To close this section, we state the Alexandrov-Zeeman Theorem, which classifies all the causal automorphisms of $\mathbb{L}^n$, for $n > 2$, and a family of counterexamples to this theorem, when $n = 2$.

Next, Section 1.4 covers the isometries of pseudo-Euclidean spaces. Those functions are essential to study curves and surfaces in terms of their geometric invariants. We study Lorentz transformations from the mathematical point of view, focusing on the group structure defined by the composition operation. To do this, we need the concepts of proper transformations and orthochronous transformations (they preserve, respectively, spatial and time orientations).

The groups $\mathrm{O}_1^{+\uparrow}(2,\mathbb{R})$ and $\mathrm{O}_1^{+\uparrow}(3,\mathbb{R})$ play a major role in our theory. We give them special attention in Section 1.5, where their complete classification is given. With this in place, we turn our attention to two classes of isometries: pure rotations and Lorentz boosts, which are very important in Physics. In the end, we provide an alternative characterization of proper orthochronous Lorentz transformations.

The final section in this chapter aims to extend the usual cross product operation in $\mathbb{R}^3$ to $\mathbb{R}^n_\nu$. We establish its basic properties and how it is related to the algebraic orientation of the space, pointing out the influence of the causal character of vectors on it. Lastly, some technical results are shown. They will be useful in the study of surface curvatures, to be studied in Chapter 3.

1.1 PSEUDO–EUCLIDEAN SPACES

1.1.1 Defining $\mathbb{R}^n_\nu$

Let $n \geq 2$ and consider the bilinear form $\langle \cdot, \cdot \rangle_\nu : \mathbb{R}^n \times \mathbb{R}^n \to \mathbb{R}$ given by:

$$\langle x, y \rangle_\nu \doteq x_1 y_1 + \cdots + x_{n-\nu} y_{n-\nu} - x_{n-\nu+1} y_{n-\nu+1} - \cdots - x_n y_n, \qquad (\dagger)$$

where $x = (x_1, \ldots, x_n)$ and $y = (y_1, \ldots, y_n)$.

Writing the canonical basis[1] of $\mathbb{R}^n$ as $\text{can} = (e_1, \ldots, e_n)$, the matrix of $\langle \cdot, \cdot \rangle_\nu$ relative to such basis is

$$\mathrm{Id}_{n-\nu,\nu} \doteq \left(\begin{array}{c|c} \mathrm{Id}_{n-\nu} & 0 \\ \hline 0 & -\mathrm{Id}_\nu \end{array} \right).$$

In short, we write $\mathrm{Id}_{n-\nu,\nu} = (\eta^\nu_{ij})^n_{i,j=1}$, where η^ν_{ij} is analogous to the Kronecker's delta[2] in this setting. Hence, if $u, v \in \mathbb{R}^n$ are column vectors, we write $\langle u, v \rangle_\nu = u^\top \mathrm{Id}_{n-\nu,\nu}\, v$. The matrix $\mathrm{Id}_{n-\nu,\nu}$ says that while $\langle \cdot, \cdot \rangle_\nu$ is not a positive-definite inner product in $\mathbb{R}^n$, it is still symmetric and non-degenerate[3].

Throughout the literature, there are some differences in terminology. For instance, in [54] a bilinear form is an *inner product* if it is bilinear, symmetric, and positive-definite; if not positive-definite, but still non-degenerate, it is called a *scalar product.* With this, we would say that $\langle \cdot, \cdot \rangle_\nu$ is just a scalar product. In general, we will refer to both cases as an *inner product*, *Euclidean* when positive-definite and *pseudo-Euclidean* when non-degenerate.

Definition 1.1.1. The *pseudo-Euclidean space of index ν*, from now on denoted by $\mathbb{R}^n_\nu$, is the vector space $\mathbb{R}^n$ with its usual sum and scalar multiplication, equipped with the inner product $\langle \cdot, \cdot \rangle_\nu$ defined in ($\dagger$). In short, we write $\mathbb{R}^n_\nu \doteq (\mathbb{R}^n, \langle \cdot, \cdot \rangle_\nu)$. The main ambient space in this book is $\mathbb{L}^n \doteq \mathbb{R}^n_1$, called the *Lorentz-Minkowski space.*

Remark.

- The inner products of $\mathbb{R}^n \equiv \mathbb{R}^n_0$ and $\mathbb{L}^n$ will be denoted by $\langle \cdot, \cdot \rangle_E$ and $\langle \cdot, \cdot \rangle_L$, respectively. The product $\langle \cdot, \cdot \rangle_L$ is called *Lorentz inner product*, or *Minkowski metric.*
- It is usual in the literature to define $\langle \cdot, \cdot \rangle_\nu$ with the ν minus signs in the first ν terms of the sum ($\dagger$), instead of in the last ν terms. This is no more than an axis permutation.
- The space $\mathbb{L}^4$ is the simplest model where the Special Relativity theory may be developed, that is, a gravity-free spacetime. Its points are usually called *events.*

[1] This notation will be consistent through this chapter.

[2] $\delta_{ij} = 1$, if $i = j$, and 0 if $i \neq j$.

[3] That is, if $\langle v, u \rangle = 0$ for all u, then $v = \mathbf{0}$.

Definition 1.1.2. The *pseudo-Euclidean norm* is the map $\|\cdot\|_\nu : \mathbb{R}^n_\nu \to \mathbb{R}$ given by $\|\boldsymbol{v}\|_\nu \doteq \sqrt{|\langle \boldsymbol{v}, \boldsymbol{v}\rangle_\nu|}$. We say that $\boldsymbol{v}$ is a *unit vector* if $\|\boldsymbol{v}\|_\nu = 1$.

It is important to notice that "norm" in the above definition is an abuse of language, since $\|\cdot\|_\nu$ is not a norm: there are non-zero vectors such that its pseudo-Euclidean norm vanishes and vectors for which the triangular inequality does not hold. As examples in $\mathbb{L}^3$, we have

$$\|(1,0,-1)\|_L = 0 \text{ and } \|(1,1,-2)\|_L = \sqrt{2} \geq 0 = \|(1,0,-1)\|_L + \|(0,1,-1)\|_L.$$

Despite these differences when compared to the Euclidean norm $\|\cdot\|_E$, the pseudo-Euclidean norms satisfy $\|\lambda \boldsymbol{v}\|_\nu = |\lambda|\|\boldsymbol{v}\|_\nu$ for all $\boldsymbol{v} \in \mathbb{R}^n_\nu$ and $\lambda \in \mathbb{R}$.

1.1.2 The causal character of a vector in $\mathbb{R}^n_\nu$

Definition 1.1.3 (Causal Character or Causal Type)**.** A vector $\boldsymbol{v} \in \mathbb{R}^n_\nu$ is:

(i) *spacelike*, if $\langle \boldsymbol{v}, \boldsymbol{v}\rangle_\nu > 0$, or $\boldsymbol{v} = \boldsymbol{0}$;

(ii) *timelike*, if $\langle \boldsymbol{v}, \boldsymbol{v}\rangle_\nu < 0$;

(iii) *lightlike*, if $\langle \boldsymbol{v}, \boldsymbol{v}\rangle_\nu = 0$ and $\boldsymbol{v} \neq \boldsymbol{0}$.

For any non-lightlike vector $\boldsymbol{v} \neq \boldsymbol{0}$, we set its *indicator* $\epsilon_{\boldsymbol{v}}$ to be the sign of $\langle \boldsymbol{v}, \boldsymbol{v}\rangle_\nu$, that is, $\epsilon_{\boldsymbol{v}} = 1$ if $\boldsymbol{v}$ is spacelike, and $\epsilon_{\boldsymbol{v}} = -1$ if $\boldsymbol{v}$ is timelike.

Remark.

- For $\nu = 1$, the nomenclature comes from Physics. Such motivations will be further explained in Section 1.3. In that context, lightlike vectors and the vector $\boldsymbol{0}$ are called *null vectors* (we won't use that convention).
- For our convenience we will consider all the vectors in $\mathbb{R}^n$ to be spacelike. In particular, any non-zero vector in $\mathbb{R}^n$ has indicator 1.

Example 1.1.4. In $\mathbb{L}^3$ we have:

(1) $(4,-1,0)$ is spacelike;

(2) $(1,2,3)$ is timelike;

(3) $(3,4,5)$ is lightlike.

Remark. The causal type of a vector remains the same, up to a change of sign in any of its entries. Furthermore, $\boldsymbol{v}$ and $\lambda\boldsymbol{v}$ share the same causal type for any $\boldsymbol{v} \in \mathbb{R}^n_\nu$ and $\lambda \in \mathbb{R} \setminus \{0\}$.

It is important to notice that the trichotomy property of the usual order relation in $\mathbb{R}$ ensures that each vector in $\mathbb{R}^n_\nu$ has one and only one of the three causal types above. To provide some geometric intuition for $\mathbb{L}^2$ and $\mathbb{L}^3$ (where we can actually draw something), we need to know the locus of all the vectors with a fixed causal type:

(1) if $\boldsymbol{v} = (x,y) \in \mathbb{L}^2$ and $c \in \mathbb{R}$, then $\langle \boldsymbol{v}, \boldsymbol{v}\rangle_L = c \iff x^2 - y^2 = c$, which is the equation of:

- the bisectors of the coordinated axes, if $c = 0$;

- a hyperbola with foci on the x-axis, if $c > 0$;
- a hyperbola with foci on the y-axis, if $c < 0$;

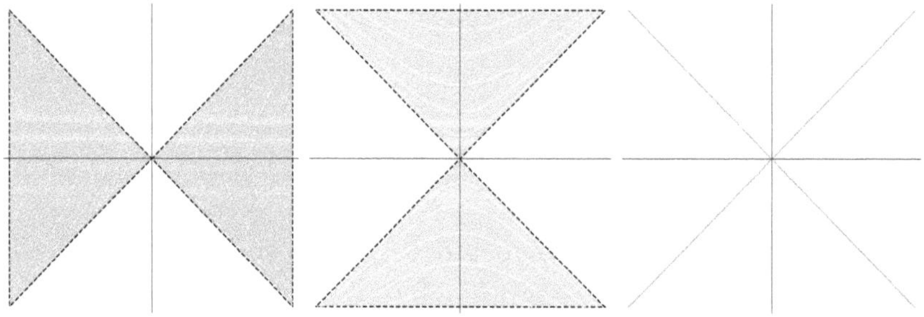

Figure 1.1: Causal type loci in $\mathbb{L}^2$.

(2) if $v = (x,y,z) \in \mathbb{L}^3$ and $c \in \mathbb{R}$, then $\langle v, v\rangle_L = c \iff x^2+y^2-z^2 = c$, which is the equation of:

- a cone, if $c = 0$;
- a one-sheeted hyperboloid, if $c > 0$;
- a two-sheeted hyperboloid, if $c < 0$.

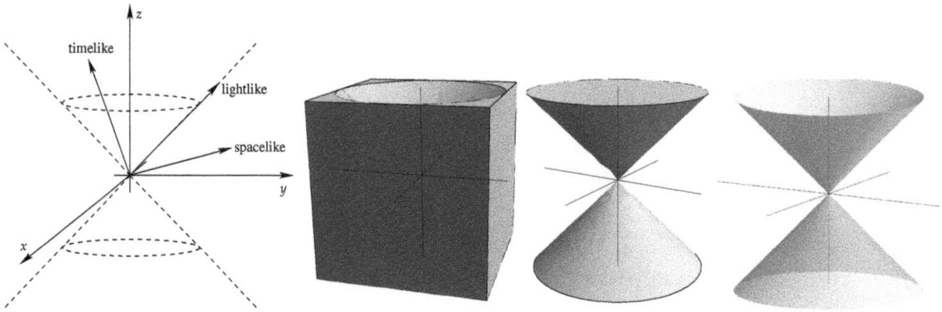

Figure 1.2: Causal type loci in $\mathbb{L}^3$.

The set of all lightlike vectors is called the *lightcone* (centered at $\mathbf{0}$). In particular, the lightcone in $\mathbb{L}^2$ is just the bisectors of the axes. In Section 1.3 we will discuss some interpretations and applications of such concepts in Special Relativity.

1.2 SUBSPACES OF $\mathbb{R}^n_\nu$

The definition of a vector subspace does not take inner products into account. Thus, subspaces of $\mathbb{L}^3$ are precisely the trivial ones (the origin and $\mathbb{L}^3$ itself), as well as the straight lines and planes containing the $\mathbf{0}$ vector. We can define the causal type of subspaces in $\mathbb{L}^n$ and, as a first motivation, we recall from Linear Algebra the:

Proposition 1.2.1. *If V is a finite-dimensional vector space with a Euclidean inner product, then, for any subspace $U \subseteq V$, we have*

$$\dim V = \dim U + \dim U^\perp \quad \textit{and} \quad V = U \oplus U^\perp.$$

We won't bother proving this at this moment, as very soon we will see what happens for pseudo-Euclidean vector spaces, where in general just the equation for the dimensions holds. We have a proof for this fact in Proposition 1.2.14. In any case, this situation leads us to discuss orthogonality in $\mathbb{L}^n$. As in Euclidean vector spaces, we have:

Definition 1.2.2 (Orthogonality).

(i) Two vectors $\boldsymbol{u}, \boldsymbol{v} \in \mathbb{R}^n_\nu$ are *pseudo-orthogonal* if $\langle \boldsymbol{u}, \boldsymbol{v} \rangle_\nu = 0$. In this case we write $\boldsymbol{u} \perp_\nu \boldsymbol{v}$.

(ii) An arbitrary basis $\mathscr{B}$ of $\mathbb{R}^n_\nu$ is *pseudo-orthogonal* if its vectors are pairwise pseudo-orthogonal.

(iii) A pseudo-orthogonal basis $\mathscr{B}$ of $\mathbb{R}^n_\nu$ is *pseudo-orthonormal* if $\|\boldsymbol{v}\|_\nu = 1$ for each $\boldsymbol{v} \in \mathscr{B}$.

(iv) Let $S \subseteq \mathbb{R}^n_\nu$ be any set. The subspace of $\mathbb{R}^n_\nu$ *pseudo-orthogonal to* S is defined as

$$S^\perp \doteq \{\boldsymbol{v} \in \mathbb{R}^n_\nu \mid \langle \boldsymbol{u}, \boldsymbol{v} \rangle_\nu = 0, \text{ for all } \boldsymbol{u} \in S\}.$$

If $\nu = 1$, we say that pseudo-orthogonal vectors are *Lorentz–orthogonal*.

Remark.

- In the above definition, if $S = \{\boldsymbol{v}\}$ is a singleton, we just write $\boldsymbol{v}^\perp$ instead of $\{\boldsymbol{v}\}^\perp$.
- If $\boldsymbol{u}$ and $\boldsymbol{v}$ are orthogonal with respect to the usual Euclidean inner product, we write $\boldsymbol{u} \perp_E \boldsymbol{v}$. Similarly, we write $\boldsymbol{u} \perp_L \boldsymbol{v}$ in Lorentz-Minkowski space. In any case, if the context is clear enough, we avoid any mention to the ambient space, simply writing $\boldsymbol{u} \perp \boldsymbol{v}$.
- Given any basis $\mathscr{B} = (\boldsymbol{u}_1, \ldots, \boldsymbol{u}_n)$ of $\mathbb{L}^n$, where $\boldsymbol{u}_n$ is timelike, then $\mathscr{B}$ is orthonormal if and only if $\langle \boldsymbol{u}_i, \boldsymbol{u}_j \rangle_L = \eta^1_{ij}$, for all $1 \leq i, j \leq n$.

To define the causal character of subspaces in $\mathbb{L}^n$, we recall the following general definition:

Definition 1.2.3. Let V be any real vector space and $B : V \times V \to \mathbb{R}$ a symmetric bilinear form. We say that

(i) B is *positive-definite* if $B(\boldsymbol{u}, \boldsymbol{u}) > 0$, for all $\boldsymbol{u} \neq \boldsymbol{0}$;

(ii) B is *negative-definite* if $-B$ is positive-definite;

(iii) B is *non-degenerate* if $B(\boldsymbol{u}, \boldsymbol{v}) = 0$ for all $\boldsymbol{v}$ implies $\boldsymbol{u} = \boldsymbol{0}$;

(iv) B is *indefinite* if there exists $\boldsymbol{u}, \boldsymbol{v} \in V$ such that $B(\boldsymbol{u}, \boldsymbol{u}) > 0$ and $B(\boldsymbol{v}, \boldsymbol{v}) < 0$.

Definition 1.2.4 (Causal Character). A vector subspace $\{\boldsymbol{0}\} \neq U \subseteq \mathbb{L}^n$ is:

(i) *spacelike*, if $\langle \cdot, \cdot \rangle_L|_U$ is positive-definite;

(ii) *timelike*, if $\langle \cdot, \cdot \rangle_L|_U$ is negative-definite, or indefinite and non-degenerate;

(iii) *lightlike*, if $\langle \cdot, \cdot \rangle_L|_U$ is degenerate.

Remark.

- In this setting, the restriction of $\langle \cdot, \cdot \rangle_L$ to U being positive-definite implies that all vectors in U are spacelike, although being negative-definite means that U is a straight line. If degenerate, then U has lightlike vectors, but no timelike ones. See Exercise 1.2.2.
- If $\dim U = 0$ then, by definition, U is spacelike.
- We define the causal character of an affine subspace as the causal type of its vector subspace counterpart.
- If $\nu > 1$, a subspace may not have a well-defined causal type. For instance, consider the subspace spanned by three orthogonal vectors in $\mathbb{R}^4_2$, one of each causal type.

Definition 1.2.5. Let $U \subseteq \mathbb{R}^n_\nu$ be an m-dimensional subspace and $\mathcal{B} = (\boldsymbol{u}_1, \cdots, \boldsymbol{u}_m)$ a basis for U. The *Gram matrix* of $\langle \cdot, \cdot \rangle_\nu \big|_U$ with respect to $\mathcal{B}$ is

$$G_{U,\mathcal{B}} \doteq \big(\langle \boldsymbol{u}_i, \boldsymbol{u}_j \rangle_\nu \big)_{1 \leq i,j \leq m} = \begin{pmatrix} \langle \boldsymbol{u}_1, \boldsymbol{u}_1 \rangle_\nu & \cdots & \langle \boldsymbol{u}_1, \boldsymbol{u}_m \rangle_\nu \\ \vdots & \ddots & \vdots \\ \langle \boldsymbol{u}_m, \boldsymbol{u}_1 \rangle_\nu & \cdots & \langle \boldsymbol{u}_m, \boldsymbol{u}_m \rangle_\nu \end{pmatrix}.$$

If the context makes the basis or subspace clear enough we write $G_{U,\mathcal{B}}$ just as G_U or G.

Lemma 1.2.6. *Let $\mathcal{B} = (\boldsymbol{u}_1, \ldots, \boldsymbol{u}_m)$ and $\mathcal{C} = (\boldsymbol{v}_1, \ldots, \boldsymbol{v}_m)$ be linearly independent subsets of $\mathbb{R}^n_\nu$ such that $U \doteq \operatorname{span} \mathcal{B} = \operatorname{span} \mathcal{C}$. Then,*

$$G_{U,\mathcal{B}} = A^\top G_{U,\mathcal{C}} A,$$

where $A = (a_{ij})_{1 \leq i,j \leq n} \in \operatorname{Mat}(n, \mathbb{R})$ is such that $\boldsymbol{u}_j = \sum_{i=1}^m a_{ij} \boldsymbol{v}_i$. In particular, $\det G_{U,\mathcal{B}} = (\det A)^2 \det G_{U,\mathcal{C}}$, and $\det G_{U,\mathcal{B}} \neq 0$ if and only if $\det G_{U,\mathcal{C}} \neq 0$.

Proof: It suffices to note that:

$$\langle \boldsymbol{u}_i, \boldsymbol{u}_j \rangle_\nu = \left\langle \sum_{k=1}^m a_{ki} \boldsymbol{v}_k, \sum_{\ell=1}^m a_{\ell j} \boldsymbol{v}_\ell \right\rangle_\nu = \sum_{k,\ell=1}^m a_{ki} a_{\ell j} \langle \boldsymbol{v}_k, \boldsymbol{v}_\ell \rangle_\nu,$$

which is precisely the (i,j)-entry in the matrix $A^\top G_{U,\mathcal{C}} A$. The remaining claim follows from applying $\det$ to both sides of $G_{U,\mathcal{B}} = A^\top G_{U,\mathcal{C}} A$. ☐

Proposition 1.2.7. *Let $\boldsymbol{u}_1, \ldots, \boldsymbol{u}_m \in \mathbb{R}^n_\nu$ be vectors such that the matrix $(\langle \boldsymbol{u}_i, \boldsymbol{u}_j \rangle_\nu)_{1 \leq i,j \leq m}$ is invertible. Then $\{\boldsymbol{u}_1, \ldots, \boldsymbol{u}_m\}$ is linearly independent.*

Proof: Let $a_1, \ldots, a_m \in \mathbb{R}$ such that $\sum_{i=1}^m a_i \boldsymbol{u}_i = \boldsymbol{0}$. Applying $\langle \cdot, \boldsymbol{u}_j \rangle_\nu$ to both sides we obtain $\sum_{i=1}^m a_i \langle \boldsymbol{u}_i, \boldsymbol{u}_j \rangle_\nu = 0$ for all $1 \leq j \leq m$. In matrix form, this is written as:

$$\underbrace{\begin{pmatrix} \langle \boldsymbol{u}_1, \boldsymbol{u}_1 \rangle_\nu & \cdots & \langle \boldsymbol{u}_1, \boldsymbol{u}_m \rangle_\nu \\ \vdots & \ddots & \vdots \\ \langle \boldsymbol{u}_m, \boldsymbol{u}_1 \rangle_\nu & \cdots & \langle \boldsymbol{u}_m, \boldsymbol{u}_m \rangle_\nu \end{pmatrix}}_{\text{invertible}} \begin{pmatrix} a_1 \\ \vdots \\ a_m \end{pmatrix} = \begin{pmatrix} 0 \\ \vdots \\ 0 \end{pmatrix} \implies \begin{pmatrix} a_1 \\ \vdots \\ a_m \end{pmatrix} = \begin{pmatrix} 0 \\ \vdots \\ 0 \end{pmatrix},$$

hence $\{\boldsymbol{u}_1, \ldots, \boldsymbol{u}_m\}$ is linearly independent. ☐

Remark. The converse of the above result does not hold. The Gram matrix of a light ray[4] is the zero matrix (0), but if $\boldsymbol{v} \neq \mathbf{0}$, then $\{\boldsymbol{v}\}$ is linearly independent. We will see soon that, under convenient conditions, Lemma 1.2.6 is a partial converse of the previous result (see Proposition 1.2.21).

In what follows, we will study subspaces of $\mathbb{L}^3$, in terms of the causal type of their vectors. After that, we will present some results relating orthogonal complements and causal characters, which hold in $\mathbb{L}^n$.

From Definition 1.2.4, the origin and $\mathbb{L}^n$ are spacelike and timelike, respectively.

Proposition 1.2.8. *Let $r \subseteq \mathbb{L}^n$ be a straight line passing through the origin. Then the causal type of r is the same of any vector giving its direction.*

Proof: See Exercise 1.2.3. □

Now we provide a criterion to know the causal character of a plane Π in $\mathbb{L}^3$ in terms of the coefficients of its general equation, written as $\Pi : ax + by + cz = 0$. If $c = 0$, the plane contains the vector $\boldsymbol{e}_3$, hence it is timelike. If $c \neq 0$ and $\boldsymbol{u} = (x, y, z) \in \Pi$, we have $z = \frac{-ax-by}{c}$ and

$$\begin{aligned}
\langle \boldsymbol{u}, \boldsymbol{u} \rangle_L &= x^2 + y^2 - \left(\frac{-ax - by}{c}\right)^2 \\
&= x^2 + y^2 - \frac{a^2}{c^2}x^2 - \frac{2ab}{c^2}xy - \frac{b^2}{c^2}y^2 \\
&= \left(1 - \frac{a^2}{c^2}\right)x^2 - \frac{2ab}{c^2}xy + \left(1 - \frac{b^2}{c^2}\right)y^2.
\end{aligned}$$

In matrix notation:

$$\langle \boldsymbol{u}, \boldsymbol{u} \rangle_L = \begin{pmatrix} x & y \end{pmatrix} \begin{pmatrix} 1 - \dfrac{a^2}{c^2} & -\dfrac{ab}{c^2} \\ -\dfrac{ab}{c^2} & 1 - \dfrac{b^2}{c^2} \end{pmatrix} \begin{pmatrix} x \\ y \end{pmatrix},$$

and the matrix

$$G_\Pi \doteq \begin{pmatrix} 1 - \dfrac{a^2}{c^2} & -\dfrac{ab}{c^2} \\ -\dfrac{ab}{c^2} & 1 - \dfrac{b^2}{c^2} \end{pmatrix} \tag{‡}$$

is the Gram matrix of $\langle \cdot, \cdot \rangle_L \big|_\Pi$ in the basis $\left((1, 0, -\frac{b}{c}), (0, 1, -\frac{a}{c})\right)$ of Π.

To move on, we need the following:

Definition 1.2.9. A symmetric matrix $A \in \mathrm{Mat}(n, \mathbb{R})$ is

(i) *semi-positive* (resp. *positive-definite*), if $\langle A\boldsymbol{u}, \boldsymbol{u} \rangle_E \geq 0$ (resp. $\langle A\boldsymbol{u}, \boldsymbol{u} \rangle_E > 0$) for all $\boldsymbol{u} \neq \mathbf{0}$.

(ii) *semi-negative* (resp. *negative-definite*), if $-A$ is semi-positive (resp. $-A$ is positive-definite);

(iii) *indefinite*, if it is none of the above.

[4] A straight line in the direction of a lightlike vector.

The following criterion characterizes semi-positive and positive-definite matrices:

Theorem 1.2.10 (Sylvester's Criterion). *A real symmetric matrix is positive-definite if and only if all of its* principal minors *are positive. In other words, if* $A = (a_{ij})_{i,j=1}^{n}$ *then* A *is positive-definite if and only if* $\Delta_k \doteq \det\left((a_{ij})_{i,j=1}^{k}\right) > 0$ *for all* $1 \leq k \leq n$.

Proof: See [32, p. 439] or follow the steps in Exercise 1.2.8. □

Theorem 1.2.11. *Let* $\Pi \subseteq \mathbb{L}^3$ *be a plane,* $\Pi : ax + by + cz = 0$*, such that* $c \neq 0$*. If* G_Π *is its Gram matrix, as in (‡), the following hold:*

(i) $\det G_\Pi > 0 \iff \Pi$ *is spacelike;*

(ii) $\det G_\Pi < 0 \implies \Pi$ *is timelike;*

(iii) $\det G_\Pi = 0 \iff \Pi$ *is lightlike.*

Proof: Let us calculate $\det G_\Pi$:

$$\begin{aligned}\det G_\Pi &= \left(1 - \frac{a^2}{c^2}\right)\left(1 - \frac{b^2}{c^2}\right) - \left(-\frac{ab}{c^2}\right)^2 \\ &= 1 - \frac{b^2}{c^2} - \frac{a^2}{c^2} + \frac{a^2b^2}{c^4} - \frac{a^2b^2}{c^4} \\ &= 1 - \left(\frac{a^2 + b^2}{c^2}\right)\end{aligned}$$

Also notice that $\operatorname{tr} G_\Pi = 2 - \left(\frac{a^2+b^2}{c^2}\right) = 1 + \det G_\Pi$. Since G_Π is a 2×2 matrix, its characteristic polynomial is:

$$\begin{aligned}c_{G_\Pi}(t) &= t^2 - \operatorname{tr} G_\Pi \, t + \det G_\Pi \\ &= t^2 - (1 + \det G_\Pi)\, t + \det G_\Pi,\end{aligned}$$

whose roots are 1 and $\det G_\Pi$. Hence, if $\det G_\Pi > 0$, all the eigenvalues of G_Π are positive and $\langle \cdot, \cdot \rangle_L|_\Pi$ is positive-definite: Π is spacelike. If $\det G_\Pi < 0$, $\langle \cdot, \cdot \rangle_L|_\Pi$ is indefinite, then Π is timelike. Finally, if $\det G_\Pi = 0$, $\langle \cdot, \cdot \rangle_L|_\Pi$ is semi-positive, then Π is lightlike. □

We now establish a result that provides a geometric interpretation for causal character of planes.

Theorem 1.2.12. *Let* $\Pi \subseteq \mathbb{L}^3$ *be a plane of general equation* $\Pi : ax + by + cz = 0$ *and* $\boldsymbol{n} = (a, b, c)$ *be its Euclidean normal vector. Then:*

(i) Π *is spacelike* $\iff$ $\boldsymbol{n}$ *is timelike;*

(ii) Π *is timelike* $\iff$ $\boldsymbol{n}$ *is spacelike;*

(iii) Π *is lightlike* $\iff$ $\boldsymbol{n}$ *is lightlike.*

Proof: Suppose $c = 0$. Then, as before, Π is timelike and $\boldsymbol{n}$ is $(a, b, 0)$, a spacelike vector. Now, let $c \neq 0$ and G_Π be the Gram matrix of $\langle \cdot, \cdot \rangle_L|_\Pi$. For each causal type:

(i) Π is spacelike $\iff 0 < \det G_\Pi = 1 - \left(\frac{a^2+b^2}{c^2}\right)$

$\iff \langle \boldsymbol{n}, \boldsymbol{n} \rangle_L = a^2 + b^2 - c^2 < 0$

$\iff \boldsymbol{n}$ is timelike;

(ii) Π is timelike $\iff 0 > \det G_\Pi = 1 - \left(\frac{a^2+b^2}{c^2}\right)$

$$\iff \langle \boldsymbol{n}, \boldsymbol{n} \rangle_L = a^2 + b^2 - c^2 > 0$$

$$\iff \boldsymbol{n} \text{ is spacelike;}$$

(iii) Π is lightlike $\iff 0 = \det G_\Pi = 1 - \left(\frac{a^2+b^2}{c^2}\right)$

$$\iff \langle \boldsymbol{n}, \boldsymbol{n} \rangle_L = a^2 + b^2 - c^2 = 0$$

$$\iff \boldsymbol{n} \text{ is lightlike.}$$

□

Remark. In the statement we could replace "Euclidean normal" with "Lorentz-normal". Why?

Example 1.2.13.

(1) The plane $\Pi_1 : 3x - 4y + 5z = 0$ is lightlike, since its Euclidean normal vector, $(3, -4, 5)$, is lightlike.

(2) The plane $\Pi_2 : x - 2y + 5z = 0$ is spacelike, since its Euclidean normal vector, $(1, -2, 5)$, is timelike.

(3) The line $L : \mathbf{X}(t) = t(1, 2, 3),\ t \in \mathbb{R}$ is timelike, since $(1, 2, 3)$ is a timelike vector.

The following images correspond to the examples above:

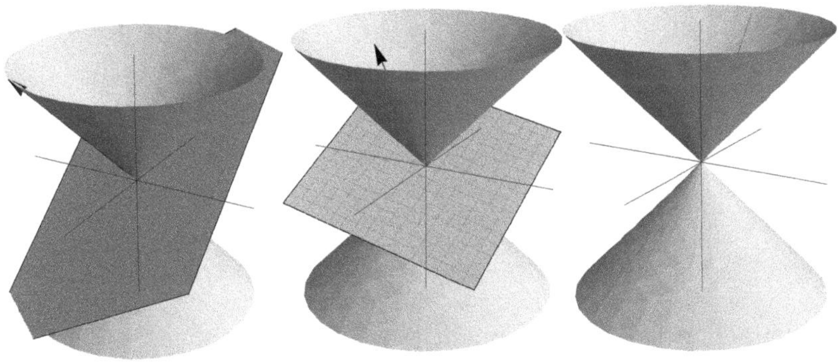

Figure 1.3: Subspaces of $\mathbb{L}^3$.

Remark. Every lightlike plane is tangent to the lightcone, the spacelike planes are "outside" of that cone and timelike planes cut it in two "lightrays", that is, lightlike lines.

Those particular results about planes in $\mathbb{L}^3$ suggest the analysis of the Gram matrix of the restriction of $\langle \cdot, \cdot \rangle_L$ to subspaces of $\mathbb{L}^n$, in order to achieve similar results in higher dimensions. We aim for such results in what follows.

Proposition 1.2.14. *Let $U \subseteq \mathbb{R}^n_\nu$ be a vector subspace. Then*

$$\dim U + \dim U^\perp = n \quad \textit{and} \quad U = (U^\perp)^\perp.$$

Proof: Let U^* be the dual space of U and consider the map $\Phi : \mathbb{R}^n_\nu \to U^*$ given by $\Phi(x) = \langle x, \cdot \rangle\big|_U$. This map is linear, surjective (since $\langle \cdot, \cdot \rangle_\nu$ is non-degenerate), and its kernel is $U^\perp$. Since $\dim U = \dim U^*$, the dimension formula follows from the rank-nullity theorem. Said formula applied twice also says that $\dim U = \dim (U^\perp)^\perp$, so $U \subseteq (U^\perp)^\perp$ implies $U = (U^\perp)^\perp$. □

From this, we have the:

Corollary 1.2.15. *Let $U \subseteq \mathbb{R}^n_\nu$ be a subspace. Then $U \oplus U^\perp = \mathbb{R}^n_\nu$ if and only if U is non-degenerate (i.e., $\langle \cdot, \cdot \rangle_\nu\big|_U$ is non-degenerate). It also follows that U is non-degenerate if and only if $U^\perp$ is also non-degenerate.*

Proof: From

$$\dim(U + U^\perp) = \dim U + \dim U^\perp - \dim(U \cap U^\perp) = n - \dim(U \cap U^\perp)$$

it follows that $U + U^\perp = \mathbb{R}^n_\nu$ if and only if $U \cap U^\perp = \{\mathbf{0}\}$, which in turn is equivalent to $\langle \cdot, \cdot \rangle_\nu\big|_U$ being non-degenerate. □

In the same way, we have the following relation between causal types and orthogonality:

Corollary 1.2.16. *A vector subspace $U \subseteq \mathbb{L}^n$ is spacelike if and only if $U^\perp$ is timelike.*

Proof: Assume that U is spacelike. So U is non-degenerate and we write $\mathbb{L}^n = U \oplus U^\perp$. Then $U^\perp$ must necessarily contain a timelike vector because $\mathbb{L}^n$ does — more precisely, take a timelike $\boldsymbol{v} \in \mathbb{L}^n$ and write $\boldsymbol{v} = \boldsymbol{x} + \boldsymbol{y}$ with $\boldsymbol{x} \in U$ and $\boldsymbol{y} \in U^\perp$, so that the relation $\langle \boldsymbol{x}, \boldsymbol{x} \rangle_L + \langle \boldsymbol{y}, \boldsymbol{y} \rangle_L = \langle \boldsymbol{v}, \boldsymbol{v} \rangle_L < 0$ with $\langle \boldsymbol{x}, \boldsymbol{x} \rangle_L \geq 0$ forces $\boldsymbol{y} \in U^\perp$ to be timelike. Conversely, assume now that U is timelike, and take $\boldsymbol{u} \in U$ timelike. Since $U^\perp \subseteq \boldsymbol{u}^\perp$, it suffices now to show that $\boldsymbol{u}^\perp$ is spacelike. We know again from Corollary 1.2.15 that $\boldsymbol{u}^\perp$ is not lightlike, and if we have $\boldsymbol{v} \in \boldsymbol{u}^\perp$ timelike, the plane spanned by $\boldsymbol{u}$ and $\boldsymbol{v}$ in $\mathbb{L}^n$ has dimension 2 while being negative-definite, which is impossible. □

The following is a restatement of Corollary 1.2.15:

Corollary 1.2.17. *If $U \subseteq \mathbb{L}^n$ is a lightlike subspace, so is $U^\perp$.*

Remark. In the definition of orthogonality we avoid saying "orthogonal complement" since, unlike what happens for the Euclidean case, for a subspace $U \subseteq \mathbb{L}^n$, we do not necessarily have $U + U^\perp = \mathbb{L}^n$. As an example, consider the subspace $U \subseteq \mathbb{L}^3$ given by $U = \{(x, y, z) \in \mathbb{L}^3 \mid y = z\}$. Then $U^\perp = \operatorname{span}\{(0, 1, 1)\} \subseteq U$.

Sylvester's criterion from Theorem 1.2.10 alone is not enough to decide the causal character of a subspace $U \subseteq \mathbb{L}^n$ in terms of the Gram matrix of $\langle \cdot, \cdot \rangle_L\big|_U$. To remedy this we have the following stronger version of that criterion:

Theorem 1.2.18 (Sylvester on steroids). *Let $A \in \operatorname{Mat}(n, \mathbb{R})$ be a symmetric matrix and Δ_k its k-th order leading principal minor determinant. Then:*

(i) *A is positive-definite (resp. semi-positive) if and only if $\Delta_k > 0$ (resp. $\Delta_k \geq 0$) for all $1 \leq k \leq n$;*

(ii) *A is negative-definite (resp. semi-negative) if and only if $(-1)^k \Delta_k > 0$ (resp. $(-1)^k \Delta_k \geq 0$) for all $1 \leq k \leq n$;*

(iii) *A is indefinite if and only if one of the following holds:*

- *there exists an even number $1 \leq k \leq n$ such that $\Delta_k < 0$, or;*
- *there exist distinct odd numbers $1 \leq k, \ell \leq n$ $\Delta_k > 0$ and $\Delta_\ell < 0$.*

Proof:

(i) This is exactly Theorem 1.2.10.

(ii) It follows directly from applying Theorem 1.2.10 to $-A$, bearing in mind that the k-th leading principal minor of $-A$ is $(-1)^k \Delta_k$.

(iii) Suppose that none of the given conditions hold, that is, all of the leading principal minors of even order are non-negative and all of the leading principal minors of odd order are simultaneously non-negative or simultaneously non-positive. If the odd minors are non-negative then, from item (i), A is semi-positive; if they are non-positive, from item (ii), A is semi-negative.

For the converse, start supposing that first condition holds. By item (i), A is not semi-positive and, from item (ii), A can't be semi-negative. Hence A is indefinite. If the second condition holds and A is semi-positive, minors of order k and ℓ are both positive, and if A is negative, both are negative. Hence A must be again indefinite.

□

The following example illustrates the above criterion to decide the causal type of a vector subspace $U \subseteq \mathbb{L}^n$ by analyzing the Gram matrix of $\langle \cdot, \cdot \rangle_L\big|_U$.

Example 1.2.19. Consider $\mathbb{L}^4$.

(1) Let $U \doteq \text{span}\{v_1, v_2, v_3\}$, where

$$v_1 = (2,0,0,1), \quad v_2 = (0,1,2,0) \quad \text{and} \quad v_3 = (-1,-2,0,1).$$

The Gram matrix of $\langle \cdot, \cdot \rangle_L\big|_U$, relative to the basis (v_1, v_2, v_3) is

$$G = \begin{pmatrix} 3 & 0 & -3 \\ 0 & 5 & -2 \\ -3 & -2 & 4 \end{pmatrix},$$

whose leading principal minors are, respectively, $\Delta_1 = 3$, $\Delta_2 = 15$ and $\Delta_3 = 3$. By Sylvester's Criterion, U is spacelike.

(2) For $U \doteq \text{span}\{v_1, v_2, v_3\}$, with

$$v_1 = (1,-1,1,1), \quad v_2 = (0,1,-1,-1) \quad \text{and} \quad v_3 = (0,0,1,1),$$

the Gram matrix of $\langle \cdot, \cdot \rangle_L\big|_U$, relative to the basis (v_1, v_2, v_3) is

$$G = \begin{pmatrix} 2 & -1 & 0 \\ -1 & 1 & 0 \\ 0 & 0 & 0 \end{pmatrix}.$$

The leading principal minors are $\Delta_1 = 2$, $\Delta_2 = 1$ and $\Delta_3 = 0$. Since there are no negative minors, G is neither indefinite nor negative-definite, hence U is not timelike, that is, it must be spacelike or lightlike. As $\Delta_3 = 0$, U is not positive-definite, so it must be lightlike.

(3) Finally, if $U \doteq \operatorname{span}\{v_1, v_2, v_3\}$, where

$$v_1 = (2,1,0,1), \quad v_2 = (0,1,1,-1) \quad \text{and} \quad v_3 = (1,0,0,-1),$$

the Gram matrix of $\langle \cdot, \cdot \rangle_L \big|_U$, relative to the basis (v_1, v_2, v_3) is

$$G = \begin{pmatrix} 4 & 2 & 3 \\ 2 & 1 & -1 \\ 3 & -1 & 0 \end{pmatrix}.$$

Now $\Delta_1 = 4$, $\Delta_2 = 0$ and $\Delta_3 = -25$. Since Δ_1 and Δ_3 have opposite signs, G is indefinite and U is timelike.

In particular, Corollaries 1.2.16 and 1.2.17 provide a generalization of Theorem 1.2.12 to hyperplanes in $\mathbb{L}^n$:

Theorem 1.2.20. *Let $\Pi \subseteq \mathbb{L}^n$ be a hyperplane of general equation*

$$\Pi: \ a_1 x_1 + \cdots + a_n x_n = 0$$

and $\boldsymbol{n} = (a_1, \ldots, a_n)$ its Euclidean normal vector. Then

(i) *Π is spacelike $\iff$ $\boldsymbol{n}$ is timelike;*

(ii) *Π is timelike $\iff$ $\boldsymbol{n}$ is spacelike;*

(iii) *Π is lightlike $\iff$ $\boldsymbol{n}$ is lightlike;*

Remark. Just as in Theorem 1.2.12, we can replace "Euclidean normal" by "Lorentz-normal".

Now we are ready to state a converse for Proposition 1.2.7 (p. 6).

Proposition 1.2.21. *Let $\{u_1, \ldots, u_m\} \subseteq \mathbb{R}^n_\nu$ be a linearly independent set of vectors whose span is non-degenerate. Then the Gram matrix of such vectors is invertible.*

Proof: Let $\mathcal{B} = (u_1, \ldots, u_m)$ and $U = \operatorname{span} \mathcal{B}$. From Proposition 1.2.15 (p. 10), since U is non-degenerate, we may write $\mathbb{R}^n_\nu = U \oplus U^\perp$. If $\mathcal{C}$ is any basis for $U^\perp$, then the (ordered) union $\mathcal{B} \cup \mathcal{C}$ is a basis for $\mathbb{R}^n_\nu$, and the Gram matrix $G_{\mathbb{R}^n_\nu, \mathcal{B} \cup \mathcal{C}}$ is block-diagonal, and thus

$$\det G_{U,\mathcal{B}} \det G_{U^\perp,\mathcal{C}} = \det G_{\mathbb{R}^n_\nu, \mathcal{B} \cup \mathcal{C}} \neq 0$$

implies that $\det G_{U,\mathcal{B}} \neq 0$, as required.

□

Theorem 1.2.22. *Let $\{\mathbf{0}\} \neq U \subseteq \mathbb{R}^n_\nu$ be a non-degenerate vector subspace. Then U has a Lorentz-orthonormal basis.*

Proof: By induction on $k = \dim U$. From non-degeneracy of U, there exists $\boldsymbol{v} \in U$ such that $\langle \boldsymbol{v}, \boldsymbol{v} \rangle_\nu \neq 0$, so $\boldsymbol{v}/\|\boldsymbol{v}\|_\nu$ is unit. With that in mind, it is enough to show that for any orthonormal subset $S \subseteq U$, with less than k vectors, we can add another vector in such a manner that the resulting set is also orthonormal. Since S spans a non-degenerate subspace, its orthogonal complement is non-degenerate and non-trivial. It then suffices to add a unit vector in $S^\perp$ to S (and such a vector exists by the above argument). □

The previous result can also be stated and proved by an algorithm:

Theorem 1.2.23 (Gram-Schmidt Orthogonalization Process)**.** *For any linearly independent vectors $v_1, \ldots, v_k \in \mathbb{R}^n_\nu$ such that for all $i \in \{1, \ldots, k\}$ we have that* $\operatorname{span}(v_1, \ldots, v_i)$ *is non-degenerate (i.e., not lightlike), there exists $\widetilde{v}_1, \ldots, \widetilde{v}_k \in \mathbb{R}^n_\nu$ pairwise Lorentz-orthogonal such that*

$$\operatorname{span}\{v_1, \ldots, v_k\} = \operatorname{span}\{\widetilde{v}_1, \ldots, \widetilde{v}_k\}.$$

Proof: Again by induction on k. For $k = 1$ just take $\widetilde{v}_1 = v_1$. Now, assume that the result is valid for some k, that is, suppose that there exists $\widetilde{v}_1, \ldots, \widetilde{v}_k \in \mathbb{R}^n_\nu$ with the stated properties.

None of the $\widetilde{v}_i$ is lightlike. In fact, if at least one of them were lightlike, we would have $\widetilde{v}_i$ orthogonal to all of the previous vectors, from $\widetilde{v}_1$ to $\widetilde{v_{i-1}}$, and even to itself, so that $\operatorname{span}\{v_1, \ldots, v_i\} = \operatorname{span}\{\widetilde{v}_1, \ldots, \widetilde{v}_i\}$ would be degenerate, contradicting our assumption. So we can set:

$$\widetilde{v_{k+1}} \doteq v_{k+1} - \sum_{i=1}^{k} \frac{\langle v_{k+1}, \widetilde{v}_i \rangle_\nu}{\langle \widetilde{v}_i, \widetilde{v}_i \rangle_\nu} \widetilde{v}_i.$$

It is clear that $\operatorname{span}\{v_1, \ldots, v_{k+1}\} = \operatorname{span}\{\widetilde{v}_1, \ldots, \widetilde{v_{k+1}}\}$, and if $j \in \{1, \ldots, k\}$, we have:

$$\begin{aligned}
\langle \widetilde{v_{k+1}}, \widetilde{v}_j \rangle_\nu &= \left\langle v_{k+1} - \sum_{i=1}^{k} \frac{\langle v_{k+1}, \widetilde{v}_i \rangle_\nu}{\langle \widetilde{v}_i, \widetilde{v}_i \rangle_\nu} \widetilde{v}_i, \widetilde{v}_j \right\rangle_\nu \\
&= \langle v_{k+1}, \widetilde{v}_j \rangle_\nu - \sum_{i=1}^{k} \frac{\langle v_{k+1}, \widetilde{v}_i \rangle_\nu}{\langle \widetilde{v}_i, \widetilde{v}_i \rangle_\nu} \langle \widetilde{v}_i, \widetilde{v}_j \rangle_\nu \\
&= \langle v_{k+1}, \widetilde{v}_j \rangle_\nu - \frac{\langle v_{k+1}, \widetilde{v}_j \rangle_\nu}{\langle \widetilde{v}_j, \widetilde{v}_j \rangle_\nu} \langle \widetilde{v}_j, \widetilde{v}_j \rangle_\nu \\
&= \langle v_{k+1}, \widetilde{v}_j \rangle_\nu - \langle v_{k+1}, \widetilde{v}_j \rangle_\nu = 0.
\end{aligned}$$

□

Remark. The expression for $\widetilde{v_{k+1}}$ can also be written as

$$\widetilde{v_{k+1}} \doteq v_{k+1} - \sum_{i=1}^{k} \epsilon_{\widetilde{v}_i} \frac{\langle v_{k+1}, \widetilde{v}_i \rangle_\nu}{\|\widetilde{v}_i\|_\nu^2} \widetilde{v}_i,$$

which may be more familiar, up to sign adjustments.

Example 1.2.24. Consider the vectors

$$v_1 = (1, 2, 1, 0), \quad v_2 = (0, 2, 0, -1) \quad \text{and} \quad v_3 = (0, 0, 0, 1)$$

in $\mathbb{L}^4$. Straightforward computations show that they are linearly independent, $\operatorname{span}\{v_1\}$ and $\operatorname{span}\{v_1, v_2\}$ are spacelike, and $U \doteq \operatorname{span}\{v_1, v_2, v_3\}$ is timelike. Theorem 1.2.23 then ensures the existence of a Lorentz-orthogonal basis for U from the given vectors. Taking $\widetilde{v}_1 = v_1$, do:

$$\widetilde{v}_2 = (0, 2, 0, -1) - \frac{4}{6}(1, 2, 1, 0) = \left(-\frac{2}{3}, \frac{2}{3}, \frac{-2}{3}, -1\right),$$

and

$$\widetilde{v}_3 = (0,0,0,1) - \frac{0}{6}(1,2,1,0) - \frac{1}{1/3}\left(-\frac{2}{3}, \frac{2}{3}, \frac{-2}{3}, -1\right) = (2,-2,2,4).$$

The vectors $\widetilde{v}_1, \widetilde{v}_2$ and $\widetilde{v}_3$ are pairwise Lorentz–orthogonal, spanning the same hyperplane in $\mathbb{L}^4$ as the original vectors.

Theorem 1.2.25. *Let $u_1, \ldots, u_{\nu+1} \in \mathbb{R}^n_\nu$ be pairwise orthogonal lightlike vectors. Then $(u_1, \ldots, u_{\nu+1})$ is linearly dependent.*

Proof: Splitting $\mathbb{R}^n_\nu$ as $\mathbb{R}^n_\nu = \mathbb{R}^{n-\nu} \times \mathbb{R}^\nu_\nu$, let $(e_i)_{i=1}^n$ be the canonical basis of $\mathbb{R}^n_\nu$. Write

$$u_j = x_j + \sum_{i=1}^{\nu} a_{ij} e_{n-\nu+i}, \qquad 1 \leq j \leq \nu + 1,$$

according to the above split (with all the x_j being spacelike and orthogonal to each timelike vector in the canonical basis). It follows that

$$0 = \langle u_i, u_j \rangle_\nu = \langle x_i, x_j \rangle_\nu - \sum_{k=1}^{\nu} a_{ki} a_{kj} \implies \langle x_i, x_j \rangle_\nu = \sum_{k=1}^{\nu} a_{ki} a_{kj}, \; 1 \leq i,j \leq \nu + 1.$$

Consider now the linear map $A\colon \mathbb{R}^{\nu+1} \to \mathbb{R}^\nu$, whose matrix in the canonical bases has the above a_{ij} as entries, and take $b = (b_1, \ldots, b_{\nu+1}) \neq \mathbf{0}$ in $\ker A$ (this is possible for dimension reasons). We have

$$\begin{aligned}
\left\langle \sum_{i=1}^{\nu+1} b_i x_i, \sum_{j=1}^{\nu+1} b_j x_j \right\rangle_\nu &= \sum_{i,j=1}^{\nu+1} b_i b_j \langle x_i, x_j \rangle_\nu \\
&= \sum_{i,j=1}^{\nu+1} b_i b_j \sum_{k=1}^{\nu} a_{ki} a_{kj} \\
&= (Ab)^\top (Ab) = 0.
\end{aligned}$$

Since the linear combination $\sum_{j=1}^{\nu+1} b_j x_j$ is spacelike, the preceding computation shows that $\sum_{j=1}^{\nu+1} b_j x_j = \mathbf{0}$. The entries in b also testify that $(u_1, \ldots, u_{\nu+1})$ is linearly dependent. In fact:

$$\sum_{j=1}^{\nu+1} b_j u_j = \sum_{j=1}^{\nu+1} b_j x_j + \sum_{i=1}^{\nu} \left(\sum_{j=1}^{\nu+1} b_j a_{ij} \right) e_{n-\nu+i} = \mathbf{0},$$

as the second term also vanishes, since $b \in \ker A$. □

In particular, we have the:

Corollary 1.2.26. *Two lightlike vectors in $\mathbb{L}^n$ are Lorentz-orthogonal if and only if they are proportional.*

Corollary 1.2.27. *If $U \subseteq \mathbb{L}^n$ is a lightlike subspace, then* $\dim(U \cap U^\perp) = 1$.

Proof: Let $u, v \in U \cap U^\perp$. Then $\langle u, v \rangle_L = 0$, hence u and v are linearly dependent, and therefore $\dim(U \cap U^\perp) \leq 1$. On the other hand, if U is lightlike, we have $\dim U > 0$. In this way take a lightlike vector (non-zero, by definition) $v \in U$. Then $\langle v, v \rangle_L = 0$ and $v \in U^\perp$, showing that $\dim(U \cap U^\perp) \geq 1$. □

Remark. The result above does not hold in $\mathbb{R}^n_\nu$, with $\nu > 1$. As an example, take any subspace containing two pairwise orthogonal and linearly independent lightlike vectors. Since its orthogonal complement contains itself, we have $\dim(U \cap U^\perp) \geq 2$.

Corollary 1.2.28. *Let $u, v \in \mathbb{L}^n$ be linearly independent lightlike vectors. Then $u+v$ and $u-v$ are not lightlike and have distinct causal types.*

Proof: Note that $\langle u \pm v, u \pm v\rangle_L = \pm 2\langle u, v\rangle_L \neq 0$. □

In particular, any degenerate subspace $U \subseteq \mathbb{L}^n$ contains only one lightray.

Proposition 1.2.29. *Let $U \subseteq \mathbb{L}^n$ be a degenerate subspace. Then U admits an orthogonal basis.*

Proof: Let $(v_1, \ldots, v_k) \subseteq \mathbb{L}^n$ be a basis of U such that v_k is a lightlike vector (therefore the remaining vectors are spacelike). Applying the Gram-Schmidt process to $\{v_1, \ldots, v_{k-1}\}$ we get $k-1$ pairwise orthogonal spacelike vectors. Such vectors are orthogonal to v_k.

In fact, for any (unit) spacelike vector $u \in U$ such that $\langle u, v_k\rangle_L \neq 0$, we may take $\lambda \in \mathbb{R}$ satisfying

$$\langle u + \lambda v_k, u + \lambda v_k\rangle_L = 1 + 2\lambda\langle u, v_k\rangle_L < 0,$$

a contradiction with the degeneracy of U. □

Remark. The preceding result still holds for subspaces of $\mathbb{R}^n_\nu$, but this requires a slightly more elaborate proof (see [31]).

Lemma 1.2.30. *Let $U \subseteq \mathbb{R}^n_\nu$ to be a vector subspace. If U admits an orthonormal basis, then it is non-degenerate.*

Proof: Let $(v_1, \ldots, v_m)$ be an orthonormal basis for U and $x \in U$. Then $x = \sum_{i=1}^m x_i v_i$. Suppose that $\langle x, v_j\rangle_\nu = 0$ for all $j = 1, \ldots, m$. Then, since $\epsilon_j = \langle v_j, v_j\rangle_\nu \in \{-1, 1\}$, we have $\epsilon_j x_j = 0$, leading to $x_j = 0$ and $x = \mathbf{0}$. This means that U is non-degenerate. □

Lemma 1.2.31. *If $U \subseteq \mathbb{L}^n$ is a lightlike hyperplane, then $U^\perp \subseteq U$.*

Proof: Let $u_1, \ldots, u_n \in \mathbb{L}^n$ be vectors such that $U = \operatorname{span}\{u_1, \ldots, u_{n-1}\}$ and with $U^\perp = \operatorname{span}\{u_n\}$. Since U is lightlike, it holds that $\dim(U \cap U^\perp) = 1$. Hence

$$\dim(U + U^\perp) = \dim U + \dim U^\perp - \dim(U \cap U^\perp) = n - 1,$$

as $\dim U = n-1$ and $\dim U^\perp = 1$. However, $U + U^\perp = \operatorname{span}\{u_1, \ldots, u_n\}$, and from linear independence of $\{u_1, \ldots, u_{n-1}\}$, it follows that $u_n \in U$. □

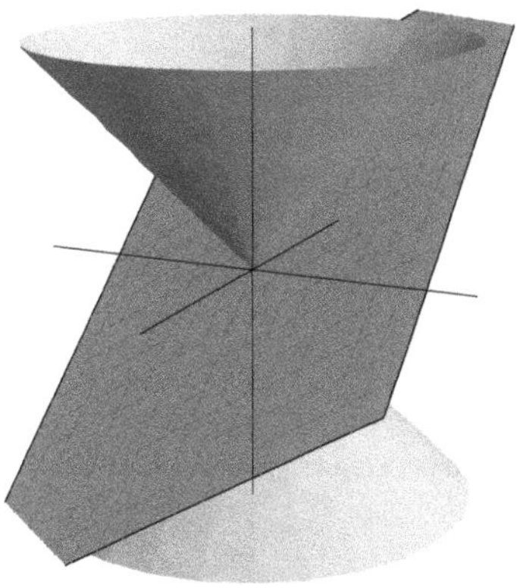

Figure 1.4: A lightlike vector contained in its own orthogonal "complement".

Remark. The above result no longer holds if $\dim U < n-1$ in $\mathbb{L}^n$. Consider the following counter-example: let

$$U = \operatorname{span}\{(0,0,1,1),(1,0,0,0)\} \quad \text{and} \quad U^\perp = \operatorname{span}\{(0,1,0,0),(0,0,1,1)\}$$

in $\mathbb{L}^4$. Note that $(0,1,0,0) \notin U$.

Lemma 1.2.32 (Orthonormal expansion). *Let $(\boldsymbol{u}_1,\ldots,\boldsymbol{u}_n)$ be a pseudo-orthonormal basis for $\mathbb{R}^n_\nu$. Then, every $\boldsymbol{v} \in \mathbb{R}^n_\nu$ can be written as*

$$\boldsymbol{v} = \sum_{i=1}^n \epsilon_{\boldsymbol{u}_i} \langle \boldsymbol{v}, \boldsymbol{u}_i\rangle_\nu \boldsymbol{u}_i.$$

Proof: Start writing $\boldsymbol{v} = \sum_{j=1}^n v_j \boldsymbol{u}_j$, for suitable $v_1,\ldots,v_n \in \mathbb{R}$ and suppose that the last ν vectors in the given basis are timelike. Applying $\langle \cdot, \boldsymbol{u}_i\rangle_\nu$ to the previous equality we have:

$$\langle \boldsymbol{v}, \boldsymbol{u}_i\rangle_\nu = \left\langle \sum_{j=1}^n v_j \boldsymbol{u}_j, \boldsymbol{u}_i \right\rangle_\nu = \sum_{j=1}^n v_j \langle \boldsymbol{u}_j, \boldsymbol{u}_i\rangle_\nu = \sum_{i=1}^n v_j \eta^\nu_{ij} = \epsilon_{\boldsymbol{u}_i} v_i.$$

Hence, $v_i = \epsilon_{\boldsymbol{u}_i}\langle \boldsymbol{v}, \boldsymbol{u}_i\rangle_\nu$. □

Corollary 1.2.33. *In $\mathbb{L}^n$ there are no pairwise orthogonal timelike vectors, as well as no timelike vectors orthogonal to lightlike ones. Furthermore, any orthogonal basis must contain precisely one timelike vector with the remaining ones being spacelike.*

Proof: See Exercise 1.2.13. □

Proposition 1.2.34. *Let $\boldsymbol{v} \in \mathbb{L}^n$ be a unit timelike vector. Then $\|\boldsymbol{v}\|_E \geq 1$.*

Proof: Writing $\boldsymbol{v} = (v_1,\ldots,v_n)$, we have:

$$\langle \boldsymbol{v}, \boldsymbol{v}\rangle_L = v_1^2 + \cdots + v_{n-1}^2 - v_n^2 = -1 \implies v_n^2 = 1 + v_1^2 + \cdots + v_{n-1}^2.$$

Therefore

$$\|\boldsymbol{v}\|_E^2 = v_1^2 + \cdots + v_n^2 = 1 + 2\left(v_1^2 + \cdots + v_{n-1}^2\right) \geq 1.$$

□

Remark. The previous proposition says the "Euclidean eyes" see a timelike vector larger than "Lorentzian eyes" see it ($\|\boldsymbol{v}\|_E \geq \|\boldsymbol{v}\|_L$). In particular, this justifies us drawing a timelike unit vector, Lorentz-orthogonal to a spacelike plane, with Euclidean length greater than 1.

Exercises

Exercise 1.2.1 (Warmup). Decide the causal type of the following vectors, lines, and planes of $\mathbb{L}^3$:

(a) $\boldsymbol{u} = (1,3,2)$;

(b) $\boldsymbol{v} = (3,0,-3)$;

(c) $r(t) = (-2, -1, 2) + t(0, 2, 3)$, $t \in \mathbb{R}$;

(d) $r(t) = (10, 0, 0) + t(3, 2, -1)$, $t \in \mathbb{R}$;

(e) $\Pi\colon 3x - 4y + 5z = 10$;

(f) $\Pi\colon x = 5y$.

Recall that it is also usual in the literature to adopt the convention

$$\langle \boldsymbol{x}, \boldsymbol{y} \rangle_L \doteq -x_1y_1 + x_2y_2 + x_3y_3.$$

What would be the answers in that convention?

Exercise† 1.2.2. Let $\{\mathbf{0}\} \neq U \subseteq \mathbb{L}^n$ be a vector subspace. Show that

(a) If $\langle \cdot, \cdot \rangle_L\big|_U$ is negative-definite, then U is a line.

> **Hint.** Show that if $\dim U \geq 2$, then $\langle \cdot, \cdot \rangle_L\big|_U$ cannot be negative-definite. For this, take $\{\boldsymbol{u}, \boldsymbol{v}\} \subseteq U$ linearly independent, with timelike $\boldsymbol{v}$. If $\boldsymbol{u}$ is spacelike or lightlike we are done. If $\boldsymbol{u}$ is timelike, write a "horizontal" (spacelike) vector by combining $\boldsymbol{v}$ and a convenient rescaling of $\boldsymbol{u}$.

(b) If U has a timelike vector, then U is non-degenerate. In particular, if $\langle \cdot, \cdot \rangle_L\big|_U$ is indefinite, it is also non-degenerate.

> **Hint.** If U is degenerate, take $\boldsymbol{u}, \boldsymbol{v} \in U$ such that $\boldsymbol{u}$ is lightlike, $\boldsymbol{v}$ is timelike, and $\langle \boldsymbol{u}, \boldsymbol{v} \rangle_L = 0$. On one hand, note that $\langle a\boldsymbol{u} + b\boldsymbol{v}, a\boldsymbol{u} + b\boldsymbol{v} \rangle_L \leq 0$ for all $a, b \in \mathbb{R}$ and, on the other hand, repeat the argument in the hint above to have a spacelike combination of $\boldsymbol{u}$ and $\boldsymbol{v}$.

Use this to show that:

(c) U is spacelike if and only if all of its vectors are spacelike;

(d) U is timelike if and only if has some timelike vector;

(e) U is lightlike if it has some lightlike vector but no timelike vectors.

Exercise 1.2.3. Prove Proposition 1.2.8 (p. 7).

Exercise 1.2.4. Show that a subspace $U \subseteq \mathbb{L}^n$ is lightlike if and only if its intersection with the lightcone in $\mathbb{L}^n$ is precisely a light ray through the origin.

Exercise 1.2.5. Show that if $U \subseteq \mathbb{L}^n$ is a 2-dimensional timelike subspace, then it intersects the lightcone of $\mathbb{L}^n$ in two lightrays passing through the origin.

Exercise 1.2.6. Exhibit a *Penrose basis* for $\mathbb{L}^n$, that is, a basis $\mathcal{P} = \{\boldsymbol{u}_1, \ldots, \boldsymbol{u}_n\}$ such that each $\boldsymbol{u}_i$ is lightlike and $\langle \boldsymbol{u}_i, \boldsymbol{u}_j \rangle_L = -1$ for $i \neq j$. Write down the Gram matrix $G_{\mathbb{L}^n, \mathcal{P}}$. Does this matrix depend on the Penrose basis $\mathcal{P}$?

Hint. Try to solve this for $\mathbb{L}^2$ and $\mathbb{L}^3$ first.

Exercise† 1.2.7. Let $U \subseteq \mathbb{R}^n_\nu$ be a vector subspace. An *orthogonal projection* of $\mathbb{R}^n_\nu$ onto U is a linear operator that fixes U (pointwise) and collapses $U^\perp$ onto the set $\{\mathbf{0}\}$.

(a) Show that U is non–degenerate if and only if there exists an orthogonal projection of $\mathbb{R}^n_\nu$ onto U, which is unique. If $v \in \mathbb{R}^n_\nu$, we write the orthogonal projection of v in U as $\mathrm{proj}_U v$.

(b) Show that, if $\dim U = k$ and $\{u_1, \dots, u_k\}$ is an orthonormal basis for U, then

$$\mathrm{proj}_U v = \sum_{i=1}^{k} \frac{\langle v, u_i \rangle_\nu}{\langle u_i, u_i \rangle_\nu} u_i.$$

Exercise 1.2.8. This exercise is a guide to prove Sylvester's Criterion. For this, let $A \in \mathrm{Mat}(n, \mathbb{R})$ be a symmetric matrix.

(a) A is positive-definite if and only if all of its eigenvalues are positive.

(b) If $W \subseteq \mathbb{R}^n$ is a k-dimensional subspace, then every subspace of $\mathbb{R}^n$ with dimension $m > n - k$ has non-trivial intersection with W.

Hint. Compute the dimension of the intersection.

(c) If $\langle Ax, x \rangle_E > 0$ for all non–zero x in a k-dimensional subspace $W \subseteq \mathbb{R}^n$, then A has at least k positive eigenvalues.

(d) Deduce the criterion using induction in n.

Exercise 1.2.9. Consider the following vectors in $\mathbb{L}^4$

$$v_1 = (1, 0, 2, 0), \quad v_2 = (0, 1, -1, 0) \quad \text{and} \quad v_3 = (1, 2, 0, 0).$$

The subspace $U \doteq \mathrm{span}\,\{v_1, v_2, v_3\}$ is spacelike but

$$\left(\langle v_i, v_j \rangle_L\right)_{1 \le i,j \le 3} = \begin{pmatrix} 5 & -2 & 1 \\ -2 & 2 & 2 \\ 1 & 2 & 5 \end{pmatrix}$$

has principal leading minors 5, 6 and 0. Proceeding like in Example 1.2.19 (p. 11), U would be lightlike. Explain this ostensible contradiction.

Exercise 1.2.10. Consider in $\mathbb{L}^5$ the vectors

$$v_1 = (1, 2, 0, 0, 1), \quad v_2 = (0, 0, 1, 1, 0) \quad \text{and} \quad v_3 = (1, 0, 1, 0, 0).$$

(a) Show that such vectors are linearly independent.

(b) Show that $\mathrm{span}\{v_1\}$, $\mathrm{span}\{v_1, v_2\}$ and $U \doteq \mathrm{span}\{v_1, v_2, v_3\}$ are spacelike subspaces of $\mathbb{L}^5$ (hence non-degenerate).

(c) Exhibit an orthonormal basis of $\mathbb{L}^5$ containing a basis for U.

Exercise† 1.2.11. Write the proof of Theorem 1.2.25 (p. 14) in the particular case $\nu = 1$.

Exercise 1.2.12 (Triangles of light)**.** Show that there are no pairwise linearly independent lightlike vectors $u, v, w \in \mathbb{L}^n$ such that $u + v + w = \mathbf{0}$. Generalize.

Exercise† 1.2.13. In $\mathbb{L}^n$, show that:

(a) there are no pairwise orthogonal timelike vectors;

(b) there are no timelike vectors orthogonal to lightlike ones;

(c) any orthonormal basis for $\mathbb{L}^n$ has exactly one timelike vector and the other ones are spacelike.

Hint. Use Lemma 1.2.32 (p. 16) and the item (a) above, assuming that all vectors in a orthonormal basis are spacelike.

1.3 CONTEXTUALIZATION IN SPECIAL RELATIVITY

In this section we give an interpretation of previous results in the setting of Special Relativity. We will focus the discussion on Lorentz-Minkowski space $\mathbb{L}^4$, for it is natural to consider three dimensions for space and one for time. Fixing the inertial frame given by the canonical basis, write the coordinates of an event as $\boldsymbol{p} = (x, y, z, t)$, so that

$$\langle \boldsymbol{p}, \boldsymbol{p} \rangle_L = x^2 + y^2 + z^2 - t^2.$$

In Physics, the expression above is usually written as $x^2 + y^2 + z^2 - (ct)^2$, where c is the speed of light in vacuum. In Mathematics, we usually work with *geometric units*[5], where $c = 1$.

Let $\boldsymbol{p}, \boldsymbol{q} \in \mathbb{L}^4$ be any given events. If $\boldsymbol{v} \doteq \boldsymbol{q} - \boldsymbol{p} = (\Delta x, \Delta y, \Delta z, \Delta t)$ is the vector joining $\boldsymbol{p}$ to $\boldsymbol{q}$ and $\Delta t \neq 0$, the causal type of $\boldsymbol{v}$ helps us understand the interaction between both events. Note that

$$\begin{aligned}
\langle \boldsymbol{v}, \boldsymbol{v} \rangle_L &= (\Delta x)^2 + (\Delta y)^2 + (\Delta z)^2 - (\Delta t)^2 \\
&= (\Delta t)^2 \left(\left(\frac{\Delta x}{\Delta t}\right)^2 + \left(\frac{\Delta y}{\Delta t}\right)^2 + \left(\frac{\Delta z}{\Delta t}\right)^2 - 1 \right) \\
&= (\Delta t)^2 (\|\widetilde{\boldsymbol{v}}\|_E^2 - 1) \\
&= (\Delta t)^2 (\|\widetilde{\boldsymbol{v}}\|_E + 1)(\|\widetilde{\boldsymbol{v}}\|_E - 1),
\end{aligned}$$

where $\widetilde{\boldsymbol{v}} \doteq \left(\frac{\Delta x}{\Delta t}, \frac{\Delta y}{\Delta t}, \frac{\Delta z}{\Delta t}\right) \in \mathbb{R}^3$ is the *spatial velocity vector* between the events $\boldsymbol{p}$ and $\boldsymbol{q}$. In particular, observe that $\langle \boldsymbol{v}, \boldsymbol{v} \rangle_L$ and $\|\widetilde{\boldsymbol{v}}\|_E - 1$ share the same sign. Hence,

(1) if $\boldsymbol{v}$ is timelike, then $\Delta t \neq 0$ and $\|\widetilde{\boldsymbol{v}}\|_E < 1$. When $\Delta t > 0$, the event $\boldsymbol{p}$ may influence the event $\boldsymbol{q}$, and $\Delta t < 0$ says that event $\boldsymbol{p}$ may have been influenced by $\boldsymbol{q}$ — such influences may manifest themselves through propagation of material waves;

(2) if $\boldsymbol{v}$ is lightlike and $\Delta t \neq 0$, then $\|\widetilde{\boldsymbol{v}}\|_E = 1$, hence the influence of $\boldsymbol{p}$ over $\boldsymbol{q}$ (or the opposite, depending on the sign of Δt as above), is realized as the propagation of an electromagnetic wave, such as a light signal emitted by $\boldsymbol{p}$ and received by $\boldsymbol{q}$;

(3) if $\boldsymbol{v}$ is spacelike, with $\Delta t \neq 0$, there is no influence relation between $\boldsymbol{p}$ and $\boldsymbol{q}$, since $\|\widetilde{\boldsymbol{v}}\|_E > 1$, that is, the speed required to move spatially from one event to another is greater than the speed of light. In other words, not even a photon or neutrino is fast enough to witness both events.

[5] See [54, p. 162] for details.

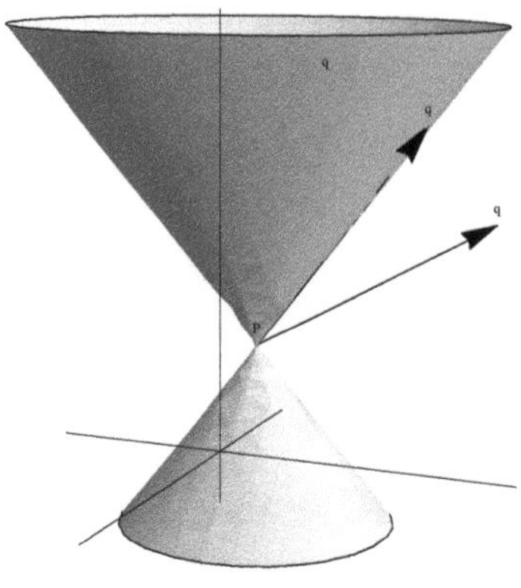

Figure 1.5: Picture of the situation above, omitting one spatial dimension.

The above considerations lead to the following definition in $\mathbb{L}^n$:

Definition 1.3.1. A timelike or lightlike vector $\boldsymbol{v} = (v_1, \ldots, v_n) \in \mathbb{L}^n$ is *future-directed* (resp. *past-directed*) if $v_n > 0$ (resp. $v_n < 0$).

Clearly $\boldsymbol{v}$ is future-directed if it is timelike or lightlike and $\langle \boldsymbol{v}, \boldsymbol{e}_n \rangle_L < 0$. Hence, we have:

Definition 1.3.2. The *lightcone* centered in $\boldsymbol{p} \in \mathbb{L}^n$ is the set

$$C_L(\boldsymbol{p}) \doteq \{\boldsymbol{q} \in \mathbb{L}^n \mid \boldsymbol{q} - \boldsymbol{p} \text{ is lightlike}\}.$$

The *timecone* centered in $\boldsymbol{p} \in \mathbb{L}^n$ is

$$C_T(\boldsymbol{p}) \doteq \{\boldsymbol{q} \in \mathbb{L}^n \mid \boldsymbol{q} - \boldsymbol{p} \text{ is timelike}\}.$$

Each of the cones $C_L(\boldsymbol{p})$ and $C_T(\boldsymbol{p})$ may be split in two disjoint components, called future/past light/timecones, denoted by $C_L^+(\boldsymbol{p}), C_L^-(\boldsymbol{p}), C_T^+(\boldsymbol{p})$ and $C_T^-(\boldsymbol{p})$. For example,

$$C_L^+(\boldsymbol{p}) \doteq \{\boldsymbol{q} \in C_L(\boldsymbol{p}) \mid \boldsymbol{q} - \boldsymbol{p} \text{ is lightlike and future-directed}\}.$$

In Physics books, when discussing Special Relativity, it is usual to consider, instead of $\mathbb{L}^4$, an arbitrary four-dimensional vector space equipped with a pseudo-Euclidean inner product of *index* 1, which is the largest dimension of a subspace for which the restriction of the inner product to it is negative-definite (for example, Exercise 1.2.2 says that $\langle \cdot, \cdot \rangle_L$ has index 1). This is done to avoid the choice of a preferred frame of reference (like the canonical basis in $\mathbb{L}^4$). Results shown so far could be written in this setting with no extra effort. In Special Relativity, using either $\mathbb{L}^4$ or any vector space in the above conditions, we are modelling a spacetime free of gravity, in the vacuum.

In General Relativity, to consider gravitation, the ambient spaces studied are no longer vector spaces, being so-called *manifolds*. In such an ambient, light rays are not necessarily straight lines. In this broader context it is possible to define, in a mathematically precise way, the future and past (chronological or absolute) of an event.

Definition 1.3.3 (Informal). Let p be a point in spacetime. The *chronological future* of p and the *absolute future* (causal) of p are, respectively,

$$\begin{aligned} I^+(p) \doteq \{&q \text{ in spacetime} \mid \text{there exists a future-directed} \\ &\text{timelike curve joining } p \text{ and } q\} \text{ and} \\ J^+(p) \doteq \{&q \text{ in spacetime} \mid \text{there exists a future-directed} \\ &\text{timelike or lightlike curve joining } p \text{ and } q\}. \end{aligned}$$

The *chronological past*, $I^-(p)$, and the *absolute past*, $J^-(p)$, are defined similarly.

We will study curves in Chapter 2, emphasizing the three-dimensional spaces $\mathbb{R}^3$ and $\mathbb{L}^3$, where most of the theory in this text will be developed. The above definitions are valid in General Relativity as well as in Special Relativity. In the latter, we have $I^{\pm}(p) = C_T^{\pm}(p)$ in $\mathbb{L}^4$, allowing us to write the:

Definition 1.3.4. Let $S \subseteq \mathbb{L}^n$ be any subset. The *chronological future* of S is the set

$$I^+(S) \doteq \bigcup_{p \in S} C_T^+(p).$$

Similarly, one defines the *chronological past* of S, $I^-(S)$.

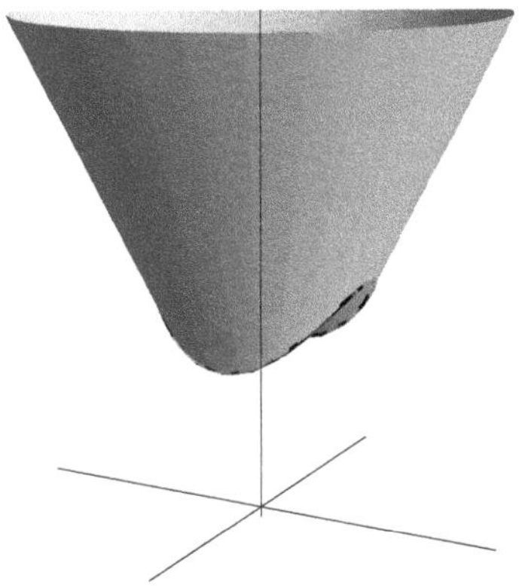

Figure 1.6: The future of a set.

With those concepts in hand, we can define the *chronological and causal orderings* in $\mathbb{L}^n$:

Definition 1.3.5. Let $p, q \in \mathbb{L}^n$. We say that p *chronologically precedes* q if $q \in C_T^+(p)$, and this is denoted by $p \ll q$. Furthermore, p *causally precedes* q if $q \in C_T^+(p) \cup C_L^+(p)$, and this will be denoted by $p \preccurlyeq q$.

In the following, we present some basic properties of future and past sets in terms of chronological precedence.

Proposition 1.3.6. *Let* $p \in \mathbb{L}^n$. *Then* $\langle u - p, v - p \rangle_L < 0$ *for all* $u, v \in C_T^+(p)$.

Proof: It follows from the definition that $u, v \in C_T^+(p)$ if and only if both $u - p$ and $v - p$ are in $C_T^+(\mathbf{0})$. Hence, without loss of generality, we may assume that $p = \mathbf{0}$. If $u, v \in C_T^+(\mathbf{0})$, recall that $\mathbb{L}^n = e_n^{\perp} \oplus \operatorname{span}\{e_n\}$ and write

$$u = x + ae_n \quad \text{and} \quad v = y + be_n,$$

for certain spacelike vectors $x, y \in \mathbb{L}^n$ and positive numbers a and b. We have

$$\langle u, u \rangle_L = \langle x + ae_n, x + ae_n \rangle_L < 0 \implies \langle x, x \rangle_L - a^2 < 0,$$

whence $\langle x, x \rangle_L < a^2$. Similarly $\langle y, y \rangle_L < b^2$. Since $\langle \cdot, \cdot \rangle_L \big|_{e_n^{\perp}}$ is a Euclidean inner product, the standard Cauchy-Schwarz inequality holds in $e_n^{\perp}$: $|\langle x, y \rangle_L| \leq \|x\|_L \|y\|_L$. Then,

$$\langle u, v \rangle_L = \langle x + ae_n, y + be_n \rangle = \langle x, y \rangle - ab \leq \|x\|_L \|y\|_L - ab < ab - ab = 0.$$

□

Remark. The previous result extends to vectors in $C_T^+(p) \cup C_L^+(p)$, but the inequality is no longer strict. When does the equality hold?

In the Lorentzian ambient space, the functions cosh and sinh play the same role that trigonometric functions do in the circle, when the inner product is Euclidean. You can recall some basic properties of such functions in Exercise 1.3.1. Now is a good moment to do so, if you are not familiar with them, since we will use those properties in the next proposition and in several of the following ones.

Proposition 1.3.7. *Let $p \in \mathbb{L}^n$. Given any two vectors $u, v \in C_T(p)$ such that $\langle u - p, v - p \rangle_L < 0$, then u and v are both in $C_T^+(p)$ or in $C_T^-(p)$.*

Proof: Just as in the previous proposition, we can assume that $p = \mathbf{0}$. Also, suppose that $u \in C_T^+(\mathbf{0})$ and $\|u\|_L = \|v\|_L = 1$. Now our aim is to prove that $v \in C_T^+(\mathbf{0})$, assuming that $\langle u, v \rangle_L < 0$. If u and v are parallel, then $v = u$ or $v = -u$ and the assumption leads to $v = u \in C_T^+(\mathbf{0})$. If $\{u, v\}$ is linearly independent, take an orthonormal basis $\{w_1, w_2\}$ for the plane spanned by u and v, where w_1 is spacelike and w_2 is timelike. Exchanging signs of the w_i if needed, we can assume that w_2 is future-directed and $\langle w_1, e_n \rangle_L \leq 0$. Write $u = a w_1 + b w_2$ for some $a, b \in \mathbb{R}$. The function $\sinh\colon \mathbb{R} \to \mathbb{R}$ is bijective and hence there exists a unique $\theta_1 \in \mathbb{R}$ such that $a = \sinh \theta_1$. Hence

$$\langle u, u \rangle_L = \sinh^2 \theta_1 - b^2 = -1,$$

and then $b^2 = \cosh^2 \theta_1$. Since u and w_2 are future-directed, the previous proposition ensures that $-b = \langle u, w_2 \rangle_L < 0$, that is, $b > 0$, whence $u = \sinh \theta_1 \, w_1 + \cosh \theta_1 \, w_2$. The same argument shows that $v = \sinh \theta_2 \, w_1 + \epsilon \cosh \theta_2 \, w_2$ for some $\theta_2 \in \mathbb{R}$ and $\epsilon \in \{-1, 1\}$. Then we have

$$\begin{aligned}
\langle u, v \rangle_L &= \sinh \theta_1 \sinh \theta_2 - \epsilon \cosh \theta_1 \cosh \theta_2 \\
&= -\epsilon \left(\cosh \theta_1 \cosh \theta_2 - \epsilon \sinh \theta_1 \sinh \theta_2 \right) \\
&= -\epsilon \left(\cosh \theta_1 \cosh(-\epsilon \theta_2) + \sinh \theta_1 \sinh(-\epsilon \theta_2) \right) \\
&= -\epsilon \cosh \left(\theta_1 - \epsilon \theta_2 \right),
\end{aligned}$$

using that cosh is an even function and sinh is an odd one. Since cosh is a positive function, $\langle u, v \rangle_L < 0$ gives $\epsilon = 1$. In this way, we have:

$$\begin{aligned}
\langle v, e_n \rangle_L &= \sinh \theta_2 \langle w_1, e_n \rangle_L + \cosh \theta_2 \langle w_2, e_n \rangle_L \\
&\leq \cosh \theta_2 \left(\langle w_1, e_n \rangle_L + \langle w_2, e_n \rangle_L \right) < 0,
\end{aligned}$$

that is, $v \in C_T^+(\mathbf{0})$, as desired. □

Remark. This result cannot be extended like we did in the remark after Proposition 1.3.6. Can you find a counterexample?

Proposition 1.3.8. *Let $p, q, r \in \mathbb{L}^n$ such that $p \ll q$ and $q \ll r$. Then $p \ll r$.*

Proof: It suffices to verify that $r - p$ is timelike and future-directed:

- To see that $r - p$ is timelike, just compute

$$\begin{aligned}
\langle r - p, r - p \rangle_L &= \langle r - q + q - p, r - q + q - p \rangle_L \\
&= \langle r - q, r - q \rangle_L + 2 \langle r - q, q - p \rangle_L + \langle q - p, q - p \rangle_L < 0,
\end{aligned}$$

  since $p \ll q$, $q \ll r$ and, from Proposition 1.3.6, we have that $\langle r - q, q - p \rangle_L < 0$ as $q - p, r - q \in C_T^+(\mathbf{0})$.

- To see that $\boldsymbol{r}-\boldsymbol{p}$ is future-directed, use again that $\boldsymbol{p} \ll \boldsymbol{q}$ and $\boldsymbol{q} \ll \boldsymbol{r}$, whence $\langle \boldsymbol{r}-\boldsymbol{p}, \boldsymbol{e}_n \rangle_L = \langle \boldsymbol{r}-\boldsymbol{q}, \boldsymbol{e}_n \rangle_L + \langle \boldsymbol{q}-\boldsymbol{p}, \boldsymbol{e}_n \rangle_L < 0$.

□

The previous proposition still holds if we replace $\ll$ by $\preccurlyeq$. See Exercise 1.3.4.

The transition from $\mathbb{R}^n$ to $\mathbb{L}^n$ affects results depending on the inner product: some fail to hold, while others undergo drastic changes. Let us explore this, starting with the:

Theorem 1.3.9 (Reverse Cauchy-Schwarz inequality). *Let $\boldsymbol{u}, \boldsymbol{v} \in \mathbb{L}^n$ be timelike vectors. Then $|\langle \boldsymbol{u}, \boldsymbol{v} \rangle_L| \geq \|\boldsymbol{u}\|_L \|\boldsymbol{v}\|_L$. Furthermore, equality holds if and only if $\boldsymbol{u}$ and $\boldsymbol{v}$ are parallel.*

Proof: Decompose $\mathbb{L}^n = \operatorname{span}\{\boldsymbol{u}\} \oplus \boldsymbol{u}^\perp$. Write $\boldsymbol{v} = \lambda \boldsymbol{u} + \boldsymbol{u}_0$, for some $\lambda \in \mathbb{R}$, and a spacelike $\boldsymbol{u}_0$ orthogonal to $\boldsymbol{u}$. On one hand we have:

$$\langle \boldsymbol{v}, \boldsymbol{v} \rangle_L = \lambda^2 \langle \boldsymbol{u}, \boldsymbol{u} \rangle_L + \langle \boldsymbol{u}_0, \boldsymbol{u}_0 \rangle_L.$$

On the other hand:

$$\begin{aligned}
\langle \boldsymbol{u}, \boldsymbol{v} \rangle_L^2 &= \langle \boldsymbol{u}, \lambda \boldsymbol{u} + \boldsymbol{u}_0 \rangle_L^2 \\
&= \lambda^2 \langle \boldsymbol{u}, \boldsymbol{u} \rangle_L^2 \\
&= \big(\langle \boldsymbol{v}, \boldsymbol{v} \rangle_L - \langle \boldsymbol{u}_0, \boldsymbol{u}_0 \rangle_L\big) \langle \boldsymbol{u}, \boldsymbol{u} \rangle_L \\
&\geq \langle \boldsymbol{v}, \boldsymbol{v} \rangle_L \langle \boldsymbol{u}, \boldsymbol{u} \rangle_L > 0,
\end{aligned}$$

using that $\boldsymbol{u}_0$ is spacelike and $\boldsymbol{u}$ is timelike. Taking square roots on both sides leads to $|\langle \boldsymbol{u}, \boldsymbol{v} \rangle_L| \geq \|\boldsymbol{u}\|_L \|\boldsymbol{v}\|_L$, as desired. Finally, note that equality holds if and only if $\langle \boldsymbol{u}_0, \boldsymbol{u}_0 \rangle_L = 0$, that is, if $\boldsymbol{u}_0 = \boldsymbol{0}$. In other words, this is the same as saying that $\boldsymbol{u}$ and $\boldsymbol{v}$ are parallel. □

Another way to prove the previous result is to adapt the proof given for the classical version, analyzing the discriminant of a certain quadratic polynomial. See how to do this on Exercise 1.3.6.

Theorem 1.3.10 (Hyperbolic Angle). *If $\boldsymbol{u}, \boldsymbol{v} \in \mathbb{L}^n$ are both future-directed or past-directed timelike vectors, there is a unique real number $\varphi \in [0, +\infty[$ such that the relation $\langle \boldsymbol{u}, \boldsymbol{v} \rangle_L = -\|\boldsymbol{u}\|_L \|\boldsymbol{v}\|_L \cosh \varphi$ holds. This φ is called the* hyperbolic angle *between the vectors $\boldsymbol{u}$ and $\boldsymbol{v}$.*

Proof: We rewrite the reverse Cauchy-Schwarz inequality as:

$$\frac{|\langle \boldsymbol{u}, \boldsymbol{v} \rangle_L|}{\|\boldsymbol{u}\|_L \|\boldsymbol{v}\|_L} \geq 1.$$

Since $\boldsymbol{u}$ and $\boldsymbol{v}$ point both to the future or to the past, Proposition 1.3.6 gives us that $\langle \boldsymbol{u}, \boldsymbol{v} \rangle_L < 0$, and then:

$$-\frac{\langle \boldsymbol{u}, \boldsymbol{v} \rangle_L}{\|\boldsymbol{u}\|_L \|\boldsymbol{v}\|_L} \geq 1.$$

The function $\cosh\colon [0, +\infty[\to [1, +\infty[$, is bijective, so there always exists a unique $\varphi \in [0, +\infty[$ such that:

$$-\frac{\langle \boldsymbol{u}, \boldsymbol{v} \rangle_L}{\|\boldsymbol{u}\|_L \|\boldsymbol{v}\|_L} = \cosh \varphi.$$

Reorganizing the expression leads to $\langle \boldsymbol{u}, \boldsymbol{v} \rangle_L = -\|\boldsymbol{u}\|_L \|\boldsymbol{v}\|_L \cosh \varphi$. □

Theorem 1.3.11 (Reverse triangle inequality). *Let $u, v \in \mathbb{L}^n$ be both future-directed or past-directed timelike vectors. Then*

$$\|u+v\|_L \geq \|u\|_L + \|v\|_L.$$

Proof: Note that $u+v$ is also timelike, and points to the future or the past, along with u and v (this is a particular case of Exercise 1.3.3). The computation is straightforward:

$$\begin{aligned}\|u+v\|_L^2 &= -\langle u+v, u+v\rangle_L \\ &= -\left(\langle u,u\rangle_L + 2\langle u,v\rangle_L + \langle v,v\rangle_L\right) \\ &= -\langle u,u\rangle_L + 2(-\langle u,v\rangle_L) - \langle v,v\rangle_L \\ &= \|u\|_L^2 + 2(-\langle u,v\rangle_L) + \|v\|_L^2 \\ &\geq \|u\|_L^2 + 2\|u\|_L\|v\|_L + \|v\|_L^2 = (\|u\|_L + \|v\|_L)^2,\end{aligned}$$

leading to $\|u+v\|_L \geq \|u\|_L + \|v\|_L$. □

Example 1.3.12 (The Twins "Paradox"). In Special Relativity, the reverse triangle inequality is used to explain the famous *Twin Paradox*. Natalia and Leticia are twins at the age of 8. Natalia is a fearless explorer and decides to start a journey through the galaxy, despite the vehement disapproval of her parents and sister. She travels in her ship at about 95% of the speed of light, for 5 years (according to her ship's calendar). After that, she gets bored and decides to return, at the same speed, arriving home another 5 years later.

There, on Earth, Natalia (18 years old) meets her sister Leticia, now married and 40 years old. How could this be?

If v is a future-directed timelike vector, connecting the events p and q, $\|v\|_L$ is interpreted as the *proper time* experienced by an observer traveling from p to q, following v. We can model this situation using the plane $\mathbb{L}^2$. The following picture shows the worldlines of Natalia and Leticia:

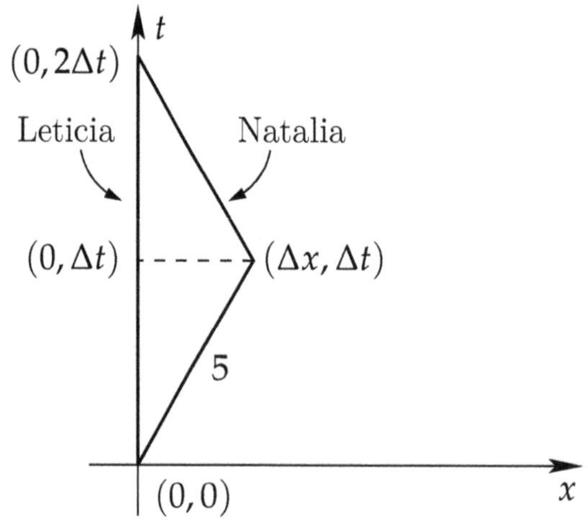

Figure 1.7: The worldlines of the sisters.

The reverse triangle inequality says that Natalia's path is shorter that Leticia's, taking less proper time to be traveled. In particular, the figure helps to effectively calculate the difference between their ages. In geometric units (where $c = 1$), the speed of Natalia's ship is 0.95. With this, the point $(\Delta x, \Delta t)$ satisfies

$$\Delta x^2 - \Delta t^2 = -25 \quad \text{and} \quad \frac{\Delta x}{\Delta t} = 0.95,$$

whence:

$$(0.95\Delta t)^2 - \Delta t^2 = -25 \implies \Delta t = \frac{5}{\sqrt{1-0.95^2}} \approx 16.$$

The symmetry gives $\|(0,32)\|_L = 32$, hence Leticia's age is $8+32 = 40$ years.

The reader might ask why this is a paradox. We could revert the analysis above and consider Natalia's ship as the reference and, in this situation, the one moving at 95% of the speed of light would be Leticia. After Natalia's return, the ages would be swapped in the above calculations.

In fact, despite the argumentation in the previous paragraph, this is not a paradox, since we cannot suppose that Natalia's ship is an inertial frame of reference, since it is subject to accelerations during the travel (at least at the departure, the arrival and in the return maneuver).

Proposition 1.3.13. *Let $\boldsymbol{v} \in \mathbb{L}^n$. Then $\langle \boldsymbol{v},\boldsymbol{v}\rangle_L = \langle \boldsymbol{v},\boldsymbol{v}\rangle_E \cos 2\theta$, where θ is the Euclidean angle between $\boldsymbol{v}$ and the hyperplane $\boldsymbol{e}_n^\perp$ (defined as the complement of the Euclidean angle between $\boldsymbol{v}$ and $\boldsymbol{e}_n$).*

Proof: Given $\boldsymbol{v} \in \mathbb{L}^n$, note the vector $\mathrm{Id}_{n-1,1}\,\boldsymbol{v}$ is just the reflection of $\boldsymbol{v}$ about the hyperplane $\boldsymbol{e}_n^\perp$. If θ is the Euclidean angle between $\boldsymbol{v}$ and the plane $\boldsymbol{e}_n^\perp$, 2θ is the Euclidean angle between $\boldsymbol{v}$ and $\mathrm{Id}_{n-1,1}\,\boldsymbol{v}$. So:

$$\begin{aligned}
\langle \boldsymbol{v},\boldsymbol{v}\rangle_L &= \langle \boldsymbol{v},\mathrm{Id}_{n-1,1}\,\boldsymbol{v}\rangle_E = \|\boldsymbol{v}\|_E\|\,\mathrm{Id}_{n-1,1}\,\boldsymbol{v}\|_E \cos 2\theta \\
&= \|\boldsymbol{v}\|_E\|\boldsymbol{v}\|_E \cos 2\theta = \|\boldsymbol{v}\|_E^2 \cos 2\theta \\
&= \langle \boldsymbol{v},\boldsymbol{v}\rangle_E \cos 2\theta.
\end{aligned}$$

□

Some consequences of this Proposition are explored in Exercises 1.3.8 and 1.3.9.

To wrap up this section, we introduce a class of transformations in $\mathbb{L}^n$, whose importance is revealed in Special Relativity: they are the ones that preserve the causal precedence $\ll$.

Definition 1.3.14 (Causal automorphism). A map $F\colon \mathbb{L}^n \to \mathbb{L}^n$ is a *causal automorphism* if F is bijective and F, as well as F^{-1}, preserve $\ll$, that is:

$$\boldsymbol{x} \ll \boldsymbol{y} \iff F(\boldsymbol{x}) \ll F(\boldsymbol{y}) \quad \text{and} \quad \boldsymbol{x} \ll \boldsymbol{y} \iff F^{-1}(\boldsymbol{x}) \ll F^{-1}(\boldsymbol{y}).$$

Remark. We do not assume linearity for causal automorphisms, and not even continuity. Those properties will be verified soon and it turns out that preserving $\ll$ is equivalent to preserving the relation $<$ defined by saying that $\boldsymbol{p} < \boldsymbol{q}$ if and only if $\boldsymbol{q} \in C_L^+(\boldsymbol{p})$, see Exercise 1.3.10.

Example 1.3.15. "Some" examples of causal automorphisms:

(1) *Positive homotheties*: for $\lambda > 0$, $H_\lambda\colon \mathbb{L}^n \to \mathbb{L}^n$ given by $H_\lambda(\boldsymbol{x}) = \lambda\boldsymbol{x}$;

(2) *Translations*: fixed $\boldsymbol{a} \in \mathbb{L}^n$, $T_{\boldsymbol{a}}\colon \mathbb{L}^n \to \mathbb{L}^n$ given by $T_{\boldsymbol{a}}(\boldsymbol{x}) = \boldsymbol{x} + \boldsymbol{a}$;

(3) *Orthochronous Lorentz transformations*: linear maps $\Lambda\colon \mathbb{L}^n \to \mathbb{L}^n$ satisfying $\langle \Lambda\boldsymbol{x}, \Lambda\boldsymbol{y}\rangle_L = \langle \boldsymbol{x},\boldsymbol{y}\rangle_L$, for any $\boldsymbol{x},\boldsymbol{y} \in \mathbb{L}^n$, such that the set of future-directed timelike vectors is fixed.

Remark. Lorentz transformations play a crucial role in the differential geometry in Lorentzian ambient spaces. They will get full attention in Section 1.4.

Finally, we present the surprising result establishing that every causal automorphism is a composition of the ones listed in the previous example:

Theorem 1.3.16 (Alexandrov-Zeeman). *Let $n \geq 3$ be a fixed integer and $F\colon \mathbb{L}^n \to \mathbb{L}^n$ be a causal automorphism. There exist a positive number $c \in \mathbb{R}_{>0}$, a vector $\boldsymbol{a} \in \mathbb{L}^n$, and an orthochronous Lorentz transformation $\Lambda\colon \mathbb{L}^n \to \mathbb{L}^n$ such that*

$$F(\boldsymbol{x}) = c\Lambda(\boldsymbol{x}) + \boldsymbol{a}, \text{ for all } \boldsymbol{x} \in \mathbb{L}^n.$$

Remark. In particular, causal automorphisms are, up to a translation, linear (hence continuous).

The proof of this result is beyond the scope of this book, but can be found in [52, Section 1.6].

Example 1.3.17. The assumption $n \geq 3$ in Theorem 1.3.16 is necessary. If $n = 2$, consider $\mathbb{L}^2$ with lightlike coordinates (u, v) given by $u \doteq x - y$ and $v \doteq x + y$ (this coordinate change takes the canonical axes on the light rays in the plane), such that

$$\langle (x,y),(x,y)\rangle_L = x^2 - y^2 = \left(\frac{v+u}{2}\right)^2 - \left(\frac{v-u}{2}\right)^2 = uv.$$

Furthermore, if (x, y) is timelike and future-directed, the condition $y > 0$ is rewritten as $v > u$. If $h\colon \mathbb{R} \to \mathbb{R}$ is any increasing diffeomorphism, define $F_h\colon \mathbb{L}^2 \to \mathbb{L}^2$ by

$$F_h(u,v) = \big(h(u), h(v)\big).$$

In terms of the original variables x and y, F_h corresponds[6] to

$$G_h(x,y) \doteq \left(\frac{h(x+y)+h(x-y)}{2}, \frac{h(x+y)-h(x-y)}{2}\right).$$

We claim that F_h (hence G_h) is a causal automorphism. Let $(u_1, v_1), (u_2, v_2) \in \mathbb{L}^2$ be vectors such that $(u_1, v_1) \ll (u_2, v_2)$. We must see that $F_h(u_1, v_1) \ll F_h(u_2, v_2)$. Our assumptions are that $(u_2 - u_1)(v_2 - v_1) < 0$ and $v_2 - v_1 > u_2 - u_1$. First, we see that

$$F_h(u_2,v_2) - F_h(u_1,v_1) = \big(h(u_2) - h(u_1), h(v_2) - h(v_1)\big)$$

is timelike. The Mean Value Theorem gives u^* between u_1 and u_2, and v^* between v_1 and v_2 such that

$$\big(h(u_2) - h(u_1)\big)\big(h(v_2) - h(v_1)\big) = \underbrace{h'(u^*)h'(v^*)}_{>0}\underbrace{(u_2 - u_1)(v_2 - v_1)}_{<0} < 0.$$

The assumption gives $u_2 - u_1 < 0$ e $v_2 - v_1 > 0$. Since h is increasing, it follows that

$$h(v_2) - h(v_1) > 0 > h(u_2) - h(u_1),$$

whence $F_h(u_1, v_1) \ll F_h(u_2, v_2)$.

Finally, notice that $F_h^{-1} = F_{h^{-1}}$ and h^{-1} is also increasing, so that we can repeat the argument with h^{-1} playing the role of h above, leading us to conclude that F_h is a causal automorphism, *for any function h satisfying the given conditions.* But we can choose h

[6]More precisely, if $\varphi\colon \mathbb{L}^2 \to \mathbb{L}^2$ is given by $\varphi(x,y) = (x - y, x + y)$, we consider $G_h \doteq \varphi^{-1} \circ F_h \circ \varphi$.

such that G_h is not of the form stated in Theorem 1.3.16. For example, if $h(t) = \sinh t$, we have

$$G_h(x, y) = (\sinh x \cosh y, \cosh x \sinh y),$$

which is not even an affine map. The surjectiveness of h is also crucial: if $h(t) = \mathrm{e}^t$, then the image of

$$G_h(x, y) = (\mathrm{e}^x \cosh y, \mathrm{e}^x \sinh y)$$

is contained in a spacelike sector of the plane $\mathbb{L}^2$.

Remark.

- Why is this argument invalid for $\mathbb{L}^3$?
- *"Baby" Alexandrov-Zeeman*: if we consider the pathological case when $n = 1$, we would have a map $F: \mathbb{L}^1 \to \mathbb{L}^1$ that is a causal automorphism if and only if it is monotonically increasing and surjective. So, there is a gap in theorem just for the case $n = 2$.

The physical meaning of this result is the following: we can recover the linear structure of Lorentz-Minkowski space from the causal relation between its events.

The first version of Theorem 1.3.16, proved by Zeeman in 1964, was a seminal work that motivated the classification of transformations in spacetimes satisfying properties analogous to $\preccurlyeq$, such as invariance of the light cone (see [1]) or invariance of the set of unit timelike vectors (see [22]).

Exercises

Exercise† 1.3.1 (Review). Let $\varphi \in \mathbb{R}$. The *hyperbolic cosine* and the *hyperbolic sine* of φ are defined by

$$\cosh \varphi \doteq \frac{\mathrm{e}^{\varphi} + \mathrm{e}^{-\varphi}}{2} \quad \text{and} \quad \sinh \varphi \doteq \frac{\mathrm{e}^{\varphi} - \mathrm{e}^{-\varphi}}{2}.$$

Check the following properties:

(a) $\cosh$ is an even function and $\sinh$ is an odd function. Furthermore, $\sinh$ is bijective.

(b) $\cosh \varphi \geq 1$, $\cosh^2 \varphi - \sinh^2 \varphi = 1$ and $\sinh \varphi \leq \cosh \varphi$ for all $\varphi \in \mathbb{R}$.

(c) $\cosh$ and $\sinh$ are differentiable, with $\cosh' = \sinh$ and $\sinh' = \cosh$.

(d) $\sinh(\varphi_1 + \varphi_2) = \sinh \varphi_1 \cosh \varphi_2 + \sinh \varphi_2 \cosh \varphi_1$ and
$\cosh(\varphi_1 + \varphi_2) = \cosh \varphi_1 \cosh \varphi_2 + \sinh \varphi_1 \sinh \varphi_2$, for all $\varphi_1, \varphi_2 \in \mathbb{R}$.

(e) Replace φ_2 by $-\varphi_2$ in (d) and use item (a) to write up formulas for $\sinh(\varphi_1 - \varphi_2)$ and $\cosh(\varphi_1 - \varphi_2)$.

(f) Make $\varphi = \varphi_1 = \varphi_2$ in (d) to state formulas for $\cosh(2\varphi)$ and $\sinh(2\varphi)$.

Exercise 1.3.2. Recall that a subset $C \subseteq \mathbb{L}^n$ is *convex* if for any $\boldsymbol{u}, \boldsymbol{v} \in C$ we have $[\boldsymbol{u}, \boldsymbol{v}] \subseteq C$, where

$$[\boldsymbol{u}, \boldsymbol{v}] \doteq \{(1-t)\boldsymbol{u} + t\boldsymbol{v} \mid 0 \leq t \leq 1\}$$

is the *straight line segment joining* $\boldsymbol{u}$ *to* $\boldsymbol{v}$. Show that, for any $\boldsymbol{p} \in \mathbb{L}^n$, $C_T^+(\boldsymbol{p})$ are $C_T^-(\boldsymbol{p})$ are convex sets.

Exercise 1.3.3. Let $p \in \mathbb{L}^n$ and $u_1, \ldots, u_k \in C_T^+(p)$. For any $\lambda_1, \ldots, \lambda_k > 0$, show that $\sum_{i=1}^{k} \lambda_i u_i \in C_T^+\left(\sum_{i=1}^{k} \lambda_i p\right)$.

Exercise 1.3.4. Show that $\preccurlyeq$ is a *transitive relation*, that is, if $p \preccurlyeq q$ and $q \preccurlyeq r$, then $p \preccurlyeq r$.

Exercise 1.3.5. A subset $C \subseteq \mathbb{L}^n$ is *$\preccurlyeq$-convex* if it satisfies the following: if $p, q \in C$ and $r \in \mathbb{L}^n$ are such that $p \preccurlyeq r \preccurlyeq q$, then $r \in C$. Also, given $S \subseteq \mathbb{L}^n$, the *$\preccurlyeq$-convex hull* of S, denoted by $H_{\preccurlyeq}(S)$, is the smallest subset $\preccurlyeq$-convex of $\mathbb{L}^n$ containing S, that is:

$$H_{\preccurlyeq}(S) \doteq \bigcap \left\{S' \subseteq \mathbb{L}^n \mid S' \text{ is } \preccurlyeq\text{-convex and } S \subseteq S'\right\}.$$

(a) Show that arbitrary intersections of $\preccurlyeq$-convex sets are $\preccurlyeq$-convex. This says that $H_{\preccurlyeq}(S)$ is well-defined, for any set $S \subseteq \mathbb{L}^n$.

(b) Show that $H_{\preccurlyeq}(S) = \{r \in \mathbb{L}^n \mid \text{there exist } p, q \in S \text{ such that } p \preccurlyeq r \preccurlyeq q\}$.

Hint. Show that the set in the right-hand side is $\preccurlyeq$-convex and contained in $H_{\preccurlyeq}(S)$.

Remark. It is possible to define a relation $\preccurlyeq$, called *observable causality* between regions in $\mathbb{L}^n$ and prove results analogous to Theorem 1.3.16 in this setting. The definition of $\preccurlyeq$-convex hull presented here is one of the first tools to achieve that. For details, see [38].

Exercise 1.3.6 (Reverse Cauchy-Schwarz inequality)**.** In this exercise we propose an adaptation for the proof of the standard Cauchy-Schwarz inequality, but for timelike vectors. Let $u, v \in \mathbb{L}^n$ be timelike vectors.

(a) If u and v are parallel to each other, prove directly that $|\langle u, v\rangle_L| = \|u\|_L\|v\|_L$, mimicking the argument in Theorem 1.3.9 (p. 23).

(b) If u and v are linearly independent, they span a timelike plane. Hence $u + tv$ may assume any causal type as we vary $t \in \mathbb{R}$. Using this, analyze the discriminant of

$$p(t) \doteq \langle u + tv, u + tv\rangle_L$$

and conclude that $|\langle u, v\rangle_L| > \|u\|_L\|v\|_L$.

Remark. The inequality is strict here, due to Exercise 1.2.5 (p. 17). Why?

(c) Can you repeat that argument when one of the vectors is lightlike? What if both are? In this setting, is there a necessary and sufficient condition for equality? Discuss.

Exercise 1.3.7 (Lorentz factor)**.** Let $p, q \in \mathbb{L}^n$ be two events and

$$v \doteq q - p = (\Delta x_1, \ldots, \Delta x_{n-1}, \Delta t)$$

the vector joining p and q. Suppose that v is timelike and future-directed (in such a way that p influences q). Show that the hyperbolic angle φ between v and e_n is determined by

$$\gamma \doteq \cosh\varphi = \frac{1}{\sqrt{1 - \|\widetilde{v}\|_E^2}},$$

where $\widetilde{v} \doteq \left(\frac{\Delta x_1}{\Delta t}, \ldots, \frac{\Delta x_{n-1}}{\Delta t}\right) \in \mathbb{R}^{n-1}$ is the spatial velocity vector associated with v. Also show that $\|\widetilde{v}\|_E = \tanh\varphi$.

Remark. The number γ is known as the *Lorentz factor* and it is used in Special Relativity for calculations involving phenomena like *time dilation* and *length contraction*, and formulas for *relativistic energy* and *relativistic linear momentum* of particles traveling at speeds close to the speed of light.

Exercise 1.3.8. Given $v \in \mathbb{L}^n$, show that $\|v\|_L = \|v\|_E \sqrt{|\cos 2\theta|}$, where θ is the Euclidean angle between v and the hyperplane $e_n^\perp$.

Exercise 1.3.9. Let $v \in \mathbb{L}^n$. Then $\|v\|_L \leq \|v\|_E$ and equality holds if and only if v is horizontal or vertical.

Remark. We use "vertical" and "horizontal" in the Euclidean sense we're used to. More precisely, u is vertical if $u \parallel e_n$ and horizontal if $u \perp e_n$. It does not matter whether you use $\perp_E$ or $\perp_L$ in this case.

Exercise 1.3.10. Recall that in $\mathbb{L}^n$ we say that $p < q$ if $q \in C_L^+(p)$. The relations $\ll$ and $<$ may be expressed in terms of each other. For this problem, given $x, y \in \mathbb{L}^n$, you may assume the following equivalences:

- $x < y \iff x \not\ll y$ and $y \ll z$ implies $x \ll z$, for any $z \in \mathbb{L}^n$.
- $x \ll y \iff x \not< y$ and there is $z \in \mathbb{L}^n$ with $x < z < y$.

Show that a bijective map $F\colon \mathbb{L}^n \to \mathbb{L}^n$ is a causal automorphism if and only if F and F^{-1} preserve $<$.

Remark. The proof of the first item you're allowed to assume above is given, for example, in [52, p. 64], while the proof for the second item is similar.

Exercise 1.3.11. In the statement of the Alexandrov-Zeeman theorem, show that the decomposition for the given causal automorphism is unique. That is, if $c_1, c_2 \in \mathbb{R}_{>0}$, $a_1, a_2 \in \mathbb{L}^n$ and Λ_1 and Λ_2 are orthochronous Lorentz transformations such that

$$c_1\Lambda_1(x) + a_1 = c_2\Lambda_2(x) + a_2,$$

for all $x \in \mathbb{L}^n$, show that $c_1 = c_2$, $a_1 = a_2$ and $\Lambda_1 = \Lambda_2$.

1.4 ISOMETRIES IN $\mathbb{R}_\nu^n$

The Euclidean norm $\|\cdot\|_E$ induces a natural distance function in $\mathbb{R}^n$ and, from the geometric point of view, functions that preserve that distance are of great interest. We would like to repeat such study in the Lorentzian ambient, but $\|\cdot\|_L$ does not induce a distance in $\mathbb{L}^n$, since $\|\cdot\|_L$ is not, in the strict sense, a norm.

Definition 1.4.1. A *Euclidean isometry* in $\mathbb{R}^n$ is a map $F\colon \mathbb{R}^n \to \mathbb{R}^n$ such that $\|F(x) - F(y)\|_E = \|x - y\|_E$, for any $x, y \in \mathbb{R}^n$. The set of all Euclidean isometries in $\mathbb{R}^n$ is denoted by $\mathrm{E}(n, \mathbb{R})$.

Remark. Euclidean isometries are also called *rigid motions in* $\mathbb{R}^n$.

To translate this to the Lorentzian ambient just replacing E by L in $\|\cdot\|$, would bring complications, due to the absolute value in the definition of $\|\cdot\|_L$. In other words, this direct translation would not be sensitive to changes in causal type (for instance, switching the axes in $\mathbb{L}^2$ would fit the bill, but axes of different causal types are geometrically and physically very distinct). However, note that F is a Euclidean isometry if and only if

$$\langle F(\boldsymbol{x}) - F(\boldsymbol{y}), F(\boldsymbol{x}) - F(\boldsymbol{y})\rangle_E = \langle \boldsymbol{x} - \boldsymbol{y}, \boldsymbol{x} - \boldsymbol{y}\rangle_E,$$

for any vectors $\boldsymbol{x}, \boldsymbol{y} \in \mathbb{R}^n$. We use this equivalence as a starting point to write the following generalization:

Definition 1.4.2 (Pseudo-Euclidean Isometry). A *pseudo-Euclidean isometry* in $\mathbb{R}^n_\nu$ is a map $F\colon \mathbb{R}^n_\nu \to \mathbb{R}^n_\nu$ such that

$$\langle F(\boldsymbol{x}) - F(\boldsymbol{y}), F(\boldsymbol{x}) - F(\boldsymbol{y})\rangle_\nu = \langle \boldsymbol{x} - \boldsymbol{y}, \boldsymbol{x} - \boldsymbol{y}\rangle_\nu,$$

for any vectors $\boldsymbol{x}, \boldsymbol{y} \in \mathbb{R}^n_\nu$. The set of all pseudo-Euclidean isometries in $\mathbb{R}^n_\nu$ is denoted by $\mathrm{E}_\nu(n, \mathbb{R})$.

Remark. For $\nu = 1$ the pseudo-Euclidean isometries are called *Poincaré transformations*. The set $\mathrm{E}_1(n, \mathbb{R})$ is often written in the literature as $\mathrm{P}(n, \mathbb{R})$.

To study such isometries it is also useful to study functions that preserve $\langle\cdot,\cdot\rangle_\nu$:

Definition 1.4.3 (Pseudo-orthogonal Transformations). A linear map $\Lambda\colon \mathbb{R}^n_\nu \to \mathbb{R}^n_\nu$ is a *pseudo-orthogonal transformation* if

$$\langle \Lambda\boldsymbol{x}, \Lambda\boldsymbol{y}\rangle_\nu = \langle \boldsymbol{x}, \boldsymbol{y}\rangle_\nu,$$

for any vectors $\boldsymbol{x}, \boldsymbol{y} \in \mathbb{R}^n_\nu$. The set of all pseudo-orthogonal transformations in $\mathbb{R}^n_\nu$ is denoted by $\mathrm{O}_\nu(n, \mathbb{R})$.

Remark. If $\nu = 1$, the pseudo-orthogonal transformations are called *Lorentz transformations*.

It follows directly from the definition that every pseudo-orthogonal transformation is a pseudo-Euclidean isometry. We will freely identify a linear transformation with its matricial representation in the canonical basis, and the vectors of $\mathbb{R}^n_\nu$ with column matrices. With that in mind, here are some examples:

Example 1.4.4. In $\mathbb{R}^n$:

(1) Translations. For each $\boldsymbol{a} \in \mathbb{R}^n$, the map $T_{\boldsymbol{a}}\colon \mathbb{R}^n \to \mathbb{R}^n$ given by $T_{\boldsymbol{a}}(\boldsymbol{x}) = \boldsymbol{x} + \boldsymbol{a}$ is an isometry. Notice that $T_{\boldsymbol{a}}$ is bijective and $(T_{\boldsymbol{a}})^{-1} = T_{-\boldsymbol{a}}$.

(2) Rotations. In $\mathbb{R}^2$, given $\theta \in [0, 2\pi[$, the map $R_\theta\colon \mathbb{R}^2 \to \mathbb{R}^2$ given by

$$R_\theta(x, y) = (x\cos\theta - y\sin\theta, x\sin\theta + y\cos\theta)$$

is an orthogonal transformation. The action of R_θ is to rotate (x, y) counter-clockwise around the origin by an angle of θ radians. It is useful to note that:

$$R_\theta(x, y) = \begin{pmatrix} \cos\theta & -\sin\theta \\ \sin\theta & \cos\theta \end{pmatrix} \begin{pmatrix} x \\ y \end{pmatrix}$$

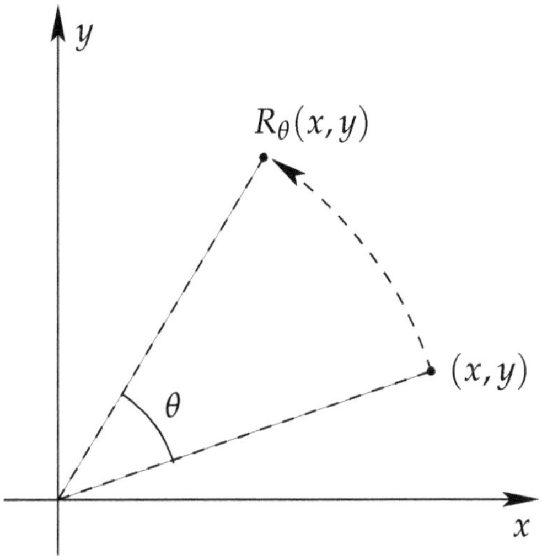

Figure 1.8: Counter-clockwise rotation of an angle θ.

(3) Rotations around coordinate axes. In $\mathbb{R}^3$, given $\theta \in [0, 2\pi[$, one can consider rotations in the form

$$\begin{pmatrix} \cos\theta & -\sin\theta & 0 \\ \sin\theta & \cos\theta & 0 \\ 0 & 0 & 1 \end{pmatrix}, \begin{pmatrix} \cos\theta & 0 & -\sin\theta \\ 0 & 1 & 0 \\ \sin\theta & 0 & \cos\theta \end{pmatrix} \text{ and } \begin{pmatrix} 1 & 0 & 0 \\ 0 & \cos\theta & -\sin\theta \\ 0 & \sin\theta & \cos\theta \end{pmatrix},$$

which are all orthogonal.

(4) Rotations around several axes. In $\mathbb{R}^{2n}$, take angles $\theta_1, \ldots, \theta_n \in [0, 2\pi[$ and consider the block matrix

$$\begin{pmatrix} \cos\theta_1 & -\sin\theta_1 & \cdots & 0 & 0 \\ \sin\theta_1 & \cos\theta_1 & \cdots & 0 & 0 \\ \vdots & \vdots & \ddots & \vdots & \vdots \\ 0 & 0 & \cdots & \cos\theta_n & -\sin\theta_n \\ 0 & 0 & \cdots & \sin\theta_n & \cos\theta_n \end{pmatrix},$$

which is another orthogonal transformation.

(5) Generalized reflections. In $\mathbb{R}^n$, given $\epsilon_1, \cdots, \epsilon_n \in \{-1, 1\}$, the map $R\colon \mathbb{R}^n \to \mathbb{R}^n$ given by

$$R(x_1, \cdots, x_n) \doteq (\epsilon_1 x_1, \cdots, \epsilon_n x_n)$$

is an orthogonal transformation. In particular, if $\epsilon_i = -1$ for a single index i, then R is the reflection on the hyperplane $x_i = 0$. When $\epsilon_i = -1$ for all i, we have $R = -\operatorname{Id}_{\mathbb{R}^n}$, the so-called *antipodal map* in $\mathbb{R}^n$.

(6) Axes permutations. In $\mathbb{R}^n$, considering a permutation[7] $\sigma \in S_n$, we have that $\Sigma\colon \mathbb{R}^n \to \mathbb{R}^n$ given by

$$\Sigma(x_1, \cdots, x_n) \doteq (x_{\sigma(1)}, \cdots, x_{\sigma(n)})$$

is an orthogonal transformation.

[7] That is, a bijection $\sigma\colon \{1, \cdots, n\} \to \{1, \cdots, n\}$. The set of all such bijections, endowed with the composition operation, is a *group* and denoted by S_n (permutations of n letters). For further details, see [23].

In analogy to the examples above, we have:

Example 1.4.5. In $\mathbb{L}^n$:

(1) Translations, as before.

(2) Hyperbolic rotations. Given any $\varphi \in \mathbb{R}$, the map $R^h_\varphi \colon \mathbb{L}^2 \to \mathbb{L}^2$ given by

$$R^h_\varphi(x,y) \doteq (x\cosh\varphi + y\sinh\varphi, x\sinh\varphi + y\cosh\varphi)$$

is a Lorentz transformation. As before, one can see R^h_φ in the form

$$R^h_\varphi(x,y) = \begin{pmatrix} \cosh\varphi & \sinh\varphi \\ \sinh\varphi & \cosh\varphi \end{pmatrix} \begin{pmatrix} x \\ y \end{pmatrix}.$$

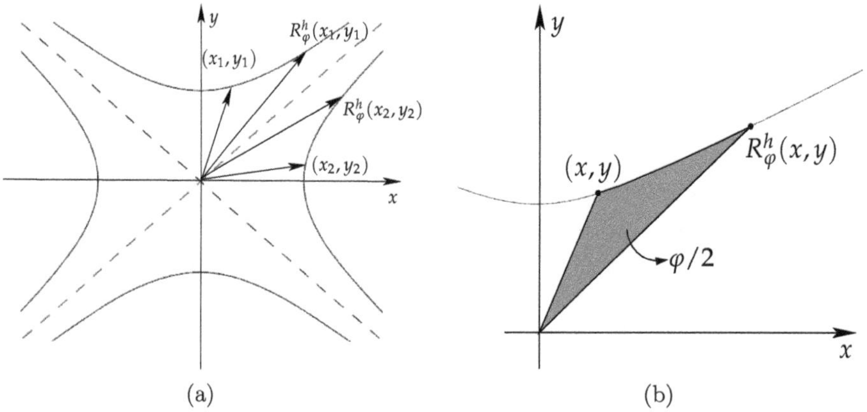

Figure 1.9: Hyperbolic rotation by angle $\varphi > 0$.

Notice that R^h_φ preserves each branch of $\{x \in \mathbb{L}^2 \mid \langle x,x\rangle_L = \pm 1\}$. See Exercise 1.4.2.

(3) Rotations around coordinate axes. Here we must take causal types into account: given $\theta \in [0, 2\pi[$ and $\varphi \in \mathbb{R}$, we can consider, among many others, the maps given by

$$\begin{pmatrix} \cos\theta & -\sin\theta & 0 \\ \sin\theta & \cos\theta & 0 \\ 0 & 0 & 1 \end{pmatrix}, \quad \begin{pmatrix} \cosh\varphi & 0 & \sinh\varphi \\ 0 & 1 & 0 \\ \sinh\varphi & 0 & \cosh\varphi \end{pmatrix} \quad \text{and} \quad \begin{pmatrix} 1 & 0 & 0 \\ 0 & \cosh\varphi & \sinh\varphi \\ 0 & \sinh\varphi & \cosh\varphi \end{pmatrix}.$$

(4) Rotations around several axes. In $\mathbb{L}^{2n}$, consider angles $\theta_1, \cdots, \theta_{n-1} \in [0, 2\pi[$ and $\varphi \in \mathbb{R}$ and the block matrix

$$\begin{pmatrix} \cos\theta_1 & -\sin\theta_1 & \cdots & 0 & 0 \\ \sin\theta_1 & \cos\theta_1 & \cdots & 0 & 0 \\ \vdots & \vdots & \ddots & \vdots & \vdots \\ 0 & 0 & \cdots & \cosh\varphi & \sinh\varphi \\ 0 & 0 & \cdots & \sinh\varphi & \cosh\varphi \end{pmatrix},$$

which is a Lorentz transformation. Is it a Lorentz transformation if the block with hyperbolic functions is not the last one?

(5) Generalized reflections, as before.

(6) Spacelike axis permutations. In $\mathbb{L}^n$, considering $\sigma \in S_{n-1}$, $\Sigma_L \colon \mathbb{L}^n \to \mathbb{L}^n$ given by

$$\Sigma_L(x_1, \cdots, x_n) \doteq \left(x_{\sigma(1)}, \cdots, x_{\sigma(n-1)}, x_n\right)$$

is a Lorentz transformation. Would Σ_L still be a Lorentz transformation if the shuffling acted on x_n?

Remark. From now on, the following natural identifications will be adopted:

$$\mathrm{O}_0(n,\mathbb{R}) \equiv \mathrm{O}(n,\mathbb{R}), \quad \mathrm{E}_0(n,\mathbb{R}) \equiv \mathrm{E}(n,\mathbb{R}) \quad \text{and} \quad \mathrm{Id}_{n-0,0} \equiv \mathrm{Id}_n\,.$$

Lemma 1.4.6. *Let $A, B \in \mathrm{Mat}(n,\mathbb{R})$ such that $\boldsymbol{x}^\top A\boldsymbol{y} = \boldsymbol{x}^\top B\boldsymbol{y}$, for any vectors $\boldsymbol{x}, \boldsymbol{y} \in \mathbb{R}^n$. Then $A = B$.*

Proof: Take $\boldsymbol{x} = \boldsymbol{e}_i$ and $\boldsymbol{y} = \boldsymbol{e}_j$, whence

$$a_{ij} = \boldsymbol{e}_i^\top A\boldsymbol{e}_j = \boldsymbol{e}_i^\top B\boldsymbol{e}_j = b_{ij},$$

for any i and j. Therefore $A = B$. □

Proposition 1.4.7. *Let $\Lambda \colon \mathbb{R}^n_\nu \to \mathbb{R}^n_\nu$ be a linear map. Then Λ is pseudo-orthogonal if and only if $\Lambda^\top \mathrm{Id}_{n-\nu,\nu}\,\Lambda = \mathrm{Id}_{n-\nu,\nu}$.*

Proof: Suppose that $\Lambda \in \mathrm{O}_\nu(n,\mathbb{R})$. Then, given any vectors $\boldsymbol{x}, \boldsymbol{y} \in \mathbb{R}^n_\nu$, it holds that $\langle \Lambda\boldsymbol{x}, \Lambda\boldsymbol{y}\rangle_\nu = \langle \boldsymbol{x}, \boldsymbol{y}\rangle_\nu$. In matrix notation this is written as

$$(\Lambda\boldsymbol{x})^\top \mathrm{Id}_{n-\nu,\nu}(\Lambda\boldsymbol{y}) = \boldsymbol{x}^\top \mathrm{Id}_{n-\nu,\nu}\,\boldsymbol{y} \implies \boldsymbol{x}^\top\Lambda^\top \mathrm{Id}_{n-\nu,\nu}\,\Lambda\boldsymbol{y} = \boldsymbol{x}^\top \mathrm{Id}_{n-\nu,\nu}\,\boldsymbol{y}.$$

Since $\boldsymbol{x}$ and $\boldsymbol{y}$ are arbitrary, the previous lemma says that $\Lambda^\top \mathrm{Id}_{n-\nu,\nu}\,\Lambda = \mathrm{Id}_{n-\nu,\nu}$. The converse is now clear. □

Corollary 1.4.8. *Let $\Lambda \in \mathrm{O}_\nu(n,\mathbb{R})$. Then $\det\Lambda \in \{-1, 1\}$. In particular, Λ is a linear isomorphism and Λ^{-1} is always defined.*

Proposition 1.4.9. *The set $\mathrm{O}_\nu(n,\mathbb{R})$ is a group, if endowed with the usual matrix multiplication. Furthermore, $\mathrm{O}_\nu(n,\mathbb{R})$ is closed under matrix transposition. Thus, $\mathrm{O}_\nu(n,\mathbb{R})$ is called a* pseudo-orthogonal group. *In particular, $\mathrm{O}(n,\mathbb{R})$ and $\mathrm{O}_1(n,\mathbb{R})$ are respectively called the* orthogonal group *and the* Lorentz group.

Proof: The verification that these sets are in fact groups is left for Exercise 1.4.3. To check closure under transposition, consider $\Lambda \in \mathrm{O}_\nu(n,\mathbb{R})$: inverting the expression $\Lambda^\top \mathrm{Id}_{n-\nu,\nu}\,\Lambda = \mathrm{Id}_{n-\nu,\nu}$ we achieve $\Lambda^{-1}\mathrm{Id}_{n-\nu,\nu}(\Lambda^\top)^{-1} = \mathrm{Id}_{n-\nu,\nu}$. Solving for $\mathrm{Id}_{n-\nu,\nu}$ in the left-hand side, we have:

$$\mathrm{Id}_{n-\nu,\nu} = \Lambda\,\mathrm{Id}_{n-\nu,\nu}(\Lambda^\top) = (\Lambda^\top)^\top \mathrm{Id}_{n-\nu,\nu}\,\Lambda^\top,$$

showing that $\Lambda^\top \in \mathrm{O}_\nu(n,\mathbb{R})$. □

Corollary 1.4.10. *If $\Lambda \in \mathrm{O}_\nu(n,\mathbb{R})$ then its columns (as well as its rows) form an orthonormal basis of $\mathbb{R}^n_\nu$.*

Proof: Considering the previous proposition, it suffices to verify the statement for the columns of Λ. In fact, noting that the i-th column of Λ is Λe_i we have

$$\langle \Lambda e_i, \Lambda e_j \rangle_\nu = \langle e_i, e_j \rangle_\nu.$$

The result follows from the fact that the canonical basis is orthonormal relative to $\langle \cdot, \cdot \rangle_\nu$, for any ν. □

Remark. In other words, the result above shows that every $\Lambda \in \mathrm{O}_\nu(n, \mathbb{R})$ maps orthonormal bases to orthonormal bases.

Note that the composition of elements of $\mathrm{O}_\nu(n, \mathbb{R})$ with translations produce elements of $\mathrm{E}_\nu(n, \mathbb{R})$. Now we prove that every map in $\mathrm{E}_\nu(n, \mathbb{R})$ is such a composition.

Proposition 1.4.11. *Let $F \in \mathrm{E}_\nu(n, \mathbb{R})$. If $F(\mathbf{0}) = \mathbf{0}$, then $F \in \mathrm{O}_\nu(n, \mathbb{R})$.*

Proof: A *polarization identity* for $\langle \cdot, \cdot \rangle_\nu$ is

$$\langle x, y \rangle_\nu = \frac{1}{2}\left(\langle x, x \rangle_\nu + \langle y, y \rangle_\nu - \langle x - y, x - y \rangle_\nu\right),$$

for any vectors $x, y \in \mathbb{R}^n_\nu$. Applying this to $F(x)$ and $F(y)$ instead of x and y, we have that:

$$\langle F(x), F(y) \rangle_\nu = \frac{1}{2}\left(\langle F(x), F(x) \rangle_\nu + \langle F(y), F(y) \rangle_\nu - \langle F(x) - F(y), F(x) - F(y) \rangle_\nu\right),$$

Since $F(\mathbf{0}) = \mathbf{0}$, it follows that

$$\langle F(x), F(x) \rangle_\nu = \langle F(x) - \mathbf{0}, F(x) - \mathbf{0} \rangle_\nu = \langle x - \mathbf{0}, x - \mathbf{0} \rangle_\nu = \langle x, x \rangle_\nu,$$

and the same holds for y. Hence, the comparison of these two formulas gives $\langle F(x), F(y) \rangle_\nu = \langle x, y \rangle_\nu$. The linearity of F then follows from Exercise 1.4.4. □

Theorem 1.4.12. *Let $F \in \mathrm{E}_\nu(n, \mathbb{R})$. Then there are a unique $a \in \mathbb{R}^n_\nu$ and $\Lambda \in \mathrm{O}_\nu(n, \mathbb{R})$ such that $F = T_a \circ \Lambda$.*

Proof: We know that $T_{-F(\mathbf{0})} \circ F \in \mathrm{E}_\nu(n, \mathbb{R})$ is an isometry taking $\mathbf{0}$ into $\mathbf{0}$, hence $T_{-F(\mathbf{0})} \circ F = \Lambda \in \mathrm{O}_\nu(n, \mathbb{R})$. Thus $F = T_{F(\mathbf{0})} \circ \Lambda$. To check the uniqueness, suppose that $T_{a_1} \circ \Lambda_1 = T_{a_2} \circ \Lambda_2$ for some $a_1, a_2 \in \mathbb{R}^n_\nu$ and $\Lambda_1, \Lambda_2 \in \mathrm{O}_\nu(n, \mathbb{R})$. Evaluating at $\mathbf{0}$ we obtain

$$a_1 = T_{a_1}(\mathbf{0}) = T_{a_1}(\Lambda_1(\mathbf{0})) = T_{a_2}(\Lambda_2(\mathbf{0})) = T_{a_2}(\mathbf{0}) = a_2.$$

Since T_{a_1} is bijective, $T_{a_1} \circ \Lambda_1 = T_{a_1} \circ \Lambda_2$ implies $\Lambda_1 = \Lambda_2$. □

Remark. In the notation above, Λ and T_a are called, respectively, the *linear part* and the *affine part* of F.

Corollary 1.4.13. *Every $F \in \mathrm{E}_\nu(n, \mathbb{R})$ is bijective.*

It is opportune to note that $\mathrm{E}_\nu(n, \mathbb{R})$ is also a group, when equipped with the composition operation. See Exercise 1.4.5. It is called the *Euclidean group* when $\nu = 0$, and the *Poincaré group* when $\nu = 1$.

The following result depends on some Differential Calculus concepts. See Appendix A if necessary.

Corollary 1.4.14. *Let $F \in \mathrm{E}_\nu(n, \mathbb{R})$. Then F is differentiable and, for each $p \in \mathbb{R}^n_\nu$, $DF(p) \in \mathrm{O}_\nu(n, \mathbb{R})$.*

Proof: Notice that $F = T_a \circ \Lambda$, for some $a \in \mathbb{R}^n_\nu$ and $\Lambda \in \mathrm{O}_\nu(n,\mathbb{R})$. Hence F is differentiable as a composition of differentiable maps. Moreover, we have:

$$DF(p) = D(T_a \circ \Lambda)(p) = DT_a(\Lambda(p)) \circ D\Lambda(p) = \mathrm{id}_{\mathbb{R}^n_\nu} \circ \Lambda = \Lambda \in \mathrm{O}_\nu(n,\mathbb{R}).$$

□

Proposition 1.4.15. *Let $p, q \in \mathbb{R}^n_\nu$, and $(v_1, \cdots, v_n), (w_1, \cdots, w_n)$ be two bases of $\mathbb{R}^n_\nu$ such that $\langle v_i, v_j \rangle_\nu = \langle w_i, w_j \rangle_\nu$ for all $1 \le i, j \le n$. Then there exists a unique $F \in \mathrm{E}_\nu(n,\mathbb{R})$ such that $F(p) = q$ and $DF(p)(v_i) = w_i$ for all $1 \le i \le n$.*

Proof: Since $F = T_a \circ \Lambda$, for some vector $a \in \mathbb{R}^n_\nu$ and some linear map $\Lambda \in \mathrm{O}_\nu(n,\mathbb{R})$, it suffices to exhibit a and Λ. The hypothesis $\langle v_i, v_j \rangle_\nu = \langle w_i, w_j \rangle_\nu$ ensures that the unique linear map Λ characterized by $\Lambda v_i = w_i$, for all i, is in $\mathrm{O}_\nu(n,\mathbb{R})$. Finally, the condition $F(p) = q$ forces $a \doteq q - \Lambda p$. □

Beyond the fact that $\Lambda \in \mathrm{O}_\nu(n,\mathbb{R})$ implies $\det \Lambda = \pm 1$ (see Corollary 1.4.8), we have that if $\Lambda = (\lambda_{ij})_{1 \le i,j \le n} \in \mathrm{O}_1(n,\mathbb{R})$, then $|\lambda_{nn}| \ge 1$. This follows from Corollary 1.4.10, by observing that the column Λe_n is a unit timelike vector. On the other hand, for $C = (c_{ij})_{1 \le i,j \le n} \in \mathrm{O}(n,\mathbb{R})$ it holds that $|c_{nn}| \le 1$.

Definition 1.4.16. The group

$$\mathrm{SO}_\nu(n,\mathbb{R}) \doteq \{\Lambda \in \mathrm{O}_\nu(n,\mathbb{R}) \mid \det \Lambda = 1\}$$

is called the *special pseudo-orthogonal group.*

Remark.

- When $\nu = 1$, we have the *special Lorentz group*. We conveniently write, as before, $\mathrm{SO}_0(n,\mathbb{R}) = \mathrm{SO}(n,\mathbb{R})$ when it is convenient.
- An element of $\mathrm{SO}_1(n,\mathbb{R})$ is called a *proper Lorentz transformation.* A Poincaré transformation is also called proper when its linear part is proper.

Now, let $\mathcal{B} = (u_1, \ldots, u_n)$ be an orthonormal basis for $\mathbb{L}^n$, where u_n is timelike. Also, let B be the matrix with the vectors u_i as its columns. Then, we have:

Definition 1.4.17. An orthonormal basis $\mathcal{B} = (u_1, \ldots, u_n)$ of $\mathbb{L}^n$ is *future-oriented* if u_n is a future-directed vector. It is *past-oriented* if u_n is past-directed.

Definition 1.4.18. A Lorentz transformation $\Lambda \in \mathrm{O}_1(n,\mathbb{R})$ is *orthochronous* (that is, *preserves time orientation*) if, for any future-oriented orthonormal basis $\mathcal{B}$, the (ordered) basis formed by the columns of ΛB, denoted by $\Lambda\mathcal{B}$, is also future-oriented. The set of orthochronous Lorentz transformations is denoted by $\mathrm{O}_1^\uparrow(n,\mathbb{R})$.

Remark. A Poincaré transformation is orthochronous if its linear part is orthochronous.

Proposition 1.4.19. *The set $\mathrm{O}_1^\uparrow(n,\mathbb{R})$ is a group.*

Proof: Let $\mathcal{B}$ be a future-oriented orthonormal basis of $\mathbb{L}^n$.

- Let $\Lambda_1, \Lambda_2 \in \mathrm{O}_1^\uparrow(n,\mathbb{R})$. Then $\Lambda_2\mathcal{B}$ is future-oriented since $\Lambda_2 \in \mathrm{O}_1^\uparrow(n,\mathbb{R})$. Hence $(\Lambda_1\Lambda_2)\mathcal{B} = \Lambda_1(\Lambda_2\mathcal{B})$ is also future-oriented, since $\Lambda_1 \in \mathrm{O}_1^\uparrow(n,\mathbb{R})$. In this way $\Lambda_1\Lambda_2 \in \mathrm{O}_1^\uparrow(n,\mathbb{R})$.

- $\mathrm{Id}_n \mathscr{B} = \mathscr{B}$, so it is clear that $\mathrm{Id}_n \in \mathrm{O}_1^{\uparrow}(n, \mathbb{R})$.
- Let $\Lambda \in \mathrm{O}_1^{\uparrow}(n, \mathbb{R})$. Then $\mathscr{B} = \Lambda^{-1}(\Lambda \mathscr{B})$. Since $\Lambda \mathscr{B}$ and $\mathscr{B}$ are both future-oriented, Λ^{-1} preserves time orientation, that is, $\Lambda^{-1} \in \mathrm{O}_1^{\uparrow}(n, \mathbb{R})$.

□

Previously we noted that if $\Lambda = (\lambda_{ij})_{1 \leq i,j \leq n} \in \mathrm{O}_1(n, \mathbb{R})$, then $|\lambda_{nn}| \geq 1$, and we could have λ_{nn} being positive or negative. It is usual in the literature to declare a Lorentz transformation Λ to be orthochronous if $\lambda_{nn} \geq 1$. The equivalence between this and our definition comes in the:

Theorem 1.4.20 (Characterization of $\mathrm{O}_1^{\uparrow}(n, \mathbb{R})$). *Let $\Lambda = (\lambda_{ij})_{1 \leq i,j \leq n}$ be a Lorentz transformation. Then:*

$$\Lambda \in \mathrm{O}_1^{\uparrow}(n, \mathbb{R}) \iff \lambda_{nn} \geq 1$$

Proof: Start supposing that Λ is orthochronous. Then $\Lambda \boldsymbol{e}_n$ is future-directed, as well as $\boldsymbol{e}_n$, and hence $\lambda_{nn} \geq 1$. Conversely, suppose that $\lambda_{nn} \geq 1$. Let $(\boldsymbol{u}_1, \ldots, \boldsymbol{u}_n)$ be a future-directed orthonormal basis of $\mathbb{L}^n$. Then $\boldsymbol{u}_n$ is timelike and future-directed, the same holding for $\Lambda \boldsymbol{u}_n$. We know that $\Lambda \boldsymbol{u}_n$ is timelike, since $\Lambda \in \mathrm{O}_1(n, \mathbb{R})$. It remains to check that its last coordinate is positive. Computing the product (recall that $\Lambda^\top \boldsymbol{e}_i$ is the i-th row of Λ), we see that such coordinate is

$$\langle \Lambda^\top \boldsymbol{e}_n, \boldsymbol{u}_n \rangle_E = \langle \Lambda^\top \boldsymbol{e}_n, \mathrm{Id}_{n-1,1} \boldsymbol{u}_n \rangle_L.$$

Since $\lambda_{nn} \geq 1$, it follows that $\Lambda^\top \boldsymbol{e}_n$ is future-directed. Furthermore, the fact that $\boldsymbol{u}_n$ is future-directed implies that $\mathrm{Id}_{n-1,1} \boldsymbol{u}_n$ is past-directed. Hence

$$\langle \Lambda^\top \boldsymbol{e}_n, \mathrm{Id}_{n-1,1} \boldsymbol{u}_n \rangle_L > 0,$$

as desired. □

Remark. The previous result could be proved using topological arguments: since $\mathbb{L}^n$ is finite-dimensional, Λ is continuous; $\det \Lambda \neq 0$ ensures the non-singularity of Λ. We know that Λ^{-1} is also linear, hence Λ^{-1} is continuous. So Λ is a homeomorphism. The time-cone, $C_T(\boldsymbol{0})$, has two connected components, that are preserved (if Λ is orthochronous) or switched (if Λ is not orthochronous).

The group $\mathrm{O}_1(n, \mathbb{R})$ admits a partition in subsets according to the signs of λ_{nn} and $\det \Lambda$ or, equivalently, the signs of λ_{nn} and the determinant of the spatial part of Λ. We formalize this recalling the splitting $\mathbb{R}^n_\nu = \mathbb{R}^{n-\nu} \times \mathbb{R}^\nu_\nu$, used in the proof of Theorem 1.2.25 (p. 14). With it we write $\Lambda \in \mathrm{O}_\nu(n, \mathbb{R})$ as blocks:

$$\Lambda = \left(\begin{array}{c|c} \Lambda_S & B \\ \hline C & \Lambda_T \end{array} \right),$$

where $\Lambda_S \in \mathrm{Mat}(n-\nu, \mathbb{R})$ and $\Lambda_T \in \mathrm{Mat}(\nu, \mathbb{R})$ are, respectively, the *spatial and temporal parts of* Λ. Since Λ is non-singular and preserves causal types, one sees (by composing with appropriate projections) that both Λ_S and Λ_T are also non-singular. We have:

Theorem 1.4.21. *Let $0 < \nu < n$ and $\Lambda \in \mathrm{O}_\nu(n, \mathbb{R})$. Then* $\det \Lambda_S = \det \Lambda_T \det \Lambda$.

Proof: As always let $\mathsf{can} = (e_i)_{i=1}^n$ be the canonical basis of $\mathbb{R}^n_\nu$. Consider also the orthonormal basis of $\mathbb{R}^n_\nu$ formed by the columns of Λ, $\mathcal{B} = (\Lambda e_1, \ldots, \Lambda e_n)$. Suppose that $\Lambda = (\lambda_{ij})_{1\leq i,j\leq n}$. We "delete" the block B, defining a linear map $T\colon \mathbb{R}^n_\nu \to \mathbb{R}^n_\nu$ by

$$T(\Lambda e_j) = \begin{cases} \Lambda e_j, & \text{if } 1 \leq j \leq n-\nu \quad \text{and} \\ \sum_{i=n-\nu+1}^n \lambda_{ij} e_i, & \text{if } n-\nu < j \leq n. \end{cases}$$

One easily sees that $[\Lambda]_{\mathsf{can},\mathcal{B}} = \mathrm{Id}_n$ and

$$[T]_{\mathcal{B},\mathsf{can}} = \left(\begin{array}{c|c} \Lambda_S & 0 \\ \hline C & \Lambda_T \end{array}\right).$$

Now, we compute the matrix $[T]_{\mathcal{B}}$. The expression $T(\Lambda e_j) = \Lambda e_j$, which holds for $1 \leq j \leq n-\nu$, says that the left upper and lower blocks of $[T]_{\mathcal{B}}$ are, respectively, $\mathrm{Id}_{n-\nu}$ and 0. To compute the determinant of $[T]_{\mathcal{B}}$ by blocks, we need to know the last ν entries of $T(\Lambda e_j)$ in the basis $\mathcal{B}$, for any $n-\nu < j \leq n$. Letting $\epsilon_k \doteq \epsilon_{e_k}$, Lemma 1.2.32 (p. 16) gives

$$\begin{aligned} T(\Lambda e_j) &= \sum_{i=n-\nu+1}^n \lambda_{ij} e_i = \sum_{i=n-\nu+1}^n \lambda_{ij} \sum_{k=1}^n \epsilon_k \langle e_i, \Lambda e_k \rangle_\nu \Lambda e_k \\ &= \sum_{i=n-\nu+1}^n \sum_{k=1}^n \sum_{\ell=1}^n \epsilon_k \lambda_{ij} \lambda_{\ell k} \langle e_i, e_\ell \rangle_\nu \Lambda e_k \\ &= \sum_{k=1}^n \left(\sum_{i=n-\nu+1}^n \sum_{\ell=1}^n \epsilon_k \lambda_{ij} \lambda_{\ell k} \eta^\nu_{i\ell} \right) \Lambda e_k. \end{aligned}$$

The desired last ν entries correspond to $n-\nu < k \leq n$ and, in this setting, the entries in the right lower block of $[T]_{\mathcal{B}}$ are given by

$$\sum_{i=n-\nu+1}^n \sum_{\ell=1}^n -\lambda_{ij}\lambda_{\ell k}(-\delta_{i\ell}) = \sum_{i=n-\nu+1}^n \lambda_{ij}\lambda_{ik},$$

which are the entries of the product of $\Lambda_T^\top$ and Λ_T, up to renaming indexes, if necessary. Then

$$[T]_{\mathcal{B}} = \left(\begin{array}{c|c} \mathrm{Id}_{n-\nu} & * \\ \hline 0 & \Lambda_T^\top \Lambda_T \end{array}\right).$$

In particular, $\det T = (\det \Lambda_T)^2$. Furthermore:

$$[T\Lambda]_{\mathcal{B}} = [T]_{\mathcal{B},\mathsf{can}}[\Lambda]_{\mathsf{can},\mathcal{B}} = \left(\begin{array}{c|c} \Lambda_S & 0 \\ \hline C & \Lambda_T \end{array}\right).$$

Hence

$$(\det \Lambda_T)^2 \det \Lambda = \det T \det \Lambda = \det(T\Lambda) = \det \Lambda_T \det \Lambda_S,$$

or $\det \Lambda_S = \det \Lambda_T \det \Lambda$, as desired. □

With this we can, as stated above, index the elements of $\mathrm{O}_\nu(n,\mathbb{R})$ according to the signs of $\det\Lambda_S$ and $\det\Lambda_T$, obtaining a partition of $\mathrm{O}_\nu(n,\mathbb{R})$:

$$\begin{aligned}
\mathrm{O}_\nu^{+\uparrow}(n,\mathbb{R}) &\doteq \{\Lambda \in \mathrm{O}_\nu(n,\mathbb{R}) \mid \det\Lambda_S > 0 \text{ and } \det\Lambda_T > 0\}\\
\mathrm{O}_\nu^{+\downarrow}(n,\mathbb{R}) &\doteq \{\Lambda \in \mathrm{O}_\nu(n,\mathbb{R}) \mid \det\Lambda_S > 0 \text{ and } \det\Lambda_T < 0\}\\
\mathrm{O}_\nu^{-\uparrow}(n,\mathbb{R}) &\doteq \{\Lambda \in \mathrm{O}_\nu(n,\mathbb{R}) \mid \det\Lambda_S < 0 \text{ and } \det\Lambda_T > 0\}\\
\mathrm{O}_\nu^{-\downarrow}(n,\mathbb{R}) &\doteq \{\Lambda \in \mathrm{O}_\nu(n,\mathbb{R}) \mid \det\Lambda_S < 0 \text{ and } \det\Lambda_T < 0\}
\end{aligned}$$

The elements of $\mathrm{O}_\nu^{+\bullet}(n,\mathbb{R})$ are said to *preserve space orientation*, while a matrix in $\mathrm{O}_\nu^{\bullet\uparrow}(n,\mathbb{R})$ *preserves time orientation* (also called *orthochronous*). If $\det\Lambda > 0$, then Λ preserves the algebraic orientation of $\mathbb{R}^n_\nu$. On the other hand, if $\Lambda \in \mathrm{O}_1(n,\mathbb{R})$ and $\det\Lambda_S > 0$, Λ preserves *spatial orientation* of $\mathbb{R}^n_\nu$, that is, it preserves the orientation of its spacelike subspaces. The previous theorem shows that if $\det\Lambda_T > 0$, then both orientation preservation concepts are equivalent.

One can show that $\mathrm{O}_\nu^{+\uparrow}(n,\mathbb{R})$ is a (normal) subgroup of $\mathrm{O}_\nu(n,\mathbb{R})$ and, if $\nu = 1$, then $\mathrm{O}_1^{+\uparrow}(n,\mathbb{R})$ is called the *special orthochronous Lorentz group*. It is clear, in this case, that $\mathrm{O}_1^{+\uparrow}(n,\mathbb{R}) \doteq \mathrm{O}_1^{\uparrow}(n,\mathbb{R}) \cap \mathrm{SO}_1(n,\mathbb{R})$ is an intersection of subgroups of $\mathrm{O}_1(n,\mathbb{R})$, hence a group itself. The relevance of $\mathrm{O}_\nu^{+\uparrow}(n,\mathbb{R})$ is highlighted when we use the *four principal representatives*

$$\tau^{+\uparrow} \doteq \mathrm{Id}_n, \quad \tau^{+\downarrow} \doteq \mathrm{Id}_{n-1,1}, \quad \tau^{-\uparrow} \doteq \mathrm{Id}_{1,n-1} \text{ and } \tau^{-\downarrow} \doteq \mathrm{diag}(-1,1,\ldots,1,-1),$$

in the following:

Proposition 1.4.22. *The sets* $\mathrm{O}_\nu^{+\downarrow}(n,\mathbb{R}), \mathrm{O}_\nu^{-\uparrow}(n,\mathbb{R})$ *and* $\mathrm{O}_\nu^{-\downarrow}(n,\mathbb{R})$ *are cosets of* $\mathrm{O}_\nu^{+\uparrow}(n,\mathbb{R})$*, whose representatives are* $\tau^{+\downarrow}, \tau^{-\uparrow}$*, and* $\tau^{-\downarrow}$*. In other words,*

$$\begin{aligned}
\mathrm{O}_\nu^{+\downarrow}(n,\mathbb{R}) &= \tau^{+\downarrow}\cdot \mathrm{O}_\nu^{+\uparrow}(n,\mathbb{R})\\
\mathrm{O}_\nu^{-\uparrow}(n,\mathbb{R}) &= \tau^{-\uparrow}\cdot \mathrm{O}_\nu^{+\uparrow}(n,\mathbb{R})\\
\mathrm{O}_\nu^{-\downarrow}(n,\mathbb{R}) &= \tau^{-\downarrow}\cdot \mathrm{O}_\nu^{+\uparrow}(n,\mathbb{R}).
\end{aligned}$$

Proof: It suffices to note how the four principal representatives act on elements of $\mathrm{O}_\nu(n,\mathbb{R})$ by left multiplication (for instance, $\tau^{-\uparrow}$ reverts just the sign of first row, $\tau^{+\downarrow}$ reverts the sign of last row, etc.), and that squaring all of the principal representatives we get the identity. See Exercise 1.4.14 for more details. □

Remark. The set $G \doteq \{\tau^{+\uparrow}, \tau^{+\downarrow}, \tau^{-\uparrow}, \tau^{-\downarrow}\}$, equipped with matrix multiplication is a group. This is a straightforward computation (suggested in Exercise 1.4.15). This helps us derive a "sign rule" for indexes in the elements of G.

Exercises

Exercise 1.4.1. Follow the notation in examples 1.4.4 and 1.4.5 (p. 30 and p. 32) and for $\theta, \theta' \in [0, 2\pi[$ and $\varphi, \varphi' \in \mathbb{R}$:

(a) Compute $R_\theta(\cos\theta', \sin\theta')$ and $R^h_\varphi(\sinh\varphi', \cosh\varphi')$. Give geometrical interpretations.

(b) Compute $R_\theta \circ R_{\theta'}$ and $R^h_\varphi \circ R^h_{\varphi'}$. Use those expressions to write $(R_\theta)^{-1}$ and $(R^h_\varphi)^{-1}$ in terms of R_θ and R^h_φ. What is the angle between $\boldsymbol{u}$ and $R_\theta(\boldsymbol{u})$, and what is the hyperbolic angle between $\boldsymbol{u}$ and $R^h_\varphi(\boldsymbol{u})$, if $\boldsymbol{u}$ is a timelike vector?

(c) Let D_θ and D^h_φ be the matrices whose entries are the derivatives of the entries in R_θ and R^h_φ, respectively. Compute

$$J \doteq D_\theta \circ (R_\theta)^{-1} \quad \text{and} \quad J^h \doteq D^h_\varphi \circ (R^h_\varphi)^{-1}.$$

Remark. Notice that D_θ is a rotation, while D^h_φ is not.

(d) Show that

$$\langle J\boldsymbol{x}, \boldsymbol{y}\rangle_E = -\langle \boldsymbol{x}, J\boldsymbol{y}\rangle_E \quad \text{and} \quad \langle J^h\boldsymbol{x}, \boldsymbol{y}\rangle_L = -\langle \boldsymbol{x}, J^h\boldsymbol{y}\rangle_L,$$

for any vectors $\boldsymbol{x}, \boldsymbol{y} \in \mathbb{R}^2_\nu$.

Remark. Maps with the above property are called *skew-symmetric*. You will face them again in Exercise 1.4.12.

Exercise 1.4.2. Let $\varphi > 0$ and R^h_φ be the hyperbolic rotation defined in Example 1.4.5.

(a) Show that the map R^h_φ preserves each branch of the set $\{\boldsymbol{x} \in \mathbb{L}^2 \mid \langle \boldsymbol{x}, \boldsymbol{x}\rangle_L = \pm 1\}$, as indicated in Figure 1.9 (a) (p. 32).

(b) Let $\theta \in \mathbb{R}$. Show that the region bounded by the upper branch of hyperbola given by $x^2 - y^2 = -1$ and the line segments joining

$$(\sinh\theta, \cosh\theta) \text{ and } (\sinh(\theta+\varphi), \cosh(\theta+\varphi))$$

to the origin $(0,0)$ has area $\varphi/2$ in $\mathbb{R}^2$, according to Figure 1.9 (b).

Hint. You may find it useful to solve this for $\theta = 0$ first.

Exercise† 1.4.3. Show that:

(a) $\mathrm{Id}_n \in \mathrm{O}_\nu(n, \mathbb{R})$;

(b) if $\Lambda_1, \Lambda_2 \in \mathrm{O}_\nu(n, \mathbb{R})$, then $\Lambda_1\Lambda_2 \in \mathrm{O}_\nu(n, \mathbb{R})$;

(c) if $\Lambda \in \mathrm{O}_\nu(n, \mathbb{R})$, then $\Lambda^{-1} \in \mathrm{O}_\nu(n, \mathbb{R})$.

Use this to prove the first part of Proposition 1.4.9 (p. 33), that is, $\mathrm{O}_\nu(n, \mathbb{R})$ is a group.

Exercise† 1.4.4. In the text, we defined pseudo-orthogonal transformations as *linear* maps that preserve $\langle \cdot, \cdot \rangle_\nu$. The word "linear" is not necessary in that definition. Prove that if $\Lambda \colon \mathbb{R}^n_\nu \to \mathbb{R}^n_\nu$ preserves $\langle \cdot, \cdot \rangle_\nu$, then Λ is automatically linear.

Hint. Use that Λ takes orthonormal bases into orthonormal bases, write $\boldsymbol{v} = \sum_{i=1}^{n} a_i\, \boldsymbol{e}_i$, $\Lambda\boldsymbol{v} = \sum_{i=1}^{n} b_i\, \Lambda\boldsymbol{e}_i$ and then prove that $a_i = b_i$ for all i.

Exercise 1.4.5. Show that the set $\mathrm{E}_\nu(n, \mathbb{R})$ is a group, when equipped with composition of functions.

Exercise 1.4.6. Consider the points $p = (1,0,3,3)$ and $q = (-1,2,5,6)$ in $\mathbb{L}^4$, as well as the vectors

$$\begin{aligned} v_1 &= (1,0,3,0), & w_1 &= (1,0,5,4) \\ v_2 &= (0,2,0,-3), & w_2 &= (0,2,-4,-5) \\ v_3 &= (-1,1,0,3), & w_3 &= (-1,1,4,5) \\ v_4 &= (0,0,6,3), & w_4 &= (0,0,14,13). \end{aligned}$$

Write out the unique Poincaré transformation $F \in \mathrm{P}(4,\mathbb{R})$ such that $F(p) = q$ and $DF(p)(v_i) = w_i$, for $1 \leq i \leq 4$.

Exercise[†] 1.4.7. Let $\mathscr{B} = (u_1,\ldots,u_n)$ and $\mathscr{C} = (v_1,\ldots,v_n)$ be ordered bases of $\mathbb{L}^n$. Show that, if $G_{\mathbb{L}^n,\mathscr{B}} = G_{\mathbb{L}^n,\mathscr{C}}$, then the change of basis matrix between $\mathscr{B}$ and $\mathscr{C}$ defines a Lorentz transformation.

Exercise 1.4.8. In the product $\mathrm{O}_\nu(n,\mathbb{R}) \times \mathbb{R}^n$, define the operation $*$ by

$$(A,v) * (B,w) \doteq (AB, Aw + v).$$

With this operation, we write the product as $\mathrm{O}_\nu(n,\mathbb{R}) \ltimes \mathbb{R}^n$.

(a) Prove that $(\mathrm{Id}_n, \mathbf{0})$ is the identity for the operation $*$ (which is not commutative).

(b) Prove that $*$ is associative.

(c) Exhibit the inverse of an element (A,v) according to $*$.

Remark. The previous items show that $\mathrm{O}_\nu(n,\mathbb{R}) \ltimes \mathbb{R}^n$ is a group, called an *(outer) semi-direct product* of $\mathrm{O}_\nu(n,\mathbb{R})$ and $\mathbb{R}^n$.

(d) Prove that $\{\mathrm{Id}_n\} \times \mathbb{R}^n$ is a normal subgroup of $\mathrm{O}_\nu(n,\mathbb{R}) \ltimes \mathbb{R}^n$ and that

$$\frac{\mathrm{O}_\nu(n,\mathbb{R}) \ltimes \mathbb{R}^n}{\{\mathrm{Id}_n\} \times \mathbb{R}^n} \cong \mathrm{O}_\nu(n,\mathbb{R}).$$

Hint. Show that the projection on $\mathrm{O}_\nu(n,\mathbb{R})$ is a group epimorphism and compute its kernel.

(e) Define $\Phi\colon \mathrm{O}_\nu(n,\mathbb{R}) \ltimes \mathbb{R}^n \to \mathrm{E}_\nu(n,\mathbb{R})$ by $\Phi(A,v) = T_v \circ A$. Show that Φ is a group isomorphism.

Exercise[†] 1.4.9. Let $\Lambda \in \mathrm{O}_1(n,\mathbb{R})$ be a Lorentz transformation.

(a) Show that for any non-lightlike eigenvector, its associated eigenvalue is 1 or -1.

(b) Show that the product of any two eigenvalues associated to linearly independent lightlike eigenvectors is 1.

(c) Let $U \subseteq \mathbb{L}^n$ be an eigenspace of Λ that contains a non-lightlike eigenvector. Show that any other eigenspace is Lorentz-orthogonal to U.

Hint. Use item (a).

(d) If $U \subseteq \mathbb{L}^n$ is a vector subspace, show that U is Λ-invariant if and only if $U^\perp$ is Λ-invariant.

Exercise 1.4.10 (Householder reflections). Let $H \subseteq \mathbb{L}^n$ be a hyperplane. Suppose that H is not lightlike and take a vector $\boldsymbol{n}$ normal to H. Write $\mathbb{L}^n = H \oplus \mathrm{span}\{\boldsymbol{n}\}$ (see Exercise 1.2.7, p. 17). Then each $\boldsymbol{v} \in \mathbb{L}^n$ is written uniquely as $\boldsymbol{v}_H + \lambda\boldsymbol{n}$, for some $\lambda \in \mathbb{R}$ and $\boldsymbol{v}_H \in H$, satisfying $\langle \boldsymbol{v}_H, \boldsymbol{n}\rangle_L = 0$. Define the *reflection relative to the hyperplane* H as $R_{\boldsymbol{n}} \colon \mathbb{L}^n \to \mathbb{L}^n$, given by $R_{\boldsymbol{n}}(\boldsymbol{v}) = \boldsymbol{v}_H - \lambda\boldsymbol{n}$.

(a) Show that $R_{\boldsymbol{n}}$ is a Lorentz transformation.

Hint. Use Exercise 1.4.4 to skip some calculations.

(b) Let $\boldsymbol{u}, \boldsymbol{v} \in \mathbb{L}^n$ be vectors such that $\langle \boldsymbol{u}, \boldsymbol{u}\rangle_L = \langle \boldsymbol{v}, \boldsymbol{v}\rangle_L \neq 0$. Show that it is possible to get $\boldsymbol{u}$ from $\boldsymbol{v}$ using one or two reflections.

Hint. Deal with two different situations: if $\boldsymbol{v} - \boldsymbol{u}$ is not a lightlike vector, compute $R_{\boldsymbol{v}-\boldsymbol{u}}(\boldsymbol{v})$; if $\boldsymbol{v} - \boldsymbol{u}$ is lightlike, show that $\boldsymbol{v} + \boldsymbol{u}$ is not lightlike, and compute $R_{\boldsymbol{v}+\boldsymbol{u}}(\boldsymbol{v})$.

Exercise 1.4.11 (Cartan-Dieudonné). Let $\Lambda \in \mathrm{O}_1(n, \mathbb{R})$ be a Lorentz transformation. Then Λ is a composition of reflections relative to non-lightlike hyperplanes.

Hint. Proceed by induction on n. For the induction step, take a non-lightlike vector $\boldsymbol{v} \in \mathbb{L}^n$, apply item (b) of Exercise 1.4.10 to take $\Lambda\boldsymbol{v}$ to $\boldsymbol{v}$ using a reflection R (or composition of reflections). Then consider $(R \circ \Lambda)\big|_{\boldsymbol{v}^\perp}$ and use item (d) of Exercise 1.4.9.

Remark.

- It is possible to show that Λ is the composition of at most n reflections. This result holds also in a more general context, see [15] for details.
- On the other hand, this result is false in infinite dimensions. For instance, since $R_{\boldsymbol{n}}$ acts as the identity on H, the isometry $-\mathbf{id}$ cannot be written as a finite product of reflections.

Exercise 1.4.12. A linear map $T \colon \mathbb{R}^n_\nu \to \mathbb{R}^n_\nu$ is *skew-symmetric* if, for any $\boldsymbol{v}, \boldsymbol{w} \in \mathbb{R}^n_\nu$, we have $\langle T\boldsymbol{v}, \boldsymbol{w}\rangle_\nu = -\langle \boldsymbol{v}, T\boldsymbol{w}\rangle_\nu$. In addition to the maps presented in Exercise 1.4.1, the cross product (to be defined in Section 1.6) with a fixed vector of $\mathbb{L}^3$ also has this property.

(a) Show that for any pseudo-orthonormal basis $\mathcal{B} = (\boldsymbol{u}_1, \boldsymbol{u}_2, \ldots, \boldsymbol{u}_n)$ of $\mathbb{R}^n_\nu$, where the last ν vectors are timelike, the map T is written in blocks as:

$$[T]_{\mathcal{B}} = \left(\begin{array}{c|c} T_S & A \\ \hline A^\top & T_T \end{array}\right),$$

where $T_S \in \mathrm{Mat}(n-\nu, \mathbb{R})$ and $T_T \in \mathrm{Mat}(\nu, \mathbb{R})$ are skew-symmetric blocks, and $A \in \mathrm{Mat}((n-\nu) \times \nu, \mathbb{R})$. What is the trace of T?

(b) Show that $\ker T = (\mathrm{Im}\, T)^\perp$ and $\mathrm{Im}\, T = (\ker T)^\perp$.

Hint. Prove the first identity directly and apply $\perp$ on both sides for the second one.

(c) If λ is an eigenvalue of T and $\boldsymbol{v}$ is an eigenvector associated to it, show that $\lambda = 0$ or $\boldsymbol{v}$ is lightlike (both can happen simultaneously).

(d) If $U \subseteq \mathbb{R}^n_\nu$ is a vector subspace, show that U is T-invariant if and only if $U^\perp$ is T-invariant.

Exercise 1.4.13. Let $T\colon \mathbb{L}^n \to \mathbb{L}^n$ be a skew-symmetric linear map, as in Exercise 1.4.12 above. The map $E_T\colon \mathbb{L}^n \to \mathbb{L}^n$ given by

$$E_T = \frac{1}{4\pi}\left(\frac{1}{n}\operatorname{tr}(T^2)\operatorname{id}_{\mathbb{L}^n} - T^2\right)$$

is called the *energy-momentum map associated to T*.

(a) Show that E_T is *self-adjoint* according to $\langle\cdot,\cdot\rangle_L$, that is,

$$\langle E_T\boldsymbol{v}, \boldsymbol{w}\rangle_L = \langle \boldsymbol{v}, E_T\boldsymbol{w}\rangle_L,$$

for any $\boldsymbol{v}, \boldsymbol{w} \in \mathbb{L}^n$. Show also that $\operatorname{tr}(E_T) = 0$.

(b) Show that if $\boldsymbol{v}$ is an eigenvector of T, then $\boldsymbol{v}$ is an eigenvector of E_T (perhaps associated to another eigenvalue). "Conversely", show that if $\boldsymbol{v}$ is an eigenvector of E_T, then $\boldsymbol{v}$ is an eigenvector of T^2.

Remark. Maps with such property appear naturally in Physics. They are used to model electromagnetic fields — it is possible to define the electric ($\boldsymbol{E}$) and the magnetic ($\boldsymbol{B}$) fields associated to T. One can then study properties of T from $\boldsymbol{E}$ and $\boldsymbol{B}$. More about that on [52].

Exercise† 1.4.14. Provide details for the proof of Proposition 1.4.22 (p. 38).

Exercise 1.4.15. Consider $G \doteq \{\tau^{+\uparrow}, \tau^{+\downarrow}, \tau^{-\uparrow}, \tau^{-\downarrow}\}$, equipped with the matrix multiplication.

(a) Show that G is a group and derive a "sign rule" for indices in the elements of G.

(b) Show that G is isomorphic to the group of reflections in the diagonal of a square (which is a realization of the *Klein group*).

Exercise 1.4.16 (Isogonal transformations)**.** A linear map $T\colon \mathbb{R}^n_\nu \to \mathbb{R}^n_\nu$ is called *isogonal* if it preserves orthogonality, that is: if $\boldsymbol{x}, \boldsymbol{y} \in \mathbb{R}^n_\nu$ are vectors such that $\langle \boldsymbol{x}, \boldsymbol{y}\rangle_\nu = 0$, then $\langle T\boldsymbol{x}, T\boldsymbol{y}\rangle_\nu = 0$. Show that:

(a) In $\mathbb{R}^n_\nu$, homotheties are isogonal. In $\mathbb{R}^n$ orthogonal transformations are isogonal and, in $\mathbb{L}^n$, Lorentz transformations are isogonal. In general, compositions of isogonal transformations are isogonal.

(b) In $\mathbb{R}^n$, every isogonal transformation is the composition of an orthogonal transformation and a homothety.

Hint. If T is isogonal and $T \neq 0$, show that T is injective. See how T acts on an orthonormal basis, and define an orthogonal transformation C in the "opposite direction". Show that $C \circ T$ is a homothety (using that $C \circ T$ is also isogonal).

(c) In $\mathbb{L}^n$, every *injective* isogonal transformation is the composition of a Lorentz transformation and a homothety.

Hint. Adapt your proof for the item above, with care.

(d) Provide a counterexample to the previous item when the transformation is not injective.

Hint. What happens if the image of T is a light ray?

1.5 INVESTIGATING $O_1(2,\mathbb{R})$ AND $O_1(3,\mathbb{R})$

1.5.1 The group $O_1(2,\mathbb{R})$ in detail

In order to fix the ideas studied so far, we'll characterize the Lorentz transformations in $\mathbb{L}^2$. For this end, we'll use the properties of hyperbolic trigonometric functions mentioned in Exercise 1.3.1 (p. 27). Taking Proposition 1.4.22 (p. 38) into account, it suffices to describe $O_1^{+\uparrow}(2,\mathbb{R})$ along with its cosets.

If $\Lambda = (\lambda_{ij})_{1\leq i,j\leq n} \in O_1^{+\uparrow}(2,\mathbb{R})$, then its columns are Lorentz-orthonormal, that is,

$$\langle \Lambda e_i, \Lambda e_j\rangle_L = \eta^1_{ij} \implies \begin{cases} \lambda_{11}^2 - \lambda_{21}^2 & = & 1 \\ \lambda_{12}^2 - \lambda_{22}^2 & = & -1 \\ \lambda_{11}\lambda_{12} - \lambda_{21}\lambda_{22} & = & 0 \end{cases} \quad \text{and,}$$

in addition, $\lambda_{11}, \lambda_{22} \geq 0$. It follows, from the first two equations, that $\lambda_{11}, \lambda_{22} \geq 1$. Therefore, there are unique $t, s \in \mathbb{R}_{\geq 0}$ such that $\lambda_{11} = \cosh t$ and $\lambda_{22} = \cosh s$. Those equations also show that $|\lambda_{21}| = \sinh t$ and $|\lambda_{12}| = \sinh s$.

Since Λ is a proper transformation, it follows that $\cosh t \cosh s - \lambda_{12}\lambda_{21} = 1$, whence $0 \leq \cosh t \cosh s - 1 = \lambda_{12}\lambda_{21}$, meaning that λ_{12} and λ_{21} share the same sign or vanish simultaneously. Regardless of which sign this is, the third equation above gives

$$0 = \cosh t \sinh s - \sinh t \cosh s = \sinh(s-t) \implies s = t.$$

Hence, we have

$$\Lambda = \begin{pmatrix} \cosh t & \sinh t \\ \sinh t & \cosh t \end{pmatrix} \quad \text{or} \quad \begin{pmatrix} \cosh t & -\sinh t \\ -\sinh t & \cosh t \end{pmatrix}, \text{ for some } t > 0.$$

In a more concise way, we write

$$\Lambda = \begin{pmatrix} \cosh\varphi & \sinh\varphi \\ \sinh\varphi & \cosh\varphi \end{pmatrix}, \text{ for some } \varphi \in \mathbb{R}.$$

Such matrices clearly lie in $O_1^{+\uparrow}(2,\mathbb{R})$, leading to

$$O_1^{+\uparrow}(2,\mathbb{R}) = \left\{ \begin{pmatrix} \cosh\varphi & \sinh\varphi \\ \sinh\varphi & \cosh\varphi \end{pmatrix} \,\middle|\, \varphi \in \mathbb{R} \right\}.$$

This allows us to write the remaining cosets in $O_1(2,\mathbb{R})$: in $\mathbb{L}^2$, we have

$$\tau^{+\uparrow} = \begin{pmatrix} 1 & 0 \\ 0 & 1 \end{pmatrix}, \quad \tau^{+\downarrow} = \begin{pmatrix} 1 & 0 \\ 0 & -1 \end{pmatrix}, \quad \tau^{-\uparrow} = \begin{pmatrix} -1 & 0 \\ 0 & 1 \end{pmatrix} \text{ and } \tau^{-\downarrow} = \begin{pmatrix} -1 & 0 \\ 0 & -1 \end{pmatrix},$$

whence

$$O_1^{+\downarrow}(2,\mathbb{R}) = \tau^{+\downarrow}\cdot O_1^{+\uparrow}(2,\mathbb{R}) = \left\{ \begin{pmatrix} \cosh\varphi & \sinh\varphi \\ -\sinh\varphi & -\cosh\varphi \end{pmatrix} \,\middle|\, \varphi \in \mathbb{R} \right\},$$

$$O_1^{-\uparrow}(2,\mathbb{R}) = \tau^{-\uparrow}\cdot O_1^{+\uparrow}(2,\mathbb{R}) = \left\{ \begin{pmatrix} -\cosh\varphi & -\sinh\varphi \\ \sinh\varphi & \cosh\varphi \end{pmatrix} \,\middle|\, \varphi \in \mathbb{R} \right\},$$

$$O_1^{-\downarrow}(2,\mathbb{R}) = \tau^{-\downarrow}\cdot O_1^{+\uparrow}(2,\mathbb{R}) = \left\{ \begin{pmatrix} -\cosh\varphi & -\sinh\varphi \\ -\sinh\varphi & -\cosh\varphi \end{pmatrix} \,\middle|\, \varphi \in \mathbb{R} \right\}.$$

It is important to notice that an analogous procedure leads to a similar characterization of $\mathrm{SO}(2,\mathbb{R})$. In this case, if $C = (c_{ij})_{1\leq i,j\leq 2}$ satisfies

$$\begin{cases} c_{11}^2 + c_{21}^2 & = & 1 \\ c_{12}^2 + c_{22}^2 & = & 1 \\ c_{11}c_{12} + c_{21}c_{22} & = & 0 \\ c_{11}c_{22} - c_{12}c_{21} & = & 1 \end{cases},$$

there exists $\theta \in [0, 2\pi[$ such that

$$C = \begin{pmatrix} \cos\theta & -\sin\theta \\ \sin\theta & \cos\theta \end{pmatrix}.$$

1.5.2 The group $\mathrm{O}_1(3,\mathbb{R})$ in (a little less) detail

Just repeating the strategy used in the previous section is not enough: there we described the elements using a single parameter, while now we'll need three parameters to describe $\mathrm{O}_1^{+\uparrow}(3,\mathbb{R})$. We start with the:

Proposition 1.5.1. *Let* $\Lambda \in \mathrm{SO}_1(3,\mathbb{R})$. *Then* 1 *is an eigenvalue of* Λ.

Proof: It suffices to show that $\det(\mathrm{Id}_3 - \Lambda) = 0$. We have:

$$\begin{aligned} \det(\mathrm{Id}_3 - \Lambda) &= \det(\mathrm{Id}_{2,1}^2 - \Lambda) \\ &= \det(\mathrm{Id}_{2,1}\Lambda^\top \mathrm{Id}_{2,1}\Lambda - \Lambda) \\ &= \det(\mathrm{Id}_{2,1}\Lambda^\top \mathrm{Id}_{2,1} - \mathrm{Id}_3)\det\Lambda \\ &= \det(\mathrm{Id}_{2,1}(\Lambda^\top - \mathrm{Id}_3)\mathrm{Id}_{2,1})\det\Lambda \\ &= \det(\Lambda^\top - \mathrm{Id}_3) \\ &= -\det(\mathrm{Id}_3 - \Lambda). \end{aligned}$$

□

Remark. The previous result, and its proof, remain valid for any odd integer n.

This ensures that every element of $\mathrm{SO}_1(3,\mathbb{R})$ fixes, pointwise, some line passing through the origin of $\mathbb{L}^3$.

Definition 1.5.2. Let $\Lambda \in \mathrm{SO}_1(3,\mathbb{R})$, $\Lambda \neq \mathrm{Id}_3$, and $\boldsymbol{v}$ be an eigenvector of Λ associated to the eigenvalue 1. The map Λ is:

(i) *hyperbolic*, if $\boldsymbol{v}$ is spacelike;

(ii) *elliptic*, if $\boldsymbol{v}$ is timelike;

(iii) *parabolic*, if $\boldsymbol{v}$ is lightlike.

Remark.

- The elements of $\mathrm{SO}_1(3,\mathbb{R})$ fix pointwise exactly one direction. To wit, if there exist two distinct eigenvectors associated to 1, since $\det\Lambda = 1$, we would have that $\Lambda = \mathrm{Id}_3$ (write the matrix of Λ in the basis of its eigenvectors).
- The nomenclature above extends to Poincaré maps, considering linear parts.

As before, in order to study $O_1(3,\mathbb{R})$, it suffices to focus on $O_1^{+\uparrow}(3,\mathbb{R})$. Consider $\Lambda = (\lambda_{ij})_{1\leq i,j\leq 3} \in O_1^{+\uparrow}(3,\mathbb{R})$ and let U be the line fixed by Λ. To justify the nomenclature in the above definition we start analyzing the special cases when U is $\operatorname{span}\{e_1\}$ (hyperbolic), $\operatorname{span}\{e_3\}$ (elliptic), or $\operatorname{span}\{(0,1,1)\}$ (parabolic), that is, one line of each causal type.

(1) $U = \operatorname{span}\{e_1\}$: in this case we have $\Lambda e_1 = e_1$. Exercise 1.4.9 (p. 40) shows that $\Lambda e_2, \Lambda e_3 \in \operatorname{span}\{e_2, e_3\}$, and hence $\lambda_{12} = \lambda_{13} = 0$. So far, we can write

$$\Lambda = \begin{pmatrix} 1 & 0 & 0 \\ 0 & \lambda_{22} & \lambda_{23} \\ 0 & \lambda_{32} & \lambda_{33} \end{pmatrix}$$

but, since $\Lambda \in O_1^{+\uparrow}(3,\mathbb{R})$, we have

$$\begin{pmatrix} \lambda_{22} & \lambda_{23} \\ \lambda_{32} & \lambda_{33} \end{pmatrix} \in O_1^{+\uparrow}(2,\mathbb{R}).$$

Therefore, there exists $\varphi \in \mathbb{R}$ such that

$$\Lambda = \begin{pmatrix} 1 & 0 & 0 \\ 0 & \cosh\varphi & \sinh\varphi \\ 0 & \sinh\varphi & \cosh\varphi \end{pmatrix}.$$

(2) $U = \operatorname{span}\{e_3\}$: now we have $\Lambda e_3 = e_3$. As above, Λe_1 and Λe_2 lie in $\operatorname{span}\{e_1, e_2\}$, thus $\lambda_{31} = \lambda_{32} = 0$. So far, the matrix has the form

$$\Lambda = \begin{pmatrix} \lambda_{11} & \lambda_{12} & 0 \\ \lambda_{21} & \lambda_{22} & 0 \\ 0 & 0 & 1 \end{pmatrix},$$

but, since $\Lambda \in O_1^{+\uparrow}(3,\mathbb{R})$, we have

$$\begin{pmatrix} \lambda_{11} & \lambda_{12} \\ \lambda_{21} & \lambda_{22} \end{pmatrix} \in SO(2,\mathbb{R}).$$

Therefore, there exists $\theta \in [0, 2\pi[$ such that

$$\Lambda = \begin{pmatrix} \cos\theta & -\sin\theta & 0 \\ \sin\theta & \cos\theta & 0 \\ 0 & 0 & 1 \end{pmatrix}.$$

(3) $U = \operatorname{span}\{e_2 + e_3\}$: we shall determine $\Lambda \equiv [\Lambda]_{\text{can}}$ indirectly, calculating the matrix $[\Lambda]_{\mathcal{B}} = (\theta_{ij})_{1\leq i,j\leq 3}$ first, where $\mathcal{B} = (e_1, e_2, e_2 + e_3)$. Relative to this basis, the Lorentz product is given by

$$\langle (x_1, y_1, z_1)_{\mathcal{B}}, (x_2, y_2, z_2)_{\mathcal{B}} \rangle_L = x_1x_2 + y_1y_2 + y_1z_2 + z_1y_2,$$

for any $(x, y, z)_{\mathcal{B}} = xe_1 + ye_2 + z(e_2 + e_3)$.

Since $\Lambda(e_2 + e_3) = e_2 + e_3$, we see that the third column of $[\Lambda]_{\mathcal{B}}$ is precisely the vector e_3. Furthermore, $e_1 \perp (e_2 + e_3)$ and, from Exercise 1.4.9 (p. 40), it follows that $\Lambda e_1 \in (e_2 + e_3)^{\perp} = \operatorname{span}\{e_1, e_2 + e_3\}$. In other words, $\theta_{21} = 0$.

Expanding the determinant of $[\Lambda]_{\mathcal{B}}$ from its already known third column, we have $1 = \det[\Lambda]_{\mathcal{B}} = \theta_{11}\theta_{22}$, that is, $\theta_{22} = 1/\theta_{11}$.

So far, we have

$$[\Lambda]_{\mathcal{B}} = \begin{pmatrix} \theta_{11} & \theta_{12} & 0 \\ 0 & \theta_{11}^{-1} & 0 \\ \theta_{31} & \theta_{32} & 1 \end{pmatrix}.$$

In order to find out the remaining entries, we note that

$$\begin{aligned} 1 &= \langle e_2, e_2 + e_3 \rangle_L &&= \langle \Lambda e_2, \Lambda(e_2 + e_3) \rangle_L &&= \theta_{11}^{-1}, \\ 1 &= \langle e_2, e_2 \rangle_L &&= \langle \Lambda e_2, \Lambda e_2 \rangle_L &&= \theta_{12}^2 + 1 + 2\theta_{32}, \\ 0 &= \langle e_1, e_2 \rangle_L &&= \langle \Lambda e_1, \Lambda e_2 \rangle_L &&= \theta_{12} + \theta_{31}. \end{aligned}$$

Setting $\theta \doteq \theta_{31}$ we have

$$[\Lambda]_{\mathcal{B}} = \begin{pmatrix} 1 & -\theta & 0 \\ 0 & 1 & 0 \\ \theta & -\theta^2/2 & 1 \end{pmatrix}.$$

In the canonical basis:

$$\begin{aligned} \Lambda = [\Lambda]_{\text{can}} &= [\text{Id}_{\mathbb{L}^3}]_{\mathcal{B},\text{can}} [\Lambda]_{\mathcal{B}} [\text{Id}_{\mathbb{L}^3}]_{\mathcal{B},\text{can}}^{-1} \\ &= \begin{pmatrix} 1 & 0 & 0 \\ 0 & 1 & 1 \\ 0 & 0 & 1 \end{pmatrix} \begin{pmatrix} 1 & -\theta & 0 \\ 0 & 1 & 0 \\ \theta & -\theta^2/2 & 1 \end{pmatrix} \begin{pmatrix} 1 & 0 & 0 \\ 0 & 1 & 1 \\ 0 & 0 & 1 \end{pmatrix}^{-1} \\ &= \begin{pmatrix} 1 & -\theta & \theta \\ \theta & 1 - \theta^2/2 & \theta^2/2 \\ \theta & -\theta^2/2 & 1 + \theta^2/2 \end{pmatrix}. \end{aligned}$$

Theorem 1.5.3. *Let $\Lambda \in \mathrm{O}_1^{+\uparrow}(3, \mathbb{R})$. Then Λ is similar to one of the three matrices described above.*

Proof: Let U be the line fixed by Λ. One (and only one) of the following cases holds:

(1) If Λ is hyperbolic, let $u_1 \in U$ be a unit vector. Extend $\{u_1\}$ to an orthonormal basis (u_1, u_2, u_3) of $\mathbb{L}^3$, with u_3 being timelike and future-directed[8]. Let $P \in \mathrm{O}_1^{\uparrow}(3, \mathbb{R})$ be such that $Pe_i = u_i$, for $1 \leq i \leq 3$. Then $P^{-1}\Lambda P \in \mathrm{O}_1^{+\uparrow}(3, \mathbb{R})$ satisfies

$$P^{-1}\Lambda P e_1 = P^{-1}\Lambda u_1 = P^{-1}u_1 = e_1.$$

Hence there exists $\varphi \in \mathbb{R}$ such that

$$\Lambda = P \begin{pmatrix} 1 & 0 & 0 \\ 0 & \cosh\varphi & \sinh\varphi \\ 0 & \sinh\varphi & \cosh\varphi \end{pmatrix} P^{-1}.$$

[8]In fact, it turns out that the time direction of u_3 is irrelevant, since $\mathrm{O}_1^{+\uparrow}(3, \mathbb{R})$ is a normal subgroup of $\mathrm{O}_1(3, \mathbb{R})$.

(2) If Λ is elliptic, let $u_3 \in U$ be a future-directed unit vector. Extend $\{u_3\}$ to an orthonormal basis (u_1, u_2, u_3) of $\mathbb{L}^3$. As before, take $P \in \mathrm{O}_1^{\uparrow}(3,\mathbb{R})$ such that $Pe_i = u_i$, for $1 \leq i \leq 3$. Then $P^{-1}\Lambda P \in \mathrm{O}_1^{+\uparrow}(3,\mathbb{R})$ satisfies

$$P^{-1}\Lambda P e_3 = P^{-1}\Lambda u_3 = P^{-1}u_3 = e_3.$$

Then there exists $\theta \in [0, 2\pi[$ such that

$$\Lambda = P \begin{pmatrix} \cos\theta & -\sin\theta & 0 \\ \sin\theta & \cos\theta & 0 \\ 0 & 0 & 1 \end{pmatrix} P^{-1}.$$

(3) If Λ is parabolic, take $u \in U$ of the form $u = (a, b, 1)$. Since $a^2 + b^2 = 1$, we have $a = \cos\theta_0$ and $b = \sin\theta_0$ for some $\theta_0 \in [0, 2\pi[$. Consider

$$P = \begin{pmatrix} \cos\left(\frac{\pi}{2} - \theta_0\right) & \sin\left(\frac{\pi}{2} - \theta_0\right) & 0 \\ -\sin\left(\frac{\pi}{2} - \theta_0\right) & \cos\left(\frac{\pi}{2} - \theta_0\right) & 0 \\ 0 & 0 & 1 \end{pmatrix}.$$

Once more, $P^{-1}\Lambda P \in \mathrm{O}_1^{+\uparrow}(3,\mathbb{R})$ satisfies

$$P^{-1}\Lambda P(e_2 + e_3) = P^{-1}\Lambda u = P^{-1}u = e_2 + e_3.$$

Hence there exists $\theta \in \mathbb{R}$ such that

$$\Lambda = P \begin{pmatrix} 1 & -\theta & \theta \\ \theta & 1 - \theta^2/2 & \theta^2/2 \\ \theta & -\theta^2/2 & 1 + \theta^2/2 \end{pmatrix} P^{-1}.$$

□

1.5.3 Rotations and boosts

Now we study two classes of Lorentz transformations that allow us to classify all of those which are proper and orthochronous. We start with the:

Definition 1.5.4 (Pure Rotation). A *pure rotation* in $\mathbb{L}^n$ is a Lorentz transformation $\Lambda = (\lambda_{ij})_{1 \leq i,j \leq n} \in \mathrm{O}_1^{+\uparrow}(n,\mathbb{R})$ that is *time fixing*, that is, $\lambda_{nn} = 1$, and therefore is written as

$$\left(\begin{array}{ccc|c} & & & 0 \\ & \Lambda_S & & \vdots \\ & & & 0 \\ \hline 0 & \cdots & 0 & 1 \end{array}\right),$$

with $\Lambda_S \in \mathrm{SO}(n-1,\mathbb{R})$.

If $n = 4$, every pure rotation is determined by a vector in $\mathbb{R}^3$ and an angle. More precisely:

Theorem 1.5.5. *For any transformation $A \in \mathrm{SO}(3,\mathbb{R})$ there exists a positive orthonormal basis $\mathcal{B} = (w_1, w_2, w_3)$ and an angle $\theta \in [0, 2\pi[$ such that*

$$[A]_{\mathcal{B}} = \begin{pmatrix} \cos\theta & -\sin\theta & 0 \\ \sin\theta & \cos\theta & 0 \\ 0 & 0 & 1 \end{pmatrix}.$$

The direction of w_3 is called the axis of rotation *of A.*

Proof: First we show that 1 is an eigenvector of A:

$$\begin{aligned}\det(\mathrm{Id}_3 - A) &= \det(AA^\top - A)\\ &= \det A \det(A^\top - \mathrm{Id}_3)\\ &= \det\left((A^\top - \mathrm{Id}_3)^\top\right)\\ &= \det(A - \mathrm{Id}_3)\\ &= -\det(\mathrm{Id}_3 - A).\end{aligned}$$

Hence, there exists a unit vector $\boldsymbol{w}_3$, such that $A\boldsymbol{w}_3 = \boldsymbol{w}_3$. Let $\boldsymbol{w}_1, \boldsymbol{w}_2 \in \mathbb{R}^3$ be vectors such that $(\boldsymbol{w}_1, \boldsymbol{w}_2, \boldsymbol{w}_3)$ is a positive orthonormal basis of $\mathbb{R}^3$. Relative to this basis, we have

$$[A]_{\mathcal{B}} = \begin{pmatrix} a_{11} & a_{12} & 0 \\ a_{21} & a_{22} & 0 \\ 0 & 0 & 1 \end{pmatrix},$$

where $(a_{ij})_{1\leq i,j\leq 2} \in \mathrm{SO}(2,\mathbb{R})$. Right after our study of $\mathrm{O}_1(2,\mathbb{R})$ we saw that there exists $\theta \in [0, 2\pi[$ such that

$$[A]_{\mathcal{B}} = \begin{pmatrix} \cos\theta & -\sin\theta & 0 \\ \sin\theta & \cos\theta & 0 \\ 0 & 0 & 1 \end{pmatrix},$$

as desired. □

With this in place, we proceed to consider a vector $\boldsymbol{v} \in \mathbb{R}^{n-1}$ as a spacelike vector in $\mathbb{L}^n = \mathbb{R}^{n-1} \times \mathbb{R}$, identifying $(v_1, \ldots, v_{n-1})$ in $\mathbb{R}^{n-1}$ with $(v_1, \ldots, v_{n-1}, 0)$ in $\mathbb{R}^{n-1} \times \{0\}$. Then we have the:

Definition 1.5.6 (Lorentz boost). A *pure boost in the direction of a unit vector* $\boldsymbol{v} \in \mathbb{R}^{n-1}$ is a Lorentz transformation $\Lambda \in \mathrm{O}_1^{+\uparrow}(n,\mathbb{R})$ that fixes pointwise the subspace $\boldsymbol{v}^\perp \subseteq \mathbb{R}^{n-1}$.

It is possible to describe such boosts, in arbitrary dimensions, in terms of $\boldsymbol{v}$ and a real parameter (the boost intensity). We'll do this when $n = 3$, as an example. The difficulty in generalizing this only lies in the larger number of equations involved. The symmetry of such equations will be made clear in what follows.

If $\Lambda = (\lambda_{ij})_{1\leq i,j\leq 3}$ is a pure boost in the direction of $\boldsymbol{v} = (v_1, v_2) \in \mathbb{R}^2$ then, by Exercise 1.4.9 (p. 40), the orthogonal complement (in $\mathbb{L}^3$) of $\boldsymbol{v}^\perp \times \{0\}$ is Λ-invariant. This complement is precisely $\operatorname{span}\{\boldsymbol{v}, \boldsymbol{e}_3\}$. In particular,

$$\Lambda\boldsymbol{v} = \gamma\boldsymbol{v} + \delta\boldsymbol{e}_3 \quad \text{and} \quad \Lambda(-v_2, v_1, 0) = (-v_2, v_1, 0),$$

with $\gamma^2 - \delta^2 = 1$, since $\langle \Lambda\boldsymbol{v}, \Lambda\boldsymbol{v}\rangle_L = \langle \boldsymbol{v}, \boldsymbol{v}\rangle_L = 1$.

Writing out the entries in the above relations we have

$$\lambda_{11}v_1 + \lambda_{12}v_2 = \gamma v_1 \tag{1.5.1}$$
$$\lambda_{21}v_1 + \lambda_{22}v_2 = \gamma v_2 \tag{1.5.2}$$
$$\lambda_{31}v_1 + \lambda_{32}v_2 = \delta \tag{1.5.3}$$
$$\lambda_{11}v_2 - \lambda_{12}v_1 = v_2 \tag{1.5.4}$$
$$-\lambda_{21}v_2 + \lambda_{22}v_1 = v_1 \tag{1.5.5}$$
$$-\lambda_{31}v_2 + \lambda_{32}v_1 = 0. \tag{1.5.6}$$

Solving the system formed by Equations (1.5.1) and (1.5.4) we have

$$\lambda_{11} = 1 + (\gamma - 1)v_1^2 \quad \text{and} \quad \lambda_{12} = (\gamma - 1)v_1v_2.$$

In the same way, Equations (1.5.2) and (1.5.5) lead to

$$\lambda_{21} = (\gamma - 1)v_1 v_2 \quad \text{and} \quad \lambda_{22} = 1 + (\gamma - 1)v_2^2.$$

Hence we have $\gamma = \det \Lambda_S > 0$. Now, using that

$$\langle \Lambda e_3, \gamma v + \delta e_3 \rangle_L = \langle e_3, v \rangle_L = 0 \text{ and}$$
$$\langle \Lambda e_3, (-v_2, v_1, 0) \rangle_L = \langle e_3, (-v_2, v_1, 0) \rangle_L = 0$$

we get

$$\lambda_{13}\gamma v_1 + \lambda_{23}\gamma v_2 - \delta\lambda_{33} = 0 \tag{1.5.7}$$
$$-\lambda_{13}v_2 + \lambda_{23}v_1 = 0. \tag{1.5.8}$$

From Equations (1.5.7) and (1.5.8), we write λ_{13} and λ_{23} in terms of the remaining parameters:

$$\lambda_{13} = \frac{\delta\lambda_{33}v_1}{\gamma} \quad \text{and} \quad \lambda_{23} = \frac{\delta\lambda_{33}v_2}{\gamma}.$$

The relation $\langle \Lambda e_3, \Lambda e_3 \rangle_L = -1$ can be read as

$$\left(\frac{\delta\lambda_{33}v_1}{\gamma}\right)^2 + \left(\frac{\delta\lambda_{33}v_2}{\gamma}\right)^2 - \lambda_{33}^2 = -1.$$

From this, using that $v_1^2 + v_2^2 = 1$ and $\lambda_{33} > 0$, we have $\lambda_{33} = \gamma$. Replacing $\boldsymbol{v}$ with $-\boldsymbol{v}$ if needed, we may suppose $\delta \leq 0$. Such a hypothesis is physically reasonable since, once $\Lambda e_3 = \delta \boldsymbol{v} + \gamma e_3$ (check this), we have that an observer subject to a pure boost in the direction of $\boldsymbol{v}$ starts to see the event $(0,0,1)$ (its immediate future) with a negative component (precisely δ) in the boost direction. See Figure 1.10 below:

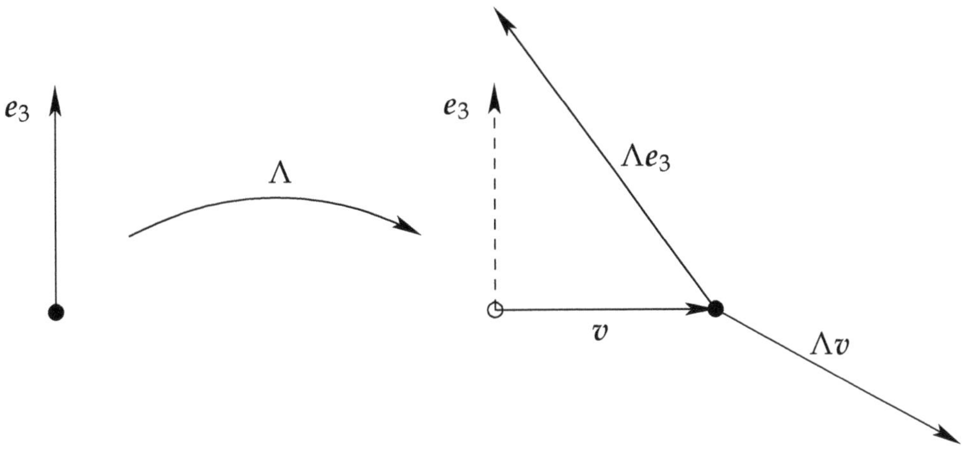

Figure 1.10: A Lorentz boost in the direction of $\boldsymbol{v}$.

Then we can proceed with $\delta = -\sqrt{-1+\gamma^2}$. From $\langle \Lambda e_i, \Lambda e_3 \rangle_L = 0$ we have

$$(1 + (\gamma - 1)v_1^2)\delta v_1 + (\gamma - 1)v_1 v_2 \delta v_2 - \lambda_{31}\gamma = 0 \implies \lambda_{31} = -v_1\sqrt{-1+\gamma^2}.$$

Similarly, $\lambda_{32} = -v_2\sqrt{-1+\gamma^2}$. Therefore,

$$\Lambda = \begin{pmatrix} 1 + (\gamma-1)v_1^2 & (\gamma-1)v_1 v_2 & -v_1\sqrt{-1+\gamma^2} \\ (\gamma-1)v_1 v_2 & 1 + (\gamma-1)v_2^2 & -v_2\sqrt{-1+\gamma^2} \\ -v_1\sqrt{-1+\gamma^2} & -v_2\sqrt{-1+\gamma^2} & \gamma \end{pmatrix}.$$

In arbitrary dimension, the expression of a boost in the direction of a unit vector $v = (v_1, \ldots, v_{n-1})$ is

$$B(v,\gamma) \doteq \left(\begin{array}{ccc|c} & \mathrm{Id}_{n-1} + (\gamma - 1)vv^\top & & \begin{matrix} -v_1\sqrt{-1+\gamma^2} \\ \vdots \\ -v_{n-1}\sqrt{-1+\gamma^2} \end{matrix} \\ \hline -v_1\sqrt{-1+\gamma^2} & \cdots & -v_{n-1}\sqrt{-1+\gamma^2} & \gamma \end{array}\right).$$

The steps to check the previous claim are the following:

1. Write $\Lambda v \in \operatorname{span}\{v, e_n\}$ in the form $\Lambda v = \gamma v + \delta e_n$, where we assume $\delta \leq 0$.

2. Take $\{u_1, \ldots, u_{n-2}\}$, a basis (orthonormal, for ease) of $v^\perp$. Solve the linear system given by $\Lambda u_i = u_i$, $1 \leq i \leq n-2$, and the entries of $\Lambda v = \gamma v + \delta e_n$ to write the block $\mathrm{Id}_{n-1} + (\gamma - 1)vv^\top$. By now we have $\gamma = \det \Lambda_S > 0$.

3. Solve the system given by the equations $\langle \Lambda u_i, \Lambda e_n \rangle_L = 0$, with $1 \leq i \leq n-2$ and $\langle \Lambda v, \Lambda e_n \rangle_L = 0$ to find out that $\lambda_{in} = -v_i\sqrt{-1+\gamma^2}$, with $1 \leq i \leq n-1$.

4. Use $\langle \Lambda e_n, \Lambda e_n \rangle_L = -1$, $\|v\| = 1$ and $\lambda_{nn} > 0$ to get $\lambda_{nn} = \gamma$.

5. Finally, solve the system given by $\langle \Lambda e_i, \Lambda e_n \rangle_L = 0$, with $1 \leq i \leq n-1$, to get $\lambda_{ni} = -v_i\sqrt{-1+\gamma^2}$.

It is convenient to reparametrize the matrix above, choosing a hyperbolic angle $\varphi \geq 0$ such that $\gamma = \cosh \varphi$, and then setting $B_\angle(v, \varphi) \doteq B(v, \gamma)$. In this form, we can express some properties of such boosts in a more natural way. See Exercises 1.5.3, 1.5.4, 1.5.5, and 1.5.6.

To close this section we state the following classification for the elements of $\mathrm{O}_1^{+\uparrow}(4, \mathbb{R})$:

Theorem 1.5.7. *Every Lorentz transformation $\Lambda \in \mathrm{O}_1^{+\uparrow}(4, \mathbb{R})$ has a unique decomposition as a product of a pure rotation R and a pure boost B:*

$$\Lambda = BR.$$

Such decomposition can also be written in the reverse order:

$$\Lambda = R\widetilde{B},$$

where $\widetilde{B}$ is a pure boost in another direction. Furthermore, there exists a pair of pure rotations R_1 and R_2, and a number $\varphi \geq 0$ such that

$$\Lambda = R_1 B_\angle(e_1, \varphi) R_2.$$

The latter decomposition is also unique, except when Λ itself is already a pure rotation.

The proof of this result relies on the representation of $\mathrm{O}_1^{+\uparrow}(4, \mathbb{R})$ in a group of complex matrices and on a particular decomposition of them. This exceeds the scope of this text and we refer to [33] for details.

Exercises

Exercise 1.5.1. We have seen that the map $\Lambda\colon \mathbb{R} \to \mathrm{O}_1^{+\uparrow}(2,\mathbb{R})$ given by:

$$\Lambda(\theta) \doteq \begin{pmatrix} \cosh\theta & \sinh\theta \\ \sinh\theta & \cosh\theta \end{pmatrix}$$

is surjective. Complete the proof that Λ is a group isomorphism and conclude that any two matrices in $\mathrm{O}_1^{+\uparrow}(2,\mathbb{R})$ commute.

Exercise 1.5.2. In Physics, it is usual to adopt the following convention for the product in $\mathbb{L}^4$:

$$\langle (t_1,x_1,y_1,z_1),(t_2,x_2,y_2,z_2)\rangle_L \doteq t_1t_2 - x_1x_2 - y_1y_2 - z_1z_2.$$

Assume this convention *only* for this exercise. Consider

$$\mathrm{Herm}(2,\mathbb{C}) \doteq \{A \in \mathrm{Mat}(2,\mathbb{C}) \mid A^\dagger \doteq \overline{A}^\top = A\},$$

the set of all *Hermitian* complex matrices of order 2.

(a) Let $\Phi\colon \mathbb{L}^4 \to \mathrm{Mat}(2,\mathbb{C})$ be given by

$$\Phi(t,x,y,z) = \begin{pmatrix} t+x & y+iz \\ y-iz & t-x \end{pmatrix}.$$

Check that Φ is linear, injective, and its image is $\mathrm{Herm}(2,\mathbb{C})$.

(b) Show that $\langle v,v\rangle_L = \det\Phi(v)$, for all $v \in \mathbb{L}^4$.

(c) Given $M \in \mathrm{Mat}(2,\mathbb{C})$, consider $\Psi_M\colon \mathrm{Herm}(2,\mathbb{C}) \to \mathrm{Herm}(2,\mathbb{C})$ defined by $\Psi_M(A) = M^\dagger AM$. Show that Ψ_M is linear, its image in fact lies in $\mathrm{Herm}(2,\mathbb{C})$, and if $\det M \in \mathbb{S}^1 = \{z \in \mathbb{C} \mid |z| = 1\}$, then Ψ_M preserves determinants.

(d) For any $M \in \mathrm{Mat}(2,\mathbb{C})$ with $\det M \in \mathbb{S}^1$, let $\Lambda_M\colon \mathbb{L}^4 \to \mathbb{L}^4$ be the *conjugation map* given by $\Lambda_M = \Phi^{-1} \circ \Psi_M \circ \Phi$. Show that $\Lambda_M \in \mathrm{O}_1(4,\mathbb{R})$.

Remark.

- In this convention, the definitions of "spacelike" and "timelike" are swapped, but the Lorentz transformations are the same as in our convention.
- For $1 \leq i \leq 4$, the matrices $\sigma_i \doteq \Phi(e_i)$ are called *Pauli matrices*.
- Not all Lorentz transformations in $\mathbb{L}^4$ can be obtained in this way. Can you find one of them?

Exercise 1.5.3. Determine, as in the text, the general form of a given Lorentz boost $\Lambda \in \mathrm{O}_1^{+\uparrow}(4,\mathbb{R})$ in the direction of a unit vector $\boldsymbol{v} \in \mathbb{R}^3$. In particular, for $\boldsymbol{v} = \boldsymbol{e}_1$, and $v \in \mathbb{R}_{\geq 0}$ determined by the relation $\gamma = 1/\sqrt{1-v^2}$, and $(x',y',z',t') \doteq \Lambda(x,y,z,t)$, show that

$$\begin{cases} x' = \gamma(x - vt) \\ y' = y \\ z' = z \\ t' = \gamma(t - vx), \end{cases}$$

where γ is defined as in the text.

Exercise 1.5.4. Let $\boldsymbol{v} \in \mathbb{R}^{n-1}$ be a unit vector and consider two Lorentz boosts $B_\angle(\boldsymbol{v},\varphi_1), B_\angle(\boldsymbol{v},\varphi_2) \in \mathrm{O}_1^{+\uparrow}(n,\mathbb{R})$ in the direction of $\boldsymbol{v}$. Show that

$$B_\angle(\boldsymbol{v},\varphi_1)B_\angle(\boldsymbol{v},\varphi_2) = B_\angle(\boldsymbol{v},\varphi_1+\varphi_2).$$

In particular, conclude that $B_\angle(\boldsymbol{v},\varphi_1)$ and $B_\angle(\boldsymbol{v},\varphi_2)$ commute. Furthermore, show that $B_\angle(\boldsymbol{v},\varphi)^{-1} = B_\angle(\boldsymbol{v},-\varphi)$.

Hint. Use the matrix product definition and analyze four cases, thinking of the block division of a general boost. It is not as complicated as it seems.

Exercise 1.5.5. If $\boldsymbol{v},\boldsymbol{w} \in \mathbb{R}^{n-1}$ are two given (spatial) unit vectors and $B_\angle(\boldsymbol{v},\varphi_1), B_\angle(\boldsymbol{w},\varphi_2) \in \mathrm{O}_1^{+\uparrow}(n,\mathbb{R})$ are the Lorentz boosts in the directions of $\boldsymbol{v}$ and $\boldsymbol{w}$, respectively, then, in general, $B_\angle(\boldsymbol{v},\varphi_1)$ and $B_\angle(\boldsymbol{w},\varphi_2)$ do not commute. Give an example of this.

Hint. You can find examples even when $\varphi_1 = \varphi_2$.

Exercise 1.5.6. In the same way we parametrized Lorentz boosts using an "angle" φ, it is usual to parametrize them in terms of the applied speed, $v = \tanh\varphi$ (see Exercise 1.3.7, p. 28). For a unit vector $\boldsymbol{v} \in \mathbb{R}^{n-1}$, denote by $B_{\mathrm{sp}}(\boldsymbol{v},v)$ the Lorentz boost in the direction of $\boldsymbol{v}$ with speed v (more precisely, $B_{\mathrm{sp}}(\boldsymbol{v},v) \doteq B_\angle(\boldsymbol{v},\varphi)$).

(a) Show that

$$B_{\mathrm{sp}}(\boldsymbol{v},v_1)B_{\mathrm{sp}}(\boldsymbol{v},v_2) = B_{\mathrm{sp}}\left(\boldsymbol{v},\frac{v_1+v_2}{1+v_1v_2}\right).$$

Hint. Use Exercise 1.5.4 and the hyperbolic trigonometric identities.

(b) Define in $]-1,1[$ the *relativistic addition of speeds* operation:

$$v_1 \oplus v_2 \doteq \frac{v_1+v_2}{1+v_1v_2}.$$

Show that $]-1,1[$ equipped with this operation is an (abelian) group. Since we use geometric units, where the speed of light is $c=1$, the numbers $v_1,v_2 \in\,]-1,1[$ may be seen as speeds with direction.

(c) Show that $\tanh\colon \mathbb{R} \to\,]-1,1[$ is a group isomorphism.

Exercise 1.5.7 (Margulis Invariant). Let $F \in \mathrm{P}(3,\mathbb{R})$ be a hyperbolic Poincaré transformation, written as $F(\boldsymbol{x}) = \Lambda\boldsymbol{x} + \boldsymbol{w}$, with $\Lambda \in \mathrm{O}_1^{+\uparrow}(3,\mathbb{R})$ and $\boldsymbol{w} \in \mathbb{L}^3$.

(a) Show that Λ has three positive eigenvectors, $1/\lambda < 1 < \lambda$. Conclude from Exercise 1.4.9 (p. 40), the eigenspaces associated to the eigenvalues λ and $1/\lambda$ are light rays.

(b) Let $\boldsymbol{v}_\lambda, \boldsymbol{v}_{1/\lambda}$ be lightlike and future-directed eigenvectors associated to the eigenvalues λ and $1/\lambda$, respectively, and $\boldsymbol{v}_1$ be a unit eigenvector associated to 1 such that the basis $\mathcal{B} = (\boldsymbol{v}_\lambda, \boldsymbol{v}_1, \boldsymbol{v}_{1/\lambda})$ is positive. Show that F fixes a single affine straight line that is parallel to the eigenvector $\boldsymbol{v}_1$, acting as a translation in this line. In other words, show that there exist $\boldsymbol{p} \in \mathbb{L}^3$ and $\alpha_F \in \mathbb{R}$ such that

$$F(\boldsymbol{p}+t\boldsymbol{v}_1) = \boldsymbol{p}+t\boldsymbol{v}_1+\alpha_F\boldsymbol{v}_1, \quad \text{for all } t \in \mathbb{R}.$$

The number α_F is the *Margulis invariant* of F.

Hint. Solve a system for the coordinates of p in the basis $\mathcal{B}$.

Remark. The choice of v_1 in a way that the basis is positive is necessary to make α_F well-defined. Changing the sign of v_1 would change the sign of α_F.

(c) Show that $\alpha_F = \langle w, v_1 \rangle_L$. Use this to show that if $F_1, F_2 \in \mathrm{P}(3,\mathbb{R})$ are hyperbolic and conjugated by an element of $\mathrm{O}_1(3,\mathbb{R})$, it holds that $\alpha_{F_1} = \alpha_{F_2}$, justifying the name "invariant".

(d) Show that for all $n > 0$, we have $\alpha_{F^n} = n\alpha_F$.

(e) Discuss the orientation of the eigenbasis of Λ^{-1} in terms of the orientation of $\mathcal{B}$. Use this to show that $\alpha_{F^n} = n\alpha_F$ holds, even for $n < 0$.

1.6 CROSS PRODUCT IN $\mathbb{R}^n_\nu$

In a first Analytic Geometry class, one learns the concept of *cross product* between two vectors in space $\mathbb{R}^3$: if $u = (u_1, u_2, u_3)$ and $v = (v_1, v_2, v_3)$, we have

$$u \times v = \begin{vmatrix} e_1 & e_2 & e_3 \\ u_1 & u_2 & u_3 \\ v_1 & v_2 & v_3 \end{vmatrix},$$

where the above determinant is formal, and used to find the components of the cross product relative to the canonical basis. This is applied, among other things, to understand orthogonality relations and to compute volumes. In this section, we will see how to extend this definition to pseudo-Euclidean spaces of arbitrary (finite) dimension, again emphasizing the particular features of the case $n = 3$.

Definition 1.6.1. The *cross product* of $v_1, \ldots, v_{n-1} \in \mathbb{R}^n_\nu$, according to $\langle \cdot, \cdot \rangle_\nu$, is the unique vector $v_1 \times \cdots \times v_{n-1} \doteq v \in \mathbb{R}^n_\nu$ such that

$$\langle v, x \rangle_\nu = \det(x, v_1, \cdots, v_{n-1}), \text{ for all } x \in \mathbb{R}^n_\nu.$$

Remark.

- The existence and uniqueness of the cross product $v_1 \times \cdots \times v_{n-1}$ are guaranteed by Riesz's Lemma applied to the particular linear functional $f\colon \mathbb{R}^n_\nu \to \mathbb{R}$ given by $f(x) = \det(x, v_1, \cdots, v_{n-1})$.
- In $\mathbb{R}^n$ we will write $\times_E$, and in $\mathbb{L}^n$ we will write $\times_L$. In particular, for $n = 3$ we will write $u \times_E v$, and similarly in $\mathbb{L}^3$.
- It is also usual to use the symbol $\wedge$ instead of $\times$. For example, one could write $u \wedge_L v$ for the Lorentzian cross product of $u, v \in \mathbb{L}^3$.

The above definition is not very efficient to explicitly compute cross products. The proposition below not only solves this issue, but also shows that the above definition is indeed an extension of the usual cross product in $\mathbb{R}^3$:

Proposition 1.6.2. *Let* $\mathcal{B} = (\boldsymbol{u}_i)_{i=1}^n$ *be a positive and orthonormal basis for* $\mathbb{R}^n_\nu$ *and* $\boldsymbol{v}_j = \sum_{i=1}^n v_{ij}\boldsymbol{u}_i \in \mathbb{R}^n_\nu$, *for* $1 \leq j \leq n-1$, *be given vectors. Writing* $\epsilon_i \doteq \epsilon_{\boldsymbol{u}_i}$ *for the indicators of the elements in* $\mathcal{B}$, *we have that:*

$$\boldsymbol{v}_1 \times \cdots \times \boldsymbol{v}_{n-1} = \begin{vmatrix} \epsilon_1\boldsymbol{u}_1 & \cdots & \epsilon_n\boldsymbol{u}_n \\ v_{11} & \cdots & v_{n1} \\ \vdots & \ddots & \vdots \\ v_{1,n-1} & \cdots & v_{n,n-1} \end{vmatrix}.$$

Proof: For simplicity, write $\boldsymbol{v} = \boldsymbol{v}_1 \times \cdots \times \boldsymbol{v}_{n-1}$. By Lemma 1.2.32 (p. 16), we have that $\boldsymbol{v} = \sum_{i=1}^n \epsilon_i\langle \boldsymbol{v}, \boldsymbol{u}_i\rangle_\nu \boldsymbol{u}_i$, but from the definition of cross product, each component is given by

$$\langle \boldsymbol{v}, \boldsymbol{u}_i\rangle_\nu = (-1)^{i+1}\begin{vmatrix} v_{11} & \cdots & v_{i-1,1} & v_{i+1,1} & \cdots & v_{n1} \\ \vdots & \ddots & \vdots & \vdots & \ddots & \vdots \\ v_{1,n-1} & \cdots & v_{i-1,n-1} & v_{i+1,n-1} & \cdots & v_{n,n-1} \end{vmatrix}.$$

With this, it suffices to recognize the orthonormal expansion of $\boldsymbol{v}$ as the expansion of the (formal) determinant given in the statement of this result via the first row. □

Remark. In particular, if $\boldsymbol{u} = (u_1, u_2, u_3)$ and $\boldsymbol{v} = (v_1, v_2, v_3)$ are vectors in $\mathbb{L}^3$, we have that

$$\boldsymbol{u} \times_L \boldsymbol{v} = \begin{vmatrix} \boldsymbol{e}_1 & \boldsymbol{e}_2 & -\boldsymbol{e}_3 \\ u_1 & u_2 & u_3 \\ v_1 & v_2 & v_3 \end{vmatrix}.$$

Next, we will register a few general properties of the cross product, which directly follow from the usual properties of the determinant function:

Proposition 1.6.3. *Let* $\boldsymbol{v}_1, \ldots, \boldsymbol{v}_{n-1} \in \mathbb{R}^n_\nu$ *and* $\lambda \in \mathbb{R}$. *Then the cross product is:*

(i) alternating, *that is, switching two of its entries changes the sign of the result. In particular, for* $n = 3$, *the cross product is skew-symmetric* ($\boldsymbol{v}_1 \times \boldsymbol{v}_2 = -\boldsymbol{v}_2 \times \boldsymbol{v}_1$);

(ii) such that $\boldsymbol{v}_1 \times \cdots \times \boldsymbol{v}_{n-1} = \boldsymbol{0}$ *if and only if* $\{\boldsymbol{v}_1, \ldots, \boldsymbol{v}_{n-1}\}$ *is linearly dependent;*

(iii) multilinear, *that is, linear in each one of its entries. In particular, for* $n = 3$, *we have:*

$$(\boldsymbol{v}_1 + \lambda\boldsymbol{u}) \times \boldsymbol{v}_2 = \boldsymbol{v}_1 \times \boldsymbol{v}_2 + \lambda(\boldsymbol{u} \times \boldsymbol{v}_2)$$
$$\boldsymbol{v}_1 \times (\boldsymbol{v}_2 + \lambda\boldsymbol{u}) = \boldsymbol{v}_1 \times \boldsymbol{v}_2 + \lambda(\boldsymbol{v}_1 \times \boldsymbol{u}),$$

for all $\boldsymbol{u} \in \mathbb{R}^3_\nu$ *and* $\lambda \in \mathbb{R}$;

(iv) orthogonal to each one of its entries:

$$\langle \boldsymbol{v}_1 \times \cdots \times \boldsymbol{v}_{n-1}, \boldsymbol{v}_i\rangle_\nu = 0,$$

for each $1 \leq i \leq n-1$;

(v) for $n = 3$, cyclic, *that is, it satisfies the cyclic identity:*

$$\langle \boldsymbol{v}_1 \times \boldsymbol{v}_2, \boldsymbol{v}_3\rangle_\nu = \langle \boldsymbol{v}_1, \boldsymbol{v}_2 \times \boldsymbol{v}_3\rangle_\nu.$$

Remark.

- Recall that a permutation $\sigma \in S_k$ may be decomposed as the composition of a sequence of permutations which only move two elements (these permutations are called *transpositions*). We say that $\mathrm{sgn}(\sigma) = 1$ if this decomposition has an even number of transpositions, and -1 if it has an odd number of transpositions. Even though this decomposition in terms of transpositions is not unique, the parity of the number of transpositions is an invariant of σ, so that $\mathrm{sgn}(\sigma)$ is indeed well-defined. For more details, see [23].

- With this terminology, saying that the cross product is alternating is equivalent to saying that given $v_1, \ldots, v_{n-1} \in \mathbb{R}^n_\nu$, we have that

$$v_{\sigma(1)} \times \cdots \times v_{\sigma(n-1)} = \mathrm{sgn}(\sigma)\, v_1 \times \cdots \times v_{n-1},$$

  for each $\sigma \in S_{n-1}$.

Some additional properties of the cross product for $n = 3$ are stated in Exercise 1.6.4. Furthermore, note that (visually), the Lorentzian product of $v_1 \ldots, v_{n-1} \in \mathbb{L}^n$ is the Euclidean product of these same vectors, but reflected on the hyperplane $x_n = 0$. More precisely:

$$\mathrm{Id}_{n-1,1}(v_1 \times_L \cdots \times_L v_{n-1}) = v_1 \times_E \cdots \times_E v_{n-1}.$$

The verification of this fact is left as Exercise 1.6.5. Moreover, we may analyze the causal type of the cross product in terms of the causal type of the space spanned by the vectors who enter the product. The following proposition is nothing more than a reformulation of Theorem 1.2.20 (p. 12):

Proposition 1.6.4. *Let $v_1, \ldots, v_{n-1} \in \mathbb{L}^n$ be linearly independent and consider the hyperplane $S \doteq \mathrm{span}\,\{v_1, \ldots, v_{n-1}\}$. Then:*

(i) $v_1 \times \cdots \times v_{n-1}$ is spacelike $\iff$ S is timelike;

(ii) $v_1 \times \cdots \times v_{n-1}$ is timelike $\iff$ S is spacelike;

(iii) $v_1 \times \cdots \times v_{n-1}$ is lightlike $\iff$ S is lightlike.

Proposition 1.6.5. *Let $u_1, \ldots, u_{n-1}, v_1, \ldots, v_{n-1} \in \mathbb{R}^n_\nu$. Then we have that*

$$\langle u_1 \times \cdots \times u_{n-1}, v_1 \times \cdots \times v_{n-1}\rangle_\nu = (-1)^\nu \det\big((\langle u_i, v_j\rangle_\nu)_{1\leq i,j\leq n-1}\big).$$

Proof: If $(u_i)_{i=1}^{n-1}$ or $(v_j)_{j=1}^{n-1}$ is linearly dependent, there is nothing to do. Suppose that both of them are linearly independent. Since both sides of the proposed equality are linear in each one of the $2n-2$ present variables, and both the cross product and the determinant function are alternate, we may assume without loss of generality that $u_k = e_{i_k}$ and $v_\ell = e_{j_\ell}$, where $(e_i)_{i=1}^n$ is the canonical basis for $\mathbb{R}^n_\nu$ and

$$1 \leq i_1 < \cdots < i_{n-1} \leq n \quad \text{and} \quad 1 \leq j_1 < \cdots < j_{n-1} \leq n.$$

We will proceed with the analysis by cases, in terms of the indices i^* and j^* omitted in each of the $(n-1)$-uples of indices being considered.

- If $i^* \neq j^*$, both sides equal zero. To wit, the left side equals $\langle e_{i^*}, e_{j^*}\rangle_\nu = 0$ and the determinant on the right side has the i^*-th row and j^*-th column consisting only of zeros.

- If $1 \leq i^* = j^* \leq n - \nu$, the left side is $\langle e_{i^*}, e_{i^*} \rangle_\nu = 1$, while the right side is $(-1)^\nu \det \mathrm{Id}_{n-1,\nu} = (-1)^\nu(-1)^\nu = 1$.
- If $n - \nu < i^* = j^* \leq n$, the left side is $\langle e_{i^*}, e_{i^*} \rangle_\nu = -1$, while the right side is $(-1)^\nu \det \mathrm{Id}_{n-1,\nu-1} = (-1)^\nu(-1)^{\nu-1} = -1$.

□

Corollary 1.6.6 (Lagrange's Identities). *Let* $u, v \in \mathbb{R}^3_\nu$. *Then:*

$$\|u \times_E v\|_E^2 = \|u\|_E^2 \|v\|_E^2 - \langle u, v \rangle_E^2,$$
$$\langle u \times_L v, u \times_L v \rangle_L = -\langle u, u \rangle_L \langle v, v \rangle_L + \langle u, v \rangle_L^2.$$

The orientation of the bases chosen for $\mathbb{R}^n_\nu$ will be very important for the development of the theory to be presented in the following chapters. We recall that the canonical basis for $\mathbb{R}^n_\nu$ is, by convention, positive.

If $v_1, \ldots, v_{n-1} \in \mathbb{R}^n_\nu$ are linearly independent, don't span a lightlike hyperplane, and $v = v_1 \times \cdots \times v_{n-1}$, then $\mathcal{B} = (v_1, \ldots, v_{n-1}, v)$ is a basis for $\mathbb{R}^n_\nu$ and it is natural to wonder whether such basis is positive or negative. The answer comes from the sign of the determinant containing those vectors (be it on rows or columns). We have that

$$\det(v_1, \ldots, v_{n-1}, v) = (-1)^{n-1} \det(v, v_1, \ldots, v_{n-1}) = (-1)^{n-1} \langle v, v \rangle_\nu$$

and, thus, positivity of the basis $\mathcal{B}$ depends not only on the parity of n, but also on the causal character of v. Explicitly: if v is spacelike, $\mathcal{B}$ is positive if n is odd, and negative if n is even; if v is timelike, $\mathcal{B}$ is positive if n is even, and negative if n is odd. Check this for the canonical basis in $\mathbb{R}^4$, for example.

In particular, for $n = 3$, we may represent all cross products between vectors in the canonical basis by the following two diagrams:

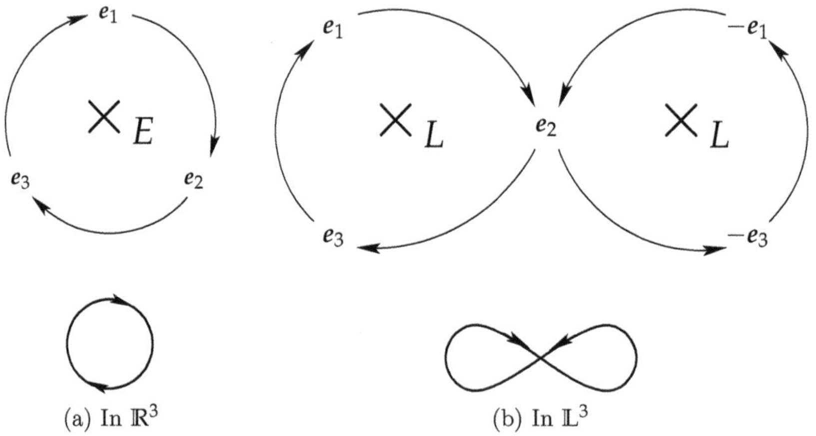

Figure 1.11: Summarizing cross products in $\mathbb{R}^3_\nu$.

The cross products are obtained by following the arrows. For example, we have that $e_2 \times_E e_3 = e_1$ and $e_1 \times_L e_2 = -e_3$. In $\mathbb{R}^3$, following the opposite direction of the given arrows will yield results with the opposite sign (since $\times_E$ is skew-symmetric), for example, $e_1 \times_E e_3 = -e_2$. But in $\mathbb{L}^3$, this does not work due to the influence of causal types: note that $e_3 \times_L e_2 = -e_1 \neq -(-e_1) = e_1$. The cross products which may not be obtained directly from the diagram 1.11b may be found out by using that $\times_L$ is skew-symmetric.

Remark. Note that the diagrams remain true if we replace the canonical basis by any positive and orthonormal basis, provided that in $\mathbb{L}^3$ the timelike vector is the last one (corresponding to e_3).

1.6.1 Completing the toolbox

We know that the trace and the determinant of a linear operator are invariant under change of basis. In Chapter 3, to define certain curvatures of a surface, we need to extend those concepts for bilinear maps. Restricting ourselves to orthonormal bases, we have the following results:

Lemma 1.6.7. *Let Z be any vector space and $B\colon \mathbb{R}^n_\nu \times \mathbb{R}^n_\nu \to Z$ be a bilinear map. So, if $(\boldsymbol{v}_i)_{i=1}^n$ and $(\boldsymbol{w}_i)_{i=1}^n$ are orthonormal bases for $\mathbb{R}^n_\nu$, we have that*

$$\sum_{i=1}^n \epsilon_{\boldsymbol{v}_i} B(\boldsymbol{v}_i, \boldsymbol{v}_i) = \sum_{i=1}^n \epsilon_{\boldsymbol{w}_i} B(\boldsymbol{w}_i, \boldsymbol{w}_i).$$

The above quantity is then called the trace of B relative to $\langle \cdot, \cdot \rangle_\nu$ *and it is denoted by* $\mathrm{tr}_{\langle \cdot, \cdot \rangle_\nu} B$.

Proof: It is possible to show that *both* considered bases have necessarily ν timelike vectors (this result is known as *Sylvester's Law of Inertia*, see [54]). With this, we may (reordering the bases if needed) assume that $\langle \boldsymbol{v}_i, \boldsymbol{v}_j \rangle_\nu = \langle \boldsymbol{w}_i, \boldsymbol{w}_j \rangle_\nu = \eta^\nu_{ij}$ and that, in particular, that $\epsilon_i \doteq \epsilon_{\boldsymbol{v}_i} = \epsilon_{\boldsymbol{w}_i}$ for $1 \leq i \leq n$. Applying Lemma 1.2.32 (p. 16) successively, we have

$$\boldsymbol{w}_j = \sum_{k=1}^n \epsilon_k \langle \boldsymbol{w}_j, \boldsymbol{v}_k \rangle_\nu \boldsymbol{v}_k = \sum_{i=1}^n \left(\sum_{k=1}^n \epsilon_i \epsilon_k \langle \boldsymbol{w}_j, \boldsymbol{v}_k \rangle_\nu \langle \boldsymbol{v}_k, \boldsymbol{w}_i \rangle_\nu \right) \boldsymbol{w}_i,$$

whence

$$\sum_{k=1}^n \epsilon_i \epsilon_k \langle \boldsymbol{w}_j, \boldsymbol{v}_k \rangle_\nu \langle \boldsymbol{v}_k, \boldsymbol{w}_i \rangle_\nu = \delta_{ij}, \quad 1 \leq i, j \leq n,$$

by linear independence of the $(\boldsymbol{w}_j)_{j=1}^n$. With this, we compute

$$\begin{aligned}
\sum_{i=1}^n \epsilon_i B(\boldsymbol{v}_i, \boldsymbol{v}_i) &= \sum_{i=1}^n B\left(\sum_{j=1}^n \epsilon_j \langle \boldsymbol{v}_i, \boldsymbol{w}_j \rangle_\nu \boldsymbol{w}_j, \sum_{k=1}^n \epsilon_k \langle \boldsymbol{v}_i, \boldsymbol{w}_k \rangle_\nu \boldsymbol{w}_k \right) \\
&= \sum_{j,k=1}^n \epsilon_k \left(\sum_{i=1}^n \epsilon_i \epsilon_j \langle \boldsymbol{w}_j, \boldsymbol{v}_i \rangle_\nu \langle \boldsymbol{v}_i, \boldsymbol{w}_k \rangle_\nu \right) B(\boldsymbol{w}_j, \boldsymbol{w}_k) \\
&= \sum_{j,k=1}^n \epsilon_k \delta_{jk} B(\boldsymbol{w}_j, \boldsymbol{w}_k) \\
&= \sum_{k=1}^n \epsilon_k B(\boldsymbol{w}_k, \boldsymbol{w}_k),
\end{aligned}$$

as wanted. □

Lemma 1.6.8. *Let $B\colon \mathbb{R}^n_\nu \times \mathbb{R}^n_\nu \to \mathbb{R}$ be a bilinear functional. If $(\boldsymbol{v}_i)_{i=1}^n$ and $(\boldsymbol{w}_i)_{i=1}^n$ are orthonormal bases for $\mathbb{R}^n_\nu$, the matrices of B relative to the given bases are related by*

$$\det\left((B(\boldsymbol{v}_i, \boldsymbol{v}_j))_{1 \leq i,j \leq n} \right) = \det\left((B(\boldsymbol{w}_i, \boldsymbol{w}_j))_{1 \leq i,j \leq n} \right).$$

The above quantity is called the determinant of B relative to $\langle \cdot, \cdot \rangle_\nu$, *and it is denoted by* $\det_{\langle \cdot, \cdot \rangle_\nu} B$.

Proof: Write, for each j, $v_j = \sum_{i=1}^{m} a_{ij} w_i$. Since $A = (a_{ij})_{1\leq i,j\leq n}$ is the change of basis matrix between orthonormal bases, the Exercise 1.4.7 (p. 40) ensures that $A \in \mathrm{O}_\nu(n,\mathbb{R})$ and so $|\det A| = 1$. This way,

$$B(v_i, v_j) = B\left(\sum_{k=1}^{n} a_{ki} w_k, \sum_{\ell=1}^{n} a_{\ell j} w_\ell\right) = \sum_{k=1}^{n} a_{ki}\left(\sum_{\ell=1}^{n} B(w_k, w_\ell) a_{\ell j}\right),$$

which, in matrix terms, is rewritten as

$$\big(B(v_i,v_j)\big)_{1\leq i,j\leq n} = A^\top \big(B(w_i,w_j)\big)_{1\leq i,j\leq n} A$$

and, thus,

$$\det\Big(\big(B(v_i,v_j)\big)_{1\leq i,j\leq n}\Big) = \det\Big(\big(B(w_i,w_j)\big)_{1\leq i,j\leq n}\Big)$$

as wanted. □

Lemma 1.6.9. *Let $v_1, \ldots, v_{n-1} \in \mathbb{R}^n_\nu$ be linearly independent, S be their linear span, and $T\colon S \to S$ be a linear map. Then:*

(i) $\displaystyle\sum_{i=1}^{n-1} (v_1 \times \cdots \times v_{i-1} \times Tv_i \times v_{i+1} \times \cdots \times v_{n-1}) = (\operatorname{tr} T)\, v_1 \times \cdots \times v_{n-1}$;

(ii) $Tv_1 \times \cdots \times Tv_{n-1} = \det(T)\, v_1 \times \cdots \times v_{n-1}$.

Proof: Write $[T]_{\mathcal{B}} = (a_{ij})_{1\leq i,j\leq n-1}$, where $\mathcal{B} = (v_1, \ldots, v_{n-1})$.

(i) We have that:

$$\begin{aligned}
&\sum_{i=1}^{n-1} (v_1 \times \cdots \times v_{i-1} \times Tv_i \times v_{i+1} \times \cdots \times v_{n-1}) = \\
&\qquad = \sum_{i=1}^{n-1}\left(v_1 \times \cdots \times v_{i-1} \times \left(\sum_{j_i=1}^{n-1} a_{j_i i} v_{j_i}\right) \times v_{i+1} \times \cdots \times v_{n-1}\right) \\
&\qquad = \sum_{i=1}^{n-1}\left(\sum_{j_i=1}^{n-1} a_{j_i i} v_1 \times \cdots \times v_{i-1} \times v_{j_i} \times v_{i+1} \times \cdots \times v_{n-1}\right) \\
&\qquad \overset{(*)}{=} \sum_{i=1}^{n-1} (a_{ii} v_1 \times \cdots \times v_{n-1}) \\
&\qquad = \operatorname{tr}([T]_{\mathcal{B}})\, v_1 \times \cdots \times v_{n-1} \\
&\qquad = \operatorname{tr}(T)\, v_1 \times \cdots \times v_{n-1},
\end{aligned}$$

where in $(*)$ we have used that if $j_i \neq i$, the cross product vanishes (one of the factors repeats).

(ii) We will use (one) definition[9] of determinant. We have:

$$
\begin{aligned}
Tv_1 \times \cdots \times Tv_{n-1} &= \left(\sum_{i_1=1}^{n-1} a_{i_1 1} v_{i_1} \right) \times \cdots \times \left(\sum_{i_{n-1}=1}^{n-1} a_{i_{n-1}, n-1} v_{i_{n-1}} \right) \\
&= \sum_{i_1,\ldots,i_{n-1}=1}^{n-1} \left(\prod_{j=1}^{n-1} a_{i_j j} \right) v_{i_1} \times \cdots \times v_{i_{n-1}} \\
&\overset{(*)}{=} \sum_{\sigma \in S_{n-1}} \left(\prod_{j=1}^{n-1} a_{\sigma(j) j} \right) v_{\sigma(1)} \times \cdots \times v_{\sigma(n-1)} \\
&= \sum_{\sigma \in S_{n-1}} \operatorname{sgn}(\sigma) \left(\prod_{j=1}^{n-1} a_{\sigma(j) j} \right) v_1 \times \cdots \times v_{n-1} \\
&= \det([T]_{\mathcal{B}}^{\top})\, v_1 \times \cdots \times v_{n-1} \\
&= \det(T)\, v_1 \times \cdots \times v_{n-1},
\end{aligned}
$$

where in $(*)$ we have used that the summand vanishes in case there is any repetition of indices, so that the summation may be reindexed with permutations. More precisely, for each non-zero term in the sum there is a unique $\sigma \in S_{n-1}$, namely, the one such that $\sigma(k) = i_k$, for each $1 \leq k \leq n-1$.

□

Exercises

Exercise 1.6.1. Consider the vectors $\boldsymbol{u} = (-1,3,1)$, $\boldsymbol{v} = (2,1,1)$, and $\boldsymbol{w} = (0,1,1)$. Prove that there is no vector $\boldsymbol{x}$ such that $\boldsymbol{x} \times_E \boldsymbol{u} = \boldsymbol{v}$ and $\boldsymbol{x} \perp_E \boldsymbol{w}$. However, there is a unique vector $\boldsymbol{x}$ satisfying $\boldsymbol{x} \times_L \boldsymbol{u} = \boldsymbol{v}$ and $\boldsymbol{x} \perp_L \boldsymbol{w}$. Determine this vector.

Exercise 1.6.2. Suppose that $\boldsymbol{u}, \boldsymbol{v} \in \mathbb{R}^3_\nu$ are linearly independent vectors. Let $\boldsymbol{w} \in \mathbb{R}^3_\nu$ be such that $\boldsymbol{w} \times \boldsymbol{u} = \boldsymbol{w} \times \boldsymbol{v} = \boldsymbol{0}$, where $\times$ is the cross product in the ambient space under consideration. Show that $\boldsymbol{w} = \boldsymbol{0}$ (that is, the result is true in both $\mathbb{R}^3$ and $\mathbb{L}^3$).

Exercise 1.6.3. Prove that if $\boldsymbol{u}, \boldsymbol{v} \in \mathbb{R}^3$ are vectors with $\langle \boldsymbol{u}, \boldsymbol{v} \rangle_E = 0$ and $\boldsymbol{u} \times_E \boldsymbol{v} = 0$, then $\boldsymbol{u}$ or $\boldsymbol{v}$ must vanish. Verify with a counter-example that the corresponding result in $\mathbb{L}^3$ is false.

Exercise 1.6.4. Let $\boldsymbol{u}, \boldsymbol{v}, \boldsymbol{w} \in \mathbb{R}^3_\nu$ be any vectors.

(a) Prove the double cross product identities:

$$
\begin{aligned}
\boldsymbol{u} \times_E (\boldsymbol{v} \times_E \boldsymbol{w}) &= \langle \boldsymbol{u}, \boldsymbol{w} \rangle_E \boldsymbol{v} - \langle \boldsymbol{u}, \boldsymbol{v} \rangle_E \boldsymbol{w} \\
\boldsymbol{u} \times_L (\boldsymbol{v} \times_L \boldsymbol{w}) &= \langle \boldsymbol{u}, \boldsymbol{v} \rangle_L \boldsymbol{w} - \langle \boldsymbol{u}, \boldsymbol{w} \rangle_L \boldsymbol{v}
\end{aligned}
$$

(b) Prove the *Jacobi identities*:

$$
\begin{aligned}
\boldsymbol{u} \times_E (\boldsymbol{v} \times_E \boldsymbol{w}) + \boldsymbol{v} \times_E (\boldsymbol{w} \times_E \boldsymbol{u}) + \boldsymbol{w} \times_E (\boldsymbol{u} \times_E \boldsymbol{v}) &= \boldsymbol{0} \\
\boldsymbol{u} \times_L (\boldsymbol{v} \times_L \boldsymbol{w}) + \boldsymbol{v} \times_L (\boldsymbol{w} \times_L \boldsymbol{u}) + \boldsymbol{w} \times_L (\boldsymbol{u} \times_L \boldsymbol{v}) &= \boldsymbol{0}
\end{aligned}
$$

[9] For more details, see [8]

Hint. Use item (a).

Exercise 1.6.5. Let $v_1, \ldots, v_{n-1} \in \mathbb{R}^n_\nu$. Show that

$$\mathrm{Id}_{n-1,1}(v_1 \times_L \cdots \times_L v_{n-1}) = v_1 \times_E \cdots \times_E v_{n-1}.$$

Exercise† 1.6.6. Verify that the diagrams given in Figure 1.11 (p. 56) indeed work.

Exercise 1.6.7. Let $B\colon \mathbb{R}^3_\nu \times \mathbb{R}^3_\nu \to \mathbb{R}^3_\nu$ be a skew-symmetric bilinear map. Show that there is a unique linear map $T\colon \mathbb{R}^3_\nu \to \mathbb{R}^3_\nu$ such that $T(v \times w) = B(v, w)$, for all $v, w \in \mathbb{R}^3_\nu$.

Exercise 1.6.8 (Representation of $\times_E$). Given any $x = (x_1, x_2, x_3) \in \mathbb{R}^3 \setminus \{\mathbf{0}\}$, we may see the cross product with x as a linear map $T = x \times_E \colon \mathbb{R}^3 \to \mathbb{R}^3$.

(a) Write the matrix of T relative to the canonical basis and check that its characteristic polynomial is $c_T(t) = t^3 + \langle x, x\rangle_E t$.

(b) Describe the eigenspace associated to 0 (i.e., $\ker T$). Regarding $\mathbb{R}^3$ inside $\mathbb{C}^3$ and considering an eigenvector $v \in \mathbb{C}^3$ associated to the eigenvalue $\mathrm{i}\|x\|_E$, compute the cross products $x \times_E \mathrm{Re}(v)$ and $x \times_E \mathrm{Im}(v)$. Conclude that $\mathrm{Re}(v)$ and $\mathrm{Im}(v)$ are both orthogonal to x. Verify also that $\mathrm{Re}(v)$ and $\mathrm{Im}(v)$ are orthogonal.

Hint. Proposition 1.6.5 (p. 55).

Remark.

- The interpretation of the eigenvalue $\mathrm{i}\|x\|_E$ is given as follows: the restriction of T to the normal plane to x, which is spanned by $\mathrm{Re}(v)$ and $\mathrm{Im}(v)$, acts as a rotation of 90° counterclockwise (inside the plane), followed by a dilation of factor $\|x\|_E$.
- One possible v is

$$v = \left(-x_1x_3 - x_2\|x\|_E\mathrm{i}, -x_2x_3 + x_1\|x\|_E\mathrm{i}, x_1^2 + x_2^2\right).$$

Exercise 1.6.9 (Representation of $\times_L$). Given any $x = (x_1, x_2, x_3) \in \mathbb{L}^3 \setminus \{\mathbf{0}\}$, we may see the cross product with x as a linear map $T = x \times_L \colon \mathbb{L}^3 \to \mathbb{L}^3$.

(a) Write the matrix of T relative to the canonical basis and check that its characteristic polynomial is $c_T(t) = t^3 - \langle x, x\rangle_L t$.

(b) Suppose that x is spacelike. The plane which has x as its normal direction is timelike and intersects the lightcone of $\mathbb{L}^3$ in two light rays. Show that T is diagonalizable and that the directions of these light rays are eigenvectors of T.

(c) Suppose that x is timelike. We have a situation similar to what was discussed in Exercise 1.6.8: if $v \in \mathbb{C}^3$ is an eigenvector associated to the eigenvalue $\mathrm{i}\|x\|_L$, compute $x \times_L \mathrm{Re}(v)$ and $x \times_L \mathrm{Im}(v)$. Conclude again that $\mathrm{Re}(v)$ and $\mathrm{Im}(v)$ are both Lorentz-orthogonal to x. Verify also that $\mathrm{Re}(v)$ and $\mathrm{Im}(v)$ are Lorentz-orthogonal.

Remark. If x is timelike, one possible v is

$$v = \left(x_1x_3 - x_2\|x\|_L\mathrm{i}, x_2x_3 + x_1\|x\|_L\mathrm{i}, x_1^2 + x_2^2\right).$$

Exercise 1.6.10. Let $\nu \in \{0, 1\}$ and $v_1, \ldots, v_{n-1} \in \mathbb{R}^n_\nu$.

(a) If θ is the (Euclidean) angle between $v_1 \times_L \cdots \times_L v_{n-1}$ (or $v_1 \times_E \cdots \times_E v_{n-1}$) and the hyperplane $e_n^\perp$, show that:

$$\langle v_1 \times_L \cdots \times_L v_{n-1}, v_1 \times_L \cdots \times_L v_{n-1} \rangle_L = \\ = \langle v_1 \times_E \cdots \times_E v_{n-1}, v_1 \times_E \cdots \times_E v_{n-1} \rangle_E \cos 2\theta.$$

(b) Show that

$$\| v_1 \times_L \cdots \times_L v_{n-1} \|_L \leq \| v_1 \times_E \cdots \times_E v_{n-1} \|_E,$$

with equality if and only if the vectors are linearly dependent, or the hyperplane spanned by $v_1, \ldots, v_{n-1}$ is horizontal or vertical.

Exercise 1.6.11. Show that $\mathrm{tr}_{\langle\cdot,\cdot\rangle} \langle \cdot, \cdot \rangle = n$ and that $\det_{\langle\cdot,\cdot\rangle} \langle \cdot, \cdot \rangle = (-1)^\nu$.

Exercise 1.6.12. Let $T \colon \mathbb{R}^n_\nu \to \mathbb{R}^n_\nu$ be a linear operator. Define a bilinear functional $B_T \colon \mathbb{R}^n_\nu \times \mathbb{R}^n_\nu \to \mathbb{R}$ by $B_T(v, w) \doteq \langle Tv, w \rangle$. Show that $\mathrm{tr}_{\langle\cdot,\cdot\rangle} B_T = \mathrm{tr}\, T$ and that $\det_{\langle\cdot,\cdot\rangle} B_T = (-1)^\nu \det T$.

Exercise 1.6.13. Let $T_1, T_2 \colon \mathbb{R}^n_\nu \to \mathbb{R}^n_\nu$ be two linear operators and consider a bilinear functional $B \colon \mathbb{R}^n_\nu \times \mathbb{R}^n_\nu \to \mathbb{R}$. Define the (T_1, T_2)-*pull-back of* B

$$(T_1, T_2)^* B \colon \mathbb{R}^n_\nu \times \mathbb{R}^n_\nu \to \mathbb{R}, \text{ by } (T_1, T_2)^* B(v, w) \doteq B(T_1 v, T_2 w).$$

Show that $\det_{\langle\cdot,\cdot\rangle}((T_1, T_2)^* B) = \det T_1 \det T_2 \det_{\langle\cdot,\cdot\rangle} B$.

CHAPTER 2

Local Theory of Curves

INTRODUCTION

In this chapter we start our study of curves in $\mathbb{R}^n$ and in $\mathbb{L}^n$.

In Section 2.1 we present the definition of a parametrized curve in Euclidean or Lorentzian spaces of arbitrary finite dimension and extend the concept of causal type, seen in Chapter 1, to curves in $\mathbb{L}^n$. We provide a wide range of examples and introduce the concept of regularity for a curve. After that, we show that there are no lightlike or timelike closed curves in $\mathbb{L}^n$, emphasizing the physical interpretation of this result in terms of the causality in $\mathbb{L}^n$. In what follows, we study two quantities naturally associated to a curve: arclength (proper time, for timelike curves) and energy. We discuss the viability and geometric interpretation of unit speed reparametrizations, show an alternative reparametrization for lightlike curves (arc-photon), and close this section with a motivation for the definition of congruence between curves.

From now on, to deal with Euclidean and Lorentzian ambient spaces simultaneously, we denote the scalar products simply by $\langle \cdot, \cdot \rangle$, omitting the subscript indexes E and L.

In Section 2.2, we particularize the results seen in the previous section to curves in the plane $\mathbb{R}^2_\nu$. We start showing that each curve is locally the graph of some smooth function, and then we characterize all the lightlike curves in the plane. With this in place, we then turn our attention to unit speed curves, defining the Frenet-Serret frame for such curves and the notion of oriented curvature: the required invariant to classify the remaining curves. After that we generalize this construction for non-unit speed curves and derive a formula for $\kappa_\alpha(t)$. We provide interpretations for the sign of the curvature, using Taylor formulae, and discuss osculating circles. The end of this section comes with Theorem 2.2.12, which characterizes all the plane curves with constant curvature and Theorem 2.2.13, which shows that the curvature is the single geometric invariant necessary to recover the trace of a plane curve, up to an isometry of $\mathbb{R}^2_\nu$.

Lastly, in Section 2.3, we study curves in space. We start with the Frenet-Serret frame, for a certain class of curves (to be called *admissible*), and we introduce the torsion of a curve which, together with the curvature, determines the trace of the curve, up to an isometry of the ambient (see Theorem 2.3.20, p. 112). In the following we show, through examples, how the Frenet-Serret frames carries geometric information of the curve, via curvature and torsion. For instance, in Proposition 2.3.11 (p. 106) we see that an admissible curve is planar if and only if its torsion vanishes. We also give conditions to the trace of an admissible curve to be contained in a sphere of the ambient space (Theorem 2.3.15, p. 108). We cannot neglect an important class of curves: the helices, which are characterized in terms of the ratio $\tau_\alpha / \kappa_\alpha$ (Lancret's Theorem, p. 110). We conclude this chapter with the so-called Cartan frame for lightlike or semi-lightlike curves, introducing a single invariant, the pseudo-torsion, for this class of curves. We use this tool to achieve results analogous to the previous ones in this context, being aware of the interpretations in each case. In particular, we obtain a lightlike version of Lancret's Theorem (p. 125), show that every semi-lightlike curve is planar, and state a version of the Fundamental Theorem of Curves (p. 125).

2.1 PARAMETRIZED CURVES IN $\mathbb{R}^n_\nu$

Throughout this chapter, $I \subseteq \mathbb{R}$ will denote a (bounded or not) open interval.

Definition 2.1.1. A *parametrized curve* is a smooth map $\boldsymbol{\alpha}\colon I \to \mathbb{R}^n_\nu$. Its image, $\boldsymbol{\alpha}(I)$, is the *trace* of $\boldsymbol{\alpha}$. For each $t \in I$, the derivative $\boldsymbol{\alpha}'(t)$ is the *velocity vector of* $\boldsymbol{\alpha}$ *in* t.

Remark. We use "smooth" meaning "infinitely differentiable", or "of class $\mathscr{C}^\infty$". Despite that requirement in the definition, most of the results presented here are valid for $\mathscr{C}^3$ or $\mathscr{C}^4$ curves. Hence we'll allow ourselves to deal only with "sufficiently differentiable" parametrizations..

In Chapter 1, we defined the causal type of vectors in $\mathbb{L}^n$, and extended this definition to its subspaces (including affine ones). Now we'll have this concept for curves in $\mathbb{L}^n$. An affine subspace naturally associated to each point of a curve is its tangent line at that point, whose causal type is determined by any spanning vector. This leads to the:

Definition 2.1.2. Let $\boldsymbol{\alpha}\colon I \to \mathbb{R}^n_\nu$ be a curve and $t_0 \in I$. Then $\boldsymbol{\alpha}$ is:

(i) *spacelike at* t_0, if $\boldsymbol{\alpha}'(t_0)$ is a spacelike vector;

(ii) *timelike at* t_0, if $\boldsymbol{\alpha}'(t_0)$ at a timelike vector;

(iii) *lightlike in* t_0, if $\boldsymbol{\alpha}'(t_0)$ at a lightlike vector.

If the causal type of $\boldsymbol{\alpha}'(t)$ is the same for all $t \in I$, this will be defined as the causal type of the curve $\boldsymbol{\alpha}$. If so, and $\boldsymbol{\alpha}$ is not lightlike, we set the *indicator or* $\boldsymbol{\alpha}$ as $\epsilon_{\boldsymbol{\alpha}} \doteq \epsilon_{\boldsymbol{\alpha}'(t)}$.

Remark.

- The continuity of $I \ni t \mapsto \langle \boldsymbol{\alpha}'(t), \boldsymbol{\alpha}'(t)\rangle \in \mathbb{R}$, implies that "being spacelike" and "being timelike" are *open properties*, that is, if $\boldsymbol{\alpha}$ is spacelike or timelike at t_0, then $\boldsymbol{\alpha}$ retains this causal type for all t in some open interval around t_0.

- If $\boldsymbol{\alpha}$ is spacelike in t_0 and timelike in t_1, the Intermediate Value Theorem ensures that there exists t_2 between t_0 and t_1 such that $\boldsymbol{\alpha}$ is lightlike at t_2. Informally, the causal type of a curve can't jump between space and time.

- Our focus here is the *local* geometry of curves, hence the remarks above allow us to assume in some proofs, possibly shrinking the domain of the curve, that $\boldsymbol{\alpha}$ has constant causal type.

- If $\boldsymbol{\alpha}$ is timelike in $\mathbb{L}^n$, we say that it is future-directed or past-directed according to time orientation of velocity vectors $\boldsymbol{\alpha}'(t)$. Again, by continuity, a timelike curve is either future-directed or past-directed (i.e., it cannot jump from one time orientation to the other).

Example 2.1.3.

(1) Points: let $\boldsymbol{p} \in \mathbb{L}^n$ and consider $\boldsymbol{\alpha}\colon \mathbb{R} \to \mathbb{L}^n$ given by $\boldsymbol{\alpha}(t) = \boldsymbol{p}$. Then $\boldsymbol{\alpha}'(t) = \boldsymbol{0}$, hence $\boldsymbol{\alpha}$ is spacelike.

(2) Straight lines: the line passing through $\boldsymbol{p} \in \mathbb{L}^n$ with direction $\boldsymbol{v} \in \mathbb{L}^n$ admits a parametrization given by $\boldsymbol{\alpha}(t) = \boldsymbol{p} + t\boldsymbol{v}$, for $t \in \mathbb{R}$. Then $\boldsymbol{\alpha}'(t) = \boldsymbol{v}$, and $\boldsymbol{\alpha}$ has the same causal type of $\boldsymbol{v}$. This agrees with the previous definition for the causal type of a straight line.

(3) Euclidean circle: consider $\boldsymbol{\alpha}\colon \mathbb{R} \to \mathbb{L}^2$, given by $\boldsymbol{\alpha}(t) = (\cos t, \sin t)$. Thinking geometrically of its image, we see that $\boldsymbol{\alpha}$ is:

- spacelike for $t \in \,]\pi/4 + k\pi, 3\pi/4 + k\pi[$;
- timelike for $t \in \,]-\pi/4 + k\pi, \pi/4 + k\pi[$;
- lightlike for $t = \pi/4 + k\pi/2$,

for all $k \in \mathbb{Z}$.

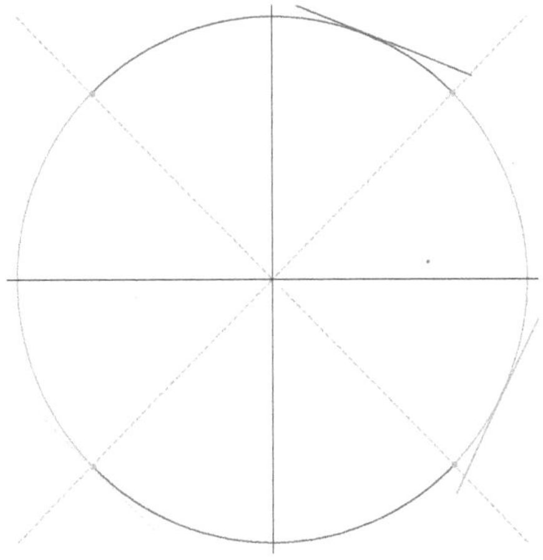

Figure 2.1: The Euclidean unit circle in $\mathbb{L}^2$.

(4) Consider the curve $\boldsymbol{\alpha} : \mathbb{R} \to \mathbb{L}^3$ given by $\boldsymbol{\alpha}(t) = (-6t, 2\sqrt{7}t, t^2)$. We have that $\boldsymbol{\alpha}'(t) = (-6, 2\sqrt{7}, 2t)$, and then: $\langle \boldsymbol{\alpha}'(t), \boldsymbol{\alpha}'(t) \rangle_L = 64 - 4t^2$. Analyzing the sign of this last expression we see that $\boldsymbol{\alpha}$ is:

- spacelike on $]-4, 4[$;
- timelike on $]-\infty, -4[\,\cup\,]4, +\infty[$;
- lightlike at $t = -4$ and $t = 4$.

(5) A Euclidean circle of radius $r > 0$, centered in $\boldsymbol{p} \in \mathbb{L}^3$. Let $\boldsymbol{\alpha}\colon \mathbb{R} \to \mathbb{L}^3$, given by $\boldsymbol{\alpha}(t) = \boldsymbol{p} + (r\cos t, r\sin t, 0)$. Then $\boldsymbol{\alpha}'(t) = (-r\sin t, r\cos t, 0)$, hence $\langle \boldsymbol{\alpha}'(t), \boldsymbol{\alpha}'(t) \rangle_L = r^2 > 0$, so $\boldsymbol{\alpha}$ is spacelike. Note that $\boldsymbol{\alpha}$ is a planar curve, contained in a spacelike plane.

(6) A hyperbola centered in $\boldsymbol{p} \in \mathbb{L}^3$ and radius $r > 0$. Let $\boldsymbol{\alpha}\colon \mathbb{R} \to \mathbb{L}^3$ be given by $\boldsymbol{\alpha}(t) = \boldsymbol{p} + (0, r\cosh t, r\sinh t)$. Now $\boldsymbol{\alpha}'(t) = (0, r\sinh t, r\cosh t)$, and $\langle \boldsymbol{\alpha}'(t), \boldsymbol{\alpha}'(t) \rangle_L = -r^2 < 0$, so $\boldsymbol{\alpha}$ is timelike. This is an example of a timelike curve lying in a timelike plane.

(7) In analogy to the previous item, if $\boldsymbol{p} \in \mathbb{L}^3$ and $r > 0$, the hyperbola $\boldsymbol{\alpha}\colon \mathbb{R} \to \mathbb{L}^3$ given by $\boldsymbol{\alpha}(t) = \boldsymbol{p} + (0, r\sinh t, r\cosh t)$ is spacelike, but lying in a timelike plane.

(8) Euclidean Helix: consider $\boldsymbol{\alpha}\colon \mathbb{R} \to \mathbb{L}^3$, given by $\boldsymbol{\alpha}(t) = (a\cos t, a\sin t, bt)$, $a, b > 0$. We give the causal type of this helix in terms of a and b.

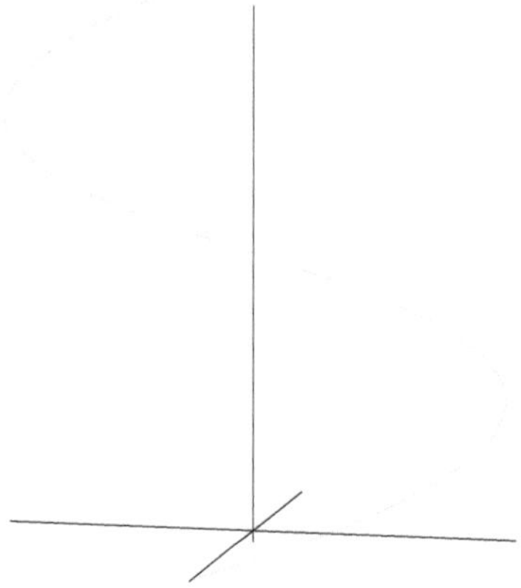

Figure 2.2: A lightlike Euclidean helix in $\mathbb{L}^3$.

We have $\boldsymbol{\alpha}'(t) = (-a\sin t, a\cos t, b)$, and then

$$\langle \boldsymbol{\alpha}'(t), \boldsymbol{\alpha}'(t) \rangle_L = a^2 - b^2 = (a+b)(a-b).$$

It follows directly that $\boldsymbol{\alpha}$ is:

- spacelike, if $a > b$;
- timelike, if $a < b$;
- lightlike, if $a = b$.

Remark. In general, a planar curve in $\mathbb{L}^3$ *doesn't have* the same causal type as the plane containing it. Take for example the x-axis and any plane containing it. One can obtain planes of each causal type by rotating this plane around the x-axis.

Definition 2.1.4. Let $\boldsymbol{\alpha}\colon I \to \mathbb{R}^n_\nu$ to be a parametrized curve. We say that $\boldsymbol{\alpha}$ is *regular* if $\boldsymbol{\alpha}'(t) \neq \mathbf{0}$, for all $t \in I$. If there is $t_0 \in I$ such that $\boldsymbol{\alpha}'(t_0) = \mathbf{0}$, we say that t_0 is a *singular point* of this parametrization.

Remark. The regularity hypothesis is reasonable, since it excludes, among other situations, the constant curve whose image is a point (as it doesn't fit our intuitive notion of a "curve").

Theorem 2.1.5. *Let $\boldsymbol{\alpha}\colon I \to \mathbb{L}^n$ be a curve. If $\boldsymbol{\alpha}$ is lightlike or timelike, then $\boldsymbol{\alpha}$ is regular.*

Proof: Write $\boldsymbol{\alpha}(t) = (x_1(t), \ldots, x_n(t))$. Then

$$x_1'(t)^2 + \cdots + x_{n-1}'(t)^2 - x_n'(t)^2 \leq 0.$$

If $x_n'(t) = 0$, then $x_i'(t) = 0$ for all $1 \leq i \leq n-1$, and $\boldsymbol{\alpha}'(t) = \mathbf{0}$. Hence $\boldsymbol{\alpha}$ is spacelike in t. Therefore $x_n'(t) \neq 0$, for all $t \in I$ and $\boldsymbol{\alpha}'(t) \neq \mathbf{0}$. □

It is sometimes interesting to consider curves defined in closed intervals $[a, b]$. In this case the map $\boldsymbol{\alpha}\colon [a, b] \to \mathbb{R}^n_\nu$ is smooth if it is the restriction of some smooth map defined in an open interval containing $[a, b]$.

Definition 2.1.6. A smooth map $\boldsymbol{\alpha}\colon [a,b] \to \mathbb{R}^n_\nu$ is a *closed parametrized curve* if $\boldsymbol{\alpha}$ and all of its derivatives agree in the boundary of its domain. In other words, for all $k \geq 0$, we have $\boldsymbol{\alpha}^{(k)}(a) = \boldsymbol{\alpha}^{(k)}(b)$.

Our first result for closed parametrized curves is:

Theorem 2.1.7. *In $\mathbb{L}^n$, there are no closed lightlike or closed timelike parametrized curves.*

Proof: Let $\boldsymbol{\alpha}\colon [a,b] \to \mathbb{L}^n$ be a curve satisfying $\boldsymbol{\alpha}(a) = \boldsymbol{\alpha}(b)$. Writing it in coordinates as $\boldsymbol{\alpha}(t) = (x_1(t), \ldots, x_n(t))$, we shall see that $\boldsymbol{\alpha}$ is spacelike (excluding the singular case) at some point. We have $x_n(a) = x_n(b)$ and, by Rolle's Theorem, there exists $t_0 \in \,]a,b[$ such that $x_n'(t_0) = 0$. Hence

$$\langle \boldsymbol{\alpha}'(t_0), \boldsymbol{\alpha}'(t_0)\rangle_L = x_1'(t_0)^2 + \cdots + x_{n-1}'(t_0)^2 \geq 0.$$

If $\langle \boldsymbol{\alpha}'(t_0), \boldsymbol{\alpha}'(t_0)\rangle_L = 0$, then $\boldsymbol{\alpha}'(t_0) = \mathbf{0}$ and $\boldsymbol{\alpha}$ is singular in t_0. If $\langle \boldsymbol{\alpha}'(t_0), \boldsymbol{\alpha}'(t_0)\rangle_L > 0$, then $\boldsymbol{\alpha}$ is spacelike in t_0. □

Remark. The existence of $\boldsymbol{\alpha}'(t)$ for all $t \in [a,b]$ is all we needed in the proof above. We didn't use $\boldsymbol{\alpha}^{(k)}(a) = \boldsymbol{\alpha}^{(k)}(b)$, for all $k \geq 1$. Give a geometric interpretation for this.

In the setting of Special Relativity, future-directed timelike curves are the worldlines of particles with positive mass, while future-directed lightlike curves model the trajectory of massless particles, like photons. In this way, Theorem 2.1.7 says that it is impossible for an observer in $\mathbb{L}^n$ to go back to the past, before they are born, to kill their own grandfather. This means that the causality of Lorentz-Minkowski is "well-behaved". Therefore, spacetime models that admit closed timelike or closed lightlike curves are not usually considered in the General Relativity context, but are of interest in Causality Theory (see, for example, [7]). Compare the above result with Exercise 1.2.12 (p. 18).

Definition 2.1.8. Let $\boldsymbol{\alpha}\colon I \to \mathbb{R}^n_\nu$ be a parametrized curve and $a, b \in I$. The *arclength of* $\boldsymbol{\alpha}$ between a and b is defined by

$$L_a^b[\boldsymbol{\alpha}] \doteq \int_a^b \|\boldsymbol{\alpha}'(t)\|_\nu \, dt.$$

When $\boldsymbol{\alpha}$ is future-directed and timelike in $\mathbb{L}^n$, its arclength is called *proper time*, denoted by $\mathrm{t}_a^b[\boldsymbol{\alpha}]$. Furthermore, an *arclength function for* $\boldsymbol{\alpha}$ is a smooth function $s\colon I \to \mathbb{R}$ written as $s(t) = L_{t_0}^t[\boldsymbol{\alpha}]$, for some $t_0 \in I$.

Remark. The proper time of a future-directed timelike parametrized curve can be seen as the time measured by a clock carried by an observer traveling along the curve.

Another quantity associated to a curve, whose importance appears in Chapter 3, is the energy:

Definition 2.1.9. Let $\boldsymbol{\alpha}\colon I \to \mathbb{R}^n_\nu$ be a parametrized curve and $a, b \in I$. The *energy of* $\boldsymbol{\alpha}$ *between* a *and* b is defined by

$$E_a^b[\boldsymbol{\alpha}] \doteq \frac{1}{2}\int_a^b \langle \boldsymbol{\alpha}'(t), \boldsymbol{\alpha}'(t)\rangle_\nu \, dt.$$

Remark. To avoid overloaded notation we don't mention the ambient space of the curve in arclength and energy, since it is given in the definition of the curve *a priori*. We also omit the boundaries a and b when the integration is over the whole interval I.

Example 2.1.10. To illustrate, we compute the arclengths of a given curve, seeing it as a curve in Euclidean space and then in Lorentz-Minkowski space.

(1) Let $\boldsymbol{\alpha}\colon \mathbb{R} \to \mathbb{R}^3$ be given by $\boldsymbol{\alpha}(t) = (a\cos t, a\sin t, bt)$. Then,

$$L_0^{2\pi}[\boldsymbol{\alpha}] = \int_0^{2\pi} \sqrt{a^2+b^2}\,\mathrm{d}t = 2\pi\sqrt{a^2+b^2}.$$

(2) If $\boldsymbol{\alpha}\colon \mathbb{R} \to \mathbb{L}^3$ is given by the expression above, we have

$$L_0^{2\pi}[\boldsymbol{\alpha}] = \int_0^{2\pi} \sqrt{|a^2-b^2|}\,\mathrm{d}t = 2\pi\sqrt{|a^2-b^2|}.$$

In particular, if $|a| = |b|$ the curve is lightlike and its length vanishes.

(3) Let $\boldsymbol{\beta} : \mathbb{R} \to \mathbb{R}^3$ be given by $\boldsymbol{\beta}(t) = (\mathrm{e}^{-t}\cos t, \mathrm{e}^{-t}\sin t, \mathrm{e}^{-t})$. Then

$$L_0^1[\boldsymbol{\beta}] = \int_0^1 \mathrm{e}^{-t}\sqrt{3}\,\mathrm{d}t = \sqrt{3}\left(1 - \frac{1}{\mathrm{e}}\right).$$

(4) If $\boldsymbol{\beta} : \mathbb{R} \to \mathbb{L}^3$ is given by the expression above,

$$L_0^1[\boldsymbol{\beta}] = \int_0^1 \mathrm{e}^{-t}\,\mathrm{d}t = 1 - \frac{1}{\mathrm{e}}.$$

Now let's compute the energy of some curves:

(5) Let $\gamma : \mathbb{R} \to \mathbb{R}^2$ be given by $\gamma(t) = (\cos t, \sin t)$, then

$$E_0^{2\pi}[\gamma] = \int_0^{2\pi} 1\,\mathrm{d}t = 2\pi.$$

(6) The same parametrization in $\mathbb{L}^2$ gives

$$E_0^{2\pi}[\gamma] = -\int_0^{2\pi} \cos(2t)\,\mathrm{d}t = 0.$$

(You would expect that, right? See Figure 2.1.)

(7) The hyperbola, parametrized by $\boldsymbol{\eta} : \mathbb{R} \to \mathbb{R}^2$, $\boldsymbol{\eta}(t) = (\sinh t, \cosh t)$ has its energy given by

$$E_0^{2\pi}[\boldsymbol{\eta}] = \int_0^{2\pi} \cosh(2t)\,\mathrm{d}t = \frac{\sinh(4\pi)}{2}.$$

(8) In $\mathbb{L}^2$, the same map $\boldsymbol{\eta}$ has energy

$$E_0^{2\pi}[\boldsymbol{\eta}] = \int_0^{2\pi} 1\,\mathrm{d}t = 2\pi.$$

The energy and arclength of a parametrized curve are related:

Proposition 2.1.11. *Let $\boldsymbol{\alpha} : I \to \mathbb{R}^n_\nu$ be a parametrized curve with constant causal type. Given any $a, b \in I$ we have*

$$L_a^b[\boldsymbol{\alpha}] \leq \sqrt{2\epsilon_{\boldsymbol{\alpha}}(b-a)E_a^b[\boldsymbol{\alpha}]}.$$

The equality holds if and only if $\boldsymbol{\alpha}$ has constant speed.

Proof: Apply the Cauchy-Schwarz inequality for the (positive-definite) inner product given by

$$\langle\langle f, g\rangle\rangle \doteq \int_a^b f(x)g(x)\,\mathrm{d}x,$$

for the functions $f(t) = \sqrt{\epsilon_{\boldsymbol{\alpha}}\langle \boldsymbol{\alpha}'(t), \boldsymbol{\alpha}'(t)\rangle}$ and $g(t) = 1$:

$$\begin{aligned} L_a^b[\boldsymbol{\alpha}] &= \langle\langle f, g\rangle\rangle \\ &\leq \|f\|\|g\| \\ &= \left(\int_a^b \epsilon_{\boldsymbol{\alpha}}\langle \boldsymbol{\alpha}'(t), \boldsymbol{\alpha}'(t)\rangle\,\mathrm{d}t\right)^{1/2} \left(\int_a^b 1\,\mathrm{d}t\right)^{1/2} \\ &= \sqrt{2\epsilon_{\boldsymbol{\alpha}} E_a^b[\boldsymbol{\alpha}]}\sqrt{b-a}. \end{aligned}$$

The equality holds if and only if the functions f and g are linearly dependent on the function space or, equivalently, $\boldsymbol{\alpha}$ has constant speed. □

Remark. The previous result is false when the causal type of the curve is not constant. We saw a counter-example not too long ago. Which one it is?

Two parametrized curves may have the same image (trace). An example is an arc of the unit circle

$$\mathbb{S}^1 = \{(x,y) \in \mathbb{R}^2 \mid x^2 + y^2 = 1\},$$

which can be parametrized by $\boldsymbol{\alpha}(t) = (\cos t, \sin t)$, $t \in \,]-\pi/2, 3\pi/2[$ and by the expression obtained in Exercise 2.1.8. This motivates the:

Definition 2.1.12. Let $\boldsymbol{\alpha} : I \to \mathbb{R}^n_\nu$ be a parametrized curve. A *reparametrization* of $\boldsymbol{\alpha}$ is another parametrized curve of the form $\boldsymbol{\beta} = \boldsymbol{\alpha} \circ h$, for some diffeomorphism $h : J \to I$ between the open intervals J and I. The function h is called a *change of parameters*.

Remark. In the notation above:

- if $\boldsymbol{\beta}$ is a reparametrization of $\boldsymbol{\alpha}$ by h, then $\boldsymbol{\alpha}$ is a reparametrization of $\boldsymbol{\beta}$ by h^{-1};
- the reparametrization is *positive* if $h' > 0$;
- $\boldsymbol{\beta}$ is regular if and only if $\boldsymbol{\alpha}$ is regular;
- $\boldsymbol{\beta}$ has the same causal type of $\boldsymbol{\alpha}$.

The interesting objects to study in Differential Geometry of curves are those intrinsic to the trace of the curve, that is, those invariant under reparametrization.

In particular, and as expected, the length of a parametrized curve is invariant under reparametrizations (see Exercise 2.1.13), but energy is not intrinsic (see Exercise 2.1.14).

One must be careful when assigning such invariant properties to the image of the curve via some parametrization. For example, the curves $\boldsymbol{\alpha}\colon\ [0, 2\pi] \to \mathbb{S}^1$ and $\boldsymbol{\beta}\colon\ [0, 4\pi] \to \mathbb{S}^1$ given by $\boldsymbol{\alpha}(t) = \boldsymbol{\beta}(t) \doteq (\cos t, \sin t)$ have the same image, but $L[\boldsymbol{\alpha}] = 2\pi$ and $L[\boldsymbol{\beta}] = 4\pi$. What would prevent us from setting the latter value as the length of the unit circle, even knowing that its length is 2π?

The images $\boldsymbol{\alpha}([0, 2\pi])$ and $\boldsymbol{\beta}([0, 4\pi])$ are the same as sets, but geometrically different in the sense that $\boldsymbol{\alpha}$ covers $\mathbb{S}^1$ once, while $\boldsymbol{\beta}$ covers it twice. The injectivity condition for the parametrization avoids artificial distortions in the "actual" length of a curve. There are reparametrizations of a given curve that carry strong geometric meaning and will simplify many expressions to be obtained in the following sections. To explore this, we need the following result:

Proposition 2.1.13. *Let $u, v : I \to \mathbb{R}^n_\nu$ be parametrized curves. Then:*

(i) $f : I \to \mathbb{R}$ *given by* $f(t) = \langle u(t), v(t)\rangle_\nu$ *is smooth and*

$$f'(t) = \langle u'(t), v(t)\rangle_\nu + \langle u(t), v'(t)\rangle_\nu;$$

(ii) $g : I \to \mathbb{R}$ *given by* $g(t) = \|u(t)\|_\nu$ *is smooth in each t such that* $\|u(t)\|_\nu \neq 0$ *and*

$$g'(t) = \frac{\epsilon_u \langle u'(t), u(t)\rangle_\nu}{\|u(t)\|_\nu};$$

(iii) If $n = 3$, the map $w : I \to \mathbb{R}^3_\nu$ given by $w(t) = u(t) \times v(t)$ is smooth and

$$w'(t) = u'(t) \times v(t) + u(t) \times v'(t).$$

Proof: Since u, v, $\langle \cdot, \cdot \rangle_\nu$ and $\times$ are smooth and the square root function is differentiable wherever it does not vanish, all of the compositions in the statement are smooth. We could achieve expressions for the derivatives expanding each object in terms of its coordinates. A more elegant way to prove this, allowing generalizations, is to use results provided in Appendix A as follows:

(i) From bilinearity of $\langle \cdot, \cdot \rangle_\nu$ we have

$$\begin{aligned} f'(t) &= Df(t)(1) = D(\langle \cdot, \cdot \rangle_\nu \circ (u, v))(t)(1) \\ &= D(\langle \cdot, \cdot \rangle_\nu)(u(t), v(t)) \circ D(u, v)(t)(1) \\ &= D(\langle \cdot, \cdot \rangle_\nu)(u(t), v(t)) \circ (Du(t), Dv(t))(1) \\ &= D(\langle \cdot, \cdot \rangle_\nu)(u(t), v(t))(u'(t), v'(t)) \\ &= \langle u'(t), v(t)\rangle_\nu + \langle u(t), v'(t)\rangle_\nu. \end{aligned}$$

(ii) Setting $v = u$ in the above,

$$g'(t) = \frac{\mathrm{d}}{\mathrm{d}t}\sqrt{\epsilon_u \langle u(t), u(t)\rangle_\nu} = \frac{2\epsilon_u \langle u'(t), u(t)\rangle_\nu}{2\sqrt{\epsilon_u \langle u(t), u(t)\rangle_\nu}} = \frac{\epsilon_u \langle u'(t), u(t)\rangle_\nu}{\|u(t)\|_\nu}.$$

(iii) Analogously to the item (i), bilinearity of $\times$ gives

$$\begin{aligned} w'(t) &= Dw(t)(1) = D(\times \circ (u, v))(t)(1) \\ &= D(\times)(u(t), v(t)) \circ D(u, v)(t)(1) \\ &= D(\times)(u(t), v(t))(u'(t), v'(t)) \\ &= u'(t) \times v(t) + u(t) \times v'(t). \end{aligned}$$

□

Remark. Item (iii) above can be generalized to the cross product in $\mathbb{R}^n_\nu$ using the formula for the total derivative of a multilinear map. Try to deduce such an expression.

Corollary 2.1.14. *Let $u, v : I \to \mathbb{R}^n_\nu$ be parametrized curves. If $\langle u, v\rangle_\nu$ is constant, then $\langle u'(t), v(t)\rangle_\nu = -\langle u(t), v'(t)\rangle_\nu$, for all $t \in I$.*

Corollary 2.1.15. *Let $u : I \to \mathbb{R}^n_\nu$ be a parametrized curve. If $\langle u, u\rangle_\nu$ is constant, then $u'(t) \perp u(t)$, for all $t \in I$ (interpret this geometrically).*

Definition 2.1.16. Let $\boldsymbol{\alpha}\colon I \to \mathbb{R}^n_\nu$ be a parametrized curve. We say that $\boldsymbol{\alpha}$ has *unit speed* if $\|\boldsymbol{\alpha}'(t)\|_\nu = 1$ for all $t \in I$, and $\boldsymbol{\alpha}$ is *parametrized by arclength* if, for all $t_0 \in I$, the arclength function from t_0 is given by $s(t) = t - t_0$.

Remark. The concepts defined above are equivalent. We ask you to verify this in Exercise 2.1.16.

Theorem 2.1.17. *Let $\boldsymbol{\alpha}\colon I \to \mathbb{R}^n_\nu$ be a non-lightlike regular curve. Then $\boldsymbol{\alpha}$ admits a unit speed reparametrization. More precisely, there exists an open interval J and a diffeomorphism $h\colon J \to I$ such that $\widetilde{\boldsymbol{\alpha}} \doteq \boldsymbol{\alpha} \circ h$ has unit speed.*

Proof: Let $t_0 \in I$ be fixed and $s\colon I \to \mathbb{R}$ be the arclength function from t_0. From the Fundamental Theorem of Calculus, $s'(t) = \|\boldsymbol{\alpha}'(t)\|_\nu > 0$. Therefore s is an increasing diffeomorphism over its image $J \doteq s(I) \subseteq \mathbb{R}$. Hence we consider its inverse map, $h\colon J \to I$ (also increasing). Setting $\widetilde{\boldsymbol{\alpha}} = \boldsymbol{\alpha} \circ h$ and identifying $s = s(t)$, we have:

$$\|\widetilde{\boldsymbol{\alpha}}'(s)\|_\nu = \|\boldsymbol{\alpha}'(h(s))h'(s)\|_\nu = \|\boldsymbol{\alpha}'(h(s))\|_\nu h'(s) = s'(h(s))h'(s) = (s \circ h)'(s) = 1.$$

□

Remark.

- The relation $\widetilde{\boldsymbol{\alpha}} = \boldsymbol{\alpha} \circ h$ implies $\boldsymbol{\alpha}(t) = \widetilde{\boldsymbol{\alpha}}(s(t))$ for all $t \in I$. This latter expression is more commonly used.
- For timelike curves, one may also write $\boldsymbol{\alpha}(t) = \widetilde{\boldsymbol{\alpha}}(\mathrm{t}(t))$.
- The above reparametrizations are not unique. In fact, they depend on the choice of t_0, but the change of parameters between two unit speed reparametrizations is necessarily an affine function. Details in Exercise 2.1.19.

For lightlike curves it is impossible to obtain unit speed reparametrizations. Intuitively, the proper time in any arc of a photon's worldline is zero, hence it can't be used as a parameter. The best we get is:

Theorem 2.1.18 (Arc-photon). *Let $\boldsymbol{\alpha}\colon I \to \mathbb{L}^n$ be a lightlike parametrized curve such that $\|\boldsymbol{\alpha}''(t)\|_L \neq 0$ for all $t \in I$. Then $\boldsymbol{\alpha}$ admits an* arc-photon *reparametrization, that is, there exists an open interval J and a diffeomorphism $h\colon J \to I$ such that $\widetilde{\boldsymbol{\alpha}} \doteq \boldsymbol{\alpha} \circ h$ satisfies $\|\widetilde{\boldsymbol{\alpha}}''(\phi)\|_L = 1$ for all $\phi \in J$.*

Proof: Such a function h must verify $\widetilde{\boldsymbol{\alpha}}(\phi) = \boldsymbol{\alpha}(h(\phi))$. Differentiating this twice we get

$$\widetilde{\boldsymbol{\alpha}}''(\phi) = \boldsymbol{\alpha}''(h(\phi))h'(\phi)^2 + \boldsymbol{\alpha}'(h(\phi))h''(\phi).$$

Since $\boldsymbol{\alpha}$ is lightlike, $\boldsymbol{\alpha}''(t)$ is orthogonal to $\boldsymbol{\alpha}'(t)$ and the condition $\|\boldsymbol{\alpha}''(t)\|_L \neq 0$ tells us that $\boldsymbol{\alpha}''(t)$ is spacelike, for all $t \in I$. Hence, applying $\langle\cdot,\cdot\rangle_L$ to both sides of the above equation, we have

$$1 = \langle \boldsymbol{\alpha}''(h(\phi)), \boldsymbol{\alpha}''(h(\phi))\rangle_L h'(\phi)^4,$$

so that $h'(\phi) = \|\boldsymbol{\alpha}''(h(\phi))\|_L^{-1/2}$: a first order ODE for the function h. For fixed numbers $\phi_0 \in J$ and $t_0 \in I$, one shows that this equation has a single solution satisfying $h(\phi_0) = t_0$. Once the existence of h is ensured, we use it to define $\widetilde{\boldsymbol{\alpha}}$ with the desired properties. □

Remark. As in the previous remark, arc-photon reparametrizations are not unique and the change of parameters between any two of them is affine. See Exercise 2.1.20.

Example 2.1.19.

(1) In $\mathbb{R}^3$ the parametrized helix $\boldsymbol{\alpha} : \mathbb{R} \to \mathbb{R}^3$, $\boldsymbol{\alpha}(t) = (a\cos t, a\sin t, bt)$, has the function $s(t) = t\sqrt{a^2+b^2}$ as an arclength. That is, $t = s(t)/\sqrt{a^2+b^2}$, hence

$$\widetilde{\boldsymbol{\alpha}}(s) = \left(a\cos\left(\frac{s}{\sqrt{a^2+b^2}}\right), a\sin\left(\frac{s}{\sqrt{a^2+b^2}}\right), \frac{bs}{\sqrt{a^2+b^2}}\right)$$

is a unit speed reparametrization of $\boldsymbol{\alpha}$.

(2) In $\mathbb{L}^3$ consider the parametrized curve $\boldsymbol{\alpha} :]0,+\infty[\to \mathbb{L}^3$ given by

$$\boldsymbol{\alpha}(t) = \left((t^2-2)\sin t + 2t\cos t, (2-t^2)\cos t + 2t\sin t, \frac{\sqrt{10}}{3}t^3\right).$$

Then $\boldsymbol{\alpha}'(t) = (t^2\cos t, t^2\sin t, \sqrt{10}t^2)$ so that $\boldsymbol{\alpha}$ is timelike amd

$$\mathfrak{t}(t) = \int_0^t 3\xi^2\,d\xi = t^3.$$

Hence $t = \sqrt[3]{\mathfrak{t}(t)}$. This way,

$$\widetilde{\boldsymbol{\alpha}}(\mathfrak{t}) = \left((\sqrt[3]{\mathfrak{t}^2}-2)\sin\sqrt[3]{\mathfrak{t}} + 2\sqrt[3]{\mathfrak{t}}\cos\sqrt[3]{\mathfrak{t}}, (2-\sqrt[3]{\mathfrak{t}^2})\cos\sqrt[3]{\mathfrak{t}} + 2\sqrt[3]{\mathfrak{t}}\sin\sqrt[3]{\mathfrak{t}}, \frac{\sqrt{10}}{3}\mathfrak{t}\right)$$

is a proper time reparametrization of $\boldsymbol{\alpha}$.

(3) Consider the Euclidean helix $\boldsymbol{\alpha} : \mathbb{R} \to \mathbb{L}^3$, $\boldsymbol{\alpha}(t) = (r\cos t, r\sin t, rt)$, where $r > 0$. Note that $\boldsymbol{\alpha}$ is lightlike. We will obtain an arc-photon reparametrization of $\boldsymbol{\alpha}$: since $\boldsymbol{\alpha}''(t) = (-r\cos t, -r\sin t, 0)$ we have $\|\boldsymbol{\alpha}''(t)\|_L = r$. Then

$$h'(\phi) = \frac{1}{\sqrt{r}} \implies h(\phi) = \frac{\phi}{\sqrt{r}}.$$

Hence

$$\widetilde{\boldsymbol{\alpha}}(\phi) = \left(r\cos\left(\frac{\phi}{\sqrt{r}}\right), r\sin\left(\frac{\phi}{\sqrt{r}}\right), \sqrt{r}\phi\right).$$

To close this section, we introduce the following concept:

Definition 2.1.20 (Congruence of curves). Let $\boldsymbol{\alpha}\colon I \to \mathbb{R}^n_\nu$ and $\boldsymbol{\beta}\colon J \to \mathbb{R}^n_\nu$ be parametrized curves. We say that $\boldsymbol{\alpha}$ and $\boldsymbol{\beta}$ are *congruent* if there is a reparametrization $\widetilde{\boldsymbol{\alpha}}\colon J \to \mathbb{R}^n_\nu$ of $\boldsymbol{\alpha}$ and an isometry $F \in \mathrm{E}_\nu(n,\mathbb{R})$ such that $\boldsymbol{\beta} = F \circ \widetilde{\boldsymbol{\alpha}}$.

Since any two open intervals are diffeomorphic, we may assume that both curves are defined in the same domain I. In the notation above this means $\widetilde{\boldsymbol{\alpha}} = \boldsymbol{\alpha}$.

Congruence is one of the most important concepts in Geometry, since it allows the study of geometric quantities of an object, which do not depend on the position of the object in the ambient space.

Proposition 2.1.21. *Let $\boldsymbol{\alpha}, \boldsymbol{\beta}\colon I \to \mathbb{R}^n_\nu$ be congruent parametrized curves. Then $\langle \boldsymbol{\beta}'(t), \boldsymbol{\beta}'(t)\rangle_\nu = \langle \boldsymbol{\alpha}'(t), \boldsymbol{\alpha}'(t)\rangle_\nu$ for all $t \in I$.*

Proof: There exists $F \in \mathrm{E}_\nu(n, \mathbb{R})$ such that $\boldsymbol{\beta} = F \circ \boldsymbol{\alpha}$. We already know that $F = T_{\boldsymbol{a}} \circ \Lambda$ for some $\Lambda \in \mathrm{O}_\nu(n, \mathbb{R})$ and $\boldsymbol{a} \in \mathbb{R}^n_\nu$. Then

$$\boldsymbol{\beta}'(t) = D\boldsymbol{\beta}(t)(1) = D(F \circ \boldsymbol{\alpha})(t)(1) = DF(\boldsymbol{\alpha}(t)) \circ D\boldsymbol{\alpha}(t)(1) = \Lambda\boldsymbol{\alpha}'(t).$$

Hence $\langle \boldsymbol{\beta}'(t), \boldsymbol{\beta}'(t)\rangle_\nu = \langle \boldsymbol{\alpha}'(t), \boldsymbol{\alpha}'(t)\rangle_\nu$. □

Corollary 2.1.22. *Let $\boldsymbol{\alpha}, \boldsymbol{\beta}\colon I \to \mathbb{R}^n_\nu$ be two congruent parametrized curves. Then we have that $L[\boldsymbol{\beta}] = L[\boldsymbol{\alpha}]$ and $E[\boldsymbol{\beta}] = E[\boldsymbol{\alpha}]$.*

In terms of congruence, we raise a question: what is the minimal geometric information necessary to recover, up to a congruence, a curve in the plane or space? The answer relies on the concepts of *curvature* and *torsion*. Both are the subject of the following sections.

Exercises

In the following exercises, "curve" stands for "parametrized curve".

Exercise 2.1.1. As in Example 2.1.3 (p. 64), discuss the causal type of:

(a) the *hyperbolic helix* $\boldsymbol{\alpha}\colon \mathbb{R} \to \mathbb{L}^3$ given by $\boldsymbol{\alpha}(t) = (at, b\cosh t, b\sinh t)$, where $a, b > 0$;

(b) the curve $\boldsymbol{\alpha}\colon \mathbb{R} \to \mathbb{L}^4$ given by $\boldsymbol{\alpha}(t) = (a_0, a_1t, a_2t^2, a_3t^3)$.

(c) the curve $\boldsymbol{\alpha}\colon \mathbb{R}_{>0} \to \mathbb{L}^3$ given by $\boldsymbol{\alpha}(t) = (t, \log t^a, \log t^b)$, where $a, b > 0$.

Exercise 2.1.2. Let $\boldsymbol{\alpha}\colon I \to \mathbb{R}^n_\nu$ be a curve.

(a) Let $\nu = 0$ and $\boldsymbol{p} \in \mathbb{R}^n$ not be contained in the image of $\boldsymbol{\alpha}$. Suppose that $\boldsymbol{\alpha}(t_0)$ is the point in the image of $\boldsymbol{\alpha}$ which is closest to $\boldsymbol{p}$. Show that $\boldsymbol{\alpha}'(t_0)$ and $\boldsymbol{\alpha}(t_0) - \boldsymbol{p}$ are orthogonal.

(b) Let $\boldsymbol{v} \in \mathbb{R}^n_\nu$. Suppose $\langle \boldsymbol{\alpha}'(t), \boldsymbol{v}\rangle_\nu = 0$ for all $t \in I$ and that there exists $t_0 \in I$ such that $\boldsymbol{\alpha}(t_0)$ is orthogonal to $\boldsymbol{v}$. Show that $\boldsymbol{\alpha}(t)$ is orthogonal to $\boldsymbol{v}$ for all $t \in I$.

(c) Let $\nu = 0$ and $\boldsymbol{\beta}\colon I \to \mathbb{R}^n$ be another curve. Show that if, for all $t \in I$, the line passing through $\boldsymbol{\alpha}(t)$ and $\boldsymbol{\beta}(t)$ is orthogonal to $\boldsymbol{\alpha}$ and $\boldsymbol{\beta}$ at t, then the length of the segment joining $\boldsymbol{\alpha}(t)$ and $\boldsymbol{\beta}(t)$ is constant.

Remark. The angle between two curves at a point is, by definition, the angle between the tangent lines to the curves at that point.

Exercise 2.1.3. Show that if $\boldsymbol{\alpha}\colon I \to \mathbb{R}^n$ is a curve whose coordinates are polynomials of order at most k, then the trace of $\boldsymbol{\alpha}$ is contained in an affine subspace of $\mathbb{R}^n$ of dimension at most k.

Hint. Taylor polynomials.

Exercise 2.1.4. Let $\boldsymbol{\alpha}\colon I \to \mathbb{L}^3$ be a regular closed curve (hence spacelike). Show that if the trace of $\boldsymbol{\alpha}$ is contained in an affine plane Π, then Π is also spacelike. How can you generalize this result to $\mathbb{L}^n$?

Hint. Apply, if necessary, a Poincaré transformation and suppose that $\Pi\colon x = 0$.

Exercise 2.1.5. Compute the arclength and energy of the following curves:

(a) $\alpha\colon\,]0,2\pi[\, \to \mathbb{R}^4$, $\alpha(t) = (\cos t, \sin t, \cos t, \sin t)$;

(b) $\beta\colon\,]1,\cosh(2)[\, \to \mathbb{L}^3$, $\beta(t) = (\cos t, \sin t, t^2/2)$;

(c) $\gamma\colon \mathbb{R} \to \mathbb{L}^3$, $\gamma(t) = \left(12t - 3\sqrt{2}t^2, 9t + \dfrac{3\sqrt{2}}{2}t^2 - \dfrac{t^3}{2}, 15t - \dfrac{3\sqrt{2}}{2}t^2 + \dfrac{t^3}{2}\right)$.

Exercise 2.1.6. Let $a, b \in \mathbb{R}$, with $a > 0$ and $b < 0$. Consider the *logarithmic spiral* $\alpha\colon \mathbb{R} \to \mathbb{R}^2$ given by $\alpha(t) = (ae^{bt}\cos t, ae^{bt}\sin t)$. Show that:

(a) $\lim_{t\to+\infty} \alpha'(t) = \mathbf{0}$;

(b) $\int_0^{+\infty} \|\alpha'(t)\|_E \, dt < +\infty$. Evaluate this integral in terms of a and b, interpreting it when $b \to 0^-$;

(c) α intersects, with a constant angle, all rays starting at the origin. Curves with such a property are called *loxodromic*.

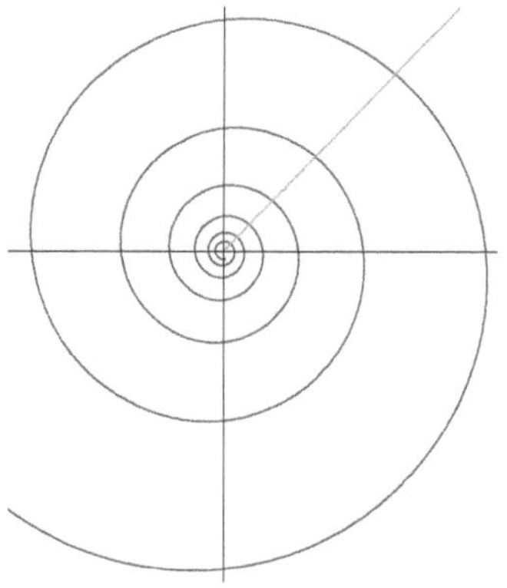

Figure 2.3: The logarithmic spiral with $a = 1$ and $b = -0.1$.

Exercise 2.1.7. Let $a, b \in \mathbb{R}$, with $a > 0$ and $b < 0$. Consider the Lorentzian analogue of the logarithmic spiral, $\beta\colon \mathbb{R} \to \mathbb{L}^2$, given by $\beta(t) = (ae^{bt}\cosh t, ae^{bt}\sinh t)$.

(a) Discuss the causal type of β in terms of a and b.

(b) Discuss the existence of $\lim_{t\to+\infty} \beta'(t)$ in terms of a and b.

(c) Show that $\int_0^{+\infty} \|\beta'(t)\|_L \, dt < +\infty$. Evaluate this integral in terms of a and b, interpreting it when $b \to 0^-$.

Exercise 2.1.8 (Rational parametrization for the circle)**.** For each $t \in \mathbb{R}$, consider the ray welling up from $(0,-1)$ passing through $(t,0)$. Let $\gamma(t)$ be the unique point in the interior of this ray contained in

$$\mathbb{S}^1 = \{(x,y) \in \mathbb{R}^2 \mid x^2 + y^2 = 1\}.$$

(a) Writing the expression for $\gamma(t)$, show that this construction defines a smooth map $\gamma\colon \mathbb{R} \to \mathbb{S}^1 \subseteq \mathbb{R}^2$ whose image omits just the point $(0,-1)$.

(b) Evaluate the limits $\lim_{t\to+\infty}\gamma(t)$ and $\lim_{t\to-\infty}\gamma(t)$. Interpret geometrically.

(c) Verify that $\int_{-\infty}^{+\infty}\|\gamma'(t)\|_E\,\mathrm{d}t = 2\pi$.

Remark. The map γ is also known as *stereographic projection from the circle, via the south pole.*

Exercise 2.1.9 (Rational parametrization for the hyperbola). For each $t \in\,]-1,1[$, consider the ray emanating from $(0,-1)$ passing through $(t,0)$. Let $\gamma(t)$ be the unique point in the interior of this ray contained in

$$\mathbb{H}^1 \doteq \{(x,y) \in \mathbb{L}^2 \mid x^2 - y^2 = -1 \text{ and } y > 0\}.$$

(a) Writing the expression of $\gamma(t)$, show that this construction defines a diffeomorphism $\gamma\colon\,]-1,1[\,\to \mathbb{H}^1 \subseteq \mathbb{L}^2$.

(b) Evaluate the limits $\lim_{t\to1}\gamma(t)$ and $\lim_{t\to-1}\gamma(t)$. Interpret geometrically.

Exercise[†] 2.1.10. Let $\boldsymbol{\alpha}\colon I \to \mathbb{R}^n$ be a curve. Show that given any $a,b \in I$ such that $a<b$, we have

$$L_a^b[\boldsymbol{\alpha}] \geq \|\boldsymbol{\alpha}(b) - \boldsymbol{\alpha}(a)\|_E,$$

with the equality holding if and only if the trace of $\boldsymbol{\alpha}\big|_{[a,b]}$ is the line segment joining $\boldsymbol{\alpha}(a)$ to $\boldsymbol{\alpha}(b)$.

Hint. Write $\boldsymbol{\alpha}(b) - \boldsymbol{\alpha}(a) = \int_a^b \boldsymbol{\alpha}'(t)\,\mathrm{d}t$, apply $\langle\cdot,\boldsymbol{\alpha}(b)-\boldsymbol{\alpha}(a)\rangle_E$ on both sides of it and use Cauchy-Schwarz.

Exercise[†] 2.1.11. Let $\boldsymbol{\alpha}\colon I \to \mathbb{L}^n$ be a future-directed timelike curve and $a,b \in I$ such that $a<b$.

(a) Show that $\boldsymbol{\alpha}(b) - \boldsymbol{\alpha}(a)$ is a future-directed timelike vector.

Hint. Write $\boldsymbol{\alpha} = (\boldsymbol{\beta}, x_n)$, where $\boldsymbol{\beta}\colon I \to \mathbb{R}^{n-1}$. Since $\boldsymbol{\alpha}$ is a future-directed timelike curve, we have the inequality $\|\boldsymbol{\beta}'(t)\|_E < x_n'(t)$ for all $t \in I$. Evaluate the quantity $\|\boldsymbol{\beta}(b) - \boldsymbol{\beta}(a)\|_E = \left\|\int_a^b \boldsymbol{\beta}'(t)\,\mathrm{d}t\right\|_E$.

Remark.

- This shows that the chronological future of a point (informally defined in Chapter 1) coincides with its future timecone, that is, $I^+(\boldsymbol{p}) = C_T^+(\boldsymbol{p})$, for all $\boldsymbol{p} \in \mathbb{L}^n$.
- The result established here holds even if we assume that $\boldsymbol{\alpha}$ is only a differentiable curve, i.e., we don't actually need $\boldsymbol{\alpha}$ to be of class $\mathscr{C}^1$. But in this case the hint is not helpful, since we can't guarantee that $\boldsymbol{\beta}'$ is integrable, so that another approach is needed. Can you think of any?

(b) Show that

$$\mathrm{t}_a^b[\boldsymbol{\alpha}] \leq \|\boldsymbol{\alpha}(b) - \boldsymbol{\alpha}(a)\|_L,$$

with equality holding if and only if the trace of $\boldsymbol{\alpha}\big|_{[a,b]}$ is the line segment joining $\boldsymbol{\alpha}(a)$ and $\boldsymbol{\alpha}(b)$. Give a physical interpretation for this and compare with the explanation for the Twins Paradox, given in Chapter 1.

Hint. Use the strategy employed in the previous exercise, with the timelike version of Cauchy-Schwarz inequality. Pay attention to the signs.

Exercise 2.1.12. The item (a) from the previous exercise has a small generalization when $n = 2$.

(a) Show that if $\boldsymbol{\alpha}\colon I \to \mathbb{R}^2_\nu$ is a curve and $a, b \in I$ are such that $\boldsymbol{\alpha}(b) \neq \boldsymbol{\alpha}(a)$, then there exists c between a and b such that $\boldsymbol{\alpha}(b) - \boldsymbol{\alpha}(a)$ and $\boldsymbol{\alpha}'(c)$ are proportional.

   Hint. Cauchy's Mean Value Theorem.

(b) Conclude that given distinct points $\boldsymbol{p}, \boldsymbol{q} \in \mathbb{L}^2$, the vector $\boldsymbol{q} - \boldsymbol{p}$ has the same causal type of *any* constant causal type curve joining $\boldsymbol{p}$ and $\boldsymbol{q}$.

Exercise† 2.1.13. Prove that the arclength of a curve is invariant under reparametrizations, that is, if $\boldsymbol{\alpha}\colon I \to \mathbb{R}^n_\nu$ and $\boldsymbol{\beta}\colon J \to \mathbb{R}^n_\nu$ are parametrized curves such that $\boldsymbol{\beta} = \boldsymbol{\alpha} \circ h$, where $h\colon J \to I$ is a diffeomorphism, then:

$$\int_J \|\boldsymbol{\beta}'(s)\|\,\mathrm{d}s = \int_I \|\boldsymbol{\alpha}'(t)\|\,\mathrm{d}t.$$

Exercise† 2.1.14. We know that arclength is invariant under reparametrizations, but this doesn't hold for the energy. Let $\boldsymbol{\alpha}\colon\ ]a,b[\ \to \mathbb{R}^n_\nu$ be a smooth curve and $k > 0$. Define $\boldsymbol{\alpha}_k\colon\ ]a/k, b/k[\ \to \mathbb{R}^n_\nu$ as $\boldsymbol{\alpha}_k(t) \doteq \boldsymbol{\alpha}(kt)$. Evaluate the energy of $\boldsymbol{\alpha}_k$ and conclude that any non-lightlike curve admits a reparametrization whose energy has an arbitrarily large absolute value. (Intuitively, we go through the trace of $\boldsymbol{\alpha}$ k times faster. Think of kinetic energy.)

Exercise 2.1.15. Let $\boldsymbol{\alpha}\colon I \to \mathbb{R}^n$ be a curve representing the trajectory of a point with mass $m > 0$, moving under the action of a conservative force field. In other words, there exists a smooth function, defined in some open subset of $\mathbb{R}^n$, $V\colon U \subseteq \mathbb{R}^n \to \mathbb{R}$, such that $\boldsymbol{\alpha}(I) \subseteq U$, satisfying $m\boldsymbol{\alpha}''(t) = -\nabla V(\boldsymbol{\alpha}(t))$ for all $t \in I$. The function V is called *potential energy*. Define the *kinetic energy* function as $T\colon \mathbb{R}^n \to \mathbb{R}$ as $T(\boldsymbol{v}) = m\|\boldsymbol{v}\|^2/2$. Show the *energy conservation principle*: along $\boldsymbol{\alpha}$ the sum of kinetic and potential energies is a constant. Compare the kinetic energy with the energy $E[\boldsymbol{\alpha}]$ defined in the text.

Exercise 2.1.16. Let $\boldsymbol{\alpha}\colon I \to \mathbb{R}^n_\nu$ be a curve. Show that $\boldsymbol{\alpha}$ has unit speed if and only if it is parametrized by arclength.

Exercise 2.1.17. Reparametrize by arclength the following curves:

(a) the *cycloid* $\gamma\colon\ ]0, 2\pi[\ \to \mathbb{R}^2$, $\gamma(t) = (t - \sin t, 1 - \cos t)$;

(b) the upper half of *Neil's parabola (a semicubic)*, given by $\boldsymbol{\eta}\colon \mathbb{R} \to \mathbb{R}^2$, $\boldsymbol{\eta}(t) = (t^2, t^3)$.

Exercise 2.1.18. Let $\boldsymbol{\alpha}\colon \mathbb{R} \to \mathbb{L}^3$ be given by

$$\boldsymbol{\alpha}(t) = (6t - t^3, 3\sqrt{2}t^2, 6t + t^3).$$

Check that $\boldsymbol{\alpha}$ is lightlike and reparametrize it by arc-photon.

Exercise† 2.1.19. Let $\boldsymbol{\alpha}\colon I \to \mathbb{R}^n_\nu$ be a curve and suppose that $\widetilde{\boldsymbol{\alpha}}_1\colon J_1 \to \mathbb{R}^n_\nu$ and $\widetilde{\boldsymbol{\alpha}}_2\colon J_2 \to \mathbb{R}^n_\nu$ are two unit speed reparametrizations of $\boldsymbol{\alpha}$, such that $\widetilde{\boldsymbol{\alpha}}_1(s_1(t)) = \widetilde{\boldsymbol{\alpha}}_2(s_2(t))$, for all $t \in I$. Show that $s_1(t) = \pm s_2(t) + a$, for some $a \in \mathbb{R}$. What is the geometric meaning of a?

Exercise 2.1.20. Let $\alpha\colon I \to \mathbb{L}^n$ be a lightlike curve, and suppose that $\widetilde{\alpha}_1\colon J_1 \to \mathbb{L}^n$ and $\widetilde{\alpha}_2\colon J_2 \to \mathbb{L}^n$ are arc-photon reparametrizations of α, such that $\widetilde{\alpha}_1(\phi_1(t)) = \widetilde{\alpha}_2(\phi_2(t))$, for all $t \in I$. Show that $\phi_1(t) = \pm\phi_2(t) + a$, for some $a \in \mathbb{R}$. What is the geometric meaning of a?

Exercise 2.1.21. Let $\alpha\colon I_1 \to \mathbb{R}^2$ and $\beta\colon I_2 \to \mathbb{R}^2$ be regular curves that *intersect transversally* at $p = \alpha(t_1^*) = \beta(t_2^*)$, i.e., such that $\{\alpha'(t_1^*), \beta'(t_2^*)\}$ is linearly independent, where $t_1^* \in I_1$ and $t_2^* \in I_2$. Let $v \in \mathbb{R}^2$ be a unit vector and define, for each $s \in \mathbb{R}$, the *perturbations of* α *in the direction of* v, $\alpha^s\colon I_1 \to \mathbb{R}^2$, by setting $\alpha^s(t) = \alpha(t) + sv$. Show that, for s sufficiently small, the traces of α^s and β intersect near p.

Hint. Use the Implicit Function Theorem for a convenient map $F\colon \mathbb{R} \times I_1 \times I_2 \to \mathbb{R}^2$.

Exercise 2.1.22. Let $\alpha\colon I \to \mathbb{R}^n$ be a regular curve and $t_0 \in I$. Show that there exists an open interval $J \subseteq I$ around t_0 such that:

(a) $\alpha\big|_J$ is injective;

(b) there exist smooth maps $F_1, \ldots, F_{n-1}\colon \mathbb{R}^n \to \mathbb{R}$ whose image $\alpha(J)$ is contained in the set
$$\{x \in \mathbb{R}^n \mid F_i(x) = 0 \text{ for all } 1 \leq i \leq n-1\}.$$

Hint. Use the Implicit Function Theorem first and, if needed, seek inspiration from Exercise A.3 (p. 329, Appendix A).

2.2 CURVES IN THE PLANE

We'll begin the study of curves in $\mathbb{R}^2_\nu$ by observing that every regular curve may be, at least locally, reparametrized as the graph of a real function. In $\mathbb{L}^2$, the causal type of the curve allows us to specify which of the coordinate axes is the domain of such graph.

Theorem 2.2.1. *Let $\alpha : I \to \mathbb{R}^2_\nu$ be a regular curve. For each $t_0 \in I$ there is an open interval J, $u_0 \in J$ and a reparametrization $\widetilde{\alpha}\colon J \to \mathbb{R}^2_\nu$ of α such that $\alpha(t_0) = \widetilde{\alpha}(u_0)$, of the form $\widetilde{\alpha}(u) = (u, f(u))$ or $\widetilde{\alpha}(u) = (f(u), u)$, for some smooth function f. Moreover, in $\mathbb{L}^2$, we ensure that the first case is always possible if α is spacelike, while the second one is always possible if α is timelike.*

Proof: Write $\alpha(t) = (x(t), y(t))$. Since α is regular, we know that $x'(t_0)$ and $y'(t_0)$ are not simultaneously zero. Suppose without loss of generality that $x'(t_0) \neq 0$. By the Inverse Function Theorem, there is an open interval J where the inverse function $x^{-1}\colon J \to x^{-1}(J) \subseteq I$ is well-defined and smooth. Define the curve $\widetilde{\alpha} \doteq \alpha \circ x^{-1}$, so that $\widetilde{\alpha}(u) = (u, y(x^{-1}(u)))$. Thus, $f \doteq y \circ x^{-1}$ and $u_0 \doteq x(t_0)$ fit the bill. In $\mathbb{L}^2$, if α is spacelike, we necessarily have that $x'(t_0) \neq 0$, while if it is timelike, we have $y'(t_0) \neq 0$. □

In the plane, it is easy to determine all the lightlike curves:

Proposition 2.2.2. *Let $\alpha\colon I \to \mathbb{L}^2$ be a lightlike curve. Then the trace of α is contained in a line parallel to one of the light rays of $\mathbb{L}^2$.*

Proof: Write $\alpha(t) = (x(t), y(t))$. Since α is lightlike, we have $x'(t)^2 - y'(t)^2 = 0$, whence $|x'(t)| = |y'(t)|$. Suppose without loss of generality that $x'(t) = y'(t)$. Then there is $c \in \mathbb{R}$ such that $x(t) = y(t) + c$, and so we have

$$\alpha(t) = (y(t) + c, y(t)) = (c, 0) + y(t)(1, 1),$$

as wanted. □

In view of the above result, we may focus our attention on curves that are not lightlike, which we may assume parametrized with unit speed (and thus with constant causal character).

The idea is, then, to associate to each point of the curve a positive orthonormal basis of $\mathbb{R}^2_\nu$. The trace of the curve in the plane is directly related to the way such basis changes along the curve. More precisely, every vector in $\mathbb{R}^2_\nu$ may be written as a linear combinations of the elements in this basis, and *in particular the derivatives of the basis vectors themselves*, and thus the coefficients of such combinations will provide geometric information about the curve. This strategy will be used again in the next section when we discuss curves in space $\mathbb{R}^3_\nu$.

Definition 2.2.3 (Frenet-Serret Dihedron)**.** Let $\alpha : I \to \mathbb{R}^2_\nu$ be a unit speed curve. The *Frenet-Serret frame* of α at $s \in I$ is

$$\mathcal{F} = (\boldsymbol{T}_\alpha(s), \boldsymbol{N}_\alpha(s)),$$

where $\boldsymbol{T}_\alpha(s) \doteq \alpha'(s)$ is the *tangent vector* to α at s, and $\boldsymbol{N}_\alpha(s)$ is the *normal vector* to α at s, characterized as the unique vector that makes the basis $\mathcal{F}$ orthonormal and positive.

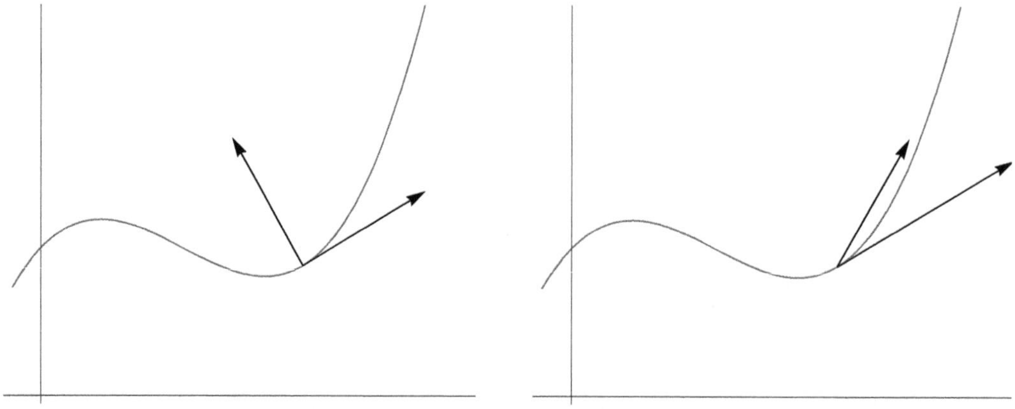

(a) In $\mathbb{R}^2$ (b) In $\mathbb{L}^2$

Figure 2.4: The Frenet-Serret dihedron in $\mathbb{R}^2_\nu$.

If $\alpha(s) = (x(s), y(s))$, then the vector $\boldsymbol{N}_\alpha(s)$ may be obtained geometrically in the following way:

- in $\mathbb{R}^2$, $\boldsymbol{N}_\alpha(s)$ is the counterclockwise rotation of angle $\pi/2$ of the vector $\boldsymbol{T}_\alpha(s)$, that is, $\boldsymbol{N}_\alpha(s) = (-y'(s), x'(s))$;
- in $\mathbb{L}^2$, $\boldsymbol{N}_\alpha(s)$ is the reflection in one of the light rays, which depend on the causal type of α. The reflection of $\boldsymbol{T}_\alpha(s)$ relative to any of these light rays produces a vector orthogonal to $\boldsymbol{T}_\alpha(s)$, and the axis of reflection is chosen to make the

obtained basis positive. If $\boldsymbol{\alpha}$ is spacelike, we perform the reflection relative to the *main light ray* ($y = x$), and thus $\boldsymbol{N}_{\alpha}(s) = (y'(s), x'(s))$, while if $\boldsymbol{\alpha}$ is timelike, we perform the reflection relative to the *secondary light ray* ($y = -x$), so that $\boldsymbol{N}_{\alpha}(s) = -(y'(s), x'(s))$.

We may deduce a general expression for the normal vector, $\boldsymbol{N}_{\alpha}(s) = (a(s), b(s))$, just from the conditions

$$\langle \boldsymbol{T}_{\alpha}(s), \boldsymbol{N}_{\alpha}(s)\rangle = 0 \quad \text{and} \quad \det(\boldsymbol{T}_{\alpha}(s), \boldsymbol{N}_{\alpha}(s)) = 1.$$

Such conditions yield the system:

$$\begin{cases} x'(s)a(s) + (-1)^{\nu}y'(s)b(s) & = 0 \\ -y'(s)a(s) + x'(s)b(s) & = 1 \end{cases}.$$

Noting that $\epsilon_{\alpha} = x'(s)^2 + (-1)^{\nu}y'(s)^2$, the solutions to the system are

$$a(s) = -\epsilon_{\alpha}(-1)^{\nu}y'(s) \quad \text{and} \quad b(s) = \epsilon_{\alpha}x'(s),$$

that is, $\boldsymbol{N}_{\alpha}(s) = \epsilon_{\alpha}\big((-1)^{\nu+1}y'(s), x'(s)\big)$. Since $\langle \boldsymbol{T}_{\alpha}(s), \boldsymbol{T}_{\alpha}(s)\rangle = \epsilon_{\alpha}$ is constant, we have that $\langle \boldsymbol{T}'_{\alpha}(s), \boldsymbol{T}_{\alpha}(s)\rangle = 0$ and thus $\boldsymbol{T}'_{\alpha}(s)$ is parallel to $\boldsymbol{N}_{\alpha}(s)$. We have the:

Definition 2.2.4 (Curvature). Let $\boldsymbol{\alpha}\colon I \to \mathbb{R}^2_{\nu}$ be a unit speed curve. The *oriented curvature of* $\boldsymbol{\alpha}$ is the function $\kappa_{\alpha}\colon I \to \mathbb{R}$ characterized by the relation

$$\boldsymbol{T}'_{\alpha}(s) = \kappa_{\alpha}(s)\boldsymbol{N}_{\alpha}(s).$$

The number $\kappa_{\alpha}(s)$ is called the *curvature of* $\boldsymbol{\alpha}$ *at* s.

Remark. Applying $\langle \cdot, \boldsymbol{N}_{\alpha}(s)\rangle$ to the relation $\boldsymbol{T}'_{\alpha}(s) = \kappa_{\alpha}(s)\boldsymbol{N}_{\alpha}(s)$ and noting that $\langle \boldsymbol{N}_{\alpha}(s), \boldsymbol{N}_{\alpha}(s)\rangle = (-1)^{\nu}\epsilon_{\alpha}$, we obtain

$$\kappa_{\alpha}(s) = (-1)^{\nu}\epsilon_{\alpha}\langle \boldsymbol{T}'_{\alpha}(s), \boldsymbol{N}_{\alpha}(s)\rangle.$$

Proposition 2.2.5 (Frenet-Serret equations). *Let* $\boldsymbol{\alpha} : I \to \mathbb{R}^2_{\nu}$ *be a unit speed curve. Then*

$$\begin{pmatrix} \boldsymbol{T}'_{\alpha}(s) \\ \boldsymbol{N}'_{\alpha}(s) \end{pmatrix} = \begin{pmatrix} 0 & \kappa_{\alpha}(s) \\ (-1)^{\nu+1}\kappa_{\alpha}(s) & 0 \end{pmatrix} \begin{pmatrix} \boldsymbol{T}_{\alpha}(s) \\ \boldsymbol{N}_{\alpha}(s) \end{pmatrix}$$

holds, for all $s \in I$.

Proof: The first equation is the definition of the curvature of $\boldsymbol{\alpha}$. For the second one, note that $\langle \boldsymbol{N}_{\alpha}(s), \boldsymbol{N}_{\alpha}(s)\rangle$ is constant, and thus $\boldsymbol{N}'_{\alpha}(s)$ is parallel to $\boldsymbol{T}_{\alpha}(s)$. By orthonormal expansion we have that

$$\boldsymbol{N}'_{\alpha}(s) = \epsilon_{\alpha}\langle \boldsymbol{T}_{\alpha}(s), \boldsymbol{N}'_{\alpha}(s)\rangle \boldsymbol{T}_{\alpha}(s).$$

But now Corollary 2.1.14 gives us that

$$\kappa_{\alpha}(s) = (-1)^{\nu}\epsilon_{\alpha}\langle \boldsymbol{T}'_{\alpha}(s), \boldsymbol{N}_{\alpha}(s)\rangle = (-1)^{\nu+1}\epsilon_{\alpha}\langle \boldsymbol{T}_{\alpha}(s), \boldsymbol{N}'_{\alpha}(s)\rangle$$

and, finally, $\boldsymbol{N}'_{\alpha}(s) = (-1)^{\nu+1}\kappa_{\alpha}(s)\boldsymbol{T}_{\alpha}(s)$. ☐

Example 2.2.6.

(1) Straight lines: let $\boldsymbol{p}, \boldsymbol{v} \in \mathbb{R}^2_\nu$, with $\|\boldsymbol{v}\| = 1$, and $\boldsymbol{\alpha}\colon \mathbb{R} \to \mathbb{R}^2_\nu$ the curve given by $\boldsymbol{\alpha}(s) = \boldsymbol{p} + s\boldsymbol{v}$. Clearly $\boldsymbol{\alpha}$ has unit speed. We then have $\boldsymbol{T}_{\boldsymbol{\alpha}}(s) = \boldsymbol{v}$ and $\boldsymbol{T}'_{\boldsymbol{\alpha}}(s) = \boldsymbol{0}$, so that $\kappa_{\boldsymbol{\alpha}} \equiv 0$.

(2) Euclidean circles: let $\boldsymbol{p} = (x_p, y_p) \in \mathbb{R}^2$, $r > 0$, and define

$$\mathbb{S}^1(\boldsymbol{p}, r) \doteq \{(x,y) \in \mathbb{R}^2 \mid (x - x_p)^2 + (y - y_p)^2 = r^2\}.$$

Consider the parametrized curve $\boldsymbol{\alpha}\colon \mathbb{R} \to \mathbb{S}^1(\boldsymbol{p}, r)$ given by

$$\boldsymbol{\alpha}(s) \doteq \left(x_p + r\cos\left(\frac{s}{r}\right), y_p + r\sin\left(\frac{s}{r}\right)\right).$$

We have that $\boldsymbol{T}_{\boldsymbol{\alpha}}(s) = \left(-\sin\left(\frac{s}{r}\right), \cos\left(\frac{s}{r}\right)\right)$. It follows from this that the normal vector is $\boldsymbol{N}_{\boldsymbol{\alpha}}(s) = \left(-\cos\left(\frac{s}{r}\right), -\sin\left(\frac{s}{r}\right)\right)$. This way, we may compute the curvature of $\boldsymbol{\alpha}$:

$$\boldsymbol{T}'_{\boldsymbol{\alpha}}(s) = \left(-\frac{1}{r}\cos\left(\frac{s}{r}\right), -\frac{1}{r}\sin\left(\frac{s}{r}\right)\right) = \frac{1}{r}\boldsymbol{N}_{\boldsymbol{\alpha}}(s),$$

and we conclude that $\kappa_{\boldsymbol{\alpha}}(s) = 1/r$, for all $s \in \mathbb{R}$.

(3) Hyperbolic lines: let $\boldsymbol{p} = (x_p, y_p) \in \mathbb{L}^2$, $r > 0$, and define

$$\mathbb{H}^1(\boldsymbol{p}, r) \doteq \{(x,y) \in \mathbb{L}^2 \mid (x - x_p)^2 - (y - y_p)^2 = -r^2 \text{ and } y > y_p\} \text{ and}$$
$$\mathbb{H}^1_-(\boldsymbol{p}, r) \doteq \{(x,y) \in \mathbb{L}^2 \mid (x - x_p)^2 - (y - y_p)^2 = -r^2 \text{ and } y < y_p\}.$$

We'll discuss here the set $\mathbb{H}^1(\boldsymbol{p}, r)$. Consider $\boldsymbol{\alpha}\colon \mathbb{R} \to \mathbb{H}^1(\boldsymbol{p}, r)$, the parametrized curve given by

$$\boldsymbol{\alpha}(s) \doteq \left(x_p + r\sinh\left(\frac{s}{r}\right), y_p + r\cosh\left(\frac{s}{r}\right)\right).$$

We have that $\boldsymbol{T}_{\boldsymbol{\alpha}}(s) = \left(\cosh\left(\frac{s}{r}\right), \sinh\left(\frac{s}{r}\right)\right)$ and thus $\epsilon_{\boldsymbol{\alpha}} = 1$. It follows from this that $\boldsymbol{N}_{\boldsymbol{\alpha}}(s) = \left(\sinh\left(\frac{s}{r}\right), \cosh\left(\frac{s}{r}\right)\right)$. This way, we may compute the curvature of $\boldsymbol{\alpha}$:

$$\boldsymbol{T}'_{\boldsymbol{\alpha}}(s) = \left(\frac{1}{r}\sinh\left(\frac{s}{r}\right), \frac{1}{r}\cosh\left(\frac{s}{r}\right)\right) = \frac{1}{r}\boldsymbol{N}_{\boldsymbol{\alpha}}(s),$$

and we conclude that $\kappa_{\boldsymbol{\alpha}}(s) = 1/r$, for all $s \in \mathbb{R}$.

(4) *de Sitter line*: let $\boldsymbol{p} = (x_p, y_p) \in \mathbb{L}^2$, $r > 0$, and define

$$\mathbb{S}^1_1(\boldsymbol{p}, r) \doteq \{(x,y) \in \mathbb{L}^2 \mid (x - x_p)^2 - (y - y_p)^2 = r^2\}.$$

Let's parametrize the right branch of $\mathbb{S}^1_1(\boldsymbol{p}, r)$ with $\boldsymbol{\alpha}\colon \mathbb{R} \to \mathbb{S}^1_1(\boldsymbol{p}, r)$ given by

$$\boldsymbol{\alpha}(s) \doteq \left(x_p + r\cosh\left(\frac{s}{r}\right), y_p + r\sinh\left(\frac{s}{r}\right)\right).$$

We have that $\boldsymbol{T}_{\boldsymbol{\alpha}}(s) = \left(\sinh\left(\frac{s}{r}\right), \cosh\left(\frac{s}{r}\right)\right)$ and thus $\epsilon_{\boldsymbol{\alpha}} = -1$. It follows from this that $\boldsymbol{N}_{\boldsymbol{\alpha}}(s) = \left(-\sinh\left(\frac{s}{r}\right), -\cosh\left(\frac{s}{r}\right)\right)$. This way, we may compute the curvature of $\boldsymbol{\alpha}$:

$$\boldsymbol{T}'_{\boldsymbol{\alpha}}(s) = \left(\frac{1}{r}\cosh\left(\frac{s}{r}\right), \frac{1}{r}\sinh\left(\frac{s}{r}\right)\right) = -\frac{1}{r}\boldsymbol{N}_{\boldsymbol{\alpha}}(s),$$

and we conclude that $\kappa_{\boldsymbol{\alpha}}(s) = -1/r$, for all $s \in \mathbb{R}$.

(5) *Catenary*[1]: one unit speed parametrization for this curve is $\alpha : \mathbb{R} \to \mathbb{R}^2$ given by

$$\alpha(s) = (\operatorname{arcsinh} s, \sqrt{1+s^2}).$$

Then $T_\alpha(s) = \left(\frac{1}{\sqrt{1+s^2}}, \frac{s}{\sqrt{1+s^2}}\right)$ and hence $N_\alpha(s) = \left(-\frac{s}{\sqrt{1+s^2}}, \frac{1}{\sqrt{1+s^2}}\right)$. So

$$T'_\alpha(s) = \left(-\frac{s}{(1+s^2)^{3/2}}, \frac{1}{(1+s^2)^{3/2}}\right) = \frac{1}{1+s^2} N_\alpha(s),$$

whence $\kappa_\alpha(s) = 1/(1+s^2) > 0$.

(6) *Cosine wave*: a unit speed parametrization for the graph of the function cos, seen in $\mathbb{L}^2$, is $\alpha\colon\ ]-1,1[\ \to \mathbb{L}^2$ given by

$$\alpha(s) = (\arcsin s, \sqrt{1-s^2}).$$

We have that

$$T_\alpha(s) = \left(\frac{1}{\sqrt{1-s^2}}, -\frac{s}{\sqrt{1-s^2}}\right) \quad \text{and} \quad N_\alpha(s) = \left(-\frac{s}{\sqrt{1-s^2}}, \frac{1}{\sqrt{1-s^2}}\right),$$

since $\epsilon_\alpha = 1$. Thus

$$T'_\alpha(s) = \left(\frac{s}{(1-s^2)^{3/2}}, -\frac{1}{(1-s^2)^{3/2}}\right) = \frac{1}{s^2-1} N_\alpha(s),$$

whence $\kappa_\alpha(s) = 1/(s^2-1) < 0$.

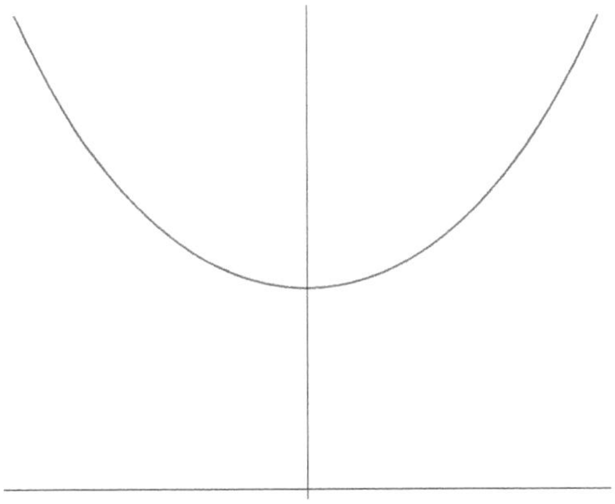

(a) The catenary in $\mathbb{R}^2$

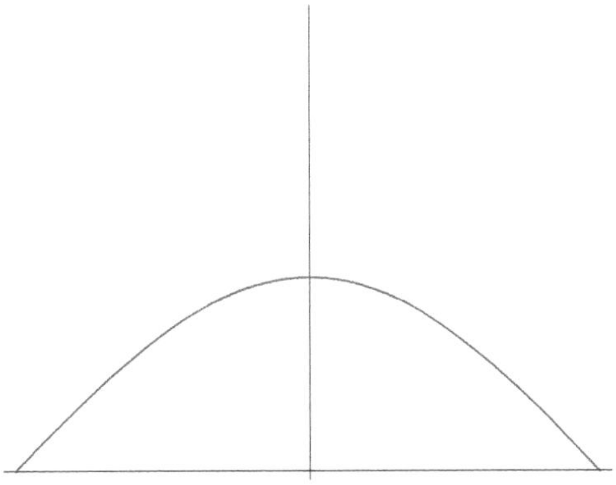

(b) The cosine wave in $\mathbb{L}^2$

Figure 2.5: Catenaries (Euclidean and Lorentzian).

Remark.

- To discuss both hyperbolic lines simultaneously we will indicate, when needed, $\mathbb{H}^1(p,r)$ by $\mathbb{H}^1_+(p,r)$. This "separation" is made for hyperbolic lines but not for the de Sitter line, since these objects have natural generalizations in higher dimensions, but it turns out that that these de Sitter spaces will be connected. In other words, it won't be divided into a "right branch" and a "left branch".

[1] The name "catenary" comes from Latin *catēna*, which means "chain". Indeed, the shape of the curve is the one of a chain being held by its endpoints, under the action only of its own weight.

- In analogy with $\mathbb{S}^1(\boldsymbol{p},r)$, $\boldsymbol{p}$ and r will also be called the *center* and the *radius* of $\mathbb{S}^1_1(\boldsymbol{p},r)$ and $\mathbb{H}^1_\pm(\boldsymbol{p},r)$. If $\boldsymbol{p}=\boldsymbol{0}$ or $r=1$, they will be omitted to simplify notation.

The above examples illustrate that our definitions so far, despite being relatively simple, are not practical for the analysis of curves which do not have unit speed. Before we extend all the *Frenet-Serret apparatus* for such curves, consider $\boldsymbol{\alpha}$ again with unit speed and assume that $\boldsymbol{\beta}$ is a reparametrization of $\boldsymbol{\alpha}$ also with unit speed. It follows from Exercise 2.1.19 (p. 76) that $\boldsymbol{\beta}(s)=\boldsymbol{\alpha}(\pm s+a)$, for some $a\in\mathbb{R}$. We have that

$$\boldsymbol{T}_{\boldsymbol{\beta}}(s)=\pm\boldsymbol{T}_{\boldsymbol{\alpha}}(\pm s+a),\quad \boldsymbol{N}_{\boldsymbol{\beta}}(s)=\pm\boldsymbol{N}_{\boldsymbol{\alpha}}(\pm s+a)\quad\text{and}\quad \boldsymbol{T}'_{\boldsymbol{\beta}}(s)=\boldsymbol{T}'_{\boldsymbol{\alpha}}(\pm s+a),$$

which gives us that $\kappa_{\boldsymbol{\beta}}(s)=\pm\kappa_{\boldsymbol{\alpha}}(\pm s+a)$. To summarize, the curvature is sensitive to the change of direction realized by the curve (such change is indicated by the sign $\pm$) and, particularly, we see that the curvature is invariant under *positive* reparametrizations.

In the theory to follow, we will make a local study of curves and thus we may assume that they are regular and with constant causal type. When the curve does not have unit speed, we have the:

Definition 2.2.7. Let $\boldsymbol{\alpha}\colon I\to\mathbb{R}^2_\nu$ be a regular curve which is not lightlike. Writing $\boldsymbol{\alpha}(t)=\tilde{\boldsymbol{\alpha}}(s(t))$, where s is an arclength function for $\boldsymbol{\alpha}$, the *tangent vector* to $\boldsymbol{\alpha}$ at t is defined by

$$\boldsymbol{T}_{\boldsymbol{\alpha}}(t)\doteq\boldsymbol{T}_{\tilde{\boldsymbol{\alpha}}}(s(t)),$$

the *normal vector* to $\boldsymbol{\alpha}$ at t is defined by

$$\boldsymbol{N}_{\boldsymbol{\alpha}}(t)\doteq\boldsymbol{N}_{\tilde{\boldsymbol{\alpha}}}(s(t))$$

and, lastly, the *curvature* of $\boldsymbol{\alpha}$ at t is defined by

$$\kappa_{\boldsymbol{\alpha}}(t)\doteq\kappa_{\tilde{\boldsymbol{\alpha}}}(s(t)).$$

Remark. The discussion preceding the above definition in fact says that the Frenet-Serret apparatus for $\boldsymbol{\alpha}$ does not depend on the arclength function chosen.

Proposition 2.2.8. *Let $\boldsymbol{\alpha}\colon I\to\mathbb{R}^2_\nu$ be a regular curve which is not lightlike. Then*

$$\kappa_{\boldsymbol{\alpha}}(t)=\frac{\det(\boldsymbol{\alpha}'(t),\boldsymbol{\alpha}''(t))}{\|\boldsymbol{\alpha}'(t)\|^3}.$$

Proof: Write $\boldsymbol{\alpha}(t)=\tilde{\boldsymbol{\alpha}}(s(t))$ with $\tilde{\boldsymbol{\alpha}}$ having unit speed, and s being an arclength function for $\boldsymbol{\alpha}$. Differentiating such relation twice, we have that

$$\boldsymbol{\alpha}'(t)=s'(t)\boldsymbol{T}_{\tilde{\boldsymbol{\alpha}}}(s(t))\quad\text{and}\quad\boldsymbol{\alpha}''(t)=s''(t)\boldsymbol{T}_{\tilde{\boldsymbol{\alpha}}}(s(t))+s'(t)^2\kappa_{\tilde{\boldsymbol{\alpha}}}(s(t))\boldsymbol{N}_{\tilde{\boldsymbol{\alpha}}}(s(t)).$$

With this:

$$\begin{aligned}\det(\boldsymbol{\alpha}'(t),\boldsymbol{\alpha}''(t))&=\det\left(s'(t)\boldsymbol{T}_{\tilde{\boldsymbol{\alpha}}}(s(t)),s''(t)\boldsymbol{T}_{\tilde{\boldsymbol{\alpha}}}(s(t))+s'(t)^2\kappa_{\tilde{\boldsymbol{\alpha}}}(s(t))\boldsymbol{N}_{\tilde{\boldsymbol{\alpha}}}(s(t))\right)\\&=s'(t)s''(t)\det(\boldsymbol{T}_{\tilde{\boldsymbol{\alpha}}}(s(t)),\boldsymbol{T}_{\tilde{\boldsymbol{\alpha}}}(s(t)))+s'(t)^3\kappa_{\tilde{\boldsymbol{\alpha}}}(s(t))\det(\boldsymbol{T}_{\tilde{\boldsymbol{\alpha}}}(s(t)),\boldsymbol{N}_{\tilde{\boldsymbol{\alpha}}}(s(t)))\\&=s'(t)^3\kappa_{\tilde{\boldsymbol{\alpha}}}(s(t)).\end{aligned}$$

Since $s'(t)=\|\boldsymbol{\alpha}'(t)\|$ and $\kappa_{\boldsymbol{\alpha}}(t)=\kappa_{\tilde{\boldsymbol{\alpha}}}(s(t))$, the result follows. □

Remark.

- The sign of the curvature of α does not depend on the ambient plane where it lies (i.e., $\mathbb{R}^2$ or $\mathbb{L}^2$).
- If $\|\alpha'(t)\| = 1$, the curvature at t is the (Euclidean) oriented area of the parallelogram spanned by $\alpha'(t)$ and $\alpha''(t)$.

Example 2.2.9. Let $a \in \mathbb{R}$, $a \neq 0$, and $\alpha\colon \mathbb{R} \to \mathbb{R}^2_\nu$ be the parametrization of a parabola, given by $\alpha(t) = (t, at^2)$. Note that α does not have unit speed, nor a constant causal character. Since $\alpha'(t) = (1, 2at)$ and $\alpha''(t) = (0, 2a)$, we have that seen in $\mathbb{L}^2$, α is

- spacelike for $|t| < 1/2|a|$;
- timelike for $|t| > 1/2|a|$;
- lightlike at $t = 1/2|a|$ and $t = -1/2|a|$.

Moreover:

$$\kappa_\alpha(t) = \frac{2a}{\left|1 + (-1)^\nu 4a^2t^2\right|^{3/2}}, \text{ if } |t| \neq \frac{1}{2|a|}.$$

See the behavior of the curvature in both ambient planes:

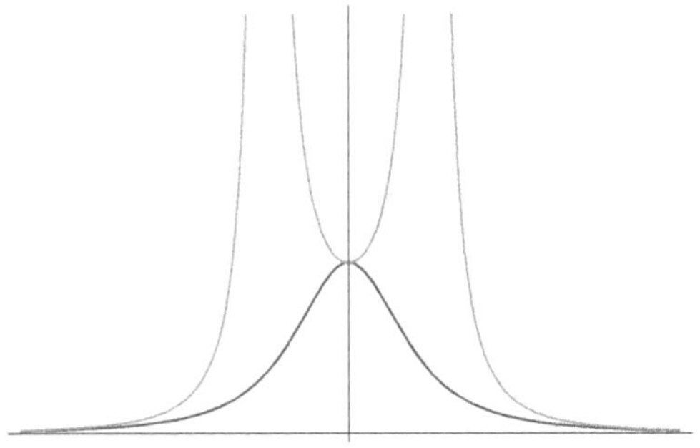

Figure 2.6: The curvatures of a parabola for $a = 1$ in both ambient planes: the bottom one in $\mathbb{R}^2$ and the upper one in $\mathbb{L}^2$.

Note that, when both are defined, the absolute value of the Lorentzian curvature is always greater than the absolute value of the Euclidean curvature. This is in fact a very general phenomenon valid for every curve, see Exercise 2.2.6. Moreover, in $\mathbb{L}^2$ we have that

$$\lim_{t \to \pm 1/2a} |\kappa_\alpha(t)| = +\infty.$$

Corollary 2.2.10. *Let $\alpha, \beta\colon I \to \mathbb{R}^2_\nu$ be two regular, non-lightlike and congruent curves. Then $|\kappa_\alpha(t)| = |\kappa_\beta(t)|$ for all $t \in I$. In particular, equality without the absolute values hold if the linear part of the isometry realizing the congruence preserves the orientation of the plane.*

Proof: Let $F = T_a \circ A \in \mathrm{E}_\nu(2, \mathbb{R})$ be such that $\beta = F \circ \alpha$. We have previously seen that $\|\beta'(t)\| = \|\alpha'(t)\|$. Moreover, the relation

$$\det(\beta'(t), \beta''(t)) = \det(A\alpha'(t), A\alpha''(t)) = \det A \det(\alpha'(t), \alpha''(t))$$

holds. Since $|\det A| = 1$, the result follows from the curvature expression given in Proposition 2.2.8. □

This result is particularly useful for a local analysis of the curve. With it, we may assume that the curve $\boldsymbol{\alpha} : I \to \mathbb{R}^2_\nu$ is not only parametrized with unit speed, but is also "well placed" in the plane. More precisely, Exercise 2.1.19 (p. 76) and Proposition 1.4.15 (p. 35) allow us to assume that $0 \in I$, $\boldsymbol{\alpha}(0) = \mathbf{0}$ and that the Frenet-Serret frame at this point is "almost"[2] the standard basis for $\mathbb{R}^2_\nu$. For a geometric interpretation of the sign of the curvature, assume that $\kappa_{\boldsymbol{\alpha}}(0) \neq 0$, and consider the Taylor formula:

$$\boldsymbol{\alpha}(s) = s\boldsymbol{\alpha}'(0) + \frac{s^2}{2}\boldsymbol{\alpha}''(0) + \boldsymbol{R}(s),$$

where $\boldsymbol{R}(s)/s^2 \to \mathbf{0}$ if $s \to 0$. Reorganizing, we have that

$$\boldsymbol{\alpha}(s) - s\boldsymbol{T}_{\boldsymbol{\alpha}}(0) = \frac{s^2}{2}\kappa_{\boldsymbol{\alpha}}(0)\boldsymbol{N}_{\boldsymbol{\alpha}}(0) + \boldsymbol{R}(s).$$

The tangent line to $\boldsymbol{\alpha}$ at $s_0 = 0$ divides $\mathbb{R}^2_\nu$ into two half-planes. Thus, the above expression says that for s small enough, the difference $\boldsymbol{\alpha}(s) - s\boldsymbol{T}_{\boldsymbol{\alpha}}(0)$ points to the same side that $\boldsymbol{N}_{\boldsymbol{\alpha}}(0)$ if $\kappa_{\boldsymbol{\alpha}}(0) > 0$, and points to the opposite side if $\kappa_{\boldsymbol{\alpha}}(0) < 0$. In any case, $\boldsymbol{\alpha}(s) - s\boldsymbol{T}_{\boldsymbol{\alpha}}(0)$ points in the same direction as $\boldsymbol{\alpha}''(0)$, which gives the direction to which the curve deviates from its tangent on that point. This justifies the name *acceleration vector* usually given to $\boldsymbol{\alpha}''(0)$ (see Exercise 2.2.8).

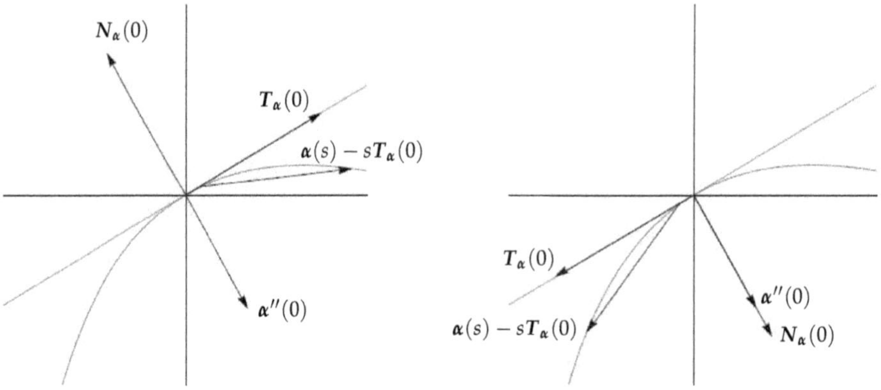

(a) $\kappa_{\boldsymbol{\alpha}}(0) < 0$ (b) $\kappa_{\boldsymbol{\alpha}}(0) > 0$

Figure 2.7: Interpretation of the *local canonical form* for a planar curve.

Following this train of thought, we may approximate the curve in a neighborhood of $\boldsymbol{\alpha}(s_0)$ with another curve having constant curvature equal to $\kappa_{\boldsymbol{\alpha}}(0)$, with the same tangent vector as $\boldsymbol{\alpha}$ at s_0.

We have seen in Example 2.2.6 that, for $\boldsymbol{p} \in \mathbb{R}^2_\nu$ and $r > 0$, $\mathbb{S}^1(\boldsymbol{p}, r)$, $\mathbb{S}^1_1(\boldsymbol{p}, r)$ and $\mathbb{H}^1_\pm(\boldsymbol{p}, r)$ are curves in the plane with constant curvature equal to $1/r$, up to sign. With this in mind, we are interested in curves of this type with the so-called *radius of curvature* equal to $r = 1/|\kappa_{\boldsymbol{\alpha}}(s_0)|$. Let's find the *center of curvature* $\boldsymbol{p}$: we know that $\boldsymbol{\alpha}(s_0) - \boldsymbol{p}$ is normal to $\boldsymbol{\alpha}$ at s_0, and so there is $\lambda \in \mathbb{R}$ such that $\boldsymbol{\alpha}(s_0) - \boldsymbol{p} = \lambda \boldsymbol{N}_{\boldsymbol{\alpha}}(s_0)$. Up to a positive reparametrization, we may assume that $\kappa_{\boldsymbol{\alpha}}(s_0) > 0$. Taking into account that we seek a curve with the same causal type as $\boldsymbol{\alpha}$ at s_0, we have that

$$\langle \boldsymbol{\alpha}(s_0) - \boldsymbol{p}, \boldsymbol{\alpha}(s_0) - \boldsymbol{p} \rangle = \frac{(-1)^\nu \epsilon_{\boldsymbol{\alpha}}}{\kappa_{\boldsymbol{\alpha}}(s_0)^2} \implies |\lambda| = \frac{1}{\kappa_{\boldsymbol{\alpha}}(s_0)}.$$

[2]If the curve is timelike, one needs to consider $(\boldsymbol{e}_2, -\boldsymbol{e}_1)$ instead.

To decide the sign of λ, we note that in $\mathbb{R}^2$, the vectors $\boldsymbol{\alpha}(s_0) - \boldsymbol{p}$ and $\boldsymbol{N}_{\boldsymbol{\alpha}}(s_0)$ point in opposite directions, while in $\mathbb{L}^2$ they point in the same direction.

With r and $\boldsymbol{p}$ determined, we call this curve the *osculating circle* of $\boldsymbol{\alpha}$ at s_0. This blatant abuse of terminology in $\mathbb{L}^2$ is justified by the fact that both $\mathbb{S}^1_1(\boldsymbol{p}, r)$ and $\mathbb{H}^1(\boldsymbol{p}, r) \cup \mathbb{H}^1_-(\boldsymbol{p}, r)$ are the locus of points which are (Lorentzian) equidistant to a given fixed center, sharing the same geometric definition of a circle in $\mathbb{R}^2$. This analogy will be further emphasized in Theorem 2.2.12 to come.

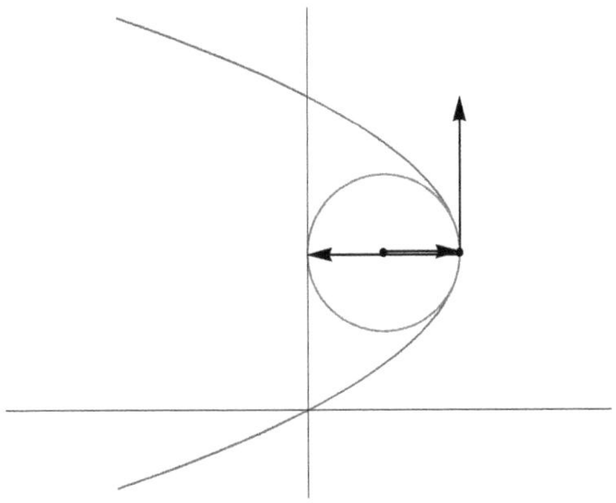

(a) Osculating circle of a parabola in $\mathbb{R}^2$

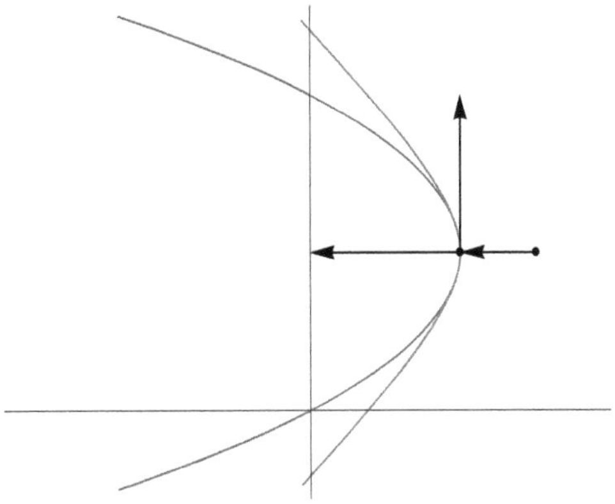

(b) Osculating de Sitter to a parabola in $\mathbb{L}^2$

Figure 2.8: Justifying the sign of λ.

That is, we have $\lambda = (-1)^{\nu+1}/\kappa_{\boldsymbol{\alpha}}(s_0)$, and thus

$$\boldsymbol{p} = \boldsymbol{\alpha}(s_0) + \frac{(-1)^{\nu}}{\kappa_{\boldsymbol{\alpha}}(s_0)} \boldsymbol{N}_{\boldsymbol{\alpha}}(s_0).$$

Moreover, note that $\boldsymbol{p}$ depends on s_0 in the above construction. Allowing s_0 to range over an interval where $\kappa_{\boldsymbol{\alpha}}$ does not vanish, we obtain a new curve describing the motion of the centers of curvature of $\boldsymbol{\alpha}$. Such curve is called the *evolute* of $\boldsymbol{\alpha}$ (see Exercise 2.2.2).

Now, let's show that the osculating circle is, among all circles tangent to $\boldsymbol{\alpha}$ at $\boldsymbol{\alpha}(s_0)$, the one that better approximates it. Denote, for each $r > 0$, by $C_{r,\nu}$ the "circle" of radius r in $\mathbb{R}^2_\nu$ with this property. The branch of $C_{r,\nu}$ containing $\boldsymbol{\alpha}(s_0)$ divides the plane $\mathbb{R}^2_\nu$ into two connected components. Assuming that branch to be parametrized with positive curvature, its *interior* is the connected component of $\mathbb{R}^2_\nu$ with the removed branch for which its normal vector at $\boldsymbol{\alpha}(s_0)$ points to (and the *exterior* is the remaining component).

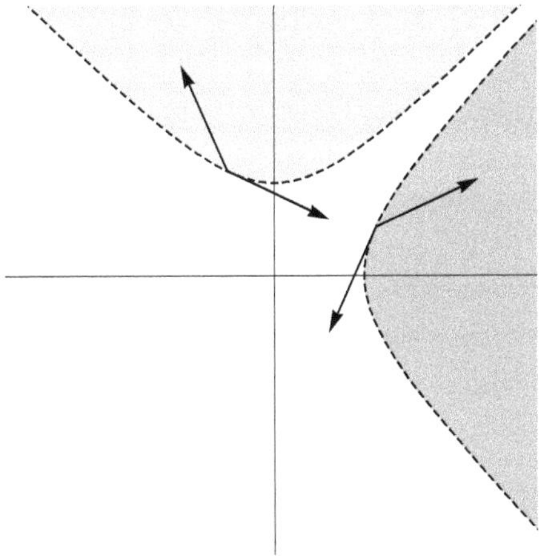

Figure 2.9: Interior of $\mathbb{H}^1$ and of a branch of $\mathbb{S}^1_1$ in $\mathbb{L}^2$.

With this, let's say that $C_{r,\nu}$ is *too curved (resp., slightly curved)* at $\alpha(s_0)$ if there is $\epsilon > 0$ such that $\alpha(\,]s_0 - \epsilon, s_0 + \epsilon[\,)$ does not intersect the interior (resp., exterior) of the branch of $C_{r,\nu}$ which is tangent to α at $\alpha(s_0)$.

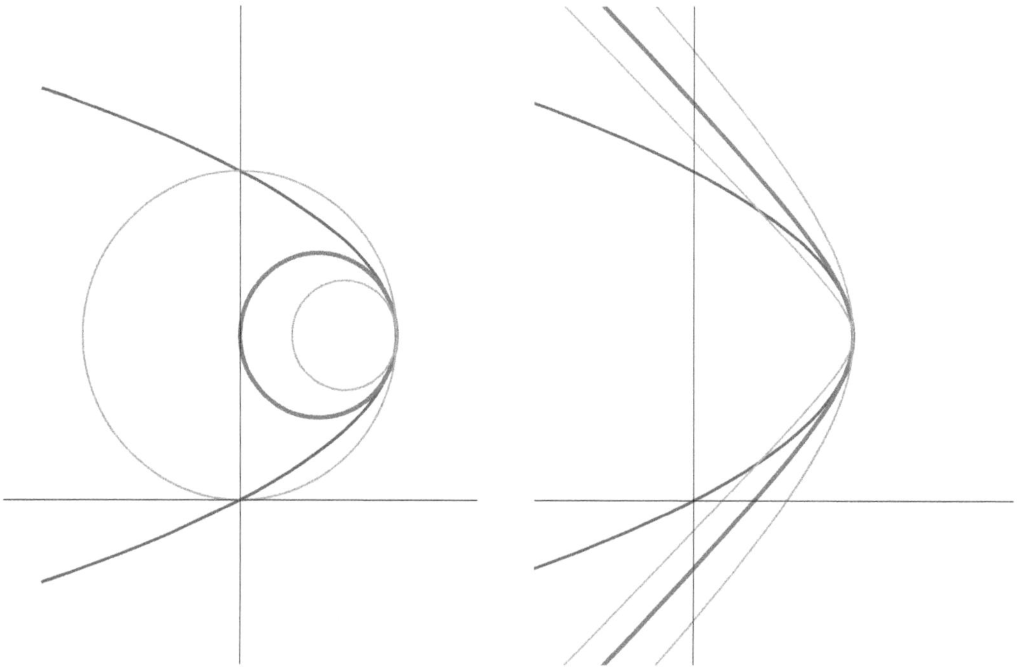

(a) Candidates for osculating circles in $\mathbb{R}^2$ (b) Candidates for osculating hyperbolas in $\mathbb{L}^2$

Figure 2.10: Circles $C_{r,\nu}$ too curved or slightly curved for a parabola in $\mathbb{R}^2_\nu$

Proposition 2.2.11. *With the notation of the above discussion, if* $r_0 = 1/\kappa_\alpha(s_0)$, *we have that* $C_{r,\nu}$ *is too curved at* $\alpha(s_0)$ *if* $r < r_0$, *and slightly curved if* $r > r_0$.

Proof: Consider $f : I \to \mathbb{R}$ given by

$$f(s) \doteq \langle \alpha(s) - \alpha(s_0) + r(-1)^{\nu+1}\mathbf{N}_\alpha(s_0), \alpha(s) - \alpha(s_0) + r(-1)^{\nu+1}\mathbf{N}_\alpha(s_0)\rangle.$$

Note that $f(s_0) = (-1)^\nu \epsilon_\alpha r^2$ and that

$$\begin{aligned} f'(s) &= 2\langle \mathbf{T}_\alpha(s), \alpha(s) - \alpha(s_0) + r(-1)^{\nu+1}\mathbf{N}_\alpha(s_0)\rangle \\ f''(s) &= 2\left(\langle \mathbf{T}'_\alpha(s), \alpha(s) - \alpha(s_0) + r(-1)^{\nu+1}\mathbf{N}_\alpha(s_0)\rangle + \epsilon_\alpha\right), \end{aligned}$$

whence $f'(s_0) = 0$ and $f''(s_0) = 2\epsilon_\alpha(1 - r/r_0)$. Let's then classify the critical point s_0 of f in terms of ϵ_α and r.

If $r < r_0$ and $\epsilon_\alpha = 1$, we have that s_0 is a local minimum for f. Thus, in $\mathbb{R}^2$, we have that $f(s) \geq f(s_0) = r^2$ for every s sufficiently close to s_0. In $\mathbb{L}^2$, we'll have that $f(s) \geq f(s_0) = -r^2$. In the case where $\epsilon_\alpha = -1$, we have that s_0 is a local maximum for f, so that $f(s) \leq -r^2$ for every s sufficiently close to s_0. In any case, we conclude that $C_{r,\nu}$ is too curved at $\alpha(s_0)$. Similarly, if $r > r_0$, then $C_{r,\nu}$ is slightly curved at $\alpha(s_0)$. □

Remark. The above result shows that the osculating circle is the one with greatest curvature among all the slightly curved circles $C_{r,\nu}$. Equivalently, it is the one with smallest curvature among all the too curved circles $C_{r,\nu}$.

Instead of only dealing with osculating circles, we could also consider approximations by polynomial curves. Despite having simple parametrizations, these curves have the disadvantage of not having constant curvature, which makes it more difficult to obtain a similar result to the above in this new setting. We will give expressions for *osculating parabolas* via Taylor expansions, and we'll compute their curvatures. In Exercise 2.2.17 we'll point out how to do this for cubics. We have two situations to consider:

- If $\boldsymbol{\alpha}$ is spacelike, assuming that $\boldsymbol{T}_{\boldsymbol{\alpha}}(0) = (1,0)$ and $\boldsymbol{N}_{\boldsymbol{\alpha}}(0) = (0,1)$, we have that the osculating parabola of $\boldsymbol{\alpha}$ at $s_0 = 0$ is given by

$$\boldsymbol{\beta}(s) \doteq \boldsymbol{\alpha}(s) - R(s) = \left(s, \frac{\kappa_{\boldsymbol{\alpha}}(0)}{2}s^2\right).$$

  By Example 2.2.9 we have

$$\kappa_{\boldsymbol{\beta}}(s) = \frac{\kappa_{\boldsymbol{\alpha}}(0)}{\left|1 + (-1)^{\nu}\kappa_{\boldsymbol{\alpha}}(0)^2 s^2\right|^{3/2}}$$

  and, in particular, $\kappa_{\boldsymbol{\beta}}(0) = \kappa_{\boldsymbol{\alpha}}(0)$, as expected. Moreover, if s is sufficiently small, we have that $|\kappa_{\boldsymbol{\beta}}(s)| \leq |\kappa_{\boldsymbol{\alpha}}(0)|$ if $\nu = 0$, while $|\kappa_{\boldsymbol{\beta}}(s)| \geq |\kappa_{\boldsymbol{\alpha}}(0)|$ if $\nu = 1$.

- If $\boldsymbol{\alpha}$ is timelike, assuming that $\boldsymbol{T}_{\boldsymbol{\alpha}}(0) = (0,1)$ and $\boldsymbol{N}_{\boldsymbol{\alpha}}(0) = (-1,0)$, we have that the osculating parabola of $\boldsymbol{\alpha}$ at $s_0 = 0$ is given by

$$\boldsymbol{\beta}(s) \doteq \boldsymbol{\alpha}(s) - R(s) = \left(-\frac{\kappa_{\boldsymbol{\alpha}}(0)}{2}s^2, s\right).$$

  Similarly, we have that

$$\kappa_{\boldsymbol{\beta}}(s) = \frac{\kappa_{\boldsymbol{\alpha}}(0)}{\left|1 - \kappa_{\boldsymbol{\alpha}}(0)^2 s^2\right|^{3/2}}$$

  and, as before, $\kappa_{\boldsymbol{\beta}}(0) = \kappa_{\boldsymbol{\alpha}}(0)$. Now, for s sufficiently small, we have the inequality $|\kappa_{\boldsymbol{\beta}}(s)| \geq |\kappa_{\boldsymbol{\alpha}}(0)|$.

This indeed shows that curves with constant curvature are the most adequate ones for good local approximations. We use $\mathbb{S}^1(\boldsymbol{p},r)$, $\mathbb{S}^1_1(\boldsymbol{p},r)$ and $\mathbb{H}^1_{\pm}(\boldsymbol{p},r)$ as such models. The following theorem tells us that no other choice was possible:

Theorem 2.2.12 (Curves with constant curvature). *Let $\boldsymbol{\alpha}\colon I \to \mathbb{R}^2_\nu$ be a non-lightlike regular curve. Suppose that the curvature of $\boldsymbol{\alpha}$ is a constant κ. Then:*

(i) if $\kappa = 0$, the trace of $\boldsymbol{\alpha}$ is contained in a straight line;

(ii) if $\kappa \neq 0$ in $\mathbb{R}^2$, there is $\boldsymbol{p}$ such that the trace of $\boldsymbol{\alpha}$ is contained in $\mathbb{S}^1(\boldsymbol{p}, 1/|\kappa|)$;

(iii) if $\kappa \neq 0$ in $\mathbb{L}^2$, there is $\boldsymbol{p}$ such that the trace of $\boldsymbol{\alpha}$ is contained in

- *$\mathbb{H}^1_{\pm}(\boldsymbol{p}, 1/|\kappa|)$, if $\boldsymbol{\alpha}$ is spacelike;*
- *$\mathbb{S}^1_1(\boldsymbol{p}, 1/|\kappa|)$, if $\boldsymbol{\alpha}$ is timelike.*

Proof: We may assume that $\boldsymbol{\alpha}$ has unit speed. From the Frenet-Serret dihedron it follows that:

- if $\kappa = 0$, then $\boldsymbol{\alpha}''(s) = \boldsymbol{T}'_{\boldsymbol{\alpha}}(s) = \boldsymbol{0}$, and so there exist $\boldsymbol{p}, \boldsymbol{v} \in \mathbb{R}^2_\nu$ such that $\boldsymbol{\alpha}(s) = \boldsymbol{p} + s\boldsymbol{v}$, for all $s \in I$;

- if $\kappa \neq 0$ we consider, motivated by evolutes, the candidate to center

$$p(s) = \alpha(s) + \frac{(-1)^\nu}{\kappa} N_\alpha(s),$$

which satisfies

$$p'(s) = T_\alpha(s) + \frac{(-1)^\nu}{\kappa}(-1)^{\nu+1}\kappa T_\alpha(s) = 0,$$

whence $p(s) = p \in \mathbb{R}^2_\nu$, for all $s \in I$. So:

$$\langle \alpha(s) - p, \alpha(s) - p \rangle = \frac{(-1)^\nu \epsilon_\alpha}{\kappa^2},$$

concluding the proof of the theorem.

□

Remark. The above results confirm yet another intuition for the curvature: it is a measure of how much a curve deviates from being a straight line. We know that in $\mathbb{R}^2$, the shortest distance between two points is a line and, informally, the curve with this property is called a *geodesic*. This would also justify calling our oriented curvature the *geodesic curvature* instead. We will formally study geodesics in Chapter 3.

Lastly, we'll answer the question posed at the end of Section 2.1 for planar curves as a natural extension of the previous result when the curvature is not a constant.

Theorem 2.2.13 (Fundamental Theorem of Plane Curves). *Given $p, v \in \mathbb{R}^2_\nu$, with v unit, a continuous function $\kappa : I \to \mathbb{R}$ and $s_0 \in I$, there is a unique unit speed curve $\alpha : I \to \mathbb{R}^2_\nu$ such that $\kappa_\alpha(s) = \kappa(s)$ for all $s \in I$, $\alpha(s_0) = p$, and $\alpha'(s_0) = v$. Moreover, if two unit speed curves have the same curvature and causal type, they're congruent.*

Proof: Write $p = (x_0, y_0)$ and $v = (a, b)$. We have three situations to consider, depending on the ambient space and causal type of the curve:

- In $\mathbb{R}^2$, define $\alpha\colon I \to \mathbb{R}^2$ by

$$\alpha(s) \doteq \left(x_0 + \int_{s_0}^{s} \cos(\theta(\xi) + \theta_0)\, d\xi, y_0 + \int_{s_0}^{s} \sin(\theta(\xi) + \theta_0)\, d\xi \right),$$

where $\theta(\xi) = \int_{s_0}^{\xi} \kappa(\tau)\, d\tau$, and θ_0 is such that $\cos\theta_0 = a$ and $\sin\theta_0 = b$ (for instance, $\theta_0 = \arctan(b/a)$ if $a \neq 0$, or $\pm\pi/2$ else). Of course that $\alpha(s_0) = p$. And since

$$\alpha'(s) = (\cos(\theta(s) + \theta_0), \sin(\theta(s) + \theta_0)),$$

we have that α has unit speed and, by construction, $\alpha'(s_0) = v$. Thus

$$N_\alpha(s) = (-\sin(\theta(s) + \theta_0), \cos(\theta(s) + \theta_0))$$

gives us that $\kappa_\alpha(s) = \theta'(s) = \kappa(s)$, for all $s \in I$. For uniqueness, suppose that $\beta\colon I \to \mathbb{R}^2$ given by $\beta(s) = (\tilde{x}(s), \tilde{y}(s))$ is another unit speed curve such that $\kappa_\beta(s) = \kappa(s)$, for all $s \in I$. Then β also satisfies the Frenet-Serret system:

$$\begin{cases} x''(s) = -\kappa(s) y'(s) \\ y''(s) = \kappa(s) x'(s) \end{cases}$$

for all $s \in I$. If $\beta(s_0) = p$ and $\beta'(s_0) = v$, the existence and uniqueness theorem from the theory of ordinary differential equations gives us that $\alpha = \beta$.

- In $\mathbb{L}^2$, to obtain a spacelike curve, define $\boldsymbol{\alpha}\colon I \to \mathbb{L}^2$ by

$$\boldsymbol{\alpha}(s) \doteq \left(x_0 + \int_{s_0}^{s} \cosh(\varphi(\xi) + \varphi_0)\,\mathrm{d}\xi, y_0 + \int_{s_0}^{s} \sinh(\varphi(\xi) + \varphi_0)\,\mathrm{d}\xi\right),$$

where $\varphi(\xi) = \int_{s_0}^{\xi} \kappa(\tau)\,\mathrm{d}\tau$, and φ_0 is such that $\cosh\varphi_0 = a$ and $\sinh\varphi_0 = b$ (here we're assuming without loss of generality that $a > 0$; the case $a < 0$ being similar). Again, we have that $\boldsymbol{\alpha}(s_0) = \boldsymbol{p}$. Since

$$\boldsymbol{\alpha}'(s) = (\cosh(\varphi(s) + \varphi_0), \sinh(\varphi(s) + \varphi_0)),$$

we have that $\boldsymbol{\alpha}$ is spacelike, has unit speed and, by construction, satisfies $\boldsymbol{\alpha}'(s_0) = \boldsymbol{v}$. Thus

$$\boldsymbol{N}_{\boldsymbol{\alpha}}(s) = (\sinh(\varphi(s) + \varphi_0), \cosh(\varphi(s) + \varphi_0))$$

gives us that $\kappa_{\boldsymbol{\alpha}}(s) = \varphi'(s) = \kappa(s)$, for all $s \in I$. For uniqueness, suppose that $\boldsymbol{\beta}\colon I \to \mathbb{L}^2$ given by $\boldsymbol{\beta}(s) = (\tilde{x}(s), \tilde{y}(s))$ is another unit speed spacelike curve such that $\kappa_{\boldsymbol{\beta}}(s) = \kappa(s)$, for all $s \in I$. Then $\boldsymbol{\beta}$ also satisfies the Frenet-Serret system:

$$\begin{cases} x''(s) = \kappa(s)y'(s) \\ y''(s) = \kappa(s)x'(s) \end{cases}$$

for all $s \in I$. As in the previous case, if $\boldsymbol{\beta}(s_0) = \boldsymbol{p}$ and $\boldsymbol{\beta}'(s_0) = \boldsymbol{v}$, the existence and uniqueness theorem from the theory of ordinary differential equations gives us that $\boldsymbol{\alpha} = \boldsymbol{\beta}$.

- In $\mathbb{L}^2$, to obtain a timelike curve, define $\boldsymbol{\alpha}\colon I \to \mathbb{L}^2$ by

$$\boldsymbol{\alpha}(s) \doteq \left(x_0 + \int_{s_0}^{s} \sinh(\varphi(\xi) + \varphi_0)\,\mathrm{d}\xi, y_0 + \int_{s_0}^{s} \cosh(\varphi(\xi) + \varphi_0)\,\mathrm{d}\xi\right),$$

where $\varphi(\xi) = -\int_{s_0}^{\xi} \kappa(\tau)\,\mathrm{d}\tau$, and φ_0 is such that $\sinh\varphi_0 = a$ and $\cosh\varphi_0 = b$ (here we're assuming without loss of generality that $b > 0$; the case $b < 0$ is similar). By the third time, we have $\boldsymbol{\alpha}(s_0) = \boldsymbol{p}$. Since

$$\boldsymbol{\alpha}'(s) = (\sinh(\varphi(s) + \varphi_0), \cosh(\varphi(s) + \varphi_0)),$$

we have that $\boldsymbol{\alpha}$ is timelike, has unit speed and, by construction, satisfies $\boldsymbol{\alpha}'(s_0) = \boldsymbol{v}$. Then

$$\boldsymbol{N}_{\boldsymbol{\alpha}}(s) = (-\cosh(\varphi(s) + \varphi_0), -\sinh(\varphi(s) + \varphi_0))$$

gives us that $\kappa_{\boldsymbol{\alpha}}(s) = -\varphi'(s) = \kappa(s)$, for all $s \in I$. The argument for uniqueness is the same as in the previous two cases, with the system:

$$\begin{cases} x''(s) = -\kappa(s)y'(s) \\ y''(s) = -\kappa(s)x'(s). \end{cases}$$

Lastly, assume that $\boldsymbol{\beta}, \boldsymbol{\gamma}\colon I \to \mathbb{R}^2_\nu$ are both unit speed curves with same causal type and curvature. Once chosen $s_0 \in I$, Proposition 1.4.15 (p. 35) gives us a transformation $F \in \mathrm{E}_\nu(2, \mathbb{R})$ satisfying $F(\boldsymbol{\beta}(s_0)) = \boldsymbol{\gamma}(s_0)$, $DF(\boldsymbol{\beta}(s_0))(\boldsymbol{T}_{\boldsymbol{\beta}}(s_0)) = \boldsymbol{T}_{\boldsymbol{\gamma}}(s_0)$ and $DF(\boldsymbol{\beta}(s_0))(\boldsymbol{N}_{\boldsymbol{\beta}}(s_0)) = \boldsymbol{N}_{\boldsymbol{\gamma}}(s_0)$. By the uniqueness argued in each case above, we have $F \circ \boldsymbol{\beta} = \boldsymbol{\gamma}$. □

Exercises

Exercise 2.2.1. Let $\alpha\colon I \to \mathbb{R}^2_\nu$ be a unit speed curve.

(a) Assume that $\nu = 0$ and, for each $s \in I$, let $\theta(s)$ be the Euclidean angle between the vector $(1,0)$ and $T_\alpha(s)$. Show that $\kappa_\alpha(s) = \theta'(s)$.

(b) Assume that $\nu = 1$, that α is a future-directed timelike curve and, for each $t \in I$, let $\varphi(t)$ be the hyperbolic angle between the vector $(0,1)$ and $T_\alpha(t)$. Show that $\kappa_\alpha(t) = -\varphi'(t)$.

(c) State and prove a result similar to item (b) for spacelike curves in $\mathbb{L}^2$.

Exercise 2.2.2 (Evolutes)**.** Let $\alpha\colon I \to \mathbb{R}^2_\nu$ be a unit speed curve with non-vanishing curvature. We define the *evolute* of α, $\beta = \mathrm{ev}_\alpha\colon I \to \mathbb{R}^2_\nu$ as

$$\beta(s) \doteq \alpha(s) + \frac{(-1)^\nu}{\kappa_\alpha(s)} N_\alpha(s).$$

Observe that β does not necessarily have unit speed, despite its parameter being denoted by s.

(a) Fix $s_0 \in I$. Show that β is regular at s_0 if and only if $\kappa'_\alpha(s_0) \neq 0$.

(b) Under the above assumptions, given $s \in I$, show that the tangent vector to the evolute at s is parallel to the normal vector to α at s. In particular, conclude that in $\mathbb{L}^2$, β and α have opposite causal types.

(c) ([63]) Suppose that $\nu = 1$. Show that α and β never intersect.

Hint. Assume that they do intersect and use item (a) from Exercise 2.1.12 to obtain a contradiction.

Exercise 2.2.3 (Four singularities, [63])**.** Let $\alpha\colon [a,b] \to \mathbb{L}^2$ be a closed curve. Show that the set of points on which α is lightlike is the union of at least four non-empty closed and disjoint subsets of the plane.

Hint. Regard α inside $\mathbb{R}^2$ instead. Since α is closed, the Euclidean normal map $N_\alpha\colon [a,b] \to \mathbb{S}^1$ is surjective. Consider the inverse images of the intersections of the two light rays with the circle $\mathbb{S}^1$.

Exercise 2.2.4 ([59])**.** Let $\alpha\colon I \to \mathbb{R}^2_\nu$ be a unit speed regular curve. For each $n \geq 1$, let $a_n, b_n\colon I \to \mathbb{R}$ be the functions satisfying

$$\alpha^{(n)}(s) = a_n(s) T_\alpha(s) + b_n(s) N_\alpha(s),$$

for all $s \in I$, where $\alpha^{(n)}$ denotes the n-th derivative of α. Show that for all $n \geq 1$ the generalized Frenet-Serret equations

$$\begin{pmatrix} a_{n+1}(s) \\ b_{n+1}(s) \end{pmatrix} = \begin{pmatrix} a'_n(s) \\ b'_n(s) \end{pmatrix} + \begin{pmatrix} 0 & (-1)^{\nu+1}\kappa_\alpha(s) \\ \kappa_\alpha(s) & 0 \end{pmatrix} \begin{pmatrix} a_n(s) \\ b_n(s) \end{pmatrix}$$

hold.

Exercise 2.2.5. Let $a, b \in \mathbb{R}_{>0}$.

(a) Obtain the values of t for which the curvature of the ellipse $\boldsymbol{\alpha}\colon \mathbb{R} \to \mathbb{R}^2$ given by

$$\boldsymbol{\alpha}(t) = (a\cos t, b\sin t)$$

is maximum and minimum.

Hint. Thinking geometrically, there's a natural guess to be made. Check whether your computations support such a guess.

Remark. We'll say that $s_0 \in I$ is a *vertex* of $\boldsymbol{\alpha}$ if $\kappa'_{\boldsymbol{\alpha}}(s_0) = 0$. The *Four Vertex Theorem* states that every closed, simple and convex curve in $\mathbb{R}^2$ has at least four vertices. For more details, see [17]. Moreover, there is a sharper version of this theorem without the convexity assumption, see [14].

(b) Repeat the discussion for the Lorentzian analogues, $\boldsymbol{\beta}_1, \boldsymbol{\beta}_2\colon \mathbb{R} \to \mathbb{L}^2$ given by

$$\boldsymbol{\beta}_1(t) = (a\cosh t, b\sinh t) \quad \text{and} \quad \boldsymbol{\beta}_2(t) = (a\sinh t, b\cosh t).$$

Do the curvatures still attain a maximum and a minimum value?

Exercise† 2.2.6. Let $\boldsymbol{\alpha}\colon I \to \mathbb{R}^2_\nu$ be a non-lightlike regular curve. Denote by $\kappa_{\boldsymbol{\alpha},E}$ and $\kappa_{\boldsymbol{\alpha},L}$ the curvatures of $\boldsymbol{\alpha}$ when seen in $\mathbb{R}^2$ and $\mathbb{L}^2$, respectively. Show that $|\kappa_{\boldsymbol{\alpha},E}(t)| \leq |\kappa_{\boldsymbol{\alpha},L}(t)|$ for all $t \in I$. Moreover, give an example of a curve satisfying $\kappa_{\boldsymbol{\alpha},L}(t) < \kappa_{\boldsymbol{\alpha},E}(t)$ for all $t \in I$.

Exercise 2.2.7. Give an example of a regular curve $\boldsymbol{\alpha}\colon I \to \mathbb{L}^2$ which is lightlike at a single instant $t_0 \in I$, satisfying

$$\lim_{t \to t_0^-} \kappa_{\boldsymbol{\alpha}}(t) = -\infty \quad \text{and} \quad \lim_{t \to t_0^+} \kappa_{\boldsymbol{\alpha}}(t) = +\infty.$$

Hint. Try some cubic similar to the one in Figure 2.4 (p. 78).

Exercise† 2.2.8. Let $\boldsymbol{\alpha}\colon I \to \mathbb{R}^2_\nu$ be a non-lightlike regular curve. Show, for all $t \in I$, that $\|\boldsymbol{T}'_{\boldsymbol{\alpha}}(t)\| = \|\boldsymbol{N}'_{\boldsymbol{\alpha}}(t)\|$. In particular, check that if $\boldsymbol{\alpha}$ has unit speed, then $|\kappa_{\boldsymbol{\alpha}}(s)| = \|\boldsymbol{\alpha}''(s)\|$, for all $s \in I$.

Exercise 2.2.9. It is usually convenient to also study curves as the set of zeros of a smooth function. For example, find a parametrized curve whose trace is the solution set of the equation $x^3 + y^3 = 3axy$, where $a \in \mathbb{R}$. This curve is called the *Folium of Descartes*.

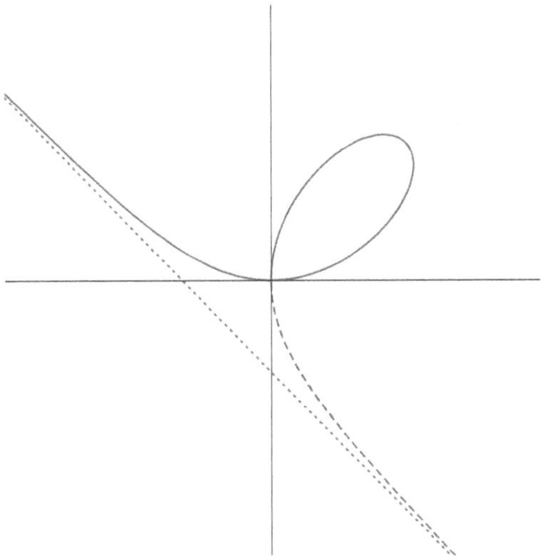

Figure 2.11: The Folium of Descartes for $a = 1/3$.

Hint. Try $y = tx$, that is, take as the parameter t the tangent of the Euclidean angle formed between the x-axis and the position vector (x,y).

Exercise† 2.2.10. Exercise 2.1.22 (p. 77) says that a curve in $\mathbb{R}^2$ is locally represented as the solution set of an equation of the form $F(x,y) = 0$, for some smooth function $F\colon \mathbb{R}^2 \to \mathbb{R}$. That is, $\boldsymbol{\alpha}\colon I \to \mathbb{R}^2$ satisfies $F(\boldsymbol{\alpha}(t)) = 0$ for all $t \in I$ (perhaps reducing the domain I). It is possible to express the curvature of $\boldsymbol{\alpha}$ in terms of the function F. Show that

$$|\kappa_{\boldsymbol{\alpha}}(t)| = \frac{\left|\operatorname{Hess} F(\boldsymbol{\alpha}(t))(R_{\pi/2}\nabla F(\boldsymbol{\alpha}(t)))\right|}{\|\nabla F(\boldsymbol{\alpha}(t))\|^3}.$$

Hint.

- Recall (from Appendix A) that for a smooth function $f\colon \mathbb{R}^n \to \mathbb{R}^k$, we identify the second total derivative $D(Df)(\boldsymbol{p})$ with the symmetric bilinear map given by $D^2f(\boldsymbol{p})(\boldsymbol{v},\boldsymbol{w}) = \sum_{i,j=1}^n v_i w_j F_{x_i x_j}(\boldsymbol{p})$. The *Hessian* of f at $\boldsymbol{p}$ is then defined by $\operatorname{Hess} f(\boldsymbol{p})(\boldsymbol{v}) \doteq D^2 f(\boldsymbol{p})(\boldsymbol{v},\boldsymbol{v})$.
- Write the Frenet-Serret frame of $\boldsymbol{\alpha}$ in terms of the gradient of F and use that $s'(t)\kappa_{\boldsymbol{\alpha}}(t) = \langle \boldsymbol{T}'_{\boldsymbol{\alpha}}(t), \boldsymbol{N}_{\boldsymbol{\alpha}}(t)\rangle$.

Remark. The absolute values are necessary since we could work with $-F$ instead of F; the correct sign is decided by analyzing whether $\nabla F(\boldsymbol{\alpha}(t))$ points in the same direction as $\boldsymbol{\alpha}'(t)$ or not.

Exercise† 2.2.11. Note that the result seen in Exercise 2.1.22 (p. 77) is independent of the scalar product used, and thus it still holds in $\mathbb{L}^2$. However, to repeat the discussion made in the previous exercise, we need the correct notion of a gradient in $\mathbb{L}^2$. Bearing in mind that for a smooth function $F\colon \mathbb{R}^2_\nu \to \mathbb{R}$ the gradient $\nabla F(\boldsymbol{p})$ is nothing more than the vector associated via $\langle\cdot,\cdot\rangle_E$ to the linear functional $DF(\boldsymbol{p})$ with Riesz's Lemma, we define

$$\nabla_L F(\boldsymbol{p}) \doteq \left(\frac{\partial F}{\partial x}(\boldsymbol{p}), -\frac{\partial F}{\partial y}(\boldsymbol{p})\right).$$

With this, if $\boldsymbol{\alpha}\colon I \to \mathbb{L}^2$ is a non-lightlike regular curve satisfying $F(\boldsymbol{\alpha}(t)) = 0$ for all $t \in I$, show that:

$$|\kappa_{\boldsymbol{\alpha}}(t)| = \frac{\left|\operatorname{Hess} F(\boldsymbol{\alpha}(t))(\mathfrak{f}\nabla_L F(\boldsymbol{\alpha}(t)))\right|}{\|\nabla_L F(\boldsymbol{\alpha}(t))\|_L^3},$$

where the "flip" operator $\mathfrak{f}\colon \mathbb{L}^2 \to \mathbb{L}^2$ is given by $\mathfrak{f}(x,y) = (y,x)$. Will the causal type of $\boldsymbol{\alpha}$ influence the correct choice of sign? Test the formula for $\mathbb{S}^1_1$ and $\mathbb{H}^1$ using $F(x,y) = x^2 - y^2 \pm 1$.

Exercise 2.2.12 (Graphs). Let $f\colon I \to \mathbb{R}$ be a smooth function and $\boldsymbol{\alpha}, \boldsymbol{\beta}\colon I \to \mathbb{R}^2_\nu$ be given by $\boldsymbol{\alpha}(t) = (t, f(t))$ and $\boldsymbol{\beta}(t) = (f(t), t)$. Determine, when possible, $\kappa_{\boldsymbol{\alpha}}(t)$ and $\kappa_{\boldsymbol{\beta}}(t)$. In particular, check that if $t_0 \in I$ is a critical point for f, we have $\boldsymbol{\alpha}''(t_0) = (0, \kappa_{\boldsymbol{\alpha}}(t_0))$ and $\boldsymbol{\beta}''(t_0) = (-\kappa_{\boldsymbol{\beta}}(t_0), 0)$. Since every curve is locally given as a graph, interpret again the signs of the curvatures in terms of the concavity of f.

Exercise† 2.2.13 (Polar Coordinates). Recall that a point $(x,y) \in \mathbb{R}^2$ may also be represented by a pair $(r,\theta) \in \mathbb{R}_{\geq 0} \times [0, 2\pi[$, where r is the distance between (x,y) and the origin, and θ is the (oriented) angle formed between $\boldsymbol{e}_1 = (1,0)$ and (x,y). That is, we have:

$$x = r\cos\theta \quad \text{and} \quad y = r\sin\theta.$$

Moreover, if $(x,y) \neq (0,0)$, such representation is unique. The variables r and θ are called the *polar coordinates* of the pair (x,y).

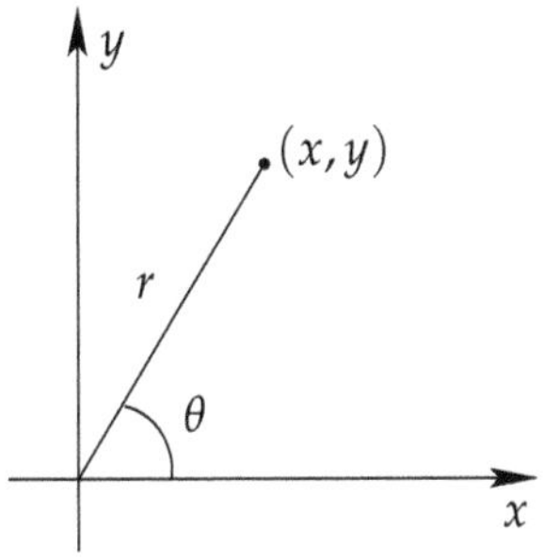

Figure 2.12: Polar coordinates in $\mathbb{R}^2$.

Curves in the plane may be expressed in the form $r = r(\theta)$, using θ as a parameter ranging over a certain interval $[\theta_0, \theta_1]$.

(a) Explicitly write the parametrization $\boldsymbol{\alpha}(\theta)$ for a curve given in *polar representation* $r = r(\theta)$. Verify that

$$\boldsymbol{\alpha}'(\theta) = R_\theta(r'(\theta), r(\theta)) \text{ and } \boldsymbol{\alpha}''(\theta) = R_\theta(r''(\theta) - r(\theta), 2r'(\theta)),$$

where R_θ denotes a counterclockwise Euclidean rotation by θ, as in Example 1.4.4 (p. 30).

(b) Show that the arclength of $\boldsymbol{\alpha}$ between θ_0 and θ_1 is given by

$$L[\boldsymbol{\alpha}] = \int_{\theta_0}^{\theta_1} \sqrt{r(\theta)^2 + r'(\theta)^2}\, d\theta.$$

(c) Assuming that $\boldsymbol{\alpha}$ is regular, show that the curvature of $\boldsymbol{\alpha}$ is given by

$$\kappa_{\boldsymbol{\alpha}}(\theta) = \frac{2r'(\theta)^2 - r(\theta)r''(\theta) + r(\theta)^2}{\left(r(\theta)^2 + r'(\theta)^2\right)^{3/2}}.$$

(d) Use the previous items to compute the length and the curvature of the *cardioid*[3] defined by $r = 1 - \cos\theta$, for $\theta \in [0, 2\pi[$. On which points is the curvature minimal?

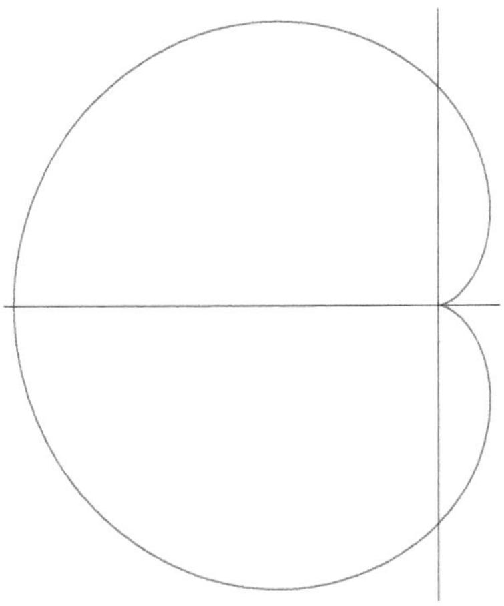

Figure 2.13: Cardioid in the Euclidean plane.

[3]The name "cardioid" comes precisely from the heart-like shape the curve has.

Exercise† 2.2.14 (Rindler Coordinates). Each point $(x,y) \in \mathbb{L}^2$ with $|y| < x$ may be represented by a pair $(\rho,\varphi) \in \mathbb{R}_{\geq 0} \times \mathbb{R}$, in an analogous way to polar coordinates in $\mathbb{R}^2$, by setting

$$x = \rho\cosh\varphi \quad \text{and} \quad y = \rho\sinh\varphi.$$

The variables ρ and φ are called the *Rindler coordinates* of the pair (x,y).

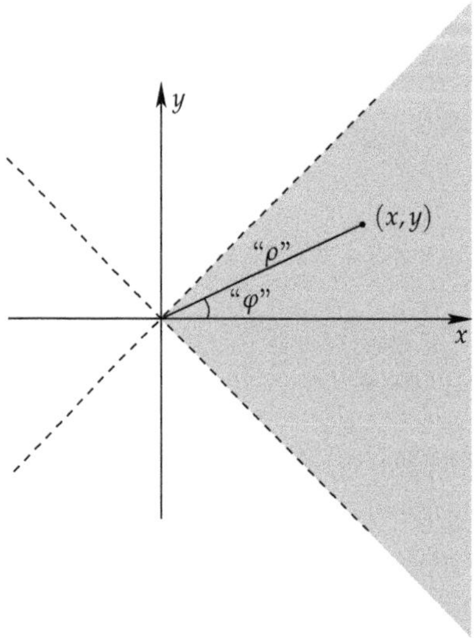

Figure 2.14: Rindler coordinates in $\mathbb{L}^2$.

Curves in this region of the plane $\mathbb{L}^2$, known as the *Rindler wedge*, may be expressed in the form $\rho = \rho(\varphi)$, using φ as a parameter ranging over a certain interval $[\varphi_0,\varphi_1]$.

(a) Explicitly write the parametrization $\boldsymbol{\alpha}(\varphi)$ for a curve given in *Rindler representation* $\rho = \rho(\varphi)$. Verify that

$$\boldsymbol{\alpha}'(\varphi) = R^h_\varphi(\rho'(\varphi),\rho(\varphi)) \quad \text{and} \quad \boldsymbol{\alpha}''(\varphi) = R^h_\varphi(\rho''(\varphi)+\rho(\varphi), 2\rho'(\varphi)),$$

where R^h_φ denotes a hyperbolic rotation by φ, as in Example 1.4.5 (p. 32).

(b) Show that the arclength of $\boldsymbol{\alpha}$ between φ_0 and φ_1 is given by

$$L[\boldsymbol{\alpha}] = \int_{\varphi_0}^{\varphi_1} \sqrt{|\rho(\varphi)^2 - \rho'(\varphi)^2|}\,d\varphi.$$

(c) Assuming that $\boldsymbol{\alpha}$ is regular and non-lightlike, show that the curvature of $\boldsymbol{\alpha}$ is given by

$$\kappa_{\boldsymbol{\alpha}}(\varphi) = \frac{2\rho'(\varphi)^2 - \rho(\varphi)\rho''(\varphi) - \rho(\varphi)^2}{|\rho(\varphi)^2 - \rho'(\varphi)^2|^{3/2}}.$$

(d) One may consider a Lorentzian version of the cardioid defined in Exercise 2.2.13: use the previous items to compute the length and curvature of the curve defined by $\rho = -1+\cosh\varphi$, for $\varphi \in \mathbb{R}\setminus\{0\}$. Are there points where the curvature is maximum or minimum? Compare with the cardioid.

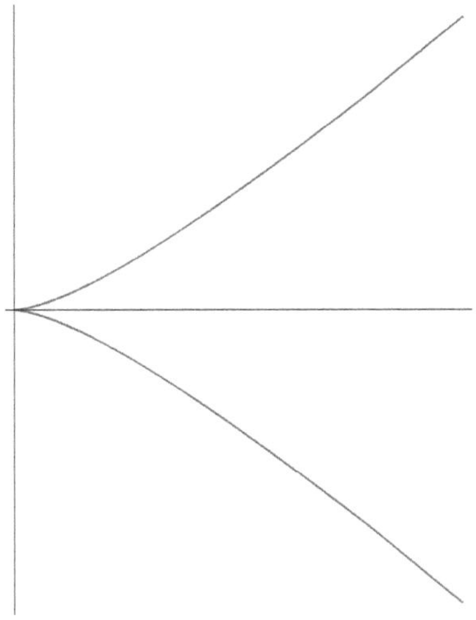

Figure 2.15: Cardioid in the Lorentzian plane.

Exercise† 2.2.15 (Curves in the complex plane)**.** We may regard curves in $\mathbb{R}^2$ as maps $\alpha\colon I \to \mathbb{C}$, written in the form $\alpha(t) = x(t) + \mathrm{i}y(t)$. Suppose that α is regular. Observe that if $z_1 = x_1 + \mathrm{i}y_1$ and $z_2 = x_2 + \mathrm{i}y_2$ are identified with the vectors $\boldsymbol{z}_1 = (x_1, y_1)$ and $\boldsymbol{z}_2 = (x_2, y_2)$, we have that $\langle \boldsymbol{z}_1, \boldsymbol{z}_2 \rangle_E = \mathrm{Re}(z_1\overline{z_2})$.

(a) Show that

$$\frac{\mathrm{d}}{\mathrm{d}t}\left(\frac{\alpha'(t)}{|\alpha'(t)|}\right) = \mathrm{i}\kappa_\alpha(t)\alpha'(t).$$

(b) Show that

$$\kappa_\alpha(t) = -\frac{\mathrm{Im}(\alpha'(t)\overline{\alpha''(t)})}{|\alpha'(t)|^3},$$

and use this to compute the curvature of the Archimedes Spiral, parametrized by $\alpha(t) = (a + bt)\mathrm{e}^{\mathrm{i}t}$.

(c) Suppose that $\alpha\colon I \to \mathbb{C}$ is given by $\alpha(\theta) = r(\theta)\mathrm{e}^{\mathrm{i}\theta}$ for some positive function r, and use item (b) to recover the formula for $\kappa_\alpha(\theta)$ seen in Exercise 2.2.13.

Exercise† 2.2.16 (A glimpse of the future)**.** In this exercise we will see how to repeat the results from the previous exercise for curves in $\mathbb{L}^2$. Consider the set of *split-complex numbers*

$$\mathbb{C}' \doteq \{x + \mathrm{h}y \mid x, y \in \mathbb{R} \text{ and } \mathrm{h}^2 = 1\},$$

with operations similar to $\mathbb{C}$. Split-complex numbers encode the geometry of $\mathbb{L}^2$ in the same way that the complex numbers encode the geometry of $\mathbb{R}^2$. We will study $\mathbb{C}'$ in more detail in Chapter 4, when we discuss surfaces with zero mean curvature in $\mathbb{L}^3$.

Given $w = x + \mathrm{h}y \in \mathbb{C}'$, define its *split-conjugate* as $\overline{w} \doteq x - \mathrm{h}y$, and its *split-complex absolute value* as $|w| \doteq \sqrt{|x^2 - y^2|}$. We'll denote the projections onto the coordinate axes, just like in $\mathbb{C}$, by Re and Im. Note that identifying $w_1 = x_1 + \mathrm{h}y_1$ and $w_2 = x_2 + \mathrm{h}y_2$ with the vectors $\boldsymbol{w}_1 = (x_1, y_1)$ and $\boldsymbol{w}_2 = (x_2, y_2)$, we have that $\langle \boldsymbol{w}_1, \boldsymbol{w}_2 \rangle_L = \mathrm{Re}(w_1\overline{w_2})$.

We may regard curves in $\mathbb{L}^2$ as maps $\alpha\colon I \to \mathbb{C}'$ written as $\alpha(t) = x(t) + \mathrm{h}y(t)$. Suppose that α is regular and non-lightlike.

(a) Show that

$$\frac{\mathrm{d}}{\mathrm{d}t}\left(\frac{\alpha'(t)}{|\alpha'(t)|}\right) = \epsilon_\alpha \mathrm{h}\kappa_\alpha(t)\alpha'(t).$$

(b) Show that

$$\kappa_{\alpha}(t) = -\frac{\mathrm{Im}(\alpha'(t)\overline{\alpha''(t)})}{|\alpha'(t)|^3},$$

and use this to compute the curvature of a Lorentzian version of the Archimedes Spiral. Defining, for $t \in \mathbb{R}$, the *split-complex exponential* by

$$e^{ht} \doteq \cosh t + h \sinh t,$$

this curve may be expressed as $\alpha(t) = (a + bt)e^{ht}$.

(c) Suppose that $\alpha\colon I \to \mathbb{C}'$ has the form $\alpha(\varphi) = \rho(\varphi)e^{h\varphi}$ for some positive function ρ, and use item (b) to recover the formula for $\kappa_{\alpha}(\varphi)$ seen in Exercise 2.2.14.

Remark. "Naively" assume that the usual rules from Calculus hold in the split-complex setting. We will formalize all of this in due time.

Exercise 2.2.17 (Osculating cubic)**.** Let $\alpha\colon I \to \mathbb{R}^2_\nu$ be a unit speed curve. Assume that $0 \in I$ and $\alpha(0) = \mathbf{0}$. Write

$$\alpha(s) = s\alpha'(0) + \frac{s^2}{2}\alpha''(0) + \frac{s^3}{6}\alpha'''(0) + \mathbf{R}(s),$$

where $\mathbf{R}(s)/s^3 \to 0$ as $s \to 0$. Define $\beta\colon I \to \mathbb{R}^2_\nu$ by setting $\beta(s) = \alpha(s) - \mathbf{R}(s)$.

(a) If α is spacelike, assume in addition that $\mathbf{T}_{\alpha}(0) = (1,0)$ and $\mathbf{N}_{\alpha}(0) = (0,1)$. Compute the curvature $\kappa_{\beta}(s)$ and compare it to $\kappa_{\alpha}(0)$.

(b) If α is timelike, repeat the argument used in item (a) above, now assuming that $\mathbf{T}_{\alpha}(0) = (0,1)$ and $\mathbf{N}_{\alpha}(0) = (-1,0)$.

Exercise 2.2.18. The definition of the evolute of a curve (given in Exercise 2.2.2) is also valid for curves with arbitrary parameter, whose curvatures never vanish. Consider the cycloid parametrized by $\alpha\colon\ ]0,2\pi[\ \to \mathbb{R}^2$, $\alpha(t) = (t - \sin t, 1 - \cos t)$. Determine its evolute $\mathbf{ev}_{\alpha}$, and show that they are congruent (that is, show that $\mathbf{ev}_{\alpha}$ is another cycloid).

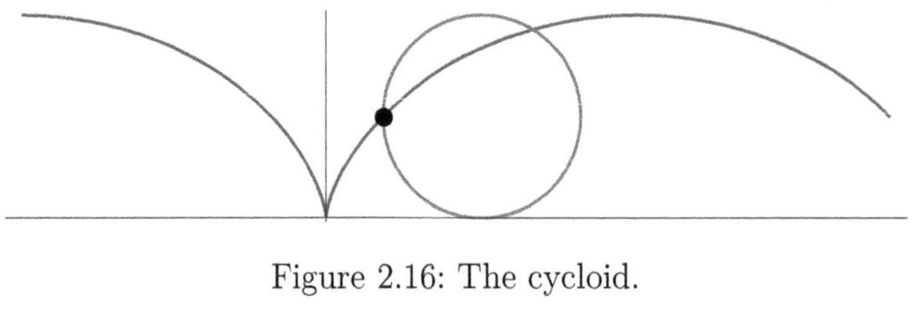

Figure 2.16: The cycloid.

Exercise 2.2.19. Determine, up to congruence, all the non-lightlike regular curves $\alpha\colon \mathbb{R} \to \mathbb{R}^2_\nu$ with the following property: for each $s \in \mathbb{R}$ there is $F_s \in \mathrm{E}_\nu(2,\mathbb{R})$ such that $F_s(\alpha(t)) = \alpha(t+s)$, for all $t \in \mathbb{R}$.

Exercise† **2.2.20.** Let $\alpha\colon I \to \mathbb{L}^2$ be a curve and consider again the "flip" operator $\mathfrak{f}\colon \mathbb{L}^2 \to \mathbb{L}^2$ given by $\mathfrak{f}(x,y) = (y,x)$.

(a) Show that α is regular if and only if $\mathfrak{f} \circ \alpha$ is also regular.

(b) Suppose that α has constant causal type. Show that α is spacelike (resp. timelike, lightlike) if and only if $\mathfrak{f} \circ \alpha$ is timelike (resp. spacelike, lightlike).

(c) Assuming that $\boldsymbol{\alpha}$ is non-lightlike and regular, show that $\kappa_{\mathfrak{f}\circ\boldsymbol{\alpha}}(t) = -\kappa_{\boldsymbol{\alpha}}(t)$ for all $t \in I$. Conclude that up to congruence and reparametrization, to every spacelike curve in $\mathbb{L}^2$ corresponds a unique timelike curve with the same curvature (and vice versa, since $\mathfrak{f} = \mathfrak{f}^{-1}$).

Exercise 2.2.21. Determine all the non-lightlike regular curves in $\mathbb{R}^2_\nu$ such that:

(a) all their tangent lines intercept in a fixed point;

(b) all their normal lines intercept in a fixed point.

Exercise 2.2.22. Determine a unit speed curve $\boldsymbol{\alpha}\colon \mathbb{R} \to \mathbb{R}^2$ such that

$$\boldsymbol{\alpha}(0) = \mathbf{0} \quad \text{and} \quad \kappa_{\boldsymbol{\alpha}}(s) = \frac{1}{1+s^2},$$

for all $s \in \mathbb{R}$.

Exercise 2.2.23. Determine all the unit speed curves $\boldsymbol{\alpha}\colon I \to \mathbb{R}^2_\nu$ such that

$$\kappa_{\boldsymbol{\alpha}}(s) = \frac{1}{as+b},$$

with $a, b \in \mathbb{R}$, $a \neq 0$.

2.3 CURVES IN SPACE

In this section, our main goal is to establish a classification result similar to Theorem 2.2.13 for curves in space. However, this time the curvature is not enough and, to this end, we'll introduce the concept of torsion.

The first thing to be done is to define the Frenet-Serret frame for a curve in space, following the guidelines outlined in the previous section: associate a positive orthonormal basis to each point of the curve. The general construction, for a non-lightlike curve, begins considering the velocity vector, and then taking a vector orthogonal to it. But in space $\mathbb{R}^3_\nu$, there is now the possibility for the "natural candidate" to normal vector to be lightlike, making unfeasible the construction of the desired orthonormal basis. We must investigate a solution around this issue.

In contrast with what was seen for planar curves, in $\mathbb{L}^3$ there are many lightlike curves that are not straight lines, which may not be neglected. It is possible to characterize those by using a single invariant (see Theorem 2.3.34, p. 125), but for this we need to introduce yet another new frame along the curve, mimicking the idea of the Frenet-Serret frame.

In view of these situations, we have many cases to discuss, depending on the causal type of the tangent and normal vectors. Moreover, we must highlight the situations where, for all t, $\boldsymbol{\alpha}''(t)$ vanishes or is lightlike. The first situation is very simple:

Proposition 2.3.1. *The trace of a curve is a straight line if and only if the curve admits a reparametrization $\boldsymbol{\alpha} : I \to \mathbb{R}^n_\nu$ such that $\boldsymbol{\alpha}''(t) = \mathbf{0}$ for all $t \in I$.*

The second situation will be addressed in Subsection 2.3.3. When one of these situations occurs in isolated points, it might not be possible to construct the desired frame in such points, and so we will not consider these (pathological) cases.

2.3.1 The Frenet-Serret trihedron

Definition 2.3.2. Let $\boldsymbol{\alpha}\colon I \to \mathbb{R}^3_\nu$ be a parametrized curve. We'll say that $\boldsymbol{\alpha}$ is *admissible*[4] if:

(i) $\boldsymbol{\alpha}$ is *biregular*, that is, $\{\boldsymbol{\alpha}'(t), \boldsymbol{\alpha}''(t)\}$ is linearly independent for all $t \in I$;

(ii) both $\boldsymbol{\alpha}'(t)$ and $\operatorname{span}\{\boldsymbol{\alpha}'(t), \boldsymbol{\alpha}''(t)\}$ are not lightlike, for each $t \in I$.

Remark.

- Every unit speed curve in $\mathbb{R}^3$ which is not a straight line is automatically admissible. This way, this definition is not a restriction for the development of the usual theory of curves developed in $\mathbb{R}^3$, but really only a device to facilitate the simultaneous treatment of several possible cases in $\mathbb{L}^3$.
- Being admissible is an intrinsic property of a curve. See Exercise 2.3.1.

As in the previous section, we will begin by focusing on unit speed curves. Note that if $\boldsymbol{\alpha}'(s)$ and $\boldsymbol{\alpha}''(s)$ are orthogonal and $\operatorname{span}\{\boldsymbol{\alpha}'(t), \boldsymbol{\alpha}''(t)\}$ is not lightlike, then $\boldsymbol{\alpha}''(s)$ cannot be lightlike. With this:

Definition 2.3.3 (Frenet-Serret Trihedron)**.** Let $\boldsymbol{\alpha}\colon I \to \mathbb{R}^3_\nu$ be an admissible unit speed curve. The *tangent vector* to $\boldsymbol{\alpha}$ at s is $\boldsymbol{T}_{\boldsymbol{\alpha}}(s) \doteq \boldsymbol{\alpha}'(s)$. The *curvature* of the curve $\boldsymbol{\alpha}$ at s is $\kappa_{\boldsymbol{\alpha}}(s) \doteq \|\boldsymbol{T}'_{\boldsymbol{\alpha}}(s)\|$. The assumptions on $\boldsymbol{\alpha}$ ensure that $\kappa_{\boldsymbol{\alpha}}(s) > 0$ for all $s \in I$, so that the unit vector pointing in the same direction as $\boldsymbol{T}'_{\boldsymbol{\alpha}}(s)$ is well defined, allowing us to define the *normal vector* to $\boldsymbol{\alpha}$ at s, by the relation $\boldsymbol{T}'_{\boldsymbol{\alpha}}(s) = \kappa_{\boldsymbol{\alpha}}(s)\boldsymbol{N}_{\boldsymbol{\alpha}}(s)$. Lastly, the *binormal vector* to $\boldsymbol{\alpha}$ at s is defined as the unique vector $\boldsymbol{B}_{\boldsymbol{\alpha}}(s)$ making the basis $(\boldsymbol{T}_{\boldsymbol{\alpha}}(s), \boldsymbol{N}_{\boldsymbol{\alpha}}(s), \boldsymbol{B}_{\boldsymbol{\alpha}}(s))$ orthonormal and positive.

Remark. The curvature here defined is fundamentally distinct from the oriented curvature defined in the previous section, as it does not encode orientation and has no sign. For a relation between them, see Exercise 2.3.3.

To express in an easier way the binormal vector in terms of $\boldsymbol{\alpha}$, the following definition is useful:

Definition 2.3.4. Let $\boldsymbol{\alpha}\colon I \to \mathbb{R}^3_\nu$ be a unit speed admissible curve. The *coindicator* of $\boldsymbol{\alpha}$ is defined as $\eta_{\boldsymbol{\alpha}} \doteq \epsilon_{\boldsymbol{N}_{\boldsymbol{\alpha}}(s)}$.

For space curves, we have the advantage of being able to express the binormal vector in the form $\boldsymbol{B}_{\boldsymbol{\alpha}}(s) = \lambda \boldsymbol{T}_{\boldsymbol{\alpha}}(s) \times \boldsymbol{N}_{\boldsymbol{\alpha}}(s)$ for some $\lambda \in \mathbb{R}$. Since we want the obtained basis to be positive, we have

$$\det\left(\boldsymbol{T}_{\boldsymbol{\alpha}}(s), \boldsymbol{N}_{\boldsymbol{\alpha}}(s), \boldsymbol{B}_{\boldsymbol{\alpha}}(s)\right) = 1 \implies \lambda \det\left(\boldsymbol{T}_{\boldsymbol{\alpha}}(s), \boldsymbol{N}_{\boldsymbol{\alpha}}(s), \boldsymbol{T}_{\boldsymbol{\alpha}}(s) \times \boldsymbol{N}_{\boldsymbol{\alpha}}(s)\right) = 1.$$

It follows from the orientability analysis done in Section 1.6 that $\lambda = (-1)^\nu \epsilon_{\boldsymbol{\alpha}} \eta_{\boldsymbol{\alpha}}$, and thus $\boldsymbol{B}_{\boldsymbol{\alpha}}(s) = (-1)^\nu \epsilon_{\boldsymbol{\alpha}} \eta_{\boldsymbol{\alpha}} \boldsymbol{T}_{\boldsymbol{\alpha}}(s) \times \boldsymbol{N}_{\boldsymbol{\alpha}}(s)$. Let's see that such $\boldsymbol{B}_{\boldsymbol{\alpha}}(s)$ is indeed a unit vector. To wit, we have:

$$\langle \boldsymbol{B}_{\boldsymbol{\alpha}}(s), \boldsymbol{B}_{\boldsymbol{\alpha}}(s)\rangle = \langle \boldsymbol{T}_{\boldsymbol{\alpha}}(s) \times \boldsymbol{N}_{\boldsymbol{\alpha}}(s), \boldsymbol{T}_{\boldsymbol{\alpha}}(s) \times \boldsymbol{N}_{\boldsymbol{\alpha}}(s)\rangle \stackrel{(*)}{=} (-1)^\nu \det\begin{pmatrix} \epsilon_{\boldsymbol{\alpha}} & 0 \\ 0 & \eta_{\boldsymbol{\alpha}} \end{pmatrix} = (-1)^\nu \epsilon_{\boldsymbol{\alpha}} \eta_{\boldsymbol{\alpha}},$$

where in $(*)$ we have used Proposition 1.6.5 (p. 55). With this we are almost ready to present the Frenet-Serret equations. Recall that the idea behind those equations is to write the derivatives $\boldsymbol{T}'_{\boldsymbol{\alpha}}(s), \boldsymbol{N}'_{\boldsymbol{\alpha}}(s)$ and $\boldsymbol{B}'_{\boldsymbol{\alpha}}(s)$ as linear combinations of the Frenet-Serret frame. We have the:

---

[4]It is reasonably usual in Mathematics to call any object "admissible" if it has the convenient properties for the development of the theory.

Definition 2.3.5 (Torsion). Let $\boldsymbol{\alpha}\colon I \to \mathbb{R}^3_\nu$ be a unit speed admissible curve. The *torsion* of $\boldsymbol{\alpha}$ at s, denoted by $\tau_{\alpha}(s)$, is the component of $\boldsymbol{N}'_{\alpha}(s)$ in the direction of $\boldsymbol{B}_{\alpha}(s)$.

Theorem 2.3.6 (Frenet-Serret Equations). *Let $\boldsymbol{\alpha}\colon I \to \mathbb{R}^3_\nu$ be a unit speed admissible curve. Then*

$$\begin{pmatrix} \boldsymbol{T}'_{\alpha}(s) \\ \boldsymbol{N}'_{\alpha}(s) \\ \boldsymbol{B}'_{\alpha}(s) \end{pmatrix} = \begin{pmatrix} 0 & \kappa_{\alpha}(s) & 0 \\ -\epsilon_{\alpha}\eta_{\alpha}\kappa_{\alpha}(s) & 0 & \tau_{\alpha}(s) \\ 0 & (-1)^{\nu+1}\epsilon_{\alpha}\tau_{\alpha}(s) & 0 \end{pmatrix} \begin{pmatrix} \boldsymbol{T}_{\alpha}(s) \\ \boldsymbol{N}_{\alpha}(s) \\ \boldsymbol{B}_{\alpha}(s) \end{pmatrix}$$

holds, for all $s \in I$.

Proof: The first equation is immediate by definition. For the remaining ones, the idea is to apply corollaries 2.1.14 and 2.1.15 (p. 70) for the orthonormal expansions of $\boldsymbol{N}'_{\alpha}(s)$ and $\boldsymbol{B}'_{\alpha}(s)$. To begin, we have

$$\begin{aligned} \boldsymbol{N}'_{\alpha}(s) &= \epsilon_{\alpha}\langle \boldsymbol{N}'_{\alpha}(s), \boldsymbol{T}_{\alpha}(s)\rangle \boldsymbol{T}_{\alpha}(s) + (-1)^{\nu}\epsilon_{\alpha}\eta_{\alpha}\langle \boldsymbol{N}'_{\alpha}(s), \boldsymbol{B}_{\alpha}(s)\rangle \boldsymbol{B}_{\alpha}(s) \\ \boldsymbol{B}'_{\alpha}(s) &= \epsilon_{\alpha}\langle \boldsymbol{B}'_{\alpha}(s), \boldsymbol{T}_{\alpha}(s)\rangle \boldsymbol{T}_{\alpha}(s) + \eta_{\alpha}\langle \boldsymbol{B}'_{\alpha}(s), \boldsymbol{N}_{\alpha}(s)\rangle \boldsymbol{N}_{\alpha}(s). \end{aligned}$$

From the first Frenet-Serret equation and the definition of torsion, we have that

$$\kappa_{\alpha}(s) = \eta_{\alpha}\langle \boldsymbol{T}'_{\alpha}(s), \boldsymbol{N}_{\alpha}(s)\rangle \quad \text{and} \quad \tau_{\alpha}(s) = (-1)^{\nu}\epsilon_{\alpha}\eta_{\alpha}\langle \boldsymbol{N}'_{\alpha}(s), \boldsymbol{B}_{\alpha}(s)\rangle.$$

To determine the component of $\boldsymbol{N}'_{\alpha}(s)$ in the direction of $\boldsymbol{T}_{\alpha}(s)$, note that

$$\epsilon_{\alpha}\langle \boldsymbol{N}'_{\alpha}(s), \boldsymbol{T}_{\alpha}(s)\rangle = -\epsilon_{\alpha}\langle \boldsymbol{N}_{\alpha}(s), \boldsymbol{T}'_{\alpha}(s)\rangle = -\epsilon_{\alpha}\eta_{\alpha}\kappa_{\alpha}(s).$$

Thus, we have the second Frenet-Serret equation. To obtain the last one, note that

$$\begin{aligned} \boldsymbol{B}'_{\alpha}(s) &= (-1)^{\nu}\epsilon_{\alpha}\eta_{\alpha}\left(\boldsymbol{T}'_{\alpha}(s) \times \boldsymbol{N}_{\alpha}(s) + \boldsymbol{T}_{\alpha}(s) \times \boldsymbol{N}'_{\alpha}(s)\right) \\ &= (-1)^{\nu}\epsilon_{\alpha}\eta_{\alpha}\boldsymbol{T}_{\alpha}(s) \times \boldsymbol{N}'_{\alpha}(s), \end{aligned}$$

so that $\langle \boldsymbol{B}'_{\alpha}(s), \boldsymbol{T}_{\alpha}(s)\rangle = 0$. And, finally:

$$\eta_{\alpha}\langle \boldsymbol{B}'_{\alpha}(s), \boldsymbol{N}_{\alpha}(s)\rangle = -\eta_{\alpha}\langle \boldsymbol{B}_{\alpha}(s), \boldsymbol{N}'_{\alpha}(s)\rangle = (-1)^{\nu+1}\epsilon_{\alpha}\tau_{\alpha}(s).$$

□

Remark. It is usual in the literature for the torsion to be defined with the sign opposite to our definition above. We have made this choice for the coefficient matrix in $\mathbb{R}^3$ to be skew-symmetric. A geometric interpretation for the torsion will be given in Proposition 2.3.11 (p. 106) later.

Example 2.3.7.

(1) Consider the curve $\boldsymbol{\alpha}\colon \mathbb{R} \to \mathbb{R}^3$ given by

$$\boldsymbol{\alpha}(s) = \left(3\cos\frac{s}{5}, 3\sin\frac{s}{5}, \frac{4s}{5}\right).$$

Note that $\boldsymbol{\alpha}$ has unit speed, and thus:

$$\boldsymbol{T}_{\alpha}(s) = \frac{1}{5}\left(-3\sin\frac{s}{5}, 3\cos\frac{s}{5}, 4\right) \implies \boldsymbol{T}'_{\alpha}(s) = \frac{3}{25}\left(-\cos\frac{s}{5}, -\sin\frac{s}{5}, 0\right),$$

whence we obtain

$$\boldsymbol{N}_{\alpha}(s) = \left(-\cos\frac{s}{5}, -\sin\frac{s}{5}, 0\right) \text{ and } \kappa_{\alpha}(s) = \frac{3}{25},$$

for all $s \in \mathbb{R}$. Proceeding, we have that:

$$\boldsymbol{B}_{\alpha}(s) = \frac{1}{5}\left(4\sin\frac{s}{5}, -4\cos\frac{s}{5}, 3\right),$$

and then it directly follows that $\tau_{\alpha}(s) = 4/25$, for all $s \in \mathbb{R}$.

(2) Consider the curve $\boldsymbol{\alpha}\colon \mathbb{R} \to \mathbb{L}^3$ given by

$$\boldsymbol{\alpha}(s) = \left(\cosh\frac{s}{\sqrt{3}}, \frac{2s}{\sqrt{3}}, \sinh\frac{s}{\sqrt{3}}\right).$$

The derivatives of $\boldsymbol{\alpha}$ are:

$$\boldsymbol{\alpha}'(s) = \boldsymbol{T}_{\alpha}(s) = \frac{1}{\sqrt{3}}\left(\sinh\frac{s}{\sqrt{3}}, 2, \cosh\frac{s}{\sqrt{3}}\right)$$
$$\boldsymbol{\alpha}''(s) = \boldsymbol{T}'_{\alpha}(s) = \frac{1}{3}\left(\cosh\frac{s}{\sqrt{3}}, 0, \sinh\frac{s}{\sqrt{3}}\right).$$

Note that $\boldsymbol{\alpha}$ is unit speed spacelike and admissible. To wit, if $\Pi = \operatorname{span}\{\boldsymbol{\alpha}'(s), \boldsymbol{\alpha}''(s)\}$ then the Gram matrix

$$G_{\Pi} = \begin{pmatrix} 1 & 0 \\ 0 & 1/9 \end{pmatrix}$$

is positive-definite. We then obtain that

$$\boldsymbol{N}_{\alpha}(s) = \left(\cosh\frac{s}{\sqrt{3}}, 0, \sinh\frac{s}{\sqrt{3}}\right) \text{ and } \kappa_{\alpha}(s) = \frac{1}{3},$$

for all $s \in \mathbb{R}$. Proceeding, we have that:

$$\boldsymbol{B}_{\alpha}(s) = -\boldsymbol{T}_{\alpha}(s) \times_L \boldsymbol{N}_{\alpha}(s) = \frac{-1}{\sqrt{3}}\left(\sinh\frac{s}{\sqrt{3}}, 1, 2\cosh\frac{s}{\sqrt{3}}\right),$$

and then $\tau_{\alpha}(s) = -2/3$, for all $s \in \mathbb{R}$.

It follows from Exercise 2.1.19 (p. 76), that if $\boldsymbol{\beta}$ is a positive reparametrization of $\boldsymbol{\alpha}$, both with unit speed, then $\boldsymbol{\beta}(s) = \boldsymbol{\alpha}(s+a)$, for some $a \in \mathbb{R}$. Hence $\kappa_{\beta}(s) = \kappa_{\alpha}(s+a)$ and $\tau_{\beta}(s) = \tau_{\alpha}(s+a)$ (see Exercise 2.3.4). With this in mind, we now extend the Frenet-Serret apparatus for admissible curves not necessarily having unit speed.

Definition 2.3.8. Let $\boldsymbol{\alpha}\colon I \to \mathbb{R}^3_\nu$ be an admissible curve and s be an arclength function for $\boldsymbol{\alpha}$. Write $\boldsymbol{\alpha}(t) = \tilde{\boldsymbol{\alpha}}(s(t))$. The *tangent, normal and binormal vectors* to $\boldsymbol{\alpha}$ at t are defined by

$$\boldsymbol{T}_{\alpha}(t) \doteq \boldsymbol{T}_{\tilde{\alpha}}(s(t)), \quad \boldsymbol{N}_{\alpha}(t) \doteq \boldsymbol{N}_{\tilde{\alpha}}(s(t)) \quad \text{and} \quad \boldsymbol{B}_{\alpha}(t) \doteq \boldsymbol{B}_{\tilde{\alpha}}(s(t))$$

and, finally, the *curvature* and the *torsion* of $\boldsymbol{\alpha}$ at t are defined by

$$\kappa_{\alpha}(t) \doteq \kappa_{\tilde{\alpha}}(s(t)) \quad \text{and} \quad \tau_{\alpha}(t) \doteq \tau_{\tilde{\alpha}}(s(t)).$$

Proposition 2.3.9. *Let $\boldsymbol{\alpha}\colon I \to \mathbb{R}^3_\nu$ be an admissible curve. Given $t \in I$, the formulas*

$$\kappa_{\boldsymbol{\alpha}}(t) = \frac{\|\boldsymbol{\alpha}'(t) \times \boldsymbol{\alpha}''(t)\|}{\|\boldsymbol{\alpha}'(t)\|^3} \quad \textit{and} \quad \tau_{\boldsymbol{\alpha}}(t) = \frac{\det\left(\boldsymbol{\alpha}'(t), \boldsymbol{\alpha}''(t), \boldsymbol{\alpha}'''(t)\right)}{\|\boldsymbol{\alpha}'(t) \times \boldsymbol{\alpha}''(t)\|^2}$$

hold.

Proof: For the curvature, we need two derivatives. As done in the previous section, we have:

$$\boldsymbol{\alpha}'(t) = s'(t)\boldsymbol{T}_{\boldsymbol{\alpha}}(t) \quad \text{and} \quad \boldsymbol{\alpha}''(t) = s''(t)\boldsymbol{T}_{\boldsymbol{\alpha}}(t) + s'(t)^2\kappa_{\boldsymbol{\alpha}}(t)\boldsymbol{N}_{\boldsymbol{\alpha}}(t).$$

With this, we have

$$\begin{aligned} \boldsymbol{\alpha}'(t) \times \boldsymbol{\alpha}''(t) &= \left(s'(t)\boldsymbol{T}_{\boldsymbol{\alpha}}(t)\right) \times \left(s''(t)\boldsymbol{T}_{\boldsymbol{\alpha}}(t) + s'(t)^2\kappa_{\boldsymbol{\alpha}}(t)\boldsymbol{N}_{\boldsymbol{\alpha}}(t)\right) \\ &= s'(t)^3\kappa_{\boldsymbol{\alpha}}(t)\boldsymbol{T}_{\boldsymbol{\alpha}}(t) \times \boldsymbol{N}_{\boldsymbol{\alpha}}(t). \end{aligned}$$

Recalling that the curvature is always positive, that $\boldsymbol{T}_{\boldsymbol{\alpha}}(t) \times \boldsymbol{N}_{\boldsymbol{\alpha}}(t)$ is a unit vector, and that $s'(t) = \|\boldsymbol{\alpha}'(t)\|$, taking $\|\cdot\|$ of both sides of the above identity allows us to solve for $\kappa_{\boldsymbol{\alpha}}(t)$ and obtain

$$\kappa_{\boldsymbol{\alpha}}(t) = \frac{\|\boldsymbol{\alpha}'(t) \times \boldsymbol{\alpha}''(t)\|}{\|\boldsymbol{\alpha}'(t)\|^3}.$$

As for the torsion, we do not necessarily need $\boldsymbol{\alpha}'''(t)$, but only its component in the direction of $\boldsymbol{B}_{\boldsymbol{\alpha}}(t)$, namely, $s'(t)^3\kappa_{\boldsymbol{\alpha}}(t)\tau_{\boldsymbol{\alpha}}(t)\boldsymbol{B}_{\boldsymbol{\alpha}}(t)$ (verify). This way (already discarding terms due to linear dependences), we have that:

$$\begin{aligned} \det\left(\boldsymbol{\alpha}'(t), \boldsymbol{\alpha}''(t), \boldsymbol{\alpha}'''(t)\right) &= \det\left(s'(t)\boldsymbol{T}_{\boldsymbol{\alpha}}(t), s'(t)^2\kappa_{\boldsymbol{\alpha}}(t)\boldsymbol{N}_{\boldsymbol{\alpha}}(t), s'(t)^3\kappa_{\boldsymbol{\alpha}}(t)\tau_{\boldsymbol{\alpha}}(t)\boldsymbol{B}_{\boldsymbol{\alpha}}(t)\right) \\ &= s'(t)^6\kappa_{\boldsymbol{\alpha}}(t)^2\tau_{\boldsymbol{\alpha}}(t)\det\left(\boldsymbol{T}_{\boldsymbol{\alpha}}(t), \boldsymbol{N}_{\boldsymbol{\alpha}}(t), \boldsymbol{B}_{\boldsymbol{\alpha}}(t)\right) \\ &= s'(t)^6\kappa_{\boldsymbol{\alpha}}(t)^2\tau_{\boldsymbol{\alpha}}(t), \end{aligned}$$

whence

$$\tau_{\boldsymbol{\alpha}}(t) = \frac{\det\left(\boldsymbol{\alpha}'(t), \boldsymbol{\alpha}''(t), \boldsymbol{\alpha}'''(t)\right)}{\|\boldsymbol{\alpha}'(t)\|^6\kappa_{\boldsymbol{\alpha}}(t)^2} = \frac{\det\left(\boldsymbol{\alpha}'(t), \boldsymbol{\alpha}''(t), \boldsymbol{\alpha}'''(t)\right)}{\|\boldsymbol{\alpha}'(t) \times \boldsymbol{\alpha}''(t)\|^2},$$

by the expression previously obtained for $\kappa_{\boldsymbol{\alpha}}(t)$. □

Remark. Note that if we had defined the torsion with the opposite sign, not only would we have an annoying negative sign in the above expression, but also in $\mathbb{L}^3$ we would have to consider the causal type of $\boldsymbol{\alpha}$ as well. In other words, our definition has the advantage of directly incorporating everything in $\tau_{\boldsymbol{\alpha}}(t)$.

Example 2.3.10 (Viviani's window)**.** Consider the curve $\boldsymbol{\alpha} : [0, 4\pi] \to \mathbb{R}^3_\nu$ given by:

$$\boldsymbol{\alpha}(t) = \left(1 + \cos t, \sin t, 2\sin\left(\frac{t}{2}\right)\right).$$

The curve $\boldsymbol{\alpha}$ is called *Viviani's window*. This curve has an interesting property: consider the cylinder over the xy-plane, centered at $(1, 0, 0)$, with radius 1, and the Euclidean sphere of radius 2 centered at the origin. Then the trace of $\boldsymbol{\alpha}$ always lies in the intersection of the cylinder with the sphere.

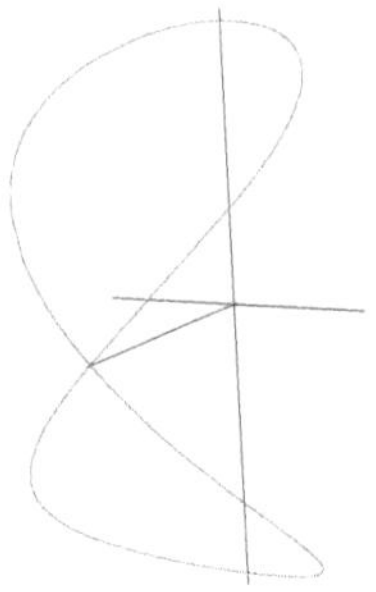

Figure 2.17: The trajectory of $\boldsymbol{\alpha}$.

To determine its curvature and torsion we need the following derivatives:

$$\boldsymbol{\alpha}'(t) = \left(-\sin t, \cos t, \cos\left(\frac{t}{2}\right)\right),$$
$$\boldsymbol{\alpha}''(t) = \left(-\cos t, -\sin t, -\frac{1}{2}\sin\left(\frac{t}{2}\right)\right), \text{ and}$$
$$\boldsymbol{\alpha}'''(t) = \left(\sin t, -\cos t, -\frac{1}{4}\cos\left(\frac{t}{2}\right)\right).$$

Since $\langle \boldsymbol{\alpha}'(t), \boldsymbol{\alpha}'(t)\rangle_L = 1 - \cos^2\left(\frac{t}{2}\right) \geq 0$, we have that $\boldsymbol{\alpha}$ is lightlike at $t = 0, 2\pi$, and 4π, and spacelike elsewhere. To apply the formulas in the previous proposition, we still need to check whether the curve is admissible: the first two components of $\boldsymbol{\alpha}'(t)$ and $\boldsymbol{\alpha}''(t)$ show that they are linearly independent, and Sylvester's Criterion (Theorem 1.2.18, p. 10) shows that $\operatorname{span}\{\boldsymbol{\alpha}'(t), \boldsymbol{\alpha}''(t)\}$ is spacelike for $t \notin \{0, 2\pi, 4\pi\}$. Excluding these points, and using the cross product and norm in each ambient space, we have

$$\kappa_{\alpha,E}(t) = \sqrt{\frac{3\cos t + 13}{(3+\cos t)^3}} \quad \text{and} \quad \tau_{\alpha,E}(t) = \frac{6\cos\left(\frac{t}{2}\right)}{13 + 3\cos t},$$
$$\kappa_{\alpha,L}(t) = \frac{\sqrt{3}}{1-\cos t} \quad \text{and} \quad \tau_{\alpha,L}(t) = \cot\left(\frac{t}{2}\right)\csc\left(\frac{t}{2}\right),$$

where the second subscript index denotes the ambient space where the calculation was done. We may now compare the graphs of the curvatures and torsions:

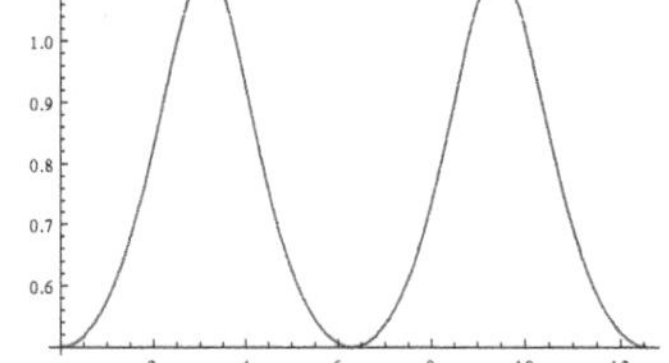

(a) The curvature of $\boldsymbol{\alpha}$ in $\mathbb{R}^3$

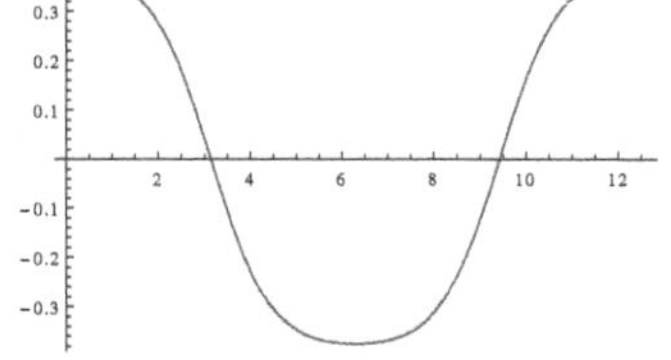

(b) The torsion of $\boldsymbol{\alpha}$ in $\mathbb{R}^3$

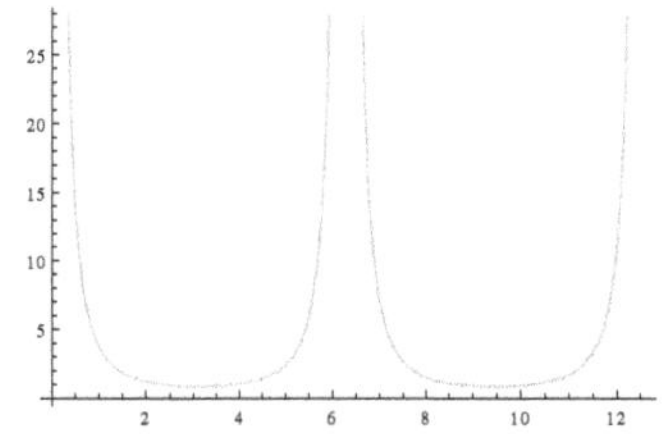

(c) The curvature of $\boldsymbol{\alpha}$ in $\mathbb{L}^3$

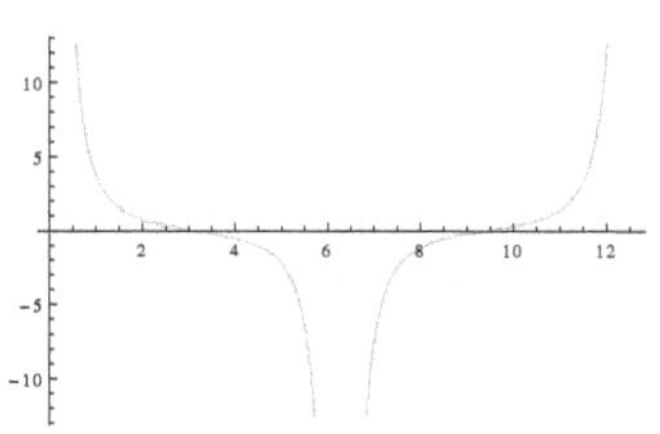

(d) The torsion of $\boldsymbol{\alpha}$ in $\mathbb{L}^3$

Figure 2.18: Curvature and torsion of Viviani's window in $\mathbb{R}^3_\nu$.

Note that, if $t_0 = 0$, 2π, or 4π (where the curve is lightlike), we have that

$$\lim_{t\to t_0} \kappa_{\alpha,L}(t) = \lim_{t\to p} \tau_{\alpha,L}(t) = \pm\infty.$$

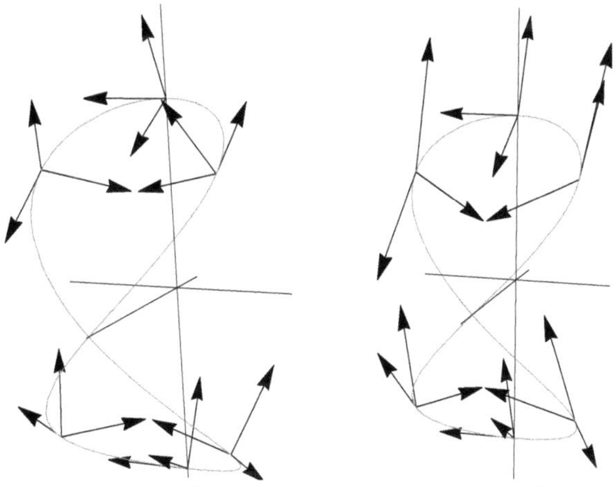

(a) In $\mathbb{R}^3$ (b) In $\mathbb{L}^3$

Figure 2.19: Frenet-Serret frames for Viviani's window in $\mathbb{R}^3_\nu$.

Exercises

Exercise† 2.3.1. Show that the property of being admissible is invariant under reparametrization. More precisely, if $\alpha\colon I \to \mathbb{R}^3_\nu$ is admissible and $h\colon J \to I$ is a diffeomorphism, show that $\alpha \circ h\colon J \to \mathbb{R}^3_\nu$ is also admissible.

Exercise 2.3.2. Consider $\alpha\colon \mathbb{R} \to \mathbb{R}^3$ given by

$$\alpha(t) = \begin{cases} (t, 0, \mathrm{e}^{-1/t^2}), \text{ if } t > 0, \\ (t, \mathrm{e}^{-1/t^2}, 0), \text{ if } t < 0 \text{ and} \\ (0,0,0), \text{if } t = 0. \end{cases}$$

Show that α is regular, but not biregular.

Exercise 2.3.3. We may look at "copies" of $\mathbb{R}^2$ and $\mathbb{L}^2$ inside $\mathbb{L}^3$, by identifying $\mathbb{R}^2$ with the plane xy, and $\mathbb{L}^2$ with the planes xz or yz. Consider the projections $\pi_E, \pi_L\colon \mathbb{R}^3_\nu \to \mathbb{R}^2_\nu$ given by

$$\pi_E(x,y,z) = (x,y) \quad \text{and} \quad \pi_L(x,y,z) = (y,z).$$

Suppose that $\alpha\colon I \to \mathbb{R}^3_\nu$ is a unit speed curve.

(a) Show that if $\alpha(s) = (x(s), y(s), 0)$ is regarded as a curve in $\mathbb{R}^3$, then its curvature satisfies $\kappa_\alpha(s) = |\kappa_{\pi_E\circ\alpha}(s)|$ for all $s \in I$.

(b) Show that if $\alpha(s) = (0, y(s), z(s))$ is regarded as a curve in $\mathbb{L}^3$, then its curvature satisfies $\kappa_\alpha(s) = |\kappa_{\pi_L\circ\alpha}(s)|$ for all $s \in I$.

Exercise† 2.3.4. Verify that the Frenet-Serret Trihedron, as well as curvature and torsion, are invariant under a positive reparametrization between unit speed curves.

Exercise 2.3.5. Show that the curvature and torsion of an admissible curve are invariant under isometries whose linear part preserves the orientation of space.

Exercise 2.3.6. A *Laguerre transformation* in $\mathbb{L}^3$ is a map $H\colon \mathbb{L}^3 \to \mathbb{L}^3$ of the form $H(\boldsymbol{x}) = a\Lambda\boldsymbol{x} + \boldsymbol{v}$ for certain $a \in \mathbb{R} \setminus \{0\}$ and $\boldsymbol{v} \in \mathbb{L}^3$, where $\Lambda \in \mathrm{O}_1(3,\mathbb{R})$ is a Lorentz transformation. If $\boldsymbol{\alpha}\colon I \to \mathbb{L}^3$ is an admissible curve, show that $H \circ \boldsymbol{\alpha}$ is also admissible, and express its curvature and torsion in terms of the curvature and torsion of $\boldsymbol{\alpha}$.

Exercise 2.3.7 ([59])**.** Let $\boldsymbol{\alpha}\colon I \to \mathbb{R}^3$ be a unit speed admissible curve, and for every $n \geq 1$, let $a_n, b_n, c_n\colon I \to \mathbb{R}$ be the functions satisfying

$$\boldsymbol{\alpha}^{(n)}(s) = a_n(s)\boldsymbol{T}_{\boldsymbol{\alpha}}(s) + b_n(s)\boldsymbol{N}_{\boldsymbol{\alpha}}(s) + c_n(s)\boldsymbol{B}_{\boldsymbol{\alpha}}(s),$$

for all $s \in I$. Show that for every $n \geq 1$ the generalized Frenet-Serret equations hold:

$$\begin{pmatrix} a_{n+1}(s) \\ b_{n+1}(s) \\ c_{n+1}(s) \end{pmatrix} = \begin{pmatrix} a_n'(s) \\ b_n'(s) \\ c_n'(s) \end{pmatrix} + \begin{pmatrix} 0 & -\epsilon_{\boldsymbol{\alpha}}\eta_{\boldsymbol{\alpha}}\kappa_{\boldsymbol{\alpha}}(s) & 0 \\ \kappa_{\boldsymbol{\alpha}}(s) & 0 & (-1)^{\nu+1}\epsilon_{\boldsymbol{\alpha}}\tau_{\boldsymbol{\alpha}}(s) \\ 0 & \tau_{\boldsymbol{\alpha}}(s) & 0 \end{pmatrix}\begin{pmatrix} a_n(s) \\ b_n(s) \\ c_n(s) \end{pmatrix}.$$

Exercise[†] 2.3.8. Let $\boldsymbol{\alpha}\colon I \to \mathbb{R}^3_\nu$ be a unit speed admissible curve and $\boldsymbol{\beta}\colon I \to \mathbb{R}^3_\nu$ its *tangent indicatrix*, that is, $\boldsymbol{\beta}(s) = \boldsymbol{T}_{\boldsymbol{\alpha}}(s)$ (which does not necessarily have unit speed). Assuming that $\boldsymbol{\beta}$ is also admissible, show that

(a) $\kappa_{\boldsymbol{\beta}}(s)^2 = \dfrac{|(-1)^\nu\eta_{\boldsymbol{\alpha}}\kappa_{\boldsymbol{\alpha}}(s)^2 + \tau_{\boldsymbol{\alpha}}(s)^2|}{\kappa_{\boldsymbol{\alpha}}(s)^2}$ and;

(b) $\tau_{\boldsymbol{\beta}}(s) = \epsilon_{\boldsymbol{\alpha}}\eta_{\boldsymbol{\alpha}}\dfrac{\kappa_{\boldsymbol{\alpha}}(s)\tau_{\boldsymbol{\alpha}}'(s) - \kappa_{\boldsymbol{\alpha}}'(s)\tau_{\boldsymbol{\alpha}}(s)}{\kappa_{\boldsymbol{\alpha}}(s)\,|(-1)^\nu\eta_{\boldsymbol{\alpha}}\kappa_{\boldsymbol{\alpha}}(s)^2 + \tau_{\boldsymbol{\alpha}}(s)^2|}$

What happens when the tangent indicatrix collapses to a point?

Exercise 2.3.9. Let $\boldsymbol{\alpha}\colon I \to \mathbb{R}^3_\nu$ be a unit speed admissible curve and $\boldsymbol{\beta}\colon I \to \mathbb{R}^3_\nu$ its *binormal indicatrix*, that is, $\boldsymbol{\beta}(s) = \boldsymbol{B}_{\boldsymbol{\alpha}}(s)$ (which does not necessarily have unit speed). Assuming that $\boldsymbol{\beta}$ is also admissible, show that

(a) $\kappa_{\boldsymbol{\beta}}(s)^2 = \dfrac{|(-1)^\nu\eta_{\boldsymbol{\alpha}}\kappa_{\boldsymbol{\alpha}}(s)^2 + \tau_{\boldsymbol{\alpha}}(s)^2|}{\tau_{\boldsymbol{\alpha}}(s)^2}$ and;

(b) $\tau_{\boldsymbol{\beta}}(s) = (-1)^{\nu+1}\eta_{\boldsymbol{\alpha}}\dfrac{\kappa_{\boldsymbol{\alpha}}(s)\tau_{\boldsymbol{\alpha}}'(s) - \kappa_{\boldsymbol{\alpha}}'(s)\tau_{\boldsymbol{\alpha}}(s)}{\tau_{\boldsymbol{\alpha}}(s)\,|(-1)^\nu\eta_{\boldsymbol{\alpha}}\kappa_{\boldsymbol{\alpha}}(s)^2 + \tau_{\boldsymbol{\alpha}}(s)^2|}$.

What happens when the binormal indicatrix degenerates to a point?

Exercise 2.3.10 (Darboux vector)**.** If a rigid body moves along a curve $\boldsymbol{\alpha}$ in $\mathbb{R}^3$ with unit speed, its motion consists of a translation and a rotation, both along the curve. The rotation is described by the so-called *Darboux vector* associated to $\boldsymbol{\alpha}$, denoted by $\boldsymbol{\omega}_{\boldsymbol{\alpha}}$. We will study the situation in both ambient spaces simultaneously. Suppose that $\boldsymbol{\alpha}\colon I \to \mathbb{R}^3_\nu$ is admissible and has unit speed.

(a) Knowing that

$$\boldsymbol{\omega}_{\boldsymbol{\alpha}}(s) \times \boldsymbol{T}_{\boldsymbol{\alpha}}(s) = \boldsymbol{T}_{\boldsymbol{\alpha}}'(s),\ \boldsymbol{\omega}_{\boldsymbol{\alpha}}(s) \times \boldsymbol{N}_{\boldsymbol{\alpha}}(s) = \boldsymbol{N}_{\boldsymbol{\alpha}}'(s) \text{ and } \boldsymbol{\omega}_{\boldsymbol{\alpha}}(s) \times \boldsymbol{B}_{\boldsymbol{\alpha}}(s) = \boldsymbol{B}_{\boldsymbol{\alpha}}'(s),$$

show that

$$\boldsymbol{\omega}_{\boldsymbol{\alpha}}(s) = (-1)^\nu\epsilon_{\boldsymbol{\alpha}}\eta_{\boldsymbol{\alpha}}\tau_{\boldsymbol{\alpha}}(s)\boldsymbol{T}_{\boldsymbol{\alpha}}(s) + \eta_{\boldsymbol{\alpha}}\kappa_{\boldsymbol{\alpha}}(s)\boldsymbol{B}_{\boldsymbol{\alpha}}(s),$$

for all $s \in I$.

(b) In $\mathbb{R}^3$, show that $T'_\alpha(s) \times_E T''_\alpha(s) = \kappa_\alpha(s)^2 \boldsymbol{\omega}_\alpha(s)$, for all $s \in I$.

Hint. Exercise 1.6.4 (p. 59) may be useful.

Exercise 2.3.11. Let $\alpha \colon I \to \mathbb{R}^3_\nu$ be an admissible curve. We could have considered other moving frames along α, according to the following construction: for every $s \in I$, take $U_\alpha(s)$ unit and orthogonal to $T_\alpha(s)$, and then let $V_\alpha(s)$ be the vector making the basis $(T_\alpha(s), U_\alpha(s), V_\alpha(s))$ orthonormal and positive. Assume that the causal type of $U_\alpha(s)$ does not depend on s and let $\epsilon_U \doteq \epsilon_{U_\alpha(s)}$.

(a) Show, as done in the text, that $V_\alpha(s) = (-1)^\nu \epsilon_\alpha \epsilon_U T_\alpha(s) \times U_\alpha(s)$, for all $s \in I$.

(b) Define the *associated curvatures* $\omega_1, \omega_2, \omega_3 \colon I \to \mathbb{R}$ as the functions such that

$$T'_\alpha(s) = \omega_3(s) U_\alpha(s) - \omega_2(s) V_\alpha(s),$$

and $\omega_1(s)$ is the component of $U'_\alpha(s)$ in the direction of $V_\alpha(s)$. Show that

$$\begin{pmatrix} T'_\alpha(s) \\ U'_\alpha(s) \\ V'_\alpha(s) \end{pmatrix} = \begin{pmatrix} 0 & \omega_3(s) & -\omega_2(s) \\ -\epsilon_\alpha \epsilon_U \omega_3(s) & 0 & \omega_1(s) \\ (-1)^\nu \epsilon_U \omega_2(s) & (-1)^{\nu+1} \epsilon_\alpha \omega_1(s) & 0 \end{pmatrix} \begin{pmatrix} T_\alpha(s) \\ U_\alpha(s) \\ V_\alpha(s) \end{pmatrix},$$

for all $s \in I$. Note that in $\mathbb{R}^3$ the coefficient matrix is skew-symmetric.

(c) We may consider the *Darboux vector* $\boldsymbol{\omega}(s)$ *associated to the frame*, satisfying

$$\boldsymbol{\omega}(s) \times T_\alpha(s) = T'_\alpha(s), \quad \boldsymbol{\omega}(s) \times U_\alpha(s) = U'_\alpha(s) \text{ and } \boldsymbol{\omega}(s) \times V_\alpha(s) = V'_\alpha(s)$$

for all $s \in I$. Show that

$$\boldsymbol{\omega}(s) = (-1)^\nu \epsilon_\alpha \epsilon_U \omega_1(s) T_\alpha(s) + (-1)^\nu \epsilon_\alpha \epsilon_U \omega_2(s) U_\alpha(s) + \epsilon_U \omega_3(s) V_\alpha(s).$$

Remark. Note that in $\mathbb{R}^3$ the above expression becomes simply

$$\boldsymbol{\omega}(s) = \omega_1(s) T_\alpha(s) + \omega_2(s) U_\alpha(s) + \omega_3(s) V_\alpha(s),$$

which explains the seemingly random indexing of the associated curvatures.

Exercise 2.3.12. Show that for a cylindrical curve $\alpha \colon I \to \mathbb{R}^3_\nu$, given in the form $\alpha(t) = (\cos t, \sin t, z(t))$, we have, whenever it makes sense, that:

$$\kappa_\alpha(t) = \frac{\left|(-1)^\nu + z'(t)^2 + z''(t)^2\right|^{1/2}}{\left|1 + (-1)^\nu z'(t)^2\right|^{3/2}} \quad \text{and} \quad \tau_\alpha(t) = \frac{z'(t) + z'''(t)}{\left|(-1)^\nu + z'(t)^2 + z''(t)^2\right|}.$$

Exercise 2.3.13. Show that for a Lorentzian cylindrical curve $\alpha \colon I \to \mathbb{L}^3$ of the form $\alpha(t) = (x(t), \cosh t, \sinh t)$ we have, whenever it makes sense, that:

$$\kappa_\alpha(t) = \frac{\left|1 - x'(t)^2 + x''(t)^2\right|^{1/2}}{\left|-1 + x'(t)^2\right|^{3/2}} \quad \text{and} \quad \tau_\alpha(t) = \frac{x'(t) - x'''(t)}{\left|1 - x'(t)^2 + x''(t)^2\right|}.$$

Exercise 2.3.14. Let $\alpha \colon I \to \mathbb{R}^3_\nu$ be an admissible curve. Denote by $\tau_{\alpha,E}$ and $\tau_{\alpha,L}$ the torsions of α when seen in $\mathbb{R}^3$ and $\mathbb{L}^3$, respectively. Show that $|\tau_{\alpha,E}(t)| \leq |\tau_{\alpha,L}(t)|$ for all $t \in I$, and give an example showing that we cannot remove the absolute values. Also try to investigate if there is any relation between the curvatures in both ambient spaces in this context.

2.3.2 Geometric effects of curvature and torsion

With the Frenet-Serret frame for admissible curves in hand, we may now obtain geometric information about such curves in $\mathbb{R}^3_\nu$. We'll begin with a geometric interpretation for torsion:

Proposition 2.3.11. *Let $\boldsymbol{\alpha}: I \to \mathbb{R}^3_\nu$ be an admissible curve. Then the trace of $\boldsymbol{\alpha}$ is contained in a plane if and only if $\tau_{\boldsymbol{\alpha}}(t) = 0$, for all $t \in I$.*

Proof: We may assume, without loss of generality, that $\boldsymbol{\alpha}$ has unit speed. Moreover, note that if $\boldsymbol{\alpha}$ is admissible then the plane spanned by $\boldsymbol{T}_{\boldsymbol{\alpha}}(s)$ and $\boldsymbol{N}_{\boldsymbol{\alpha}}(s)$ is non-degenerate for all $s \in I$ and thus the binormal vector $\boldsymbol{B}_{\boldsymbol{\alpha}}(s)$ cannot be lightlike. If there is a plane Π in $\mathbb{R}^3_\nu$ containing the trace of $\boldsymbol{\alpha}$, then the plane spanned by $\boldsymbol{T}_{\boldsymbol{\alpha}}(s)$ and $\boldsymbol{N}_{\boldsymbol{\alpha}}(s)$ passing through $\boldsymbol{\alpha}(s)$ does not depend on s and is, in fact, the plane Π. Indeed, let $\boldsymbol{v} \in \mathbb{R}^3_\nu$ be a unit vector normal to Π. Fixed $s_0 \in I$, we have that $\langle \boldsymbol{\alpha}(s) - \boldsymbol{\alpha}(s_0), \boldsymbol{v}\rangle = 0$, for all $s \in I$. Repeatedly differentiating such expression we have that

$$\langle \boldsymbol{T}_{\boldsymbol{\alpha}}(s), \boldsymbol{v}\rangle = 0 \quad \text{and} \quad \langle \boldsymbol{T}'_{\boldsymbol{\alpha}}(s), \boldsymbol{v}\rangle = \kappa_{\boldsymbol{\alpha}}(s)\langle \boldsymbol{N}_{\boldsymbol{\alpha}}(s), \boldsymbol{v}\rangle = 0.$$

Thus $\boldsymbol{v}$ and $\boldsymbol{B}_{\boldsymbol{\alpha}}(s)$ are collinear for all $s \in I$ and so $\boldsymbol{B}_{\boldsymbol{\alpha}}(s) = \pm \boldsymbol{v}$ is a constant. From the last Frenet-Serret equation, it follows that

$$\boldsymbol{0} = \boldsymbol{B}'_{\boldsymbol{\alpha}}(s) = (-1)^{\nu+1}\epsilon_{\boldsymbol{\alpha}}\tau_{\boldsymbol{\alpha}}(s)\boldsymbol{N}_{\boldsymbol{\alpha}}(s).$$

Then $\tau_{\boldsymbol{\alpha}}(s) = 0$ for all $s \in I$.

Conversely, if $\tau_{\boldsymbol{\alpha}}(s) = 0$ for all $s \in I$, then $\boldsymbol{B}_{\boldsymbol{\alpha}}(s) = \boldsymbol{B}$. Fix $s_0 \in I$ and consider the function $f\colon I \to \mathbb{R}$ given by $f(s) = \langle \boldsymbol{\alpha}(s) - \boldsymbol{\alpha}(s_0), \boldsymbol{B}\rangle$. Geometrically, this function measures how much $\boldsymbol{\alpha}$ deviates from the plane orthogonal to $\boldsymbol{B}$ that passes through $\boldsymbol{\alpha}(s_0)$. Note that $f(s_0) = 0$ and $f'(s) = \langle \boldsymbol{T}_{\boldsymbol{\alpha}}(s), \boldsymbol{B}\rangle = 0$, whence f identically vanishes and we conclude that the trace of $\boldsymbol{\alpha}$ is contained in such a plane. □

In other words, just like the curvature measures how much the curve deviates from being a straight line, the torsion measures how much the curve deviates from being planar.

Remark. The admissibility assumption is essential. Indeed, if the curve is not admissible, it is possible to extend the definition of torsion for this case, in such a way that the curve is not planar and has zero torsion (see, for example, the curve in Exercise 2.3.2, p. 103).

Recall that in $\mathbb{R}^2_\nu$ the curves with non-zero constant curvature were congruent to $\mathbb{S}^1(\boldsymbol{p}, r)$, $\mathbb{H}^1_\pm(\boldsymbol{p}, r)$ or $\mathbb{S}^1_1(\boldsymbol{p}, r)$. Identifying $\mathbb{R}^2_\nu$ with the planes xy and yz, according to whether $\nu = 0$ or 1 (see again Exercise 2.3.3, p. 103), we have the following result for planar curves in space:

Proposition 2.3.12. *Let $\boldsymbol{\alpha}: I \to \mathbb{R}^3_\nu$ be an admissible curve. Then $\kappa_{\boldsymbol{\alpha}}(s) = \kappa$ is constant and $\tau_{\boldsymbol{\alpha}} = 0$ if and only if $\boldsymbol{\alpha}$ is congruent to a piece of $\mathbb{S}^1(1/\kappa)$, $\mathbb{H}^1_\pm(1/\kappa)$ or $\mathbb{S}^1_1(1/\kappa)$.*

Proof: If $\boldsymbol{\alpha}$ is congruent to a piece of one of the curves listed above, it follows from Exercises 2.3.3 and 2.3.5, that $\boldsymbol{\alpha}$ is planar and has non-zero constant curvature. Conversely, consider $\boldsymbol{c}: I \to \mathbb{R}$ given by

$$\boldsymbol{c}(s) = \boldsymbol{\alpha}(s) + \frac{\epsilon_{\boldsymbol{\alpha}}\eta_{\boldsymbol{\alpha}}}{\kappa}\boldsymbol{N}_{\boldsymbol{\alpha}}(s).$$

Since $\tau_{\boldsymbol{\alpha}}(s) = 0$, we have that

$$\boldsymbol{c}'(s) = \boldsymbol{T}_{\boldsymbol{\alpha}}(s) + \frac{\epsilon_{\boldsymbol{\alpha}}\eta_{\boldsymbol{\alpha}}}{\kappa}(-\epsilon_{\boldsymbol{\alpha}}\eta_{\boldsymbol{\alpha}}\kappa\boldsymbol{T}_{\boldsymbol{\alpha}}(s)) = \boldsymbol{0},$$

whence $c(s) = p$, for all $s \in I$. It follows that

$$\langle \alpha(s) - p, \alpha(s) - p \rangle = \frac{\eta_\alpha}{\kappa^2}.$$

In $\mathbb{R}^3$ or if the trace of α is contained in a spacelike plane in $\mathbb{L}^3$, this means that α is congruent to a piece of a Euclidean circle of radius $1/\kappa$ on the plane xy. If the trace of α is contained in a timelike plane in $\mathbb{L}^3$, η_α says whether α is congruent to a piece of $\mathbb{H}^1_\pm(1/\kappa)$ or $\mathbb{S}^1_1(1/\kappa)$ in the plane yz. □

Remark. An observer with "Euclidean eyes" (such as ourselves) sees a Euclidean circle in a non-horizontal spacelike plane in $\mathbb{L}^3$ as an ellipse.

Aiming to give another geometric interpretation for the torsion of an admissible curve $\alpha\colon I \to \mathbb{R}^3_\nu$, let's consider the following order 3 Taylor expansion, assuming that $0 \in I$ and $\alpha(0) = \mathbf{0}$:

$$\begin{aligned}\alpha(s) &= s\alpha'(0) + \frac{s^2}{2}\alpha''(0) + \frac{s^3}{6}\alpha'''(0) + R(s)\\ &= \left(s - \frac{s^3}{6}\epsilon_\alpha\eta_\alpha\kappa_\alpha(0)^2\right)T_\alpha(0) + \left(\frac{s^2}{2}\kappa_\alpha(0) + \frac{s^3}{6}\kappa_\alpha'(0)\right)N_\alpha(0)\\ &\quad + \frac{s^3}{6}\kappa_\alpha(0)\tau_\alpha(0)B_\alpha(0) + R(s),\end{aligned}$$

where $R(s)/s^3 \to \mathbf{0}$ as $s \to 0$. Before proceeding, a bit more terminology is in order:

Definition 2.3.13. Let $\alpha : I \to \mathbb{R}^3_\nu$ be an admissible curve and $t_0 \in I$.

(i) The *osculating plane* to α at t_0 is the plane spanned by $T_\alpha(t_0)$ and $N_\alpha(t_0)$, passing through $\alpha(t_0)$.

(ii) The *normal plane* to α at t_0 is the plane spanned by $N_\alpha(t_0)$ and $B_\alpha(t_0)$, passing through $\alpha(t_0)$.

(iii) The *rectifying plane* to α at t_0 is the plane spanned by $T_\alpha(t_0)$ and $B_\alpha(t_0)$, passing through $\alpha(t_0)$.

The coordinates of $\alpha(s) - R(s)$ relative to the frame $\mathcal{F} = \big(T_\alpha(0), N_\alpha(0), B_\alpha(0)\big)$ are

$$\alpha(s) - R(s) = \left(s - \frac{s^3}{6}\epsilon_\alpha\eta_\alpha\kappa_\alpha(0)^2, \frac{s^2}{2}\kappa_\alpha(0) + \frac{s^3}{6}\kappa_\alpha'(0), \frac{s^3}{6}\kappa_\alpha(0)\tau_\alpha(0)\right)_{\mathcal{F}}.$$

For s small enough, we then consider the projections of $\alpha(s)$ onto the normal and rectifying planes:

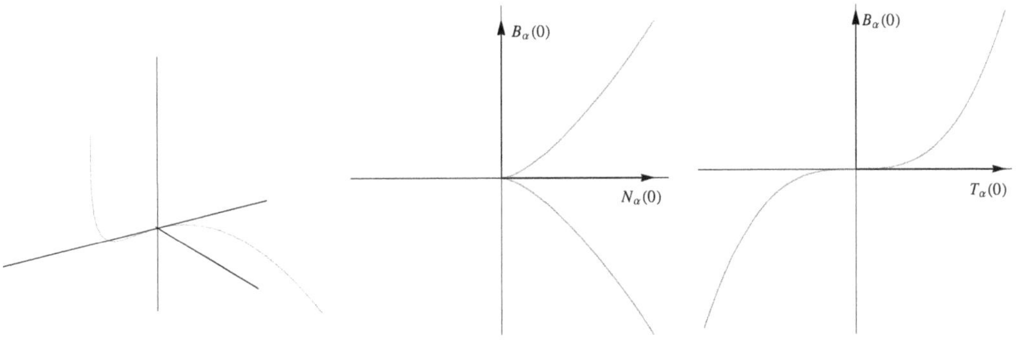

(a) The curve α (b) The normal plane (c) The rectifying plane

Figure 2.20: An interpretation for $\tau_\alpha(0) > 0$.

That is, when $\tau_\alpha(0) > 0$ we have that α crosses the osculating plane at $\alpha(0)$ in the same direction as the vector $B_\alpha(0)$ points to, and in the opposite direction if $\tau_\alpha(0) < 0$.

For the next result, we'll introduce the analogues in $\mathbb{L}^3$ of the Euclidean sphere, as the locus of points which are (Lorentzian) equidistant with a given center p.

Definition 2.3.14. Let $p \in \mathbb{L}^3$ and $r > 0$.

(i) The *de Sitter space* of center p and radius r is the set

$$\mathbb{S}_1^2(p,r) \doteq \{x \in \mathbb{L}^3 \mid \langle x-p, x-p\rangle_L = r^2\}.$$

(ii) The *hyperbolic plane* of center p and radius r is the set

$$\begin{aligned}\mathbb{H}^2(p,r) &\equiv \mathbb{H}_+^2(p,r)\\ &\doteq \{x \in \mathbb{L}^3 \mid \langle x-p, x-p\rangle_L = -r^2 \text{ and } \langle x-p, e_3\rangle_L < 0\}.\end{aligned}$$

As in the previous section, we also have

$$\mathbb{H}_-^2(p,r) \doteq \{x \in \mathbb{L}^3 \mid \langle x-p, x-p\rangle_L = -r^2 \text{ and } \langle x-p, e_3\rangle_L > 0\}.$$

Remark. In order to keep consistency with the notation adopted in the previous section, the center or radius will be omitted according to whether $p = \mathbf{0}$ or $r = 1$. The name *hyperbolic plane* will also be used in Section 4.1, when we discuss some models for Hyperbolic Geometry.

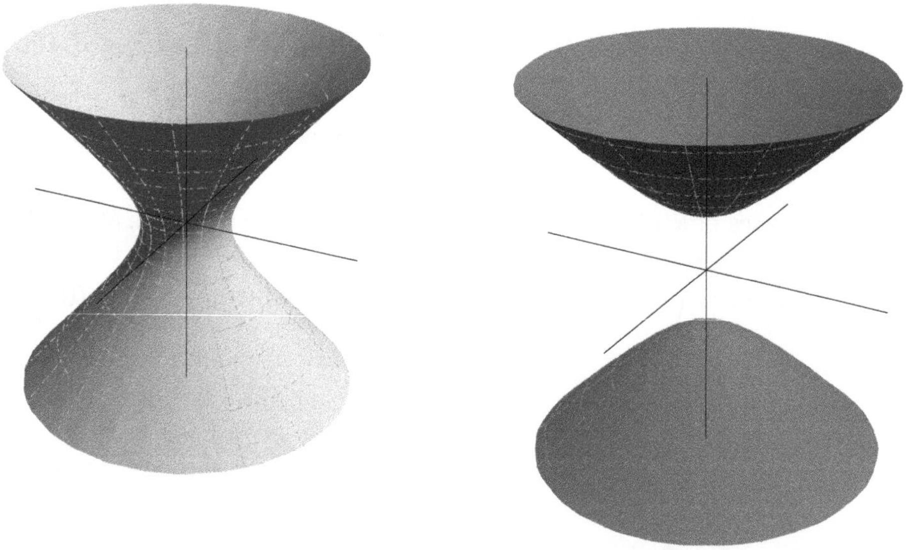

(a) The de Sitter space $\mathbb{S}_1^2$ (b) The two-sheeted hyperboloid $\mathbb{H}^2 \cup \mathbb{H}_-^2$

Figure 2.21: The analogues to Euclidean spheres in $\mathbb{L}^3$.

Theorem 2.3.15. *Let $\alpha\colon I \to \mathbb{R}_\nu^3$ be a unit speed admissible curve with non-vanishing torsion. If the trace of α is contained in $\mathbb{S}^2(p,r)$, $\mathbb{S}_1^2(p,r)$, $\mathbb{H}_\pm^2(p,r)$, or $C_L(p)$, for certain $p \in \mathbb{R}_\nu^3$ and $r > 0$, then we have that*

$$\alpha(s) - p = \frac{-\epsilon_\alpha \eta_\alpha}{\kappa_\alpha(s)} N_\alpha(s) + (-1)^{\nu+1} \frac{\eta_\alpha}{\tau_\alpha(s)} \left(\frac{1}{\kappa_\alpha(s)}\right)' B_\alpha(s)$$

and also

$$\pm r^2 = \frac{\eta_\alpha}{\kappa_\alpha(s)^2} + (-1)^\nu \epsilon_\alpha \eta_\alpha \left(\frac{1}{\tau_\alpha(s)} \left(\frac{1}{\kappa_\alpha(s)} \right)' \right)^2,$$

for all $s \in I$. On the other hand, if this last relation holds and $\kappa_\alpha(s)$ is non-constant, there is $\boldsymbol{p} \in \mathbb{R}^3_\nu$ such that the trace of $\boldsymbol{\alpha}$ is contained in $\mathbb{S}^2(\boldsymbol{p}, r)$, $\mathbb{S}^2_1(\boldsymbol{p}, r)$, $\mathbb{H}^2_\pm(\boldsymbol{p}, r)$ or $C_L(\boldsymbol{p})$.

Proof: Let's assume without loss of generality that $\boldsymbol{p} = \boldsymbol{0}$. By orthonormal expansion, we have that

$$\boldsymbol{\alpha}(s) = \epsilon_\alpha \langle \boldsymbol{\alpha}(s), \boldsymbol{T}_\alpha(s) \rangle \boldsymbol{T}_\alpha(s) + \eta_\alpha \langle \boldsymbol{\alpha}(s), \boldsymbol{N}_\alpha(s) \rangle \boldsymbol{N}_\alpha(s) + (-1)^\nu \epsilon_\alpha \eta_\alpha \langle \boldsymbol{\alpha}(s), \boldsymbol{B}_\alpha(s) \rangle \boldsymbol{B}_\alpha(s),$$

for all $s \in I$, so that our task is finding out those coefficients. Well, since $\langle \boldsymbol{\alpha}(s), \boldsymbol{\alpha}(s) \rangle$ is constant, it follows that $\langle \boldsymbol{\alpha}(s), \boldsymbol{T}_\alpha(s) \rangle = 0$. Differentiating that again we obtain

$$\langle \boldsymbol{T}_\alpha(s), \boldsymbol{T}_\alpha(s) \rangle + \langle \boldsymbol{\alpha}(s), \boldsymbol{T}_\alpha'(s) \rangle = 0 \implies \langle \boldsymbol{\alpha}(s), \boldsymbol{N}_\alpha(s) \rangle = -\frac{\epsilon_\alpha}{\kappa_\alpha(s)},$$

by the first Frenet-Serret equation. Differentiating one last time and using the second Frenet-Serret equation together with the fact that $\boldsymbol{T}_\alpha(s)$ and $\boldsymbol{N}_\alpha(s)$ are orthogonal and the relation $\langle \boldsymbol{\alpha}(s), \boldsymbol{T}_\alpha(s) \rangle = 0$ already obtained, it follows that:

$$\langle \boldsymbol{\alpha}(s), \boldsymbol{B}_\alpha(s) \rangle = -\frac{\epsilon_\alpha}{\tau_\alpha(s)} \left(\frac{1}{\kappa_\alpha(s)} \right)'.$$

Putting everything together, we finally obtain

$$\boldsymbol{\alpha}(s) = \frac{-\epsilon_\alpha \eta_\alpha}{\kappa_\alpha(s)} \boldsymbol{N}_\alpha(s) + (-1)^{\nu+1} \frac{\eta_\alpha}{\tau_\alpha(s)} \left(\frac{1}{\kappa_\alpha(s)} \right)' \boldsymbol{B}_\alpha(s),$$

and thus

$$\langle \boldsymbol{\alpha}(s), \boldsymbol{\alpha}(s) \rangle = \frac{\eta_\alpha}{\kappa_\alpha(s)^2} + (-1)^\nu \epsilon_\alpha \eta_\alpha \left(\frac{1}{\tau_\alpha(s)} \left(\frac{1}{\kappa_\alpha(s)} \right)' \right)^2,$$

as wanted.

As for the converse, let's see that the natural candidate to center is, indeed, constant. We have:

$$\begin{aligned}
&\frac{\mathrm{d}}{\mathrm{d}s} \left(\boldsymbol{\alpha}(s) + \frac{\epsilon_\alpha \eta_\alpha}{\kappa_\alpha(s)} \boldsymbol{N}_\alpha(s) + (-1)^\nu \frac{\eta_\alpha}{\tau_\alpha(s)} \left(\frac{1}{\kappa_\alpha(s)} \right)' \boldsymbol{B}_\alpha(s) \right) = \\
&= \boldsymbol{T}_\alpha(s) + \epsilon_\alpha \eta_\alpha \left(\frac{1}{\kappa_\alpha(s)} \right)' \boldsymbol{N}_\alpha(s) + \frac{\epsilon_\alpha \eta_\alpha}{\kappa_\alpha(s)} (-\epsilon_\alpha \eta_\alpha \kappa_\alpha(s) \boldsymbol{T}_\alpha(s) + \tau_\alpha(s) \boldsymbol{B}_\alpha(s)) \\
&\quad + (-1)^\nu \eta_\alpha \left(\frac{1}{\tau_\alpha(s)} \left(\frac{1}{\kappa_\alpha(s)} \right)' \right)' \boldsymbol{B}_\alpha(s) + (-1)^\nu \frac{\eta_\alpha}{\tau_\alpha(s)} \left(\frac{1}{\kappa_\alpha(s)} \right)' (-1)^{\nu+1} \epsilon_\alpha \tau_\alpha(s) \boldsymbol{N}_\alpha(s) \\
&= \eta_\alpha \left(\frac{\epsilon_\alpha \tau_\alpha(s)}{\kappa_\alpha(s)} + (-1)^\nu \left(\frac{1}{\tau_\alpha(s)} \left(\frac{1}{\kappa_\alpha(s)} \right)' \right)' \right) \boldsymbol{B}_\alpha(s).
\end{aligned}$$

Now, it remains to use the assumption to conclude that the coefficient of $\boldsymbol{B}_\alpha(s)$ in the above vanishes, which happens if and only if the term inside parenthesis vanishes. To wit, setting $\rho(s) \doteq \kappa_\alpha(s)^{-1}$ and $\sigma(s) \doteq \tau_\alpha(s)^{-1}$ to simplify the notation, we have that this term equals

$$\frac{\epsilon_\alpha \rho(s)}{\sigma(s)} + (-1)^\nu(\sigma'(s)\rho'(s) + \sigma(s)\rho''(s)).$$

On the other hand, differentiating the constant expression from the assumption gives us that

$$\begin{aligned} 0 &= \frac{\mathrm{d}}{\mathrm{d}s}\left(\eta_\alpha \rho(s)^2 + (-1)^\nu \epsilon_\alpha \eta_\alpha (\sigma(s)\rho'(s))^2\right) \\ &= 2\epsilon_\alpha \eta_\alpha \rho'(s)\sigma(s)\left(\frac{\epsilon_\alpha \rho(s)}{\sigma(s)} + (-1)^\nu(\sigma'(s)\rho'(s) + \sigma(s)\rho''(s))\right). \end{aligned}$$

Since both $\sigma(s)$ and $\rho'(s)$ are non-zero, the desired coefficient vanishes and thus

$$\boldsymbol{\alpha}(s) - \boldsymbol{p} = \frac{-\epsilon_\alpha \eta_\alpha}{\kappa_\alpha(s)} \boldsymbol{N}_\alpha(s) + (-1)^{\nu+1} \frac{\eta_\alpha}{\tau_\alpha(s)} \left(\frac{1}{\kappa_\alpha(s)}\right)' \boldsymbol{B}_\alpha(s)$$

for some $\boldsymbol{p} \in \mathbb{R}^3_\nu$. So,

$$\langle \boldsymbol{\alpha}(s) - \boldsymbol{p}, \boldsymbol{\alpha}(s) - \boldsymbol{p} \rangle = \frac{\eta_\alpha}{\kappa_\alpha(s)^2} + (-1)^\nu \epsilon_\alpha \eta_\alpha \left(\frac{1}{\tau_\alpha(s)}\left(\frac{1}{\kappa_\alpha(s)}\right)'\right)^2$$

is constant, as wanted. □

Before delivering what we have promised for this section, the Fundamental Theorem of Curves in Space, we'll give one last example of application of the theory of curves developed so far, studying a very important class of curves:

Definition 2.3.16 (Helices). A unit speed admissible curve $\boldsymbol{\alpha} : I \to \mathbb{R}^3_\nu$ is a *helix* if there is a vector $\boldsymbol{v} \in \mathbb{R}^3_\nu$, $\boldsymbol{v} \neq \boldsymbol{0}$, such that $\langle \boldsymbol{T}_\alpha(s), \boldsymbol{v} \rangle$ is a constant. Moreover, in $\mathbb{L}^3$, we'll say that the helix is:

(i) *hyperbolic*, if $\boldsymbol{v}$ is spacelike;

(ii) *elliptic*, if $\boldsymbol{v}$ is timelike;

(iii) *parabolic*, if $\boldsymbol{v}$ is lightlike.

The direction determined by the vector $\boldsymbol{v}$ is called the *helical axis* of $\boldsymbol{\alpha}$.

Remark.

- In $\mathbb{R}^3$ we may assume that $\boldsymbol{v}$ is a unit vector, and thus $\langle \boldsymbol{T}_\alpha(s), \boldsymbol{v} \rangle_E = \cos\theta(s)$ says that for a helix, the angle formed between $\boldsymbol{T}_\alpha(s)$ and $\boldsymbol{v}$ is constant.
- In general, differentiating the identity $\langle \boldsymbol{T}_\alpha(s), \boldsymbol{v} \rangle = \text{cte.}$ says that the acceleration vector of $\boldsymbol{\alpha}$ always lies in the orthogonal plane to $\boldsymbol{v}$.

We may characterize helices in terms of the ratio $\tau_\alpha/\kappa_\alpha$:

Theorem 2.3.17 (Lancret). *Let $\boldsymbol{\alpha} : I \to \mathbb{R}^3_\nu$ be a unit speed admissible curve. Then $\boldsymbol{\alpha}$ is a helix if and only if the ratio $\tau_\alpha(s)/\kappa_\alpha(s)$ is constant.*

Proof: Suppose that $\boldsymbol{\alpha}$ is a helix and that the helical axis is determined by $\boldsymbol{v}$. Setting $c \doteq \langle \boldsymbol{T}_\alpha(s), \boldsymbol{v}\rangle$ we have

$$\langle \kappa_\alpha(s)\boldsymbol{N}_\alpha(s), \boldsymbol{v}\rangle = 0 \implies \langle \boldsymbol{N}_\alpha(s), \boldsymbol{v}\rangle = 0,$$

since $\kappa_\alpha(s) \neq 0$. Differentiating again, we have

$$-\epsilon_\alpha \eta_\alpha \kappa_\alpha(s) c + \tau_\alpha(s)\langle \boldsymbol{B}_\alpha(s), \boldsymbol{v}\rangle = 0.$$

Then it suffices to verify that $\langle \boldsymbol{B}_\alpha(s), \boldsymbol{v}\rangle$ is a non-zero constant. To wit,

$$\frac{\mathrm{d}}{\mathrm{d}s}\langle \boldsymbol{B}_\alpha(s), \boldsymbol{v}\rangle = (-1)^{\nu+1}\epsilon_\alpha \tau_\alpha(s)\langle \boldsymbol{N}_\alpha(s), \boldsymbol{v}\rangle = 0.$$

If $\langle \boldsymbol{B}_\alpha(s), \boldsymbol{v}\rangle = 0$ for all s, it follows that $c = 0$, and by orthonormal expansion we conclude that $\boldsymbol{v} = \boldsymbol{0}$, contradicting the definition of a helix. Hence $\tau_\alpha(s)/\kappa_\alpha(s)$ is a constant.

Conversely, suppose that $\tau_\alpha(s) = c\kappa_\alpha(s)$, for some $c \in \mathbb{R}$. If $c = 0$, then the curve is planar and $\boldsymbol{B}_\alpha(s) = \boldsymbol{B}$ may be taken to be the vector $\boldsymbol{v}$ we seek. If $c \neq 0$, we seek a constant vector

$$\boldsymbol{v} = v_1(s)\boldsymbol{T}_\alpha(s) + v_2(s)\boldsymbol{N}_\alpha(s) + v_3(s)\boldsymbol{B}_\alpha(s)$$

such that $\langle \boldsymbol{T}_\alpha(s), \boldsymbol{v}\rangle$ is also a constant. Such condition is equivalent to $v_1(s) = v_1$ being a constant. Differentiating the expression for $\boldsymbol{v}$ yields

$$\begin{aligned}\boldsymbol{0} = &-\epsilon_\alpha \eta_\alpha \kappa_\alpha(s) v_2(s)\boldsymbol{T}_\alpha(s)\\ &+ \left(v_1\kappa_\alpha(s) + v_2'(s) + (-1)^{\nu+1}\epsilon_\alpha c\kappa_\alpha(s)v_3(s)\right)\boldsymbol{N}_\alpha(s)\\ &+ \left(c\kappa_\alpha(s)v_2(s) + v_3'(s)\right)\boldsymbol{B}_\alpha(s).\end{aligned}$$

By linear independence, we have

$$\begin{cases} 0 = -\epsilon_\alpha \eta_\alpha \kappa_\alpha(s) v_2(s) \\ 0 = v_1\kappa_\alpha(s) + v_2'(s) + (-1)^{\nu+1}\epsilon_\alpha c\kappa_\alpha(s)v_3(s), \\ 0 = c\kappa_\alpha(s)v_2(s) + v_3'(s), \end{cases}$$

and thus

$$v_2(s) = 0 \quad \text{and} \quad v_3(s) = \frac{(-1)^\nu}{c}\epsilon_\alpha v_1.$$

This means that we have found a parametrization for the helical axis of $\boldsymbol{\alpha}$, with v_1 as the real parameter. For concreteness, we may set $v_1 = 1$ and take

$$\boldsymbol{v} \doteq \boldsymbol{T}_\alpha(s) + \frac{(-1)^\nu}{c}\epsilon_\alpha \boldsymbol{B}_\alpha(s).$$

This satisfies our requirements. □

Remark.

- In particular, the proof above shows the existence of exactly one helical axis for each helix.
- If $\boldsymbol{\alpha}$ is a parabolic helix, then $\tau_\alpha(s) = \pm\kappa_\alpha(s)$. The converse holds if $\eta_\alpha = 1$.

An alternative way to express helices is in terms of their tangent indicatrices:

Definition 2.3.18. Let $\boldsymbol{\alpha} : I \to \mathbb{R}^3_\nu$ be an admissible curve. Its *tangent indicatrix* is the curve $\boldsymbol{\beta} : I \to \mathbb{R}^3_\nu$, given by $\boldsymbol{\beta}(t) = \boldsymbol{T}_{\boldsymbol{\alpha}}(t)$.

Remark.

- Note that, even when $\boldsymbol{\alpha}$ has unit speed, we cannot say the same about its tangent indicatrix.
- Similarly, one may define the normal indicatrix and the binormal indicatrix of a curve.

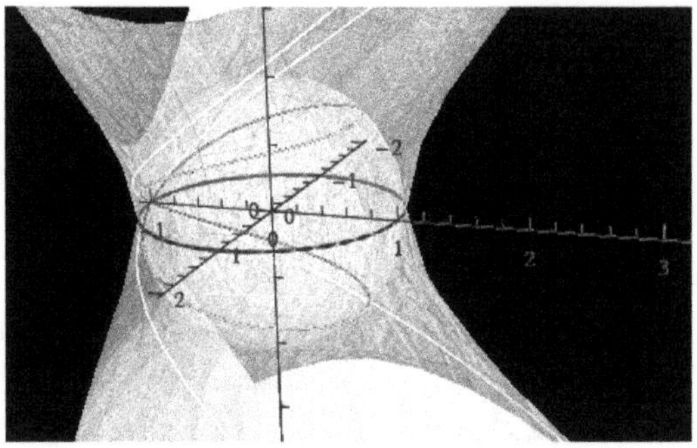

Figure 2.22: The tangent indicatrices of Viviani's window in both ambient spaces.

With this definition, Exercise 2.3.8 (p. 104) gives us the:

Proposition 2.3.19. *Let $\boldsymbol{\alpha} : I \to \mathbb{R}^3_\nu$ be a unit speed admissible curve. Then $\boldsymbol{\alpha}$ is a helix if and only if its tangent indicatrix has constant curvature.*

Remark. It also follows from Exercise 2.3.8 that the curvature of any planar tangent indicatrix is constant.

Theorem 2.3.20 (Fundamental Theorem for Admissible Curves)**.** *Let $\kappa, \tau \colon I \to \mathbb{R}$ be given continuous functions, with $\kappa \geq 0$, $\boldsymbol{p}_0 \in \mathbb{R}^3_\nu$, $s_0 \in I$, and $(\boldsymbol{T}_0, \boldsymbol{N}_0, \boldsymbol{B}_0)$ be a positive orthonormal basis for $\mathbb{R}^3_\nu$. There is a unique unit speed admissible curve $\boldsymbol{\alpha} : I \to \mathbb{R}^3_\nu$ such that:*

- $\boldsymbol{\alpha}(s_0) = \boldsymbol{p}_0$;
- $(\boldsymbol{T}_{\boldsymbol{\alpha}}(s_0), \boldsymbol{N}_{\boldsymbol{\alpha}}(s_0), \boldsymbol{B}_{\boldsymbol{\alpha}}(s_0)) = (\boldsymbol{T}_0, \boldsymbol{N}_0, \boldsymbol{B}_0)$;
- $\kappa_{\boldsymbol{\alpha}}(s) = \kappa(s)$ *and* $\tau_{\boldsymbol{\alpha}}(s) = \tau(s)$, *for all* $s \in I$.

Proof: Consider the following Initial Value Problem (IVP) in $\mathbb{R}^9$:

$$\begin{cases} \begin{pmatrix} \boldsymbol{T}'(s) \\ \boldsymbol{N}'(s) \\ \boldsymbol{B}'(s) \end{pmatrix} = \begin{pmatrix} 0 & \kappa(s) & 0 \\ -\epsilon_{\boldsymbol{T}_0}\epsilon_{\boldsymbol{N}_0}\kappa(s) & 0 & \tau(s) \\ 0 & (-1)^{\nu+1}\epsilon_{\boldsymbol{T}_0}\tau(s) & 0 \end{pmatrix} \begin{pmatrix} \boldsymbol{T}(s) \\ \boldsymbol{N}(s) \\ \boldsymbol{B}(s) \end{pmatrix} \\ \text{and} \quad (\boldsymbol{T}(s_0), \boldsymbol{N}(s_0), \boldsymbol{B}(s_0)) = (\boldsymbol{T}_0, \boldsymbol{N}_0, \boldsymbol{B}_0). \end{cases}$$

By the theorem of existence and uniqueness of solutions to systems of ordinary differential equations, there is a unique solution $(\boldsymbol{T}(s), \boldsymbol{N}(s), \boldsymbol{B}(s))$ of this IVP. We claim that such a solution forms a positive orthonormal basis for $\mathbb{R}^3_\nu$ for all $s \in I$ (and not only for s_0). Indeed, we now consider a second IVP for the curve $\boldsymbol{a} : I \to \mathbb{R}^6$:

$$\begin{cases} \boldsymbol{a}'(s) &= A(s)\boldsymbol{a}(s), \\ \boldsymbol{a}(s_0) &= (\epsilon_{T_0}, \epsilon_{N_0}, (-1)^\nu \epsilon_{T_0}\epsilon_{N_0}, 0, 0, 0) \end{cases}$$

where $A(s)$ is the following matrix of coefficients:

$$\begin{pmatrix} 0 & 0 & 0 & 2\kappa(s) & 0 & 0 \\ 0 & 0 & 0 & -2\epsilon_{T_0}\epsilon_{N_0}\kappa(s) & 0 & 2\tau(s) \\ 0 & 0 & 0 & 0 & 0 & 2(-1)^{\nu+1}\epsilon_{T_0}\tau(s) \\ -\epsilon_{T_0}\epsilon_{N_0}\kappa(s) & \kappa(s) & 0 & 0 & \tau(s) & 0 \\ 0 & 0 & 0 & (-1)^{\nu+1}\epsilon_{T_0}\tau(s) & 0 & \kappa(s) \\ 0 & (-1)^{\nu+1}\epsilon_{T_0}\tau(s) & \tau(s) & 0 & -\epsilon_{T_0}\epsilon_{N_0}\kappa(s) & 0 \end{pmatrix}.$$

If the components of $\boldsymbol{a}(s)$ are all the scalar products[5] between the frame vectors $\boldsymbol{T}(s)$, $\boldsymbol{N}(s)$, and $\boldsymbol{B}(s)$, forming the solution to the previous IVP, we conclude that the constant vector

$$\boldsymbol{a}_0 = (\epsilon_{T_0}, \epsilon_{N_0}, (-1)^\nu \epsilon_{T_0}\epsilon_{N_0}, 0, 0, 0)$$

is the unique solution for the given initial conditions, from where our claim follows. Defining then $\boldsymbol{\alpha} : I \to \mathbb{R}^3_\nu$ by

$$\boldsymbol{\alpha}(s) \doteq \boldsymbol{p}_0 + \int_{s_0}^{s} \boldsymbol{T}(\xi)\, d\xi,$$

we see that $\boldsymbol{\alpha}(s_0) = \boldsymbol{p}_0$, $\boldsymbol{\alpha}'(s) = \boldsymbol{T}(s)$, and $\boldsymbol{\alpha}''(s) = \boldsymbol{T}'(s) = \kappa(s)\boldsymbol{N}(s)$. Thus, $\boldsymbol{\alpha}$ is a unit speed curve, $\epsilon_{\boldsymbol{\alpha}} = \epsilon_{T_0}$, and $\operatorname{span}\{\boldsymbol{\alpha}'(s), \boldsymbol{\alpha}''(s)\} = \operatorname{span}\{\boldsymbol{T}(s), \boldsymbol{N}(s)\}$, hence $\boldsymbol{\alpha}$ is admissible. Moreover, $\kappa_{\boldsymbol{\alpha}}(s)\boldsymbol{N}_{\boldsymbol{\alpha}}(s) = \kappa(s)\boldsymbol{N}(s)$. Taking $\|\cdot\|$ of both sides and noting that the curvatures are positive, we obtain $\kappa_{\boldsymbol{\alpha}}(s) = \kappa(s)$, and it follows from this not only that $\boldsymbol{N}_{\boldsymbol{\alpha}}(s) = \boldsymbol{N}(s)$ for all $s \in I$, but also that $\eta_{\boldsymbol{\alpha}} = \epsilon_{N_0}$. If $\boldsymbol{T}_{\boldsymbol{\alpha}}(s) = \boldsymbol{T}(s)$ and $\boldsymbol{N}_{\boldsymbol{\alpha}}(s) = \boldsymbol{N}(s)$ we also obtain that $\boldsymbol{B}_{\boldsymbol{\alpha}}(s) = \boldsymbol{B}(s)$. Differentiating this, we have that $(-1)^{\nu+1}\epsilon_{\boldsymbol{\alpha}}\tau_{\boldsymbol{\alpha}}(s)\boldsymbol{N}_{\boldsymbol{\alpha}}(s) = (-1)^{\nu+1}\epsilon_{T_0}\tau(s)\boldsymbol{N}(s)$. From the relations established so far, we get that $\tau_{\boldsymbol{\alpha}}(s) = \tau(s)$, as wanted.

Lastly, to verify the uniqueness of $\boldsymbol{\alpha}$, let $\boldsymbol{\beta} : I \to \mathbb{R}^3_\nu$ by another unit speed admissible curve such that $\boldsymbol{\alpha}(s_0) = \boldsymbol{\beta}(s_0)$, with the same Frenet-Serret frame as $\boldsymbol{\alpha}$ at s_0, with the same curvature and torsion as $\boldsymbol{\alpha}$ for all $s \in I$. Since curvatures and torsions match, both $(\boldsymbol{T}_{\boldsymbol{\alpha}}(s), \boldsymbol{N}_{\boldsymbol{\alpha}}(s), \boldsymbol{B}_{\boldsymbol{\alpha}}(s))$ and $(\boldsymbol{T}_{\boldsymbol{\beta}}(s), \boldsymbol{N}_{\boldsymbol{\beta}}(s), \boldsymbol{B}_{\boldsymbol{\beta}}(s))$ are solutions to the same IVP, and in particular it follows that $\boldsymbol{T}_{\boldsymbol{\alpha}}(s) = \boldsymbol{T}_{\boldsymbol{\beta}}(s)$ for all $s \in I$, so that $\boldsymbol{\alpha}$ and $\boldsymbol{\beta}$ differ by a constant. Then $\boldsymbol{\alpha}(s_0) = \boldsymbol{\beta}(s_0)$ ensures that such constant is zero, and so $\boldsymbol{\alpha} = \boldsymbol{\beta}$ as wanted. □

Remark. The proof of uniqueness in the theorem above may be simplified in $\mathbb{R}^3$. See how to do this in Exercise 2.3.20.

Corollary 2.3.21. *Two unit speed admissible curves, with same curvature and torsion, whose Frenet-Serret frames have the same causal type, differ by a positive isometry of* $\mathbb{R}^3_\nu$.

Proof: Let $\boldsymbol{\alpha}, \boldsymbol{\beta} \colon I \to \mathbb{R}^3_\nu$ be two curves as in the statement above. Fix any $s_0 \in I$. Since both $\boldsymbol{\alpha}$ and $\boldsymbol{\beta}$ share the same causal type, we use Proposition 1.4.15 (p. 35) to obtain $F \in \mathrm{E}_\nu(3, \mathbb{R})$ such that $F(\boldsymbol{\alpha}(s_0)) = \boldsymbol{\beta}(s_0)$, $DF(\boldsymbol{\alpha}(s_0))(\boldsymbol{T}_{\boldsymbol{\alpha}}(s_0)) = \boldsymbol{T}_{\boldsymbol{\beta}}(s_0)$, and similarly for $\boldsymbol{N}$ and $\boldsymbol{B}$. Such an isometry F is in fact positive (since its linear part takes a positive basis of the space into another positive basis), and hence preserves torsions, so that the curves $F \circ \boldsymbol{\alpha}$ and $\boldsymbol{\beta}$ are now in the setting of Theorem 2.3.20 above. We conclude that $\boldsymbol{\beta} = F \circ \boldsymbol{\alpha}$, as wanted. □

[5]More precisely, $\boldsymbol{a} = (\langle \boldsymbol{T}, \boldsymbol{T}\rangle, \langle \boldsymbol{N}, \boldsymbol{N}\rangle, \langle \boldsymbol{B}, \boldsymbol{B}\rangle, \langle \boldsymbol{T}, \boldsymbol{N}\rangle, \langle \boldsymbol{T}, \boldsymbol{B}\rangle, \langle \boldsymbol{N}, \boldsymbol{B}\rangle)$.

Remark. Given an admissible $\alpha\colon I \to \mathbb{L}^3$ not necessarily with unit speed, some care is needed to determine the causal type of its Frenet-Serret trihedron. For example, if

$$\alpha(t) = (t^2, \sinh(t^2), \cosh(t^2)), \quad t > 0,$$

we have that $\alpha'(t)$ is always spacelike, while $\alpha''(t)$ is spacelike for $0 < t < \sqrt{2}$, lightlike for $t = \sqrt{2}$ and timelike for $t > \sqrt{2}$ (verify). Meanwhile, a unit speed reparametrization of α is

$$\widetilde{\alpha}(s) = (s/\sqrt{2}, \sinh(s/\sqrt{2}), \cosh(s/\sqrt{2})).$$

Then $\widetilde{\alpha}'(s)$ is always spacelike and $\widetilde{\alpha}''(s)$ is always timelike.

If it is not possible to concretely exhibit a unit speed reparametrization of the curve, one may analyze the Gram matrix of the osculating plane $\operatorname{span}\{\alpha'(t), \alpha''(t)\}$ to determine the causal type of the binormal vector. In this case the Gram matrix

$$\begin{pmatrix} 4t^2 & 8t \\ 8t & 8 - 16t^4 \end{pmatrix}$$

is indefinite, for all $t > 0$. Thus, the binormal vector is always spacelike and so the normal vector is always timelike.

Lastly, Theorem 2.3.20 also allows us to efficiently conclude our study of admissible helices, when both curvature and torsion are independently constant:

Corollary 2.3.22. *A helix $\alpha\colon I \to \mathbb{R}^3_\nu$ with both constant curvature and torsion is congruent, for a suitable choice of $a, b \in \mathbb{R}$, to a piece of one and only one of the following* standard helices*:*

- $\beta_1(s) = \big(a\cos(s/c), a\sin(s/c), bs/c\big)$*;*
- $\beta_2(s) = \big(a\cos(s/c), a\sin(s/c), bs/c\big)$*;*
- $\beta_3(s) = \big(bs/c, a\cosh(s/c), a\sinh(s/c)\big)$*;*
- $\beta_4(s) = \big(bs/c, a\sinh(s/c), a\cosh(s/c)\big)$*;*
- $\beta_5(s) = \big(as^2/2, a^2s^3/6, s + a^2s^3/6\big)$*;*
- $\beta_6(s) = \big(as^2/2, s - a^2s^3/6, -a^2s^3/6\big)$*,*

where β_1 is seen in $\mathbb{R}^3$, the remaining ones in $\mathbb{L}^3$, $c \doteq \sqrt{a^2+b^2}$ for β_1 and β_4, and $c \doteq \sqrt{|a^2-b^2|}$ for β_2 and β_3;

Proof: We'll denote the curvature and torsion of α, respectively, simply by κ and τ. If the helix lies in $\mathbb{R}^3$, it is congruent with β_1. Let's then focus on the remaining helices in $\mathbb{L}^3$. Recalling the proof of Lancret's Theorem, we have that a vector determining the helical axis of α is

$$v = T_\alpha(s) - \frac{\epsilon_\alpha \kappa}{\tau} B_\alpha(s),$$

whence $\langle v, v\rangle_L = \epsilon_\alpha\big(1 - \eta_\alpha\kappa^2/\tau^2\big)$. According to Exercise 2.3.25, the causal type of the curves given in the statement of this result is, in general, determined by the constants a and b. Thus, a timelike helix is:

- hyperbolic if $\kappa > |\tau|$, and thus congruent with β_3;
- elliptic if $\kappa < |\tau|$, and thus congruent with β_2; and

- parabolic if $\kappa = |\tau|$, and thus congruent with β_5.

Similarly, a spacelike helix with timelike normal is necessarily hyperbolic and thus congruent with β_4. Lastly, a spacelike helix with timelike binormal is:

- hyperbolic if $\kappa < |\tau|$, and thus congruent with β_3;
- elliptic if $\kappa > |\tau|$, and thus congruent with β_2, and
- parabolic if $\kappa = |\tau|$, and thus congruent with β_6.

□

Exercises

Exercise 2.3.15. Let $\alpha\colon I \to \mathbb{R}^3$ be a unit speed admissible curve and $s_0 \in I$. Assume that $\|\alpha(s)\| \leq R \doteq \|\alpha(s_0)\|$ for all s sufficiently close to s_0. Show that $\kappa_\alpha(s_0) \geq 1/R$. Can you get a similar result in $\mathbb{L}^3$? Which complications do arise?

Hint. Use that s_0 is a local maximum for $f\colon I \to \mathbb{R}$ given by $f(s) = \|\alpha(s)\|_E^2$.

Exercise 2.3.16. Determine all the admissible curves in $\mathbb{R}^3_\nu$ such that:

(a) all of their tangent lines intercept at a fixed point;

(b) all or their normal lines intercept at a fixed point.

Why didn't we bother writing an item (c) here for the binormal vector?

Exercise 2.3.17. Let $\alpha\colon I \to \mathbb{R}^3_\nu$ be an admissible curve. Show that if all the osculating planes of α are parallel, then α is a planar curve.

Exercise† 2.3.18. Show Proposition 2.3.19 (p. 112).

Exercise 2.3.19. Let $\alpha\colon I \to \mathbb{R}^3_\nu$ be a unit speed admissible curve. Show that α is a helix if and only if

$$\det(N_\alpha(s), N_\alpha'(s), N_\alpha''(s)) = 0, \text{ for all } s \in I.$$

Exercise† 2.3.20. In $\mathbb{R}^3$, the part of Theorem 2.3.20 (p. 112) regarding the uniqueness of the curve may be proved in other ways. In this exercise we will see two of them. Let $\alpha, \beta\colon I \to \mathbb{R}^3$ be two unit speed admissible curves and $s_0 \in I$ be such that $\alpha(s_0) = \beta(s_0)$. Suppose that α and β have the same Frenet-Serret frame at s_0, that $\kappa_\alpha(s) = \kappa_\beta(s)$, and $\tau_\alpha(s) = \tau_\beta(s)$ for all $s \in I$.

(a) Consider a deviation function $\mathscr{D}_1\colon I \to \mathbb{R}$, given by

$$\mathscr{D}_1(s) \doteq \langle T_\alpha(s), T_\beta(s)\rangle_E + \langle N_\alpha(s), N_\beta(s)\rangle_E + \langle B_\alpha(s), B_\beta(s)\rangle_E.$$

Show that, under the given assumptions, we have that $\mathscr{D}_1(s) = 3$ for all $s \in I$. Argue then that each of the terms in the definition of $\mathscr{D}_1(s)$ has to be precisely 1, and use this to deduce that the Frenet-Serret frames of α and β actually coincide at all points. Finally, conclude that $\alpha = \beta$.

Hint. The Cauchy-Schwarz inequality may be useful.

(b) An alternative argument to the one given in the above item considers a different deviation function $\mathscr{D}_2\colon I \to \mathbb{R}$ given by

$$\mathscr{D}_2(s) \doteq \|\boldsymbol{T}_\alpha(s) - \boldsymbol{T}_\beta(s)\|_E^2 + \|\boldsymbol{N}_\alpha(s) - \boldsymbol{N}_\beta(s)\|_E^2 + \|\boldsymbol{B}_\alpha(s) - \boldsymbol{B}_\beta(s)\|_E^2,$$

and then proving that, under the given assumptions, we have $\mathscr{D}_2(s) = 0$ for all $s \in I$. Use this to conclude again that $\boldsymbol{\alpha} = \boldsymbol{\beta}$.

Remark. Item (b) follows almost trivially from (a) by observing the relation $\mathscr{D}_2(s) = 6 - 2\mathscr{D}_1(s)$. Many other books just present our deviation $\mathscr{D}_2$ without mentioning our $\mathscr{D}_1$, so it is still instructive to prove item (b) without this remark.

Think a bit about the several reasons that make these arguments fail in $\mathbb{L}^3$ and understand why we had to use the theory of ordinary differential equations all the way through in the proof of Theorem 2.3.20.

Exercise 2.3.21. Let $a, b, c \in \mathbb{R}$. Show that if $2b^2 = \pm 3ac$, then the curve $\boldsymbol{\alpha}\colon \mathbb{R} \to \mathbb{R}^3$ given by $\boldsymbol{\alpha}(t) = (at, bt^2, ct^3)$ is a helix. Is there a similar condition for $\boldsymbol{\alpha}$ to be a helix when seen in $\mathbb{L}^3$?

Exercise 2.3.22. Let $\boldsymbol{\alpha}\colon I \to \mathbb{L}^3$ be a hyperbolic (resp. elliptic, parabolic) helix. If $F \in \mathrm{P}(3, \mathbb{R})$ is a Poincaré transformation, show that $F \circ \boldsymbol{\alpha}$ is also a hyperbolic (resp. elliptic, parabolic) helix.

Exercise 2.3.23. Let $\boldsymbol{\alpha}\colon I \to \mathbb{R}^3_\nu$ be a unit speed admissible curve. Suppose that $\boldsymbol{\alpha}$ is a non-parabolic helix with constant curvature. If $\boldsymbol{v} \in \mathbb{R}^3_\nu$ is a unit vector giving the direction of the helical axis, consider the projection $\overline{\boldsymbol{\alpha}}\colon I \to \mathbb{R}^3_\nu$ onto the orthogonal plane to $\boldsymbol{v}$, given by

$$\overline{\boldsymbol{\alpha}}(s) \doteq \boldsymbol{\alpha}(s) - \epsilon_{\boldsymbol{v}} \langle \boldsymbol{\alpha}(s), \boldsymbol{v} \rangle \boldsymbol{v}.$$

Show that $\overline{\boldsymbol{\alpha}}$ has constant curvature and justify (in part) the terminology adopted in Definition 2.3.16 (p. 110).

Exercise† 2.3.24. Determine admissible curves $\boldsymbol{\alpha}_1, \boldsymbol{\alpha}_2\colon \mathbb{R} \to \mathbb{L}^3$ with constant curvature and torsion verifying $\kappa = |\tau|$ and $\boldsymbol{\alpha}_1(0) = \boldsymbol{\alpha}_2(0) = \boldsymbol{0}$, such that:

(a) $\big(\boldsymbol{T}_{\alpha_1}(0), \boldsymbol{N}_{\alpha_1}(0), \boldsymbol{B}_{\alpha_1}(0)\big) = \big(\boldsymbol{e}_3, \boldsymbol{e}_1, \boldsymbol{e}_2\big)$;

(b) $\big(\boldsymbol{T}_{\alpha_2}(0), \boldsymbol{N}_{\alpha_2}(0), \boldsymbol{B}_{\alpha_2}(0)\big) = \big(\boldsymbol{e}_2, \boldsymbol{e}_1, -\boldsymbol{e}_3\big)$.

Exercise† 2.3.25. This exercise is a guide to deduce the correct choice of the coefficients a and b in the statement of Corollary 2.3.22 (p. 114). Let $a, b \in \mathbb{R}$ be non-zero. For simplicity, we'll omit the subscript indices in all of the curvatures and torsions to follow.

(a) Show that the curve $\boldsymbol{\beta}_1\colon \mathbb{R} \to \mathbb{R}^3$ given by $\boldsymbol{\beta}_1(s) = \big(a\cos(s/c), a\sin(s/c), bs/c\big)$, where $c \doteq \sqrt{a^2 + b^2}$, has unit speed and curvature and torsion given by

$$\kappa = \frac{|a|}{a^2 + b^2} \quad \text{and} \quad \tau = \frac{b}{a^2 + b^2}.$$

Also show that

$$|a| = \frac{\kappa}{\kappa^2 + \tau^2} \quad \text{and} \quad b = \frac{\tau}{\kappa^2 + \tau^2}.$$

(b) Assuming $|a| \neq |b|$, show that $\beta_2, \beta_3 \colon \mathbb{R} \to \mathbb{L}^3$ given by

$$\beta_2(s) = \left(a\cos\left(\frac{s}{c}\right), a\sin\left(\frac{s}{c}\right), \frac{bs}{c}\right),$$
$$\beta_3(s) = \left(\frac{bs}{c}, a\cosh\left(\frac{s}{c}\right), a\sinh\left(\frac{s}{c}\right)\right),$$

where $c \doteq \sqrt{|a^2 - b^2|}$, have opposite causal types, unit speed, are admissible, and have curvature and torsion given by

$$\kappa = \frac{|a|}{|a^2 - b^2|} \quad \text{and} \quad \tau = \frac{b}{|a^2 - b^2|}.$$

Also show that

$$|a| = \frac{\kappa}{|\kappa^2 - \tau^2|} \quad \text{and} \quad b = \frac{\tau}{|\kappa^2 - \tau^2|}.$$

(c) Show that $\beta_4 \colon \mathbb{R} \to \mathbb{L}^3$ given by $\beta_4(s) = (bs/c, a\sinh(s/c), a\cosh(s/c))$, where $c \doteq \sqrt{a^2 + b^2}$, has unit speed, and curvature and torsion given by

$$\kappa = \frac{|a|}{a^2 + b^2} \quad \text{and} \quad \tau = \frac{-b}{a^2 + b^2}.$$

Also show that

$$|a| = \frac{\kappa}{\kappa^2 + \tau^2} \quad \text{and} \quad b = \frac{-\tau}{\kappa^2 + \tau^2}.$$

(d) Show that $\beta_5, \beta_6 \colon \mathbb{R} \to \mathbb{L}^3$ given by

$$\beta_5(s) = \left(\frac{as^2}{2}, \frac{a^2s^3}{6}, s + \frac{a^2s^3}{6}\right),$$
$$\beta_6(s) = \left(\frac{as^2}{2}, s - \frac{a^2s^3}{6}, -\frac{a^2s^3}{6}\right),$$

have opposite causal types, unit speed, are admissible, and have curvature and torsion given by

$$\kappa = |a| \quad \text{and} \quad \tau = a.$$

Remark. The previous exercise motivates the definitions for the curves β_5 and β_6 above.

Exercise 2.3.26. Suppose that $\boldsymbol{\alpha} \colon I \to \mathbb{R}^3_\nu$ is a unit speed admissible curve with constant curvature and torsion. Determine $\boldsymbol{\alpha}$ explicitly, by solving the Frenet-Serret system.

Hint. Differentiate the second Frenet-Serret equation to obtain a second order ODE with constant coefficients involving only $\boldsymbol{N}''_{\boldsymbol{\alpha}}(s)$ and $\boldsymbol{N}_{\boldsymbol{\alpha}}(s)$ (which is then solved componentwise). Knowing $\boldsymbol{N}_{\boldsymbol{\alpha}}(s)$, integrating the first Frenet-Serret equation gives us $\boldsymbol{T}_{\boldsymbol{\alpha}}(s)$. Then integrate again to get $\boldsymbol{\alpha}(s)$.

2.3.3 Curves with degenerate osculating plane

In this subsection we will assume that all the curves are biregular. In particular, we're excluding the case where $\boldsymbol{\alpha}$ is a light ray. Our curves are assumed to have unit speed or to be parametrized by arc-photon. To begin this discussion, we have the:

Definition 2.3.23. A unit speed curve $\boldsymbol{\alpha}: I \to \mathbb{L}^3$ is called *semi-lightlike* if $\operatorname{span}\{\boldsymbol{\alpha}'(s), \boldsymbol{\alpha}''(s)\}$ is degenerate (and, thus, $\boldsymbol{\alpha}''(s)$ is lightlike) for all $s \in I$.

Remark.

- In view of this definition, we'll allow the indicator $\epsilon_{\boldsymbol{\alpha}}$ and the coindicator $\eta_{\boldsymbol{\alpha}}$ to be zero. This way, if $\boldsymbol{\alpha}$ is lightlike, we have that $(\epsilon_{\boldsymbol{\alpha}}, \eta_{\boldsymbol{\alpha}}) = (0,1)$, while if $\boldsymbol{\alpha}$ is semi-lightlike, we have $(\epsilon_{\boldsymbol{\alpha}}, \eta_{\boldsymbol{\alpha}}) = (1,0)$. This is done to treat both cases simultaneously.
- Since the arclength parameter is denoted by s and the arc-photon parameter by ϕ, we will allow ourselves to simply omit the parameter when discussing results for both cases.

The idea is, as before, to define a suitable frame adapted to the curve. In this case, an orthonormal frame would not carry geometric information about acceleration vector of the curve. We'll start redefining the tangent and normal vectors:

Definition 2.3.24. Let $\boldsymbol{\alpha}: \ : I \to \mathbb{L}^3$ be lightlike or semi-lightlike. We define the *tangent* and the *normal* to the curve by

$$\boldsymbol{T}_{\boldsymbol{\alpha}} \doteq \boldsymbol{\alpha}' \quad \text{and} \quad \boldsymbol{N}_{\boldsymbol{\alpha}} \doteq \boldsymbol{\alpha}'',$$

respectively.

To complete the basis we seek, we need to determine a third vector $\boldsymbol{B}_{\boldsymbol{\alpha}}$ (to be called again the *binormal vector*), such that basis $(\boldsymbol{T}_{\boldsymbol{\alpha}}, \boldsymbol{N}_{\boldsymbol{\alpha}}, \boldsymbol{B}_{\boldsymbol{\alpha}})$ is positive at all points of the curve.

In general, we may define the orientation of a basis $(\boldsymbol{v}, \boldsymbol{w})$ for a lightlike plane in terms of a choice of vector $\boldsymbol{n}$ which is Euclidean-normal to the plane. More precisely, let's say that $(\boldsymbol{v}, \boldsymbol{w})$ is *positive* if $(\boldsymbol{v}, \boldsymbol{w}, \boldsymbol{n})$ is a positive basis of $\mathbb{L}^3$, for $\boldsymbol{n}$ future-directed. If $\boldsymbol{v}$ is lightlike and $\boldsymbol{w}$ is unit, we have that $\boldsymbol{v} \times_L \boldsymbol{w}$ is lightlike and proportional to $\boldsymbol{v}$. Writing $\boldsymbol{v} \times_L \boldsymbol{w} = \lambda \boldsymbol{v}$ for some $\lambda \in \mathbb{R}$, we analyze the sign of λ as follows:

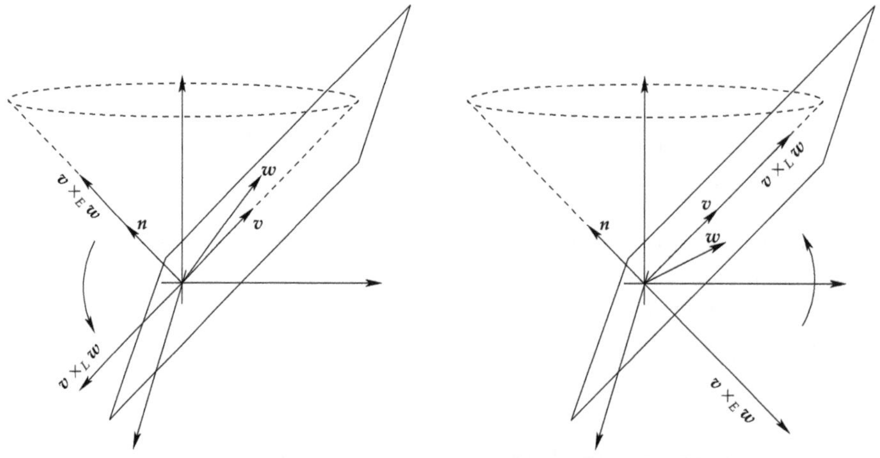

(a) $(\boldsymbol{v}, \boldsymbol{w})$ positive $(\lambda < 0)$ (b) $(\boldsymbol{v}, \boldsymbol{w})$ negative $(\lambda > 0)$

Figure 2.23: Orientations for a lightlike plane.

This way, if (v, w) is positive, then $\lambda < 0$ and, similarly, if (v, w) is negative we have $\lambda > 0$.

Back to the construction of the basis $(T_\alpha, N_\alpha, B_\alpha)$: we may assume, up to reparametrization, that the basis (T_α, N_α) is positive. In this case, to determine the vector B_α, which will be lightlike, we need to know the values of $\langle T_\alpha, B_\alpha\rangle_L$ and $\langle N_\alpha, B_\alpha\rangle_L$ as well. In view of the above definition, one of these values has to be 0 (for it to be orthogonal to the spacelike vector) and the other one -1 (so it is linearly independent with the lightlike vector). Which one will be zero and which one will be -1 naturally depends on the causal type of α. Choosing B_α such that $\langle T_\alpha, B_\alpha\rangle_L = -\eta_\alpha$ and $\langle N_\alpha, B_\alpha\rangle_L = -\epsilon_\alpha$ we treat both cases simultaneously, and then we have the:

Proposition 2.3.25. *Let $\alpha\colon I \to \mathbb{L}^3$ be lightlike or semi-lightlike. The triple $(T_\alpha, N_\alpha, B_\alpha)$ is a positive basis for $\mathbb{L}^3$.*

Proof: We must show that $\det(T_\alpha, N_\alpha, B_\alpha) > 0$. Let's treat here only the case $\epsilon_\alpha = 0$ and $\eta_\alpha = 1$ (the other case is similar and we ask you to do it in Exercise 2.3.27).

Writing $B_\alpha(\phi)$ relative to the basis $\big(T_\alpha(\phi), N_\alpha(\phi), T_\alpha(\phi) \times_E N_\alpha(\phi)\big)$, we see that the only component of $B_\alpha(\phi)$ relevant for the required determinant is the one in the direction of $T_\alpha(\phi) \times_E N_\alpha(\phi)$, and let's say that it is $\mu(\phi) T_\alpha(\phi) \times_E N_\alpha(\phi)$. Then we have

$$\det(T_\alpha(\phi), N_\alpha(\phi), B_\alpha(\phi)) = \mu(\phi) \underbrace{\det(T_\alpha(\phi), N_\alpha(\phi), T_\alpha(\phi) \times_E N_\alpha(\phi))}_{>0},$$

so that it only remains to check that $\mu(\phi) > 0$. The discussion illustrated by Figure 2.23 tells us that $T_\alpha(\phi) \times_L N_\alpha(\phi) = \lambda(\phi) T_\alpha(\phi)$ for certain $\lambda(\phi) < 0$ (since the basis $(T_\alpha(\phi), N_\alpha(\phi))$ is assumed positive). Applying $\mathrm{Id}_{2,1}$ we obtain that

$$\begin{aligned} T_\alpha(\phi) \times_E N_\alpha(\phi) = \mathrm{Id}_{2,1}\big(T_\alpha(\phi) \times_L N_\alpha(\phi)\big) = \lambda(\phi)\,\mathrm{Id}_{2,1}\, T_\alpha(\phi) \implies \\ \implies \langle T_\alpha(\phi) \times_E N_\alpha(\phi), \mathrm{Id}_{2,1}\, T_\alpha(\phi)\rangle_E < 0. \end{aligned}$$

Finally, as $T_\alpha(\phi)$ and $N_\alpha(\phi)$ are both Lorentz-orthogonal to $T_\alpha(\phi)$, we have that

$$-1 = \langle B_\alpha(\phi), T_\alpha(\phi)\rangle_L = \mu(\phi) \langle T_\alpha(\phi) \times_E N_\alpha(\phi), \mathrm{Id}_{2,1}\, T_\alpha(\phi)\rangle_E,$$

and so we conclude that $\mu(\phi) > 0$. □

So, the frame $\big(T_\alpha, N_\alpha, B_\alpha\big)$ is called the *Cartan frame* of α.

Geometrically, when the curve is lightlike, the situation is as follows: the vector $N_\alpha(\phi)$ is spacelike, and so its orthogonal complement is a timelike plane that cuts the lightcone in two light rays, one of them in the direction of $T_\alpha(\phi)$. Thus, the binormal vector will be in the direction of the remaining light ray in $N_\alpha(\phi)^\perp$, being determined by the relation $\langle B_\alpha(\phi), T_\alpha(\phi)\rangle_L = -1$. A similar interpretation holds when the curve is semi-lightlike.

Recall that to define the Frenet-Serret equations when we studied admissible curves, we have used a tool which is no longer available here: the expression of a vector in terms of the trihedron via orthonormal expansion. We'll remediate this in the following:

Lemma 2.3.26. *Let $\alpha\colon I \to \mathbb{L}^3$ and $v \in \mathbb{L}^3$. Then:*

(i) if α is lightlike, we have that

$$v = -\langle v, B_\alpha(\phi)\rangle_L T_\alpha(\phi) + \langle v, N_\alpha(\phi)\rangle_L N_\alpha(\phi) - \langle v, T_\alpha(\phi)\rangle_L B_\alpha(\phi),$$

for all $\phi \in I$;

(ii) if α is semi-lightlike, we have that

$$v = \langle v, T_\alpha(s)\rangle_L T_\alpha(s) - \langle v, B_\alpha(s)\rangle_L N_\alpha(s) - \langle v, N_\alpha(s)\rangle_L B_\alpha(s),$$

for all $s \in I$.

Remark. A possible mnemonic is: switch the position and sign only of the coefficients corresponding to lightlike directions.

Proof: We'll treat both cases at the same time, noting the relations $\epsilon_\alpha^n = \epsilon_\alpha$, $\eta_\alpha^n = \eta_\alpha$ for all $n \geq 1$, $\epsilon_\alpha \eta_\alpha = 0$ and $\epsilon_\alpha + \eta_\alpha = 1$. They follow from the only possibilities at hand being $(\epsilon_\alpha, \eta_\alpha) = (1, 0)$ and $(\epsilon_\alpha, \eta_\alpha) = (0, 1)$. Recall that we're still assuming that (T_α, N_α) is positive. That being said, write $v = aT_\alpha + bN_\alpha + cB_\alpha$. Applying all the possible products in both sides of this expression, and already organizing everything in matrix form, we have

$$\begin{pmatrix} \langle v, T_\alpha\rangle_L \\ \langle v, N_\alpha\rangle_L \\ \langle v, B_\alpha\rangle_L \end{pmatrix} = \begin{pmatrix} \epsilon_\alpha & 0 & -\eta_\alpha \\ 0 & \eta_\alpha & -\epsilon_\alpha \\ -\eta_\alpha & -\epsilon_\alpha & 0 \end{pmatrix} \begin{pmatrix} a \\ b \\ c \end{pmatrix}.$$

It follows from the observations just made that the inverse of the coefficient matrix exists, and it is itself, so that

$$\begin{pmatrix} a \\ b \\ c \end{pmatrix} = \begin{pmatrix} \epsilon_\alpha & 0 & -\eta_\alpha \\ 0 & \eta_\alpha & -\epsilon_\alpha \\ -\eta_\alpha & -\epsilon_\alpha & 0 \end{pmatrix} \begin{pmatrix} \langle v, T_\alpha\rangle_L \\ \langle v, N_\alpha\rangle_L \\ \langle v, B_\alpha\rangle_L \end{pmatrix}.$$

Particularizing, we conclude the lemma. □

Before applying the above lemma for the derivatives of the vectors in the Cartan frame, we have the:

Definition 2.3.27. Let $\alpha\colon I \to \mathbb{L}^3$ be lightlike or semi-lightlike. The *pseudo-torsion* of α is given by $\mathcal{T}_\alpha \doteq -\langle N_\alpha', B_\alpha\rangle_L$.

Remark. In the literature, this pseudo-torsion is also called the *Cartan curvature* of α.

Theorem 2.3.28. *Let $\alpha\colon I \to \mathbb{L}^3$ be lightlike or semi-lightlike. Then we have that*

$$\begin{pmatrix} T_\alpha' \\ N_\alpha' \\ B_\alpha' \end{pmatrix} = \begin{pmatrix} 0 & 1 & 0 \\ \eta_\alpha \mathcal{T}_\alpha & \epsilon_\alpha \mathcal{T}_\alpha & \eta_\alpha \\ \epsilon_\alpha & \eta_\alpha \mathcal{T}_\alpha & -\epsilon_\alpha \mathcal{T}_\alpha \end{pmatrix} \begin{pmatrix} T_\alpha \\ N_\alpha \\ B_\alpha \end{pmatrix}.$$

Remark. Explicitly, the coefficient matrices when α is lightlike or semi-lightlike are, respectively,

$$\begin{pmatrix} 0 & 1 & 0 \\ \mathcal{T}_\alpha(\phi) & 0 & 1 \\ 0 & \mathcal{T}_\alpha(\phi) & 0 \end{pmatrix} \quad \text{and} \quad \begin{pmatrix} 0 & 1 & 0 \\ 0 & \mathcal{T}_\alpha(s) & 0 \\ 1 & 0 & -\mathcal{T}_\alpha(s) \end{pmatrix}.$$

Proof: The first equation is the definition of the normal vector. For the second equation, applying the Lemma 2.3.26 regarding N_α' as a column vector, we have:

$$\mathbf{N}'_\alpha = \begin{pmatrix} \epsilon_\alpha & 0 & -\eta_\alpha \\ 0 & \eta_\alpha & -\epsilon_\alpha \\ -\eta_\alpha & -\epsilon_\alpha & 0 \end{pmatrix} \begin{pmatrix} \langle \mathbf{N}'_\alpha, \mathbf{T}_\alpha \rangle_L \\ \langle \mathbf{N}'_\alpha, \mathbf{N}_\alpha \rangle_L \\ \langle \mathbf{N}'_\alpha, \mathbf{B}_\alpha \rangle_L \end{pmatrix}$$
$$= \begin{pmatrix} \epsilon_\alpha & 0 & -\eta_\alpha \\ 0 & \eta_\alpha & -\epsilon_\alpha \\ -\eta_\alpha & -\epsilon_\alpha & 0 \end{pmatrix} \begin{pmatrix} -\eta_\alpha \\ 0 \\ -\tau_\alpha \end{pmatrix} = \begin{pmatrix} \eta_\alpha \tau_\alpha \\ \epsilon_\alpha \tau_\alpha \\ \eta_\alpha \end{pmatrix},$$

and so we obtain the second row of the coefficient matrix given in the statement. Similarly for $\mathbf{B}'_\alpha$, we have:

$$\mathbf{B}'_\alpha = \begin{pmatrix} \epsilon_\alpha & 0 & -\eta_\alpha \\ 0 & \eta_\alpha & -\epsilon_\alpha \\ -\eta_\alpha & -\epsilon_\alpha & 0 \end{pmatrix} \begin{pmatrix} \langle \mathbf{B}'_\alpha, \mathbf{T}_\alpha \rangle_L \\ \langle \mathbf{B}'_\alpha, \mathbf{N}_\alpha \rangle_L \\ \langle \mathbf{B}'_\alpha, \mathbf{B}_\alpha \rangle_L \end{pmatrix}$$
$$= \begin{pmatrix} \epsilon_\alpha & 0 & -\eta_\alpha \\ 0 & \eta_\alpha & -\epsilon_\alpha \\ -\eta_\alpha & -\epsilon_\alpha & 0 \end{pmatrix} \begin{pmatrix} \epsilon_\alpha \\ \tau_\alpha \\ 0 \end{pmatrix} = \begin{pmatrix} \epsilon_\alpha \\ \eta_\alpha \tau_\alpha \\ -\epsilon_\alpha \tau_\alpha \end{pmatrix},$$

and so we obtain the last row. □

Example 2.3.29. Let $r > 0$ and consider the curve $\alpha\colon \mathbb{R} \to \mathbb{L}^3$ given by

$$\alpha(\phi) = \left(r\cos\left(\frac{\phi}{\sqrt{r}}\right), r\sin\left(\frac{\phi}{\sqrt{r}}\right), \sqrt{r}\phi \right).$$

We have seen in Example 2.1.19 (p. 72) that α is lightlike and parametrized by arc-photon. We have

$$\mathbf{T}_\alpha(\phi) = \alpha'(\phi) = \left(-\sqrt{r}\sin\left(\frac{\phi}{\sqrt{r}}\right), \sqrt{r}\cos\left(\frac{\phi}{\sqrt{r}}\right), \sqrt{r} \right) \quad \text{and}$$
$$\mathbf{N}_\alpha(\phi) = \alpha''(\phi) = \left(-\cos\left(\frac{\phi}{\sqrt{r}}\right), -\sin\left(\frac{\phi}{\sqrt{r}}\right), 0 \right).$$

To compute $\mathbf{B}_\alpha(\phi)$, note that the cross product

$$\mathbf{T}_\alpha(\phi) \times_E \mathbf{N}_\alpha(\phi) = \left(\sqrt{r}\sin\left(\frac{\phi}{\sqrt{r}}\right), -\sqrt{r}\cos\left(\frac{\phi}{\sqrt{r}}\right), \sqrt{r} \right)$$

seen in $\mathbb{L}^3$ is lightlike and future-directed, so the basis $(\mathbf{T}_\alpha(\phi), \mathbf{N}_\alpha(\phi))$ is already positive and no reparametrization for α is required. Moreover, in this case we have something very particular: $\mathbf{T}_\alpha(\phi) \times_E \mathbf{N}_\alpha(\phi)$ is also Lorentz-orthogonal to $\mathbf{N}_\alpha(\phi)$. This means that $\mathbf{B}_\alpha(\phi)$ must be a positive multiple of the product $\mathbf{T}_\alpha(\phi) \times_E \mathbf{N}_\alpha(\phi)$. For the condition $\langle \mathbf{B}_\alpha(\phi), \mathbf{T}_\alpha(\phi) \rangle_L = -1$ to be satisfied, it suffices to take

$$\mathbf{B}_\alpha(\phi) = \left(\frac{1}{2\sqrt{r}}\sin\left(\frac{\phi}{\sqrt{r}}\right), -\frac{1}{2\sqrt{r}}\cos\left(\frac{\phi}{\sqrt{r}}\right), \frac{1}{2\sqrt{r}} \right).$$

And so, we finally have:

$$\tau_\alpha(\phi) = -\langle \mathbf{N}'_\alpha(\phi), \mathbf{B}_\alpha(\phi) \rangle_L = -\frac{1}{2r}\sin^2\left(\frac{\phi}{\sqrt{r}}\right) - \frac{1}{2r}\cos^2\left(\frac{\phi}{\sqrt{r}}\right) + 0 = -\frac{1}{2r}.$$

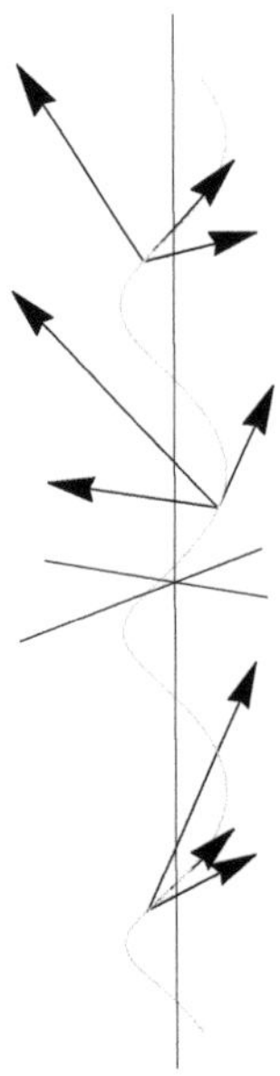

Figure 2.24: Cartan frame for α with $r = 1/4$.

The next two results illustrate some of the differences between $\mho_\alpha$ and τ_α.

Theorem 2.3.30. *The only planar lightlike curves in $\mathbb{L}^3$ are light rays.*

Proof: Of course, light rays are planar and if α is not a light ray, it admits a reparametrization with arc-photon parameter. Thus, it suffices to check that if $\alpha\colon I \to \mathbb{L}^3$ is a lightlike curve parametrized with arc-photon and $\langle \alpha(\phi) - p, v\rangle_L = 0$ for all $\phi \in I$, for certain $p, v \in \mathbb{L}^3$, then $v = \mathbf{0}$. Indeed, differentiating that expression three times we obtain

$$\langle T_\alpha(\phi), v\rangle_L = \langle N_\alpha(\phi), v\rangle_L = \mho_\alpha(\phi)\langle T_\alpha(\phi), v\rangle_L + \langle B_\alpha(\phi), v\rangle_L = 0.$$

By Lemma 2.3.26 it follows that $v = \mathbf{0}$. □

Example 2.3.31. Let $f\colon I \to \mathbb{R}$ be a smooth function with second derivative strictly positive (i.e., a strictly convex function) and consider $\alpha\colon I \to \mathbb{L}^3$ given by

$$\alpha(s) = (s, f(s), f(s)).$$

One then sees that α is semi-lightlike with

$$T_\alpha(s) = \alpha'(s) = (1, f'(s), f'(s)) \text{ and}$$
$$N_\alpha(s) = \alpha''(s) = (0, f''(s), f''(s)).$$

We have that

$$T_\alpha(s) \times_E N_\alpha(s) = (0, -f''(s), f''(s))$$

is lightlike and future-directed, so that the basis $(T_\alpha(s), N_\alpha(s))$ is positive. We seek for a lightlike vector $B_\alpha(s) = (a(s), b(s), c(s))$, orthogonal to $T_\alpha(s)$ and such that $\langle B_\alpha(s), N_\alpha(s)\rangle_L = -1$. Explicitly, we have:

$$\begin{cases} a(s)^2 + b(s)^2 - c(s)^2 & = 0 \\ a(s) + f'(s)(b(s) - c(s)) & = 0 \\ f''(s)(b(s) - c(s)) & = -1. \end{cases}$$

Substituting the third equation into the second one, it follows that $a(s) = f'(s)/f''(s)$. With this, the first equation reads

$$(b(s) - c(s))(b(s) + c(s)) = b(s)^2 - c(s)^2 = -\frac{f'(s)^2}{f''(s)^2} \implies b(s) + c(s) = \frac{f'(s)^2}{f''(s)},$$

after using the third equation again. We then obtain

$$\boldsymbol{B}_{\boldsymbol{\alpha}}(s) = \frac{1}{2f''(s)} \left(2f'(s), f'(s)^2 - 1, f'(s)^2 + 1\right).$$

Finally, we have that

$$\tau_{\boldsymbol{\alpha}}(s) = -\langle \boldsymbol{N}'_{\boldsymbol{\alpha}}(s), \boldsymbol{B}_{\boldsymbol{\alpha}}(s)\rangle_L = \frac{f'''(s)}{f''(s)}.$$

In particular, note that $\boldsymbol{\alpha}$ is contained in the plane $\Pi\colon y - z = 0$, but we may choose functions f for which the pseudo-torsion never vanishes.

The above example already says that, in general, the pseudo-torsion of a semi-lightlike curve is not a measure of how much the curve deviates from being planar. In fact, the next theorem says that the situation is much more extreme:

Theorem 2.3.32. *Every semi-lightlike is planar and contained in some lightlike plane.*

Proof: If $\boldsymbol{\alpha}\colon I \to \mathbb{L}^3$ is semi-lightlike, we seek elements $\boldsymbol{p}, \boldsymbol{v} \in \mathbb{L}^3$, with $\boldsymbol{v}$ lightlike, such that $\langle \boldsymbol{\alpha}(s) - \boldsymbol{p}, \boldsymbol{v}\rangle_L = 0$ for all $s \in I$. If this is to happen, differentiating twice we obtain $\langle \boldsymbol{N}_{\boldsymbol{\alpha}}(s), \boldsymbol{v}\rangle_L = 0$, and conclude that $\boldsymbol{v}$ must be parallel to $\boldsymbol{N}_{\boldsymbol{\alpha}}(s)$ (two orthogonal lightlike vectors in $\mathbb{L}^3$ must be proportional). Motivated by this, we seek a smooth function $\lambda\colon I \to \mathbb{R}$ such that $\boldsymbol{v} = \lambda(s)\boldsymbol{N}_{\boldsymbol{\alpha}}(s)$ is constant. This leads us to

$$\boldsymbol{0} = (\lambda'(s) + \tau_{\boldsymbol{\alpha}}(s)\lambda(s))\boldsymbol{N}_{\boldsymbol{\alpha}}(s),$$

for all $s \in I$. Define $\boldsymbol{v}$ in such a way, by taking

$$\lambda(s) = \exp\left(-\int_{s_0}^{s} \tau_{\boldsymbol{\alpha}}(\xi)\, \mathrm{d}\xi\right),$$

for some fixed $s_0 \in I$. By construction, $\boldsymbol{v}$ is constant and then we may take $\boldsymbol{p} = \boldsymbol{\alpha}(s_0)$. That done, the justification that $\boldsymbol{p}$ and $\boldsymbol{v}$ satisfy all that was required proceeds as usual: consider $f\colon I \to \mathbb{R}$ given by $f(s) = \langle \boldsymbol{\alpha}(s) - \boldsymbol{\alpha}(s_0), \boldsymbol{v}\rangle_L$. Clearly, we have $f(s_0) = 0$ and $f'(s) = \langle \boldsymbol{T}_{\boldsymbol{\alpha}}(s), \boldsymbol{v}\rangle_L = 0$, for all $s \in I$. □

In contrast to what happens for admissible curves, the sign of the pseudo-torsion does not influence how the curve crosses its osculating planes. Indeed, it $\boldsymbol{\alpha}\colon I \to \mathbb{L}^3$ and assuming that $0 \in I$ and $\boldsymbol{\alpha}(0) = \boldsymbol{0}$, we have the Taylor formula

$$\boldsymbol{\alpha}(\phi) = \phi\boldsymbol{\alpha}'(0) + \frac{\phi^2}{2}\boldsymbol{\alpha}''(0) + \frac{\phi^3}{6}\boldsymbol{\alpha}'''(0) + \boldsymbol{R}(\phi),$$

where $\boldsymbol{R}(\phi)/\phi^3 \to \boldsymbol{0}$ as $\phi \to 0$. Reorganizing, in the frame $\mathfrak{F} = (\boldsymbol{T}_{\boldsymbol{\alpha}}(0), \boldsymbol{N}_{\boldsymbol{\alpha}}(0), \boldsymbol{B}_{\boldsymbol{\alpha}}(0))$, we have that the coordinates of $\boldsymbol{\alpha}(\phi) - \boldsymbol{R}(\phi)$ are

$$\boldsymbol{\alpha}(\phi) - \boldsymbol{R}(\phi) = \left(\phi + \tau_{\boldsymbol{\alpha}}(0)\frac{\phi^3}{6}, \frac{\phi^2}{2}, \frac{\phi^3}{6}\right)_{\mathfrak{F}}.$$

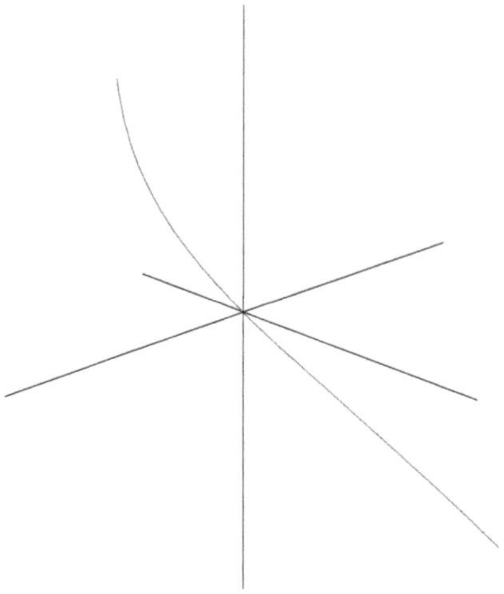

Figure 2.25: Local canonical form for lightlike $\boldsymbol{\alpha}$.

Projecting, no matter the sign of $\mho_{\boldsymbol{\alpha}}(0)$, we have:

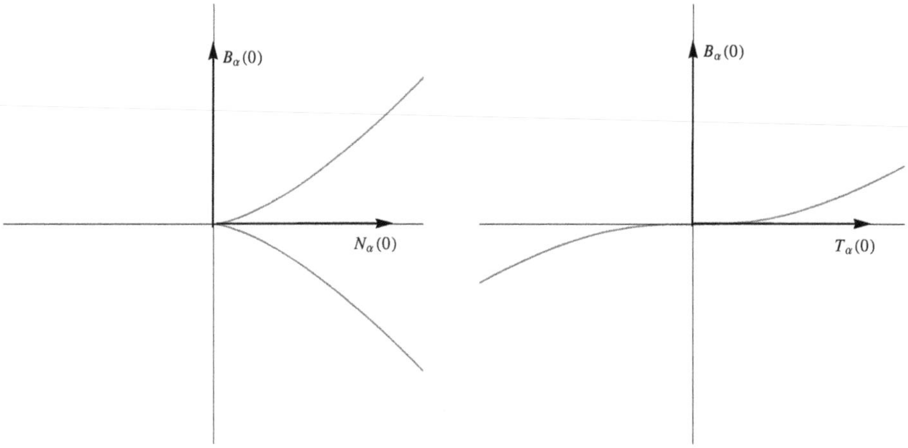

(a) Projection onto the normal plane (b) Projection in the rectifying plane

Figure 2.26: Projections onto the Cartan frame's coordinate planes.

We observe that here, even though the vectors in the Cartan frame are not pairwise orthogonal (no matter the ambient space), we have still represented as the above, since for qualitative effects, only their linear independence is relevant. Thus, we conclude that no matter the sign of the pseudo-torsion, the curve always crosses its own osculating planes in the direction of the binormal vectors.

If $\boldsymbol{\alpha}$ is semi-lightlike, in turn, we'll have

$$\boldsymbol{\alpha}(s) - \boldsymbol{R}(s) = \left(s, \frac{s^2}{2} + \mho_{\boldsymbol{\alpha}}(0)\frac{s^3}{6}, 0\right)_{\mathcal{F}},$$

which would actually allow us to predict Theorem 2.3.32 above. The only relevant projection is $\gamma \colon I \to \mathbb{R}^2$ given by

$$\gamma(s) = \left(s, \frac{s^2}{2} + \mho_{\boldsymbol{\alpha}}(0)\frac{s^3}{6}\right),$$

and it would be natural to seek a relation between the curvature of γ at 0 and the pseudo-torsion $\tau_{\alpha}(0)$. The issue, however, is that since $\boldsymbol{T}_{\alpha}(0)$ is spacelike, $\boldsymbol{N}_{\alpha}(0)$ is lightlike and they're orthogonal, to study γ we would have to consider the product given by $\langle\!\langle (x_1, x_2), (y_1, y_2) \rangle\!\rangle \doteq x_1 y_1$. Yet another caution is needed: the expression

$$\frac{\det(\gamma'(s), \gamma''(s))}{\|\gamma'(s)\|^3} = 1 + \tau_{\alpha}(0)s$$

may no longer be interpreted as the curvature of γ, since the ambient plane is degenerate. To wit, there is no reasonable notion of curvature in this setting, since every curve of the form $(s, f(s))$, where f is a smooth function, may be taken to the x-axis via the function $F\colon \mathbb{R}^2 \to \mathbb{R}^2$ given by $F(x,y) = (x, y - f(x))$. The derivative $DF(x,y)$ is a linear map, orthogonal relative to $\langle\!\langle \cdot, \cdot \rangle\!\rangle$ and, thus, F would be an "isometry" of this plane. In other words, all the graphs of smooth functions would be congruent. Since every spacelike curve may be parametrized as a graph over the x-axis and the lightlike curves are vertical lines, we see that the is no geometric invariant to associate to each curve here.

Before moving on and presenting a version of the Fundamental Theorem of Curves for lightlike and semi-lightlike curves, we will complete the comparison with the results obtained in the previous subsection by considering lightlike and semi-lightlike helices. Definition 2.3.16 (p. 110) is extended without issues to this new setting. Since every semi-lightlike is planar, it is automatically a helix. For lightlike curves, we recover Lancret's Theorem:

Theorem 2.3.33 (Lancret, lightlike version). *Let $\boldsymbol{\alpha}\colon I \to \mathbb{L}^3$ be a lightlike curve. Then $\boldsymbol{\alpha}$ is a helix if and only if $\tau_{\alpha}(\phi)$ is a constant.*

Proof: Assume that $\boldsymbol{\alpha}$ is a helix and let $\boldsymbol{v} \in \mathbb{L}^3$ be such that the $\langle \boldsymbol{T}_{\alpha}(\phi), \boldsymbol{v} \rangle_L = c \in \mathbb{R}$ is constant. Differentiating this twice we directly obtain that

$$\langle \boldsymbol{N}_{\alpha}(\phi), \boldsymbol{v} \rangle_L = \tau_{\alpha}(\phi)c + \langle \boldsymbol{B}_{\alpha}(\phi), \boldsymbol{v} \rangle_L = 0$$

for all $\phi \in I$. We claim that $c \neq 0$ and that $\langle \boldsymbol{B}_{\alpha}(\phi), \boldsymbol{v} \rangle_L$ is constant, whence $\tau_{\alpha}(\phi)$ must also be a constant. To wit, if $c = 0$ then Lemma 2.3.26 gives that $\boldsymbol{v} = \boldsymbol{0}$, contradicting the definition of a helix and, moreover, we have

$$\frac{\mathrm{d}}{\mathrm{d}\phi} \langle \boldsymbol{B}_{\alpha}(\phi), \boldsymbol{v} \rangle_L = \tau_{\alpha}(\phi) \langle \boldsymbol{N}_{\alpha}(\phi), \boldsymbol{v} \rangle_L = 0.$$

Conversely, assume that $\tau_{\alpha}(\phi) = \tau$ is a constant. If $\tau = 0$, then $\boldsymbol{v} = \boldsymbol{B}_{\alpha}(\phi)$ is a constant vector that clearly satisfies the required. And if $\tau \neq 0$, define

$$\boldsymbol{v} \doteq \boldsymbol{T}_{\alpha}(\phi) - \frac{1}{\tau} \boldsymbol{B}_{\alpha}(\phi).$$

Indeed, we have that

$$\frac{\mathrm{d}\boldsymbol{v}}{\mathrm{d}\phi} = \boldsymbol{N}_{\alpha}(\phi) - \frac{1}{\tau} \tau \boldsymbol{N}_{\alpha}(\phi) = \boldsymbol{0}$$

and $\boldsymbol{v}$ is constant. Furthermore, $\langle \boldsymbol{T}_{\alpha}(\phi), \boldsymbol{v} \rangle_L = 1/\tau$ is a constant, as desired. □

Theorem 2.3.34 (Fundamental Theorem, second version). *Let $\tau\colon I \to \mathbb{R}$ be a continuous function, $\boldsymbol{p}_0 \in \mathbb{L}^3$, $s_0, \phi_0 \in I$ and $(\boldsymbol{T}_0, \boldsymbol{N}_0, \boldsymbol{B}_0)$ a positive basis for $\mathbb{L}^3$ such that $\boldsymbol{B}_0$ is lightlike, and $(\boldsymbol{T}_0, \boldsymbol{N}_0)$ is a positive basis for a lightlike plane. Then:*

(i) if $\boldsymbol{T}_0$ is lightlike, $\boldsymbol{N}_0$ is unit, and $\langle \boldsymbol{T}_0, \boldsymbol{B}_0 \rangle_L = -1$, there is a unique lightlike curve $\boldsymbol{\alpha}\colon I \to \mathbb{L}^3$ with arc-photon parameter such that

- $\alpha(\phi_0) = p_0$;
- $(T_\alpha(\phi_0), N_\alpha(\phi_0), B_\alpha(\phi_0)) = (T_0, N_0, B_0)$;
- $\tau_\alpha(\phi) = \tau(\phi)$, *for all* $\phi \in I$;

(ii) if T_0 *is unit,* N_0 *is lightlike, and* $\langle N_0, B_0\rangle_L = -1$, *there is a unique semi-lightlike curve* $\alpha\colon I \to \mathbb{L}^3$ *such that*

- $\alpha(s_0) = p_0$;
- $(T_\alpha(s_0), N_\alpha(s_0), B_\alpha(s_0)) = (T_0, N_0, B_0)$;
- $\tau_\alpha(s) = \tau(s)$ *for all* $s \in I$;

Proof: We will focus only on the first case. As in the proof of the previous Fundamental Theorem, consider the following IVP in $\mathbb{R}^9$:

$$\begin{cases} \begin{pmatrix} T'(\phi) \\ N'(\phi) \\ B'(\phi) \end{pmatrix} = \begin{pmatrix} 0 & 1 & 0 \\ \tau(\phi) & 0 & 1 \\ 0 & \tau(\phi) & 0 \end{pmatrix} \begin{pmatrix} T(\phi) \\ N(\phi) \\ B(\phi) \end{pmatrix} \\ \text{and } \ (T(\phi_0), N(\phi_0), B(\phi_0)) = (T_0, N_0, B_0). \end{cases}$$

By the usual result regarding existence and uniqueness of solutions for systems of ordinary differential equations, there is a unique solution $\big(T(\phi), N(\phi), B(\phi)\big)$ for the IVP. We claim that the solution satisfies the conditions given in the hypothesis for all $\phi \in I$ (and not only ϕ_0), that is: $T(\phi)$ and $B(\phi)$ are always lightlike, $N(\phi)$ is unit spacelike and orthogonal to $B(\phi)$, and we still have $\langle T(\phi), B(\phi)\rangle_L = -1$. To wit, we consider the following IVP for the curve $a : I \to \mathbb{R}^6$:

$$\begin{cases} a'(\phi) &= A(\phi)a(\phi), \\ a(\phi_0) &= (0,1,0,0,-1,0) \end{cases}$$

where

$$A(\phi) = \begin{pmatrix} 0 & 0 & 0 & 2 & 0 & 0 \\ 0 & 0 & 0 & 2\tau(\phi) & 0 & 2 \\ 0 & 0 & 0 & 0 & 0 & 2\tau(\phi) \\ \tau(\phi) & 1 & 0 & 0 & 1 & 0 \\ 0 & 0 & 0 & \tau(\phi) & 0 & 1 \\ 0 & \tau(\phi) & 1 & 0 & \tau(\phi) & 0 \end{pmatrix}.$$

If the components of $a(\phi)$ are the possible scalar products[6] between the vectors $T(\phi)$, $N(\phi)$ and $B(\phi)$, solutions for the previous IVP, we conclude that the unique solution with the given initial conditions is the constant vector $a_0 = \big(0,1,0,0,-1,0\big)$, whence the claim follows.

With this done, we define

$$\alpha(\phi) \doteq p_0 + \int_{\phi_0}^{\phi} T(\xi)\,\mathrm{d}\xi.$$

Clearly we have $\alpha(\phi_0) = p_0$ and $\alpha'(\phi) = T(\phi)$, whence α is lightlike. Differentiating again, we obtain $\alpha''(\phi) = N(\phi)$, so that α has arc-photon parameter. Thus, we have that $T_\alpha(\phi) = T(\phi)$ and $N_\alpha(\phi) = N(\phi)$, while the positivity of all bases under discussion also ensures that $B_\alpha(\phi) = B(\phi)$ as well.

[6]In order, $a = \big(\langle T,T\rangle_L, \langle N,N\rangle_L, \langle B,B\rangle_L, \langle T,N\rangle_L, \langle T,B\rangle_L, \langle N,B\rangle_L\big)$.

With this, differentiating $\boldsymbol{N}_{\boldsymbol{\alpha}}(\phi) = \boldsymbol{N}(\phi)$ gives us that

$$\tau_{\boldsymbol{\alpha}}(\phi)\boldsymbol{T}_{\boldsymbol{\alpha}}(\phi) + \boldsymbol{B}_{\boldsymbol{\alpha}}(\phi) = \tau(\phi)\boldsymbol{T}(\phi) + \boldsymbol{B}(\phi),$$

and from the relations established so far it follows that $\tau_{\boldsymbol{\alpha}}(\phi) = \tau(\phi)$ for all $\phi \in I$.

The verification that such $\boldsymbol{\alpha}$ is unique is exactly the same as the one done in the proof of the Fundamental Theorem for admissible curves. □

Corollary 2.3.35. *Two curves, both lightlike or semi-lightlike whose osculating planes are positively oriented, and having the same pseudo-torsion, differ by a positive Poincaré transformation of* $\mathbb{L}^3$.

Proof: Let $\boldsymbol{\alpha}, \boldsymbol{\beta} \colon I \to \mathbb{R}^3_\nu$ be curves as in the statement above and $t_0 \in I$. Since the bases for the osculating planes are positive, the Cartan frames for both curves at t_0 satisfy the assumptions of Proposition 1.4.15 (p. 35), which gives us $F \in \mathrm{P}(3,\mathbb{R})$ such that $F(\boldsymbol{\alpha}(t_0)) = \boldsymbol{\beta}(t_0)$, $DF(\boldsymbol{\alpha}(t_0))(\boldsymbol{T}_{\boldsymbol{\alpha}}(t_0)) = \boldsymbol{T}_{\boldsymbol{\beta}}(t_0)$, and similarly for $\boldsymbol{N}$ and $\boldsymbol{B}$. Such F is in fact positive (since its linear part takes a positive basis into another positive basis), and thus preserves pseudo-torsions, so that the curves $F \circ \boldsymbol{\alpha}$ and $\boldsymbol{\beta}$ are now in the setting of Theorem 2.3.34 above. We conclude that $\boldsymbol{\beta} = F \circ \boldsymbol{\alpha}$, as wanted. □

Corollary 2.3.36. *A lightlike helix* $\boldsymbol{\alpha} \colon I \to \mathbb{L}^3$ *is congruent, for a suitable choice of* $r > 0$, *to a piece of one and only one of the following* standard helices*:*

- $\gamma_1(\phi) = \left(\sqrt{r}\phi, r\cosh(\phi/\sqrt{r}), r\sinh(\phi/\sqrt{r})\right)$;
- $\gamma_2(\phi) = \left(r\cos(\phi/\sqrt{r}), r\sin(\phi/\sqrt{r}), \sqrt{r}\phi\right)$;
- $\gamma_3(\phi) = \left(-\dfrac{\phi^3}{4} + \dfrac{\phi}{3}, \dfrac{\phi^2}{2}, -\dfrac{\phi^3}{4} - \dfrac{\phi}{3}\right)$.

Proof: We'll denote the pseudo-torsion of $\boldsymbol{\alpha}$, which we know to be constant, just by τ. From the proof of the lightlike version of Lancret's Theorem, we know that the helical axis of the curve, if $\tau \neq 0$, is given by

$$\boldsymbol{v} = \boldsymbol{T}_{\boldsymbol{\alpha}}(\phi) - \frac{1}{\tau}\boldsymbol{B}_{\boldsymbol{\alpha}}(\phi),$$

whence $\langle \boldsymbol{v}, \boldsymbol{v} \rangle_L = 2/\tau$, while we take $\boldsymbol{v} = \boldsymbol{B}_{\boldsymbol{\alpha}}(\phi)$ if $\tau = 0$ (and thus $\langle \boldsymbol{v}, \boldsymbol{v} \rangle_L = 0$). This way, we have that $\boldsymbol{\alpha}$ is

- hyperbolic if $\tau > 0$, and thus congruent with γ_1;
- elliptic if $\tau < 0$, and thus congruent with γ_2;
- parabolic if $\tau = 0$, and thus congruent with γ_3.

□

Exercises

Exercise† 2.3.27. Show Proposition 2.3.25 in the case where the curve is semi-lightlike.

Exercise 2.3.28 (More examples)**.** Determine the Cartan frame and the pseudo-torsion for the following curves:

(a) $\boldsymbol{\beta}\colon \mathbb{R} \to \mathbb{L}^3$ given by

$$\boldsymbol{\beta}(\phi) = \left(\sqrt{r}\phi, r\cosh\left(\frac{\phi}{\sqrt{r}}\right), r\sinh\left(\frac{\phi}{\sqrt{r}}\right)\right),$$

where $r > 0$.

(b) $\boldsymbol{\gamma}\colon \mathbb{R} \to \mathbb{L}^3$, $\boldsymbol{\gamma}(s) = (-e^s, s, e^s)$.

(c) $\boldsymbol{\zeta}\colon \mathbb{R}_{>0} \to \mathbb{L}^3$, $\boldsymbol{\zeta}(s) = (\sqrt{2}/2)(s - \log(1/s), s + \log(1/s), \sqrt{2}\log s)$.

Exercise† 2.3.29. Prove the Fundamental Theorem for semi-lightlike curves.

Exercise 2.3.30 ([44]). Show that every semi-lightlike curve $\boldsymbol{\alpha}\colon I \to \mathbb{L}^3$ with constant pseudo-torsion $\text{б}_{\boldsymbol{\alpha}}(s) = \text{б} \neq 0$ contained in the plane $\Pi\colon y = z$ has the form

$$\boldsymbol{\alpha}(s) = \left(\pm s + a, \frac{b}{\text{б}^2}e^{\text{б}s} + cs + d, \frac{b}{\text{б}^2}e^{\text{б}s} + cs + d\right)$$

for certain constants $a, b, c, d \in \mathbb{R}$, perhaps up to reparametrization.

CHAPTER 3

Surfaces in Space

INTRODUCTION

Now we turn our attention to surfaces in $\mathbb{R}^3_\nu$, the main goal of this text.

In Section 3.1, we introduce the concept of regular surface and present a few examples, to then classify such surfaces locally as inverse images of regular values of functions from $\mathbb{R}^3$ to $\mathbb{R}$ and graphs of functions from $\mathbb{R}^2$ to $\mathbb{R}$ (which provides a larger class of examples). In what follows, we formalize the important concept of tangent plane, seen in Calculus courses, now in our new setting. Aiming to replicate the tools from Calculus in the theory of regular surfaces, we establish the notions of smooth function and differential (through tangent planes), illustrating those with several examples. We conclude this section by discussing the notion of orientation for surfaces, which intuitively says when a surface has an "inside" and an "outside" or, equivalently, when an inhabitant of the surface is capable to decide what is "left" and what is "right".

In Section 3.2, unlike the previous section, we start to consider the influence of the ambient scalar product on the surface, that is, we start to actually study the geometry of the surface. The first step is, naturally, to define the causal type of a surface from the one of its tangent planes, and present some criteria for deciding such a causal type. Next, we'll see how to measure lengths, angles, and areas in a surface with its First Fundamental Form, which is nothing more than the restriction of the ambient scalar product to its tangent planes. We conclude the section by presenting the concept of isometry: the correct notion of equivalence, when discussing geometry of surfaces.

In Chapter 2, to study the trace of a curve in $\mathbb{R}^3_\nu$ we employed the invariants curvature and torsion, which depended not only on the tangent directions to the curve, but on its entire Frenet-Serret trihedron. In Section 3.3, to study the analogous situation for surfaces, it suffices to understand how the normal directions to the surface are changing, as this is directly related to the tangent planes. Such analysis is done by using the Gauss normal map and its differential, the Weingarten map. With those, we introduce the Second Fundamental Form and, consequently, the curvatures of a surface. We state *Gauss' Theorema Egregium* (whose proof is postponed to Section 3.7), which says that the Gaussian curvature of a surface depends only on geometric measures realized on the surface, independently on the ambient where the surface lies (be it $\mathbb{R}^3$ or $\mathbb{L}^3$).

One issue that does not occur in $\mathbb{R}^3$, but appears in the study of surfaces in $\mathbb{L}^3$, concerns the diagonalizability of the Weingarten map, which is always possible at least in the so-called umbilic points of the surface. All of this is discussed in Section 3.4, where we also provide a complete classification of all the surfaces for which all the points are umbilic. In what follows, we prove the existence of inertial parametrizations for any surface, which allows us to obtain visual interpretations for the sign of the Gaussian curvature in any given point. The mean curvature, in turn, is closely related to areas of regions in the surface, and we classify surfaces with zero mean curvature (called critical surfaces) as critical points of the area functional.

At this point, we are ready to study curves in a surface, from the point of view of its inhabitants. This begins in Section 3.5. Two important classes of curves in a surface are the asymptotic curves and lines of curvature, which may be described by certain differential equations.

In Section 3.6, our focus is on the curves on a surface which play the same role as straight lines in a plane: geodesics. For this, we introduce covariant derivatives of vector fields along curves. In a similar way to what was done in Chapter 2, we use a suitable frame (Darboux-Ribaucour Trihedron) to simultaneously study geodesics, asymptotic curves, and lines of curvature, via new invariants: geodesic curvature and geodesic torsion. Next, we introduce Christoffel symbols of a parametrization, which may be used to locally characterize geodesics in terms of a system of differential equations. Geodesics also arise naturally as solutions of a variational problem, expressed through these same differential equations.

We conclude this chapter with Section 3.7, where we state and prove the Fundamental Theorem of Surfaces, completing the comparison with the local theory of curves developed in Chapter 2. Moreover, the propositions to be proved here provide a quick proof of *Gauss' Theorema Egregium*, first stated in Section 3.3.

3.1 BASIC TOPOLOGY OF SURFACES

Initially, we present the definition of a regular surface, which does not take into account aspects relative to the ambient geometry ($\mathbb{R}^3$ or $\mathbb{L}^3$), and we study a few useful general facts for what follows. In particular, we'll need to consider functions defined only on a surface, as well as suitable notions of continuity and differentiability. For this, it is necessary to say what will be the "open subsets of the surface". Given a subset $S \subseteq \mathbb{R}^n$, recall that $A \subseteq S$ is an *open subset of* S if A is the intersection of S with an open subset of $\mathbb{R}^n$.

That being said, we need to answer the following question: what does it mean for a subset of $\mathbb{L}^n$ to be open? Recall that, in $\mathbb{R}^n$, given $\boldsymbol{p} = (p_1, \ldots, p_n)$ and $r > 0$, we define the open ball of center $\boldsymbol{p}$ and radius r as

$$B_r(\boldsymbol{p}) = \{(x_1, \ldots, x_n) \in \mathbb{R}^n \mid (x_1 - p_1)^2 + \cdots + (x_n - p_n)^2 < r^2\},$$

and with this one defines what is an open subset of $\mathbb{R}^n$, which, *a priori*, does not depend on the extra structure $\langle \cdot, \cdot \rangle_E$ or $\langle \cdot, \cdot \rangle_L$ chosen. An open ball may be expressed in terms of $\| \cdot \|_E$, but not of $\| \cdot \|_L$ (which is not a norm). Not only that, but the natural attempt to define open balls in a similar way to what was done above but using $\langle \cdot, \cdot \rangle_L$ instead does not yield a good "collection of open sets" in $\mathbb{L}^n$. Thus: *we'll take the open subsets of* $\mathbb{L}^n$ *to be the same ones as in* $\mathbb{R}^n$.

Throughout this chapter, $U \subseteq \mathbb{R}^2$ will always denote an open subset of the plane.

Definition 3.1.1 (Regular Surfaces). A *regular surface* in $\mathbb{R}^3_\nu$ is a subset $M \subseteq \mathbb{R}^3_\nu$ such that for all $\boldsymbol{p} \in M$, there are open sets $U \subseteq \mathbb{R}^2$ and $\boldsymbol{p} \in V \subseteq M$, and a mapping $\boldsymbol{x} : U \to V$ such that:

(i) $\boldsymbol{x}$ is smooth;

(ii) $\boldsymbol{x}$ is a homeomorphism;

(iii) $D\boldsymbol{x}(u, v)$ has full rank for all $(u, v) \in U$.

Under these conditions we say that, around $\boldsymbol{p}$, $\boldsymbol{x}$ is a *parametrization for* M, (u,v) are *local coordinates*, and $\boldsymbol{x}^{-1}\colon \boldsymbol{x}(U) \to U$ is a *chart*.

Remark. When convenient, we will write just $(U,\boldsymbol{x})$ instead of $\boldsymbol{x}\colon U \to \boldsymbol{x}(U) \subseteq M$.

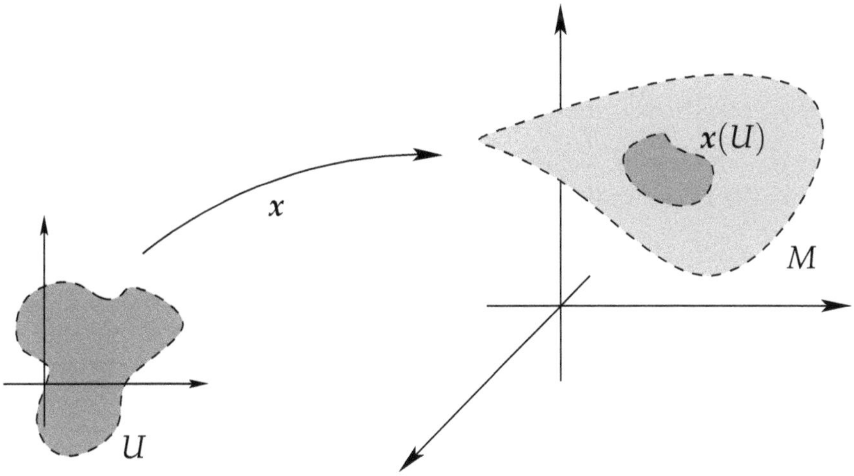

Figure 3.1: Illustrating the above definition.

Definition 3.1.2. A smooth map $\boldsymbol{x}\colon U \to \mathbb{R}^3_\nu$ is called a *regular parametrized surface* if $D\boldsymbol{x}(u,v)$ has full rank for all $(u,v) \in U$.

Remark. A regular parametrized surface is not necessarily injective, so that its image in space may have self-intersections.

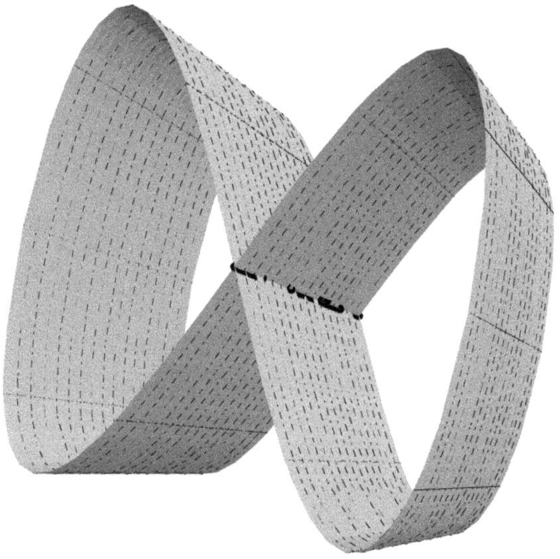

Figure 3.2: A regular parametrization with self-intersections.

In general, keeping the notation from the previous two definitions, we have that $D\boldsymbol{x}(u,v)$ having full rank is equivalent to the vectors $\boldsymbol{x}_u(u,v)$ and $\boldsymbol{x}_v(u,v)$ (which are the columns of $D\boldsymbol{x}(u,v)$) being linearly independent. This, in turn, is equivalent to any of the following conditions:

- $\boldsymbol{x}_u(u,v) \times_E \boldsymbol{x}_v(u,v) \neq \boldsymbol{0}$;
- $\boldsymbol{x}_u(u,v) \times_L \boldsymbol{x}_v(u,v) \neq \boldsymbol{0}$;
- $\|\boldsymbol{x}_u(u,v) \times_E \boldsymbol{x}_v(u,v)\|_E^2 \neq 0$;
- $\|\boldsymbol{x}_u(u,v) \times_L \boldsymbol{x}_v(u,v)\|_E^2 \neq 0$.

Fixing $u = u_0$ and letting v change (or similarly, letting u change and fixing $v = v_0$), we have the so-called *coordinate curves* of $\boldsymbol{x}$:

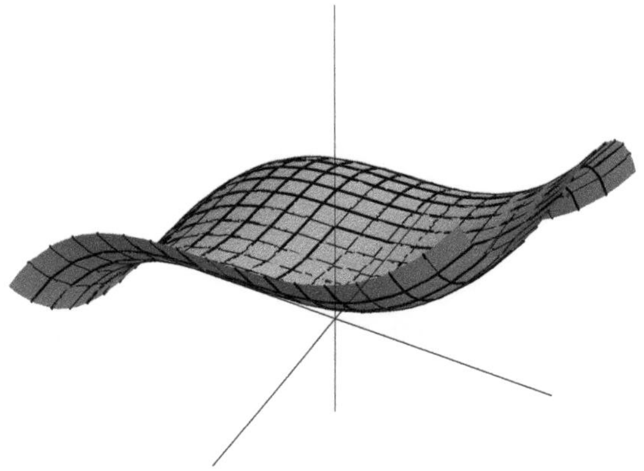

Figure 3.3: The coordinate curves of a parametrization.

The vectors $\boldsymbol{x}_u(u_0,v_0)$ and $\boldsymbol{x}_v(u_0,v_0)$ are tangent to such coordinate curves, at the point $\boldsymbol{x}(u_0,v_0)$.

The relation between the concepts listed above is summarized in the following:

Proposition 3.1.3. *Let M be a regular surface, $\boldsymbol{p} \in M$, and consider a bijective smooth map $\boldsymbol{x} : U \to \boldsymbol{x}(U) \subseteq M$ such that $\boldsymbol{p} \in \boldsymbol{x}(U)$ and $D\boldsymbol{x}(u,v)$ has full rank for all $(u,v) \in U$. Then $\boldsymbol{x}^{-1}$ is continuous, and thus $\boldsymbol{x}$ is indeed a parametrization for M around $\boldsymbol{p}$.*

Remark. In other words, forcing all the parametrizations $\boldsymbol{x}$ in Definition 3.1.1 to be homeomorphisms is redundant in the presence of the remaining conditions.

Proof: Explicitly write $\boldsymbol{x}(u,v) = (x(u,v), y(u,v), z(u,v))$, and take $(u_0,v_0) \in U$. Let's prove that $\boldsymbol{x}^{-1}$ is continuous in a neighborhood of (u_0,v_0), so that global continuity follows from (u_0,v_0) being arbitrary. Since $D\boldsymbol{x}(u,v)$ has full rank, assume without loss of generality that

$$\frac{\partial(x,y)}{\partial(u,v)}(u_0,v_0) = \begin{vmatrix} \dfrac{\partial x}{\partial u}(u_0,v_0) & \dfrac{\partial x}{\partial v}(u_0,v_0) \\ \dfrac{\partial y}{\partial u}(u_0,v_0) & \dfrac{\partial y}{\partial v}(u_0,v_0) \end{vmatrix} \neq 0.$$

So, if $\varphi : U \subseteq \mathbb{R}^2 \to \mathbb{R}^2$ is given by $\varphi(u,v) = (x(u,v), y(u,v))$, we have that:

$$\det D\varphi(u_0,v_0) = \frac{\partial(x,y)}{\partial(u,v)}(u_0,v_0) \neq 0,$$

and the Inverse Function Theorem gives us an open set $(u_0, v_0) \in V \subseteq U$ where the inverse $\varphi^{-1} : \varphi(V) \to V$ exists and is smooth. Write $\varphi^{-1}(x,y) = (u(x,y), v(x,y))$ and observe that $\varphi = \pi \circ \boldsymbol{x}$, where π is the projection in the first two coordinates. So, given $(x,y,z) \in \boldsymbol{x}(V)$, we have:

$$\varphi^{-1} \circ \pi(x,y,z) = \varphi^{-1}(x,y) = (u(x,y), v(x,y)) = \boldsymbol{x}^{-1}(x,y,z),$$

whence $\boldsymbol{x}^{-1}|_{\boldsymbol{x}(V)} = \varphi^{-1} \circ \pi$ is the composition of continuous functions, and hence continuous as well. □

Corollary 3.1.4. *Let $\boldsymbol{x}\colon U \to \mathbb{R}^3_\nu$ be a regular parametrized surface. If $\boldsymbol{x}$ is injective, then $\boldsymbol{x}(U)$ is a regular surface.*

Example 3.1.5.

(1) Planes: let $\boldsymbol{p}, \boldsymbol{w}_1, \boldsymbol{w}_2 \in \mathbb{R}^3_\nu$ be such that $\{\boldsymbol{w}_1, \boldsymbol{w}_2\}$ is linearly independent. Then $\boldsymbol{x}\colon \mathbb{R}^2 \to \mathbb{R}^3_\nu$ given by $\boldsymbol{x}(u,v) = \boldsymbol{p} + u\boldsymbol{w}_1 + v\boldsymbol{w}_2$ is an injective regular parametrized surface. Thus, the plane Π passing through $\boldsymbol{p}$ and spanned by $\boldsymbol{w}_1$ and $\boldsymbol{w}_2$ is a regular surface. Since every plane in space has this form for suitable $\boldsymbol{p}, \boldsymbol{w}_1$ and $\boldsymbol{w}_2$, we see that *all* planes are regular surfaces.

(2) Graphs: if $f\colon U \to \mathbb{R}$ is a smooth function, then its graph

$$\mathrm{gr}(f) \doteq \{(u, v, f(u,v)) \in \mathbb{R}^3_\nu \mid (u,v) \in U\}$$

is a regular surface. To wit, the map $\boldsymbol{x}\colon U \to \mathbb{R}^3_\nu$ given by $\boldsymbol{x}(u,v) = (u, v, f(u,v))$ is an injective regular parametrized surface whose image is precisely $\mathrm{gr}(f)$. A parametrization $\boldsymbol{x}$ of this form is called a *Monge parametrization*. We have similar results for graphs of functions defined in the other coordinate planes in space, $x = 0$ or $y = 0$.

(3) Each lightcone $C_L(\boldsymbol{p})$ in $\mathbb{L}^3$ is a regular surface. Recall that we have removed the cone vertex (in this case, $\boldsymbol{p}$ itself). Writing $\boldsymbol{p} = (p_1, p_2, p_3)$, we have that the cone is covered by the two parametrizations $\boldsymbol{x}_\pm\colon \mathbb{R}^2 \setminus \{(p_1, p_2)\} \to \mathbb{L}^3$, given by

$$\boldsymbol{x}_\pm(u,v) = \left(u, v, p_3 \pm \sqrt{(u-p_1)^2 + (v-p_2)^2}\right).$$

(4) The *helicoid*: consider a circular helix whose axis is the z-axis, and join its points to the z-axis by horizontal lines. One possible parametrization for the obtained set is $\boldsymbol{x}\colon\,]0,1[\, \times \mathbb{R} \to \mathbb{R}^3_\nu$, given by $\boldsymbol{x}(u,v) = (u\cos v, u \sin v, v)$. Clearly $\boldsymbol{x}$ is smooth and we have that:

$$\frac{\partial \boldsymbol{x}}{\partial u}(u,v) = (\cos v, \sin v, 0) \quad \text{and} \quad \frac{\partial \boldsymbol{x}}{\partial v}(u,v) = (-u\sin v, u\cos v, 1),$$

whence $\boldsymbol{x}$ is a regular parametrized surface (to wit, the partial derivatives are linearly independent because of their last components). Moreover, note that $\boldsymbol{x}$ is injective, whence the helicoid $\boldsymbol{x}(\,]0,1[\, \times \mathbb{R})$ is a regular surface.

Figure 3.4: The helicoid.

(5) Surfaces of Revolution: consider a smooth curve, regular and injective in the plane $y = 0$, $\boldsymbol{\alpha}\colon I \to \mathbb{R}^3_\nu$, given by $\boldsymbol{\alpha}(u) = (f(u), 0, g(u))$, and such that $f(u) > 0$ for all $u \in I$. This condition only says that the curve does not touch the z-axis. Rotating the curve around the z-axis, we obtain the *revolution surface generated by* $\boldsymbol{\alpha}$. The curves given by intersections of such surface with horizontal planes are called *parallels*, while the intersections with vertical planes passing through the origin are called *meridians*.

One standard parametrization that misses only one of the surface's meridians is $\boldsymbol{x}\colon I \times \,]0, 2\pi[\, \to \mathbb{R}^3_\nu$, given by $\boldsymbol{x}(u,v) = (f(u)\cos v, f(u)\sin v, g(u))$. One may derive such expression for $\boldsymbol{x}$ in at least two ways: fixed u, we are parametrizing circles of radius $f(u)$ inside horizontal planes of height $g(u)$, or then directly applying the rotation around the z-axis to the curve $\boldsymbol{\alpha}$, as follows:

$$\begin{pmatrix} \cos v & -\sin v & 0 \\ \sin v & \cos v & 0 \\ 0 & 0 & 1 \end{pmatrix} \begin{pmatrix} f(u) \\ 0 \\ g(u) \end{pmatrix} = \begin{pmatrix} f(u)\cos v \\ f(u)\sin v \\ g(u) \end{pmatrix}.$$

In any case, since both f and g are smooth, so is $\boldsymbol{x}$. We have:

$$\frac{\partial \boldsymbol{x}}{\partial u}(u,v) = (f'(u)\cos v, f'(u)\sin v, g'(u)) \quad \text{and}$$
$$\frac{\partial \boldsymbol{x}}{\partial v}(u,v) = (-f(u)\sin v, f(u)\cos v, 0).$$

With this:

$$\frac{\partial \boldsymbol{x}}{\partial u}(u,v) \times_E \frac{\partial \boldsymbol{x}}{\partial v}(u,v) = (-f(u)g'(u)\cos v, -f(u)g'(u)\sin v, f'(u)f(u)),$$

and so:

$$\left\| \frac{\partial \boldsymbol{x}}{\partial u}(u,v) \times_E \frac{\partial \boldsymbol{x}}{\partial v}(u,v) \right\|_E^2 = f(u)^2(f'(u)^2 + g'(u)^2) \neq 0.$$

This shows that $\boldsymbol{x}$ is a regular parametrized surface. And clearly $\boldsymbol{x}$ is injective, so that the image $\boldsymbol{x}\big(I \times \,]0, 2\pi[\,\big)$ is a regular surface.

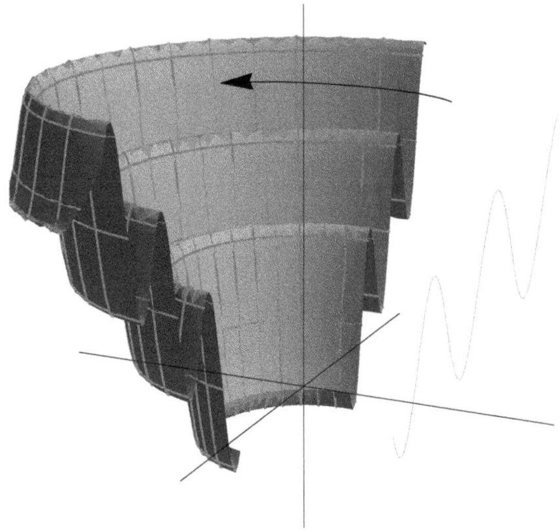

Figure 3.5: A curve and the revolution surface it generates.

For this parametrization, the parallels are the coordinate curves of the form $u = \text{cte.}$, while the meridians are the curves of the form $v = \text{cte.}$.

Besides the previous examples, many others may be exhibited implicitly via functions satisfying certain properties. That is, it is not always necessary to exhibit parametrizations to show that a set is a regular surface. We start with the:

Definition 3.1.6. Let $\Omega \subseteq \mathbb{R}^3_\nu$ be open and $f : \Omega \to \mathbb{R}$ be a smooth function. We say that $q \in \mathbb{R}^3_\nu$ is a *regular point* if $Df(q)$ is surjective[1] and a *critical value* if $Df(q)$ is not surjective. Furthermore, a number $a \in \mathbb{R}$ is called a *regular value* if $f^{-1}(\{a\})$ is non-empty and consists only of regular points.

Theorem 3.1.7 (Inverse image of a regular value)**.** *Let $\Omega \subseteq \mathbb{R}^3_\nu$ be open and $f : \Omega \to \mathbb{R}$ be smooth. If $a \in \mathbb{R}$ is a regular value for f, then $f^{-1}(\{a\})$ is a regular surface in $\mathbb{R}^3_\nu$.*

Proof: Let $p \in f^{-1}(\{a\})$. Assume without loss of generality that $f_z(p) \neq 0$ and consider the function $\varphi : \Omega \to \mathbb{R}^3$ given by $\varphi(x,y,z) = (x,y,f(x,y,z))$. Then:

$$\det D\varphi(p) = \begin{vmatrix} 1 & 0 & 0 \\ 0 & 1 & 0 \\ \dfrac{\partial f}{\partial x}(p) & \dfrac{\partial f}{\partial y}(p) & \dfrac{\partial f}{\partial z}(p) \end{vmatrix} = \frac{\partial f}{\partial z}(p) \neq 0.$$

So, the Inverse Function Theorem yields an open set $p \in V \subseteq \Omega$ for which the inverse $\varphi^{-1} : \varphi(V) \to V$ exists and is smooth. In view of the first two components of $\varphi(x,y,z)$, we have that $\varphi^{-1}(x,y,z) = (x,y,g(x,y,z))$ for some smooth function g. Restricted to $V \cap f^{-1}(\{a\})$ (which is open in $f^{-1}(\{a\})$), we have that:

$$\begin{aligned} (x,y,z) &= \varphi^{-1} \circ \varphi(x,y,z) = \varphi^{-1}(x,y,f(x,y,z)) \\ &= \varphi^{-1}(x,y,a) = (x,y,g(x,y,a)), \end{aligned}$$

whence $z = g(x,y,a)$. This way, if π denotes the projection of $\mathbb{R}^3$ in the first two

[1]In this case, this is equivalent to saying that $Df(q)$ is not the zero linear functional.

components, we may define a parametrization $x: \pi(V \cap f^{-1}(\{a\})) \to V \cap f^{-1}(\{a\})$ by $x(u,v) = (u,v,g(u,v,a))$. Note that x indeed takes values in $f^{-1}(\{a\})$, since $f(u,v,g(u,v,a)) = a$. Furthermore, it is straightforward to check that x is smooth, injective, and has full rank (hence a homeomorphism). As $p \in f^{-1}(\{a\})$ was arbitrary, we conclude that $f^{-1}(\{a\})$ is a regular surface. □

Example 3.1.8. If $p \in \mathbb{R}^3_\nu$ is a fixed point and the function $f: \mathbb{R}^3_\nu \to \mathbb{R}$ is defined by $f(q) = \langle q-p, q-p \rangle$, we have that $Df(q) = 2\langle q-p, \cdot \rangle$, which is only the zero functional when $q = p$, once $\langle \cdot, \cdot \rangle$ is non-degenerate. In particular, if $r \neq 0$, the following are regular surfaces:

- $f^{-1}(\{r^2\}) = \mathbb{S}^2(p,r)$, in $\mathbb{R}^3$;
- $f^{-1}(\{r^2\}) = \mathbb{S}^2_1(p,r)$ and $f^{-1}(\{-r^2\}) = \mathbb{H}^2_+(p,r) \cup \mathbb{H}^2_-(p,r)$, in $\mathbb{L}^3$.

Each connected component of $f^{-1}(\{-r^2\})$ is, in its own right, a regular surface (see Exercise 3.1.12).

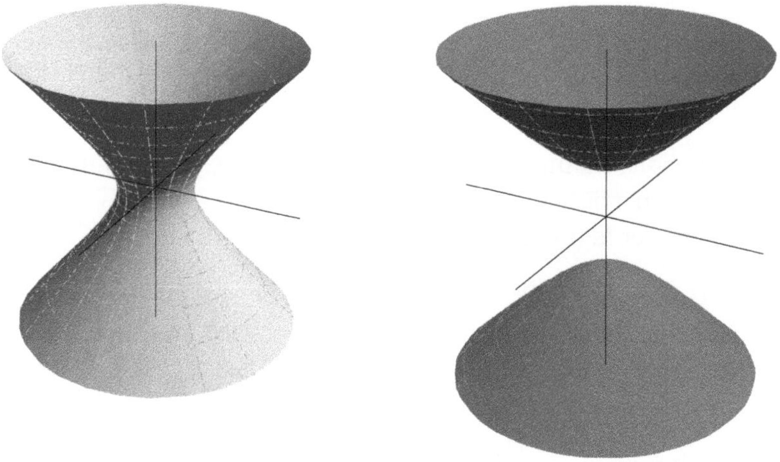

(a) The de Sitter space $\mathbb{S}^2_1$ (b) The two-sheeted hyperboloid $\mathbb{H}^2 \cup \mathbb{H}^2_-$

Figure 3.6: The analogues to Euclidean spheres, in $\mathbb{L}^3$.

Just as $\mathbb{R}^3$ is the disjoint union of the origin with spheres centered at the origin with arbitrary positive radius, $\mathbb{L}^3$ is also a disjoint union of de Sitter spaces, hyperbolic planes, the lightcone, and the origin. More precisely, we have that:

$$\mathbb{L}^3 = \left(\bigcup_{r>0} \mathbb{S}^2_1(r)\right) \cup \left(\bigcup_{r>0} \mathbb{H}^2(r)\right) \cup \left(\bigcup_{r>0} \mathbb{H}^2_-(r)\right) \cup C_L(\mathbf{0}) \cup \{\mathbf{0}\}.$$

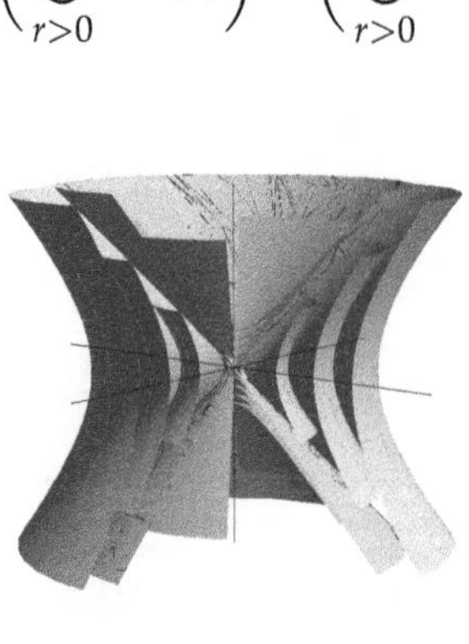

Figure 3.7: The structure of $\mathbb{L}^3$ in slices.

Among the examples of regular surfaces seen so far, graphs of smooth functions might seem a quite restricted class of examples, but in fact, in a similar fashion to what was done for curves in the last chapter, it holds that every regular surface is locally the graph of a smooth function. This is made precise in:

Proposition 3.1.9 (Local Graph). *Let $M \subseteq \mathbb{R}^3_\nu$ be a regular surface and $\boldsymbol{p} \in M$. Then, there is a Monge parametrization for M around $\boldsymbol{p}$.*

Proof: Take a parametrization $(U, \boldsymbol{x})$ for M around $\boldsymbol{p}$. Writing its components as $\boldsymbol{x}(u,v) = (x(u,v), y(u,v), z(u,v))$. By the regularity of $\boldsymbol{x}$, we may assume without loss of generality that

$$\frac{\partial(x,y)}{\partial(u,v)}(\boldsymbol{x}^{-1}(\boldsymbol{p})) \neq 0.$$

Define $\varphi : U \to \mathbb{R}^2$ by $\varphi(u,v) = (x(u,v), y(u,v))$. So we have that:

$$\det D\varphi(\boldsymbol{x}^{-1}(\boldsymbol{p})) = \frac{\partial(x,y)}{\partial(u,v)}(\boldsymbol{x}^{-1}(\boldsymbol{p})) \neq 0,$$

and the Inverse Function Theorem gives us an open set $\boldsymbol{x}^{-1}(\boldsymbol{p}) \in V \subseteq U$ where the inverse $\varphi^{-1} : \varphi(V) \to V$ exists and is smooth, of the form $\varphi^{-1}(s,t) = (u(s,t), v(s,t))$. Observe that:

$$(s,t) = \varphi \circ \varphi^{-1}(s,t) = \varphi(u(s,t), v(s,t)) = (x(u(s,t), v(s,t)), y(u(s,t), v(s,t))).$$

Now consider the map $\boldsymbol{y} \doteq \boldsymbol{x} \circ \varphi^{-1} : \varphi(V) \to \boldsymbol{x}(V) \subseteq M$. Since $\varphi|_V$ is a diffeomorphism, $\boldsymbol{y}$ is a parametrization for M around $\boldsymbol{p}$. And lastly, let's see that $\boldsymbol{y}$ is a Monge parametrization:

$$\begin{aligned}
\boldsymbol{y}(s,t) &= \boldsymbol{x} \circ \varphi^{-1}(s,t) \\
&= \boldsymbol{x}(u(s,t), v(s,t)) \\
&= (x(u(s,t), v(s,t)), y(u(s,t), v(s,t)), z(u(s,t), v(s,t))) \\
&= (s, t, z(u(s,t), v(s,t))).
\end{aligned}$$

□

Remark. We will see in the next section that it is possible to more precise with this result in $\mathbb{L}^3$ by using the causal character of M, to be properly defined there.

Regular parametrizations may be used to describe, in terms of coordinates in the plane, certain aspects of the surface. This is important because in the plane, we have available the necessary tools from Calculus and Linear Algebra to study the *local geometry* of the surface. On the other hand, we want descriptions made using different coordinates to be, in a certain sense, equivalent, and the geometric objects to be defined to be independent of the choice of parametrization. A first step towards that is the:

Theorem 3.1.10. *Let $M \subseteq \mathbb{R}^3_\nu$ be a regular surface and $(U_{\boldsymbol{x}}, \boldsymbol{x})$ and $(U_{\boldsymbol{y}}, \boldsymbol{y})$ be two parametrizations for M such that $W \doteq \boldsymbol{x}(U_{\boldsymbol{x}}) \cap \boldsymbol{y}(U_{\boldsymbol{y}})$ is non-empty. Then the* change of coordinates[2] $\boldsymbol{x}^{-1} \circ \boldsymbol{y} : \boldsymbol{y}^{-1}(W) \to \boldsymbol{x}^{-1}(W)$ *is smooth.*

[2]Also called *change of parameters.*

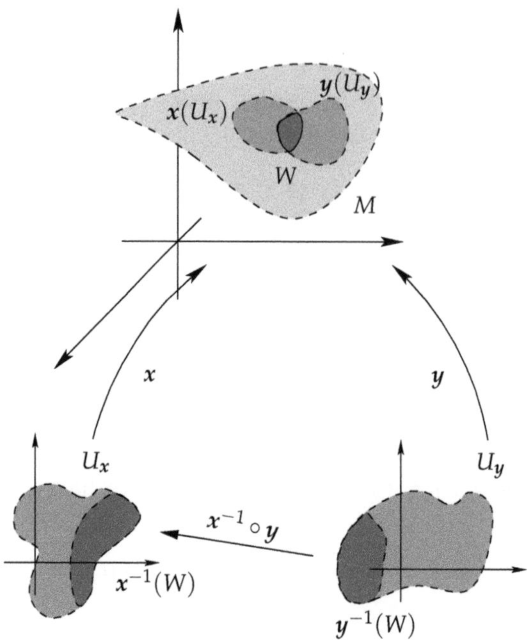

Figure 3.8: Summarizing the situation below.

Proof: Pick any point $(s_0, t_0) \in \boldsymbol{y}^{-1}(W)$. We will show that $\boldsymbol{x}^{-1} \circ \boldsymbol{y}$ is smooth in an open neighborhood of (s_0, t_0). Then global smoothness follows from (s_0, t_0) being arbitrary. Explicitly write

$$\boldsymbol{x}(u,v) = (x_1(u,v), x_2(u,v), x_3(u,v)) \quad \text{and} \quad \boldsymbol{y}(s,t) = (y_1(s,t), y_2(s,t), y_3(s,t)),$$

and let $(u_0, v_0) = \boldsymbol{x}^{-1} \circ \boldsymbol{y}(s_0, t_0)$. By the regularity of $\boldsymbol{x}$, we may assume without loss of generality that

$$\frac{\partial(x_1, x_2)}{\partial(u,v)}(u_0, v_0) \neq 0.$$

Define $\varphi \colon \boldsymbol{x}^{-1}(W) \to \mathbb{R}^2$ by $\varphi(u,v) = (x_1(u,v), x_2(u,v))$, and note that

$$\det D\varphi(u_0, v_0) = \frac{\partial(x_1, x_2)}{\partial(u,v)}(u_0, v_0) \neq 0,$$

whence the Inverse Function Theorem gives us an open set $(u_0, v_0) \in V \subseteq \boldsymbol{x}^{-1}(W)$ where the inverse function $\varphi^{-1} \colon \varphi(V) \to V$ exists and is smooth, say, written in components as $\varphi^{-1}(x_1, x_2) = (u(x_1, x_2), v(x_1, x_2))$. If π is the projection of $\mathbb{R}^3$ in the first two components, note that $\boldsymbol{y}^{-1}(\pi^{-1}(\varphi(V)))$ is an open set containing (s_0, t_0). Lastly, if $(s,t) \in \boldsymbol{y}^{-1}(\pi^{-1}(\varphi(V)))$, we have that

$$\begin{aligned}(\boldsymbol{x}^{-1} \circ \boldsymbol{y})\big|_{\boldsymbol{y}^{-1}(\pi^{-1}(\varphi(V)))}(s,t) &= \varphi^{-1}(\pi(\boldsymbol{y}(s,t))) \\ &= \varphi^{-1}(y_1(s,t), y_2(s,t)) \\ &= (u(y_1(s,t), y_2(s,t)), v(y_1(s,t), y_2(s,t)))\end{aligned}$$

is the composition of smooth maps, hence smooth as well, as wanted. $\square$

Remark. It follows from the above result that the change of coordinates is a diffeomorphism. It suffices to apply this result to its inverse as well.

We know, from Calculus, that the best linear approximation for a smooth function $f: U \subseteq \mathbb{R}^2 \to \mathbb{R}$ is represented by the tangent plane to its graph. Considering the Monge parametrization $\boldsymbol{x}: U \to \mathrm{gr}(f)$ previously seen, we have that the tangent plane to the graph of f at the point $(u_0, v_0, f(u_0, v_0))$ is spanned by the vectors

$$\frac{\partial \boldsymbol{x}}{\partial u}(u_0, v_0) = \left(1, 0, \frac{\partial f}{\partial u}(u_0, v_0)\right) \quad \text{and} \quad \frac{\partial \boldsymbol{x}}{\partial v}(u_0, v_0) = \left(0, 1, \frac{\partial f}{\partial v}(u_0, v_0)\right).$$

The next step in our discussion will be, motivated by this, to extend this notion to arbitrary regular surfaces, as follows:

Lemma 3.1.11. *Let $M \subseteq \mathbb{R}^3_\nu$ be a regular surface and $(U_{\boldsymbol{x}}, \boldsymbol{x})$ and $(U_{\boldsymbol{y}}, \boldsymbol{y})$ be two parametrizations for M such that $\boldsymbol{x}(u_0, v_0) = \boldsymbol{y}(s_0, t_0)$, for some $(u_0, v_0) \in U_{\boldsymbol{x}}$ and $(t_0, s_0) \in U_{\boldsymbol{y}}$. Then*

$$D\boldsymbol{x}(u_0, v_0)(\mathbb{R}^2) = D\boldsymbol{y}(s_0, t_0)(\mathbb{R}^2).$$

Proof: By symmetry, it suffices to show one of the inclusions. Restricting domains, it follows from Theorem 3.1.10 that $\varphi = \boldsymbol{x}^{-1} \circ \boldsymbol{y}$ is a diffeomorphism. Thus, differentiating $\boldsymbol{y} = \boldsymbol{x} \circ \varphi$ at (s_0, t_0), we have that

$$D\boldsymbol{y}(s_0, t_0) = D\boldsymbol{x}(\varphi(s_0, t_0)) \circ D\varphi(s_0, t_0) = D\boldsymbol{x}(u_0, v_0) \circ D\varphi(s_0, t_0),$$

and so we conclude that $D\boldsymbol{y}(s_0, t_0)(\mathbb{R}^2) \subseteq D\boldsymbol{x}(u_0, v_0)(\mathbb{R}^2)$, as wanted. □

Remark. With the above notation, if

$$D\varphi(s_0, t_0) = \begin{pmatrix} a & c \\ b & d \end{pmatrix},$$

applying the equality in display in the above proof to the vectors $(1, 0)$ and $(0, 1)$, respectively, gives us that

$$\frac{\partial \boldsymbol{y}}{\partial s}(s_0, t_0) = a\frac{\partial \boldsymbol{x}}{\partial u}(u_0, v_0) + b\frac{\partial \boldsymbol{x}}{\partial v}(u_0, v_0) \quad \text{and}$$
$$\frac{\partial \boldsymbol{y}}{\partial t}(s_0, t_0) = c\frac{\partial \boldsymbol{x}}{\partial u}(u_0, v_0) + d\frac{\partial \boldsymbol{x}}{\partial v}(u_0, v_0).$$

In particular, observe that

$$\frac{\partial \boldsymbol{y}}{\partial s}(s_0, t_0) \times \frac{\partial \boldsymbol{y}}{\partial t}(s_0, t_0) = \det D\varphi(s_0, t_0)\frac{\partial \boldsymbol{x}}{\partial u}(u_0, v_0) \times \frac{\partial \boldsymbol{x}}{\partial v}(u_0, v_0).$$

This lemma allows us to write the:

Definition 3.1.12 (Tangent Plane)**.** Let M be a regular surface and $\boldsymbol{p} \in M$. The *tangent plane to M at $\boldsymbol{p}$* is defined by

$$T_{\boldsymbol{p}}M \doteq D\boldsymbol{x}(u_0, v_0)(\mathbb{R}^2) = \mathrm{span}\left\{\frac{\partial \boldsymbol{x}}{\partial u}(u_0, v_0), \frac{\partial \boldsymbol{x}}{\partial v}(u_0, v_0)\right\},$$

where $(U, \boldsymbol{x})$ is any parametrization for M around $\boldsymbol{p}$, such that $\boldsymbol{x}(u_0, v_0) = \boldsymbol{p}$.

Remark.

- The tangent plane, as defined above, is a vector subspace of $\mathbb{R}^3_\nu$ and thus passes through the origin. It is usual to represent the tangent plane $T_{\boldsymbol{p}}M$ "affinely", passing through the point $\boldsymbol{p}$ instead (which will play the role of $\boldsymbol{0}$). Except in the cases where the distinction is extremely necessary, we will identify both planes.

- We have seen that regular parametrized surfaces may, in general, have self-intersections, and in such points the tangent plane (as defined above) is not well-defined. This is because if $\boldsymbol{p} = \boldsymbol{x}(u_0, v_0) = \boldsymbol{x}(u_1, v_1)$, we cannot ensure that $D\boldsymbol{x}(u_0, v_0)(\mathbb{R}^2) = D\boldsymbol{x}(u_1, v_1)(\mathbb{R}^2)$. In other words, we do not have a canonical choice to make here. In this case, we define the *tangent plane to the parametrization* $\boldsymbol{x}$ *at* (u_0, v_0) as $T_{(u_0,v_0)}\boldsymbol{x} \doteq D\boldsymbol{x}(u_0, v_0)(\mathbb{R}^2)$.

We say that $\mathscr{B}_{\boldsymbol{x}} \doteq \{\boldsymbol{x}_u(u_0, v_0), \boldsymbol{x}_v(u_0, v_0)\}$ is the *basis for* $T_{\boldsymbol{p}}M$ *associated to the parametrization* $\boldsymbol{x}$. We may also describe the tangent plane $T_{\boldsymbol{p}}M$ as the space of velocities of curves starting at $\boldsymbol{p}$, whose traces lie in the surface M. To make this idea rigorous, we start with the:

Lemma 3.1.13. *Let $M \subseteq \mathbb{R}^3_\nu$ be a regular surface, $\boldsymbol{\alpha}\colon I \to M$ be a curve, and $(U, \boldsymbol{x})$ be a parametrization for M such that $\boldsymbol{\alpha}(I) \subseteq \boldsymbol{x}(U)$. Then there are unique smooth functions $u, v : I \subseteq \mathbb{R} \to \mathbb{R}$ such that $\boldsymbol{\alpha}(t) = \boldsymbol{x}(u(t), v(t))$, for all $t \in I$.*

Proof: It suffices to consider $\boldsymbol{x}^{-1} \circ \boldsymbol{\alpha} : I \to U$. Then:

$$(u(t), v(t)) \doteq \boldsymbol{x}^{-1} \circ \boldsymbol{\alpha}(t) \implies \boldsymbol{\alpha}(t) = \boldsymbol{x}(u(t), v(t)).$$

Clearly u and v are smooth, by definition, and their uniqueness follows from $\boldsymbol{x}$ being a homeomorphism. □

Proposition 3.1.14. *Let $M \subseteq \mathbb{R}^3_\nu$ be a regular surface and $\boldsymbol{p} \in M$. Then*

$$T_{\boldsymbol{p}}M = \{\boldsymbol{\alpha}'(0) \mid \boldsymbol{\alpha} :]-\epsilon, \epsilon[\to M \text{ such that } \boldsymbol{\alpha}(0) = \boldsymbol{p}\}.$$

Proof: On one hand, if $\boldsymbol{\alpha} :]-\epsilon, \epsilon[\to M$ is a curve such that $\boldsymbol{\alpha}(0) = \boldsymbol{p}$ and $(U, \boldsymbol{x})$ is a parametrization for M with $\boldsymbol{x}(u_0, v_0) = \boldsymbol{p}$, the previous lemma allows us to write $\boldsymbol{\alpha}$ in the form $\boldsymbol{\alpha}(t) = \boldsymbol{x}(u(t), v(t))$. Note that, in particular, we have $(u(0), v(0)) = (u_0, v_0)$. Differentiating at $t = 0$, we have that

$$\boldsymbol{\alpha}'(0) = u'(0)\frac{\partial \boldsymbol{x}}{\partial u}(u_0, v_0) + v'(0)\frac{\partial \boldsymbol{x}}{\partial v}(u_0, v_0),$$

proving one of the inclusions.

On the other hand, if $\boldsymbol{x}$ is again a parametrization as before, take an arbitrary tangent vector $\boldsymbol{v}_{\boldsymbol{p}} = a\boldsymbol{x}_u(u_0, v_0) + b\boldsymbol{x}_v(u_0, v_0) \in T_{\boldsymbol{p}}M$. As U is open, there is $\epsilon > 0$ such that $(u_0 + ta, v_0 + tb) \in U$, for all $t \in]-\epsilon, \epsilon[$, so that we have a well-defined smooth curve $\boldsymbol{\alpha}\colon]-\epsilon, \epsilon[\to \boldsymbol{x}(U) \subseteq M$, given by $\boldsymbol{\alpha}(t) \doteq \boldsymbol{x}(u_0 + ta, v_0 + tb)$. Clearly $\boldsymbol{\alpha}(0) = \boldsymbol{p}$ and $\boldsymbol{\alpha}'(0) = \boldsymbol{v}_{\boldsymbol{p}}$, proving the remaining inclusion. □

Remark.

- Under the above conditions, we say that $\boldsymbol{\alpha}$ *realizes* $\boldsymbol{v}_{\boldsymbol{p}}$. When there is no risk of confusion, we omit $\boldsymbol{p}$ from $\boldsymbol{v}_{\boldsymbol{p}}$.

- This result is particularly interesting because it allows us to characterize the tangent plane without the explicit use of any parametrization.

Proposition 3.1.15. *Let $\Omega \subseteq \mathbb{R}^3$ be open, $f\colon \Omega \to \mathbb{R}$ be smooth, and $a \in \mathbb{R}$ be a regular value for f. If $M = f^{-1}(\{a\})$ and $\boldsymbol{p} \in M$, then $T_{\boldsymbol{p}}M = \ker Df(\boldsymbol{p})$.*

Proof: See Exercise 3.1.7. □

Now we move on to the study of functions defined over surfaces.

Definition 3.1.16. Let $M \subseteq \mathbb{R}^3_\nu$ be a regular surface and $f\colon M \to \mathbb{R}^k$ be a function. We'll say that f is *smooth* if, for every parametrization $(U, \boldsymbol{x})$ for M, the composition $f \circ \boldsymbol{x}\colon U \to \mathbb{R}^k$ is smooth as a map between Euclidean spaces.

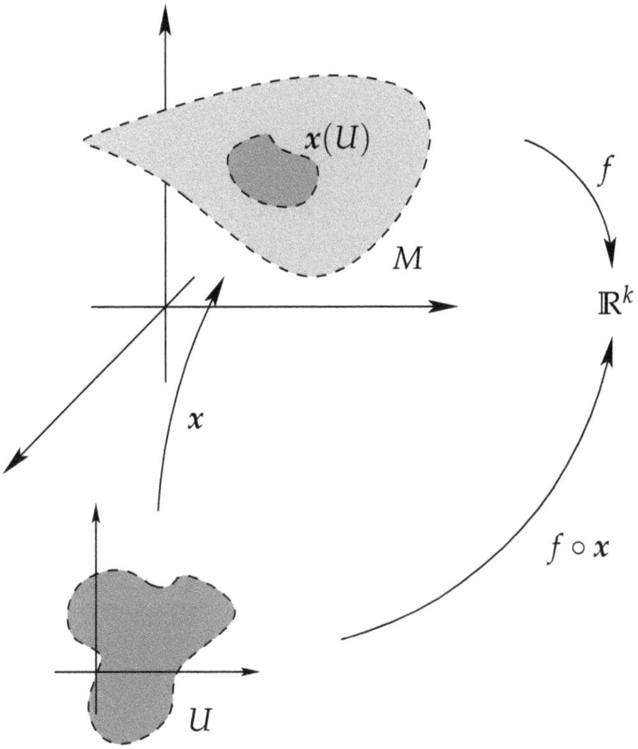

Figure 3.9: Illustrating the above definition.

Lemma 3.1.17. *Let $M \subseteq \mathbb{R}^3_\nu$ be a regular surface, $f\colon M \to \mathbb{R}^k$ be a function, and $(U_{\boldsymbol{x}}, \boldsymbol{x})$ and $(U_{\boldsymbol{y}}, \boldsymbol{y})$ be two parametrizations for M such that $W \doteq \boldsymbol{x}(U_{\boldsymbol{x}}) \cap \boldsymbol{y}(U_{\boldsymbol{y}}) \neq \varnothing$. Then $f \circ \boldsymbol{x} : \boldsymbol{x}^{-1}(W) \to \mathbb{R}^k$ is smooth if and only if $f \circ \boldsymbol{y} : \boldsymbol{y}^{-1}(W) \to \mathbb{R}^k$ is as well.*

Proof: It suffices to observe that $f \circ \boldsymbol{x} = (f \circ \boldsymbol{y}) \circ (\boldsymbol{y}^{-1} \circ \boldsymbol{x})$ and that the change of coordinates $\boldsymbol{y}^{-1} \circ \boldsymbol{x} : \boldsymbol{x}^{-1}(W) \to \boldsymbol{y}^{-1}(W)$ is a diffeomorphism. □

Remark. In view of this, to verify whether a function $f\colon M \to \mathbb{R}^k$ is smooth, it is not necessary to check the smoothness of $f \circ \boldsymbol{x}$ for all parametrizations of M, but only for enough parametrizations to cover all of M (usually a small number).

Example 3.1.18.

(1) Let $M \subseteq \mathbb{R}^3_\nu$ be a regular surface, $f, g : M \to \mathbb{R}^k$ smooth functions, and $\lambda \in \mathbb{R}$. Then $f + g$, λf and $\langle f, g \rangle$ are smooth. We denote by $\mathscr{C}^\infty(M, \mathbb{R}^k)$ the real vector space consisting of all the smooth functions from M to $\mathbb{R}^k$. When $k = 1$, we denote the real algebra $\mathscr{C}^\infty(M, \mathbb{R})$ simply by $\mathscr{C}^\infty(M)$.

(2) Let $\Omega \subseteq \mathbb{R}^3_\nu$ be open, $F\colon \Omega \to \mathbb{R}^k$ be a smooth function, and $M \subseteq \Omega$ be a regular surface. Then $F\big|_M\colon M \to \mathbb{R}^k$ is smooth.

(3) Let $\boldsymbol{x}\colon U \to \mathbb{R}^3_\nu$ be an injective regular parametrized surface. Then its inverse, $\boldsymbol{x}^{-1} : \boldsymbol{x}(U) \to U$, is smooth.

(4) Suppose that $\boldsymbol{p}_0, \boldsymbol{n} \in \mathbb{R}^3_\nu$ are given, and Π is the plane orthogonal to $\boldsymbol{n}$ and passing through $\boldsymbol{p}_0$. If $M \subseteq \mathbb{R}^3_\nu$ is a regular surface, then the *height function relative to* Π, $h\colon M \to \mathbb{R}$ given by $h(\boldsymbol{p}) = \langle \boldsymbol{p} - \boldsymbol{p}_0, \boldsymbol{n}\rangle$, is smooth. In $\mathbb{R}^3$, when $\|\boldsymbol{n}\|_E = 1$, h measures the (signed) height of points in M relative to Π.

(5) Let $M \subseteq \mathbb{R}^3_\nu$ be a regular surface and $\boldsymbol{p}_0 \in \mathbb{R}^3_\nu \setminus M$ be given. Then the *distance function to* $\boldsymbol{p}_0$, $f\colon M \to \mathbb{R}$ given by $f(\boldsymbol{p}) = \|\boldsymbol{p} - \boldsymbol{p}_0\|_E$, is smooth. Such function is not necessarily smooth when we replace $\|\cdot\|_E$ with $\|\cdot\|_L$. Why?

Definition 3.1.19. Let $M_1, M_2 \subseteq \mathbb{R}^3_\nu$ be regular surfaces. A function $f : M_1 \to M_2$ is *smooth* if, for all parametrizations $(U_1, \boldsymbol{x}_1)$ and $(U_2, \boldsymbol{x}_2)$ of M_1 and M_2 such that $f(\boldsymbol{x}_1(U_1)) \subseteq \boldsymbol{x}_2(U_2)$, the *local representation* $\boldsymbol{x}_2^{-1} \circ f \circ \boldsymbol{x}_1 : U_1 \to U_2$ is smooth as a map between Euclidean spaces.

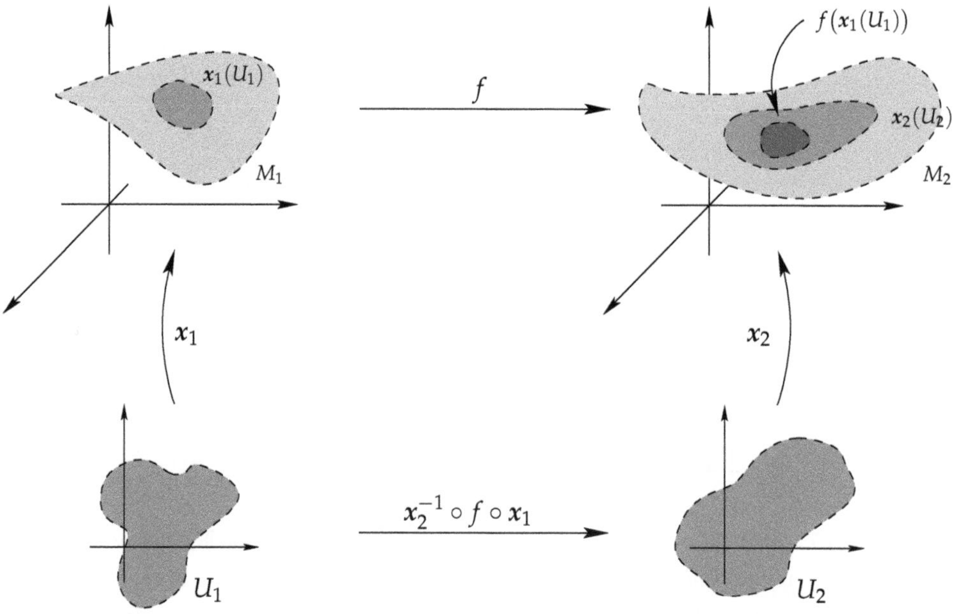

Figure 3.10: Summarizing the above definition.

Moreover, we'll say that f is a

(i) *diffeomorphism* (between the surfaces) if f is smooth and bijective, with its inverse also smooth.

(ii) *local diffeomorphism* if for every $\boldsymbol{p} \in M_1$, there is an open subset U of M_1 containing $\boldsymbol{p}$ such that the restriction $f\big|_U\colon U \to f(U)$ is a diffeomorphism.

Like in the case of functions $f : M \to \mathbb{R}^k$, we have the:

Lemma 3.1.20. *Let M_1 and M_2 be regular surfaces in $\mathbb{R}^3_\nu$, $f\colon M_1 \to M_2$ be a function, and $(U_{\boldsymbol{x}_1}, \boldsymbol{x}_1)$ and $(U_{\boldsymbol{x}_2}, \boldsymbol{x}_2)$ be parametrizations for M_1 and M_2, respectively, with $f(\boldsymbol{x}_1(U_{\boldsymbol{x}_1})) \subseteq \boldsymbol{x}_2(U_{\boldsymbol{x}_2})$. If $(U_{\boldsymbol{y}_1}, \boldsymbol{y}_1)$ and $(U_{\boldsymbol{y}_2}, \boldsymbol{y}_2)$ are further parametrizations for M_1 and M_2 with $f(\boldsymbol{y}_1(U_{\boldsymbol{y}_1})) \subseteq \boldsymbol{y}_2(U_{\boldsymbol{y}_2})$, such that $W_1 \doteq \boldsymbol{x}_1(U_{\boldsymbol{x}_1}) \cap \boldsymbol{y}_1(U_{\boldsymbol{y}_1})$ and $W_2 \doteq \boldsymbol{x}_2(U_{\boldsymbol{x}_2}) \cap \boldsymbol{y}_2(U_{\boldsymbol{y}_2})$ are both non-empty, then:*

(i) $f(W_1) \subseteq W_2$ and

(ii) the local expression $\boldsymbol{x}_2^{-1} \circ f \circ \boldsymbol{x}_1\colon \boldsymbol{x}_1^{-1}(W_1) \to \boldsymbol{x}_2^{-1}(W_2)$ is smooth if and only if $\boldsymbol{y}_2^{-1} \circ f \circ \boldsymbol{y}_1\colon \boldsymbol{y}_1^{-1}(W_1) \to \boldsymbol{y}_2^{-1}(W_2)$ is smooth as well.

Proof: The seemingly overwhelming quantity of conditions assumed on the domains and images of the parametrizations considered are necessary only to ensure that all the relevant compositions all make sense. That being understood, the proof is entirely similar to the proof given for Lemma 3.1.17, and we ask you to do it in Exercise 3.1.4. □

Example 3.1.21.

(1) For each $\boldsymbol{p} \in \mathbb{S}^2 \setminus \{(0,0,\pm 1)\}$, let $f(\boldsymbol{p}) \in \mathbb{S}^1 \times \mathbb{R}$ be the intersection of the horizontal ray starting at the z-axis passing through $\boldsymbol{p}$, with the cylinder $\mathbb{S}^1 \times \mathbb{R}$. The map $f\colon \mathbb{S}^2 \setminus \{(0,0,\pm 1)\} \to \mathbb{S}^1 \times \mathbb{R}$ so defined is known as the *Lambert cylindrical projection*. Let's see that this map is smooth, by considering the parametrizations $\boldsymbol{x}\colon \left]0,\pi\right[\times \left]0,2\pi\right[\to \mathbb{S}^2 \setminus \{(0,0,\pm 1)\}$ and $\widetilde{\boldsymbol{x}}\colon \mathbb{R} \times \left]0,2\pi\right[\to \mathbb{S}^1 \times \mathbb{R}$ given by

$$\boldsymbol{x}(u,v) \doteq (\cos u \cos v, \cos u \sin v, \sin u)$$
$$\widetilde{\boldsymbol{x}}(\widetilde{u},\widetilde{v}) \doteq (\cos \widetilde{v}, \sin \widetilde{v}, \widetilde{u}),$$

which omit a single meridian from each surface.

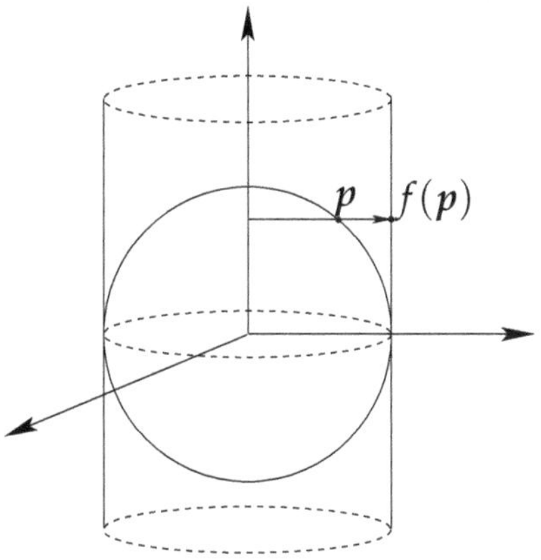

Figure 3.11: The Lambert projection.

We then see that $f(\boldsymbol{x}(u,v)) = \widetilde{\boldsymbol{x}}(\sin u, v)$, and since the map taking (u,v) to $(\sin u, v)$ is smooth, we conclude that f restricted to the image of $\boldsymbol{x}$ is smooth. To verify that f is smooth along the meridian omitted by $\boldsymbol{x}$, one repeats this argument by considering instead of $\boldsymbol{x}$ and $\widetilde{\boldsymbol{x}}$, other parametrizations $\boldsymbol{y}$ and $\widetilde{\boldsymbol{y}}$ given, for example, by $\boldsymbol{y}(u,v) = \boldsymbol{x}(u, v+\pi)$ and $\widetilde{\boldsymbol{y}}(u,v) = \widetilde{\boldsymbol{x}}(u, v+\pi)$. Note that f is not a diffeomorphism, since it is not surjective.

(2) The unit open Euclidean disk $D = \{(x,y,1) \in \mathbb{R}^3_\nu \mid x^2 + y^2 < 1\}$ is diffeomorphic to $\mathbb{H}^2$. Given $\boldsymbol{p} \in D$, let $f(\boldsymbol{p})$ be the intersection of $\mathbb{H}^2$ with the ray starting from the origin and passing through $\boldsymbol{p}$. This defines a function $f\colon D \to \mathbb{H}^2$, and we claim it is a diffeomorphism. We consider the parametrizations $\boldsymbol{x}\colon \left]0,1\right[\times \left]0,2\pi\right[\to D$ and $\widetilde{\boldsymbol{x}}\colon \mathbb{R}_{>0} \times \left]0,2\pi\right[\to \mathbb{H}^2$ given by

$$\boldsymbol{x}(u,v) \doteq (u \cos v, u \sin v, 1)$$
$$\widetilde{\boldsymbol{x}}(\widetilde{u},\widetilde{v}) \doteq (\sinh \widetilde{u} \cos \widetilde{v}, \sinh \widetilde{u} \sin \widetilde{v}, \cosh \widetilde{u}),$$

omitting, respectively, a line segment in D and a meridian in $\mathbb{H}^2$.

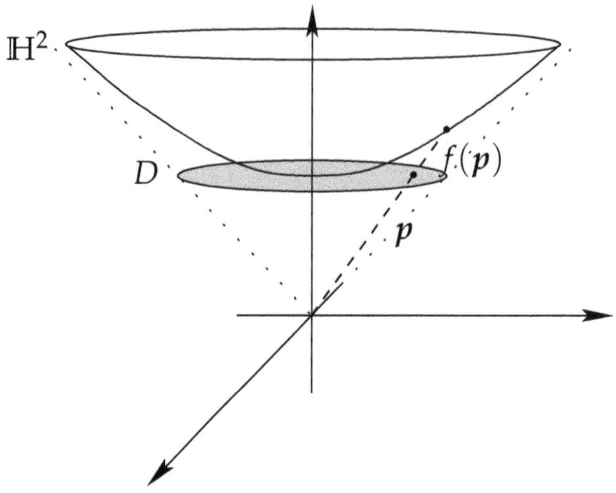

Figure 3.12: A diffeomorphism between D and $\mathbb{H}^2$.

To determine $f(\boldsymbol{x}(u,v))$, we look for $t > 0$ such that $t\boldsymbol{x}(u,v) \in \mathbb{H}^2$. Such t must satisfy

$$\langle t\boldsymbol{x}(u,v), t\boldsymbol{x}(u,v)\rangle_L = t^2u^2 - t^2 = -1,$$

whence $t = 1/\sqrt{1-u^2}$. Thus, we have

$$\begin{aligned} f(\boldsymbol{x}(u,v)) &= \left(\frac{u}{\sqrt{1-u^2}}\cos v, \frac{u}{\sqrt{1-u^2}}\sin v, \frac{1}{\sqrt{1-u^2}}\right) \\ &= \widetilde{\boldsymbol{x}}\left(\operatorname{arcsinh}\frac{u}{\sqrt{1-u^2}}, v\right). \end{aligned}$$

Since the map taking (u,v) to $\left(\operatorname{arcsinh}(u/\sqrt{1-u^2}), v\right)$ is a diffeomorphism, we conclude that f is a diffeomorphism between the images of $\boldsymbol{x}$ and $\widetilde{\boldsymbol{x}}$. Repeating the argument from the previous example, taking new parametrizations to cover what was left out by $\boldsymbol{x}$ and $\widetilde{\boldsymbol{x}}$, show that f is smooth and, in fact, a local diffeomorphism. Since f is bijective, it follows that f is a diffeomorphism.

It is natural, in the process of transferring notions of Calculus to a surface, not only to define a notion of smoothness (as we just did), but also to define what would be the "derivative" of a smooth function between surfaces. For this end, we need the following:

Lemma 3.1.22. *Let $M \subseteq \mathbb{R}^3_\nu$ be a regular surface, $\boldsymbol{p} \in M$, $\boldsymbol{v} \in T_{\boldsymbol{p}}M$ and $f\colon M \to \mathbb{R}^k$ be a smooth function. If $\boldsymbol{\alpha}\colon\,]-\epsilon,\epsilon[\, \to M$ is a curve which realizes $\boldsymbol{v}$, then $(f\circ\boldsymbol{\alpha})'(0)$ depends only on $\boldsymbol{p}$ and $\boldsymbol{v}$ (but not on $\boldsymbol{\alpha}$).*

Proof: Let $(U,\boldsymbol{x})$ be a parametrization for M around $\boldsymbol{p}$, with $\boldsymbol{p} = \boldsymbol{x}(u_0,v_0) = \boldsymbol{\alpha}(0)$, such that $\boldsymbol{x}(U)$ contains the trace of $\boldsymbol{\alpha}$ (this is possible by reducing the domain of $\boldsymbol{\alpha}$, if necessary). Write the curve in coordinates as $\boldsymbol{\alpha}(t) = \boldsymbol{x}(u(t),v(t))$, noting that $(u(0),v(0)) = (u_0,v_0) = \boldsymbol{x}^{-1}(\boldsymbol{p})$ depends only on $\boldsymbol{p}$, and that $u'(0)$ and $v'(0)$ depend only on $\boldsymbol{v}$, since

$$\boldsymbol{v} = u'(0)\frac{\partial\boldsymbol{x}}{\partial u}(\boldsymbol{x}^{-1}(\boldsymbol{p})) + v'(0)\frac{\partial\boldsymbol{x}}{\partial v}(\boldsymbol{x}^{-1}(\boldsymbol{p})).$$

With this:

$$(f\circ\boldsymbol{\alpha})'(0) = u'(0)\frac{\partial(f\circ\boldsymbol{x})}{\partial u}(\boldsymbol{x}^{-1}(\boldsymbol{p})) + v'(0)\frac{\partial(f\circ\boldsymbol{x})}{\partial v}(\boldsymbol{x}^{-1}(\boldsymbol{p}))$$

depends only on $\boldsymbol{p}$ and $\boldsymbol{v}$, as wanted. □

Remark.

- That the above calculation does not depend on the chosen parametrization is a consequence of the remark following Lemma 3.1.11.
- It is usual, in the above notation, to abbreviate

$$\frac{\partial f}{\partial u}(\boldsymbol{p}) \doteq \frac{\partial (f \circ \boldsymbol{x})}{\partial u}(\boldsymbol{x}^{-1}(\boldsymbol{p})) \quad \text{and} \quad \frac{\partial f}{\partial v}(\boldsymbol{p}) \doteq \frac{\partial (f \circ \boldsymbol{x})}{\partial v}(\boldsymbol{x}^{-1}(\boldsymbol{p})).$$

That is, *once one has a parametrization*, one may talk about partial derivatives of a function defined on a surface, which measure how a function changes along coordinate curves *of this parametrization.*

This allows us to write the:

Definition 3.1.23. Let $M \subseteq \mathbb{R}^3_\nu$ be a regular surface and $f\colon M \to \mathbb{R}^k$ be a smooth function. The *differential* of f at the point $\boldsymbol{p} \in M$ is the map $\mathrm{d}f_{\boldsymbol{p}}\colon T_{\boldsymbol{p}}M \to \mathbb{R}^k$ given by $\mathrm{d}f_{\boldsymbol{p}}(\boldsymbol{v}) = (f \circ \boldsymbol{\alpha})'(0)$, where $\boldsymbol{\alpha}$ realizes $\boldsymbol{v}$.

Remark. In the previous proof, we have seen that $(f \circ \boldsymbol{\alpha})'(0)$ is linear on the coordinates of $\boldsymbol{v}$ relative to the basis for $T_{\boldsymbol{p}}M$ associated to $\boldsymbol{x}$, $\mathcal{B}_{\boldsymbol{x}}$, whence it follows that the map $\mathrm{d}f_{\boldsymbol{p}}$ is linear.

When the image of such a function $f\colon M_1 \to \mathbb{R}^3_\nu$ is contained in another surface M_2, Proposition 3.1.14 (p. 140) says that $\mathrm{d}f_{\boldsymbol{p}}(\boldsymbol{v}) \in T_{f(\boldsymbol{p})}M_2$ for each $\boldsymbol{v} \in T_{\boldsymbol{p}}M_1$, that is, we have $\mathrm{d}f_{\boldsymbol{p}}\colon T_{\boldsymbol{p}}M_1 \to T_{f(\boldsymbol{p})}M_2$. Given parametrizations of M_1 and M_2 around $\boldsymbol{p}$ and $f(\boldsymbol{p})$, we may write the matrix of the linear operator $\mathrm{d}f_{\boldsymbol{p}}$ relative to the bases associated to these parametrizations. We have:

Proposition 3.1.24. *Let $M_1, M_2 \subseteq \mathbb{R}^3_\nu$ be regular surfaces, $\boldsymbol{p} \in M_1$ and $f\colon M_1 \to M_2$ be a smooth function. Suppose that $(U_1, \boldsymbol{x}_1)$ and $(U_2, \boldsymbol{x}_2)$ are two regular parametrizations for M_1 and M_2 around $\boldsymbol{p} = \boldsymbol{x}_1(u_0, v_0)$ and $f(\boldsymbol{p}) = \boldsymbol{x}_2(\widetilde{u}_0, \widetilde{v}_0)$, respectively, such that $f(\boldsymbol{x}_1(U_1)) \subseteq \boldsymbol{x}_2(U_2)$. Then*

$$\left[\mathrm{d}f_{\boldsymbol{p}}\right]_{\mathcal{B}_{\boldsymbol{x}_1}, \mathcal{B}_{\boldsymbol{x}_2}} = D\psi_f(u_0, v_0),$$

where $\psi_f = \boldsymbol{x}_2^{-1} \circ f \circ \boldsymbol{x}_1$ is the local representation of f.

Proof: Writing $\psi_f = (\psi_1, \psi_2)$, it suffices to note that

$$\begin{aligned} \mathrm{d}f_{\boldsymbol{p}}\left(\frac{\partial \boldsymbol{x}_1}{\partial u}(u_0, v_0)\right) &= \frac{\partial (f \circ \boldsymbol{x}_1)}{\partial u}(u_0, v_0) = \frac{\partial (\boldsymbol{x}_2 \circ \psi_f)}{\partial u}(u_0, v_0) \\ &= \frac{\partial \psi_1}{\partial u}(u_0, v_0)\frac{\partial \boldsymbol{x}_2}{\partial \widetilde{u}}(\widetilde{u}_0, \widetilde{v}_0) + \frac{\partial \psi_2}{\partial u}(u_0, v_0)\frac{\partial \boldsymbol{x}_2}{\partial \widetilde{v}}(\widetilde{u}_0, \widetilde{v}_0), \end{aligned}$$

and similarly to obtain the second column of $\left[\mathrm{d}f_{\boldsymbol{p}}\right]_{\mathcal{B}_{\boldsymbol{x}_1}, \mathcal{B}_{\boldsymbol{x}_2}}$. □

Example 3.1.25. Let's compute the differentials of the functions seen in Example 3.1.18.

(1) Let M be a regular surface, $f, g \in \mathscr{C}^\infty(M, \mathbb{R}^k)$ and $\lambda \in \mathbb{R}$. Then, for each $\boldsymbol{p} \in M$, we have that $\mathrm{d}(f+g)_{\boldsymbol{p}} = \mathrm{d}f_{\boldsymbol{p}} + \mathrm{d}g_{\boldsymbol{p}}$, $\mathrm{d}(\lambda f)_{\boldsymbol{p}} = \lambda \mathrm{d}f_{\boldsymbol{p}}$ and

$$\mathrm{d}(\langle f, g\rangle)_{\boldsymbol{p}} = \langle g(\boldsymbol{p}), \mathrm{d}f_{\boldsymbol{p}}\rangle + \langle f(\boldsymbol{p}), \mathrm{d}g_{\boldsymbol{p}}\rangle.$$

(2) Let $\Omega \subseteq \mathbb{R}^3_\nu$ be open, $F\colon \Omega \to \mathbb{R}^k$ a smooth function and $M \subseteq \Omega$ a regular surface. Then $\mathrm{d}(F|_M)_p = DF(p)|_{T_pM}$, for each $p \in M$. In particular, $\mathrm{d}(\mathrm{id}_M)_p = \mathrm{id}_{T_pM}$.

(3) If $p_0, n \in \mathbb{R}^3_\nu$ are given and $M \subseteq \mathbb{R}^3_\nu$ is a regular surface, we have seen that the height function $h\colon M \to \mathbb{R}$ relative to the plane Π (orthogonal to n and passing through p_0), given by $h(p) = \langle p - p_0, n\rangle$, is smooth. We have that $\mathrm{d}h_p = \langle \cdot, n\rangle$, for each $p \in M$.

(4) Let $M \subseteq \mathbb{R}^3$ be a regular surface and $p_0 \in \mathbb{R}^3 \setminus M$. Then $f\colon M \to \mathbb{R}$, defined by $f(p) = \|p - p_0\|_E$, has differential given by

$$\mathrm{d}f_p = \frac{\langle p - p_0, \cdot\rangle_E}{\|p - p_0\|_E}.$$

Proposition 3.1.26 (Chain rule). *Let $M_1, M_2, M_3 \subseteq \mathbb{R}^3_\nu$ be three regular surfaces, $f\colon M_1 \to M_2$ and $g\colon M_2 \to M_3$ be smooth functions. Then $g \circ f\colon M_1 \to M_3$ is smooth and, for each $p \in M_1$, we have that*

$$\mathrm{d}(g \circ f)_p = \mathrm{d}g_{f(p)} \circ \mathrm{d}f_p.$$

Proof: To verify smoothness of $g \circ f$, it suffices to write the local representation of this composition in terms of the local representations for f and g, which are smooth. Now, let $v \in T_pM_1$ and $\alpha\colon\ ]-\epsilon, \epsilon[\ \to M_1$ be a curve realizing v. Noting that $f \circ \alpha$ realizes the tangent vector $\mathrm{d}f_p(v) \in T_{f(p)}M_2$, we have that

$$\mathrm{d}(g \circ f)_p(v) = ((g \circ f) \circ \alpha)'(0) = (g \circ (f \circ \alpha))'(0) = \mathrm{d}g_{f(p)}(\mathrm{d}f_p(v)),$$

as wanted. □

Corollary 3.1.27. *Let $M_1, M_2 \subseteq \mathbb{R}^3_\nu$ be regular surfaces and $f\colon M_1 \to M_2$ be a diffeomorphism. Then, for each $p \in M_1$, $\mathrm{d}f_p$ is a linear isomorphism, whose inverse is given by $(\mathrm{d}f_p)^{-1} = \mathrm{d}(f^{-1})_{f(p)}$.*

Just like for functions between Euclidean spaces, the above corollary has the following local converse:

Theorem 3.1.28 (Inverse Function Theorem). *Let $M_1, M_2 \subseteq \mathbb{R}^3_\nu$ be two regular surfaces and $f\colon M_1 \to M_2$ be a smooth function. If $p_0 \in M_1$ is such that $\mathrm{d}f_{p_0}$ is a linear isomorphism, then there exists an open subset U of M_1 containing p_0 such that the restriction $f|_U\colon U \to f(U)$ is a diffeomorphism.*

Proof: Take parametrizations (U_1, x_1) around $p_0 = x_1(u_0, v_0)$ and (U_2, x_2) around $f(p_0)$, such that $f(x_1(U_1)) \subseteq x_2(U_2)$. Since $\mathrm{d}f_{p_0}$ is an isomorphism, the chain rule gives that $D(x_2^{-1} \circ f \circ x_1)(u_0, v_0)$ is the composition of three isomorphisms, hence an isomorphism as well. The Inverse Function Theorem for Euclidean spaces yields an open subset $V \subseteq U_1$ containing (u_0, v_0) such that $(x_2^{-1} \circ f \circ x_1)|_V\colon V \to (x_2^{-1} \circ f \circ x_1)(V)$ is a diffeomorphism. With this in place, if $U \doteq x_1(V)$, we have that U is open in M_1 and $f|_U\colon U \to f(U)$ is a diffeomorphism, whose inverse is given by the composition $x_1 \circ (x_2^{-1} \circ f \circ x_1)^{-1} \circ x_2^{-1}\colon f(U) \to U$. □

As an example of application of the Inverse Function Theorem for surfaces, we have the following result, of interesting geometric intuition:

Proposition 3.1.29. *Let $M \subseteq \mathbb{R}^3_\nu$ be a regular surface and $\boldsymbol{p}_0 \in M$. Then there are open subsets $U \subseteq M$ and $V \subseteq \boldsymbol{p}_0 + T_{\boldsymbol{p}_0}M$, containing $\boldsymbol{p}_0$ and $\boldsymbol{0}$, respectively, and a diffeomorphism $h : V \to U$ such that $h(\boldsymbol{q}) - \boldsymbol{q}$ is (Euclidean) normal to $T_{\boldsymbol{p}_0}M$, for each $\boldsymbol{q} \in V$. In other words, the surface is locally the graph of a function defined on its tangent plane.*

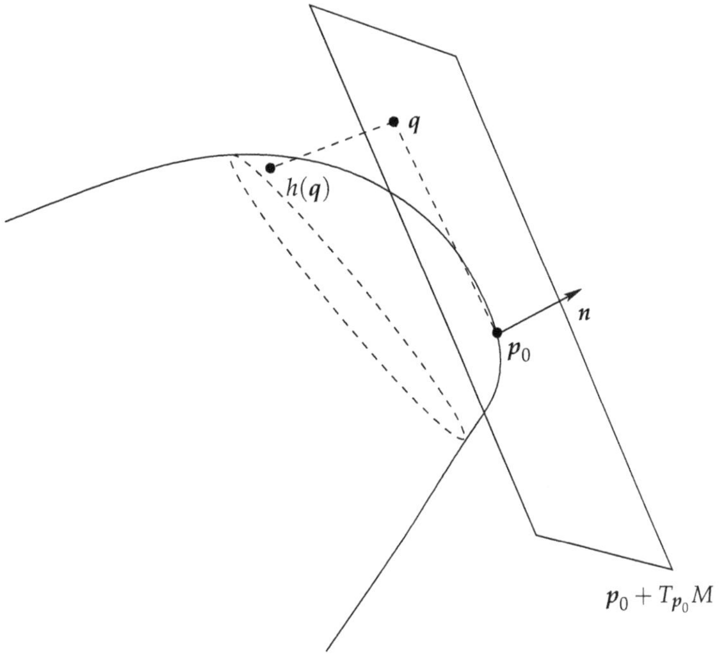

Figure 3.13: A surface as a graph over one tangent plane.

Proof: For this proof, we will use only the Euclidean inner product. Let $\boldsymbol{n} \in \mathbb{R}^3_\nu$ be a unit normal vector to $T_{\boldsymbol{p}_0}M$. Consider then the orthogonal projection $f : M \to \mathbb{R}^3$ given by

$$f(\boldsymbol{p}) = \boldsymbol{p} - \langle \boldsymbol{p} - \boldsymbol{p}_0, \boldsymbol{n}\rangle \boldsymbol{n}.$$

Since $f(\boldsymbol{p}_0) = \boldsymbol{p}_0$ and $\langle f(\boldsymbol{p}) - \boldsymbol{p}_0, \boldsymbol{n}\rangle = 0$ we have, in fact, $f : M \to \boldsymbol{p}_0 + T_{\boldsymbol{p}_0}M$. Clearly f is smooth and its differential $\mathrm{d}f_{\boldsymbol{p}} : T_{\boldsymbol{p}}M \to T_{\boldsymbol{p}_0}M$ is given by

$$\mathrm{d}f_{\boldsymbol{p}}(\boldsymbol{v}) = \boldsymbol{v} - \langle \boldsymbol{v}, \boldsymbol{n}\rangle \boldsymbol{n}.$$

In particular, as $\boldsymbol{n} \in (T_{\boldsymbol{p}_0}M)^\perp$, we have that $\mathrm{d}f_{\boldsymbol{p}_0} = \mathrm{id}_{T_{\boldsymbol{p}_0}M}$, and so the Inverse Function Theorem yields open subsets $U \subseteq M$ and $V \subseteq \boldsymbol{p}_0 + T_{\boldsymbol{p}_0}M$ containing $\boldsymbol{p}_0$ such that $f|_U : U \to V$ is a diffeomorphism, with inverse $h: V \to U$. Such h fits the bill: to wit, if $h(\boldsymbol{q}) = \boldsymbol{q}$, then $h(\boldsymbol{q}) - \boldsymbol{q} \in (T_{\boldsymbol{p}_0}M)^\perp$, trivially. Else, writing $\boldsymbol{q} = f(\boldsymbol{p}) \in V$, we have

$$\begin{aligned}\langle h(\boldsymbol{q}) - \boldsymbol{q}, \boldsymbol{q} - \boldsymbol{p}_0\rangle &= \langle \boldsymbol{p} - f(\boldsymbol{p}), \boldsymbol{q} - \boldsymbol{p}_0\rangle = 0 \quad \text{and} \\ \langle h(\boldsymbol{q}) - \boldsymbol{q}, \boldsymbol{n}\rangle &= \langle \boldsymbol{p} - f(\boldsymbol{p}), \boldsymbol{n}\rangle = \langle \boldsymbol{p} - \boldsymbol{p}_0, \boldsymbol{n}\rangle \neq 0,\end{aligned}$$

since $\boldsymbol{p} - f(\boldsymbol{p}) \in (T_{\boldsymbol{p}_0}M)^\perp$, $\boldsymbol{q} - \boldsymbol{p}_0 \in T_{\boldsymbol{p}_0}M$, and the last term does not vanish, seeing that if $h(\boldsymbol{q}) \neq \boldsymbol{q}$, then $\boldsymbol{p} \notin \boldsymbol{p}_0 + T_{\boldsymbol{p}_0}M$. We conclude that $h(\boldsymbol{q}) - \boldsymbol{q} \in (T_{\boldsymbol{p}_0}M)^\perp$ in this case as well, as wanted. □

We will conclude this section with a brief discussion regarding a notion of fundamental importance for the following sections: orientability. The basic idea is to define the orientation of a regular surface from the orientation of its tangent planes, in a similar fashion done for the orientation of lightlike planes, in Section 2.3 (Subsection 2.3.3, to be precise).

The initial definition is motivated by the remark made after the proof of Lemma 3.1.11 (p. 139). The change of basis matrix between bases associated to different parametrizations is the Jacobian matrix of the change of coordinates itself, which we know to be a diffeomorphism. Our focus is then turned to the sign of the determinant of such matrix. We have the:

Definition 3.1.30 (Orientability). Let $M \subseteq \mathbb{R}^3_\nu$ be a regular surface. We'll say that M is *orientable* if it is possible to obtain a collection $\mathfrak{O}$ of parametrizations whose images together cover M, such that for any pair of parametrizations $(U_i, \boldsymbol{x}_i)$ $(i = 1, 2)$ in $\mathfrak{O}$, with $W \doteq \boldsymbol{x}_1(U_1) \cap \boldsymbol{x}_2(U_2) \neq \varnothing$, one has that

$$\det D(\boldsymbol{x}_2^{-1} \circ \boldsymbol{x}_1)(u, v) > 0$$

for all $(u, v) \in \boldsymbol{x}_1^{-1}(W)$. We'll also say that:

(i) a parametrization $(U, \boldsymbol{x})$ is *compatible* with $\mathfrak{O}$ if $\det D(\widetilde{\boldsymbol{x}}^{-1} \circ \boldsymbol{x}) > 0$ for each parametrization $(\widetilde{U}, \widetilde{\boldsymbol{x}})$ in $\mathfrak{O}$ such that $\boldsymbol{x}(U) \cap \widetilde{\boldsymbol{x}}(\widetilde{U}) \neq \varnothing$;

(ii) the collection $\mathfrak{O}$ is an *orientation for* M;

(iii) M is *non-orientable* if it is not possible to obtain such an orientation $\mathfrak{O}$.

Proposition 3.1.31. *Let $M \subseteq \mathbb{R}^3_\nu$ be a regular surface. If M may be covered with a single parametrization, or with two parametrizations whose images have connected intersection, then M is orientable.*

Proof: If M is covered by a single parametrization, then the orientation $\mathfrak{O}$ will consist of this single parametrization only, and the only possible change of coordinates between parametrizations in $\mathfrak{O}$ is the identity map, whose derivative (itself) has positive determinant. If M is covered by two parametrizations with connected intersection, say, $(U_{\boldsymbol{x}}, \boldsymbol{x})$ and $(U_{\boldsymbol{y}}, \boldsymbol{y})$, put $W = \boldsymbol{x}(U_{\boldsymbol{x}}) \cap \boldsymbol{y}(U_{\boldsymbol{y}})$ and take $(u_0, v_0) \in \boldsymbol{x}^{-1}(W)$. By connectedness, the sign of $\det D(\boldsymbol{y}^{-1} \circ \boldsymbol{x})(u, v)$ is the same as the sign of $\det D(\boldsymbol{y}^{-1} \circ \boldsymbol{x})(u_0, v_0)$, for each $(u, v) \in \boldsymbol{x}^{-1}(W)$. We then have two possibilities:

- if $\det D(\boldsymbol{y}^{-1} \circ \boldsymbol{x})(u_0, v_0) > 0$, we may take $\mathfrak{O}$ to be simply the collection formed by $\boldsymbol{x}$ and $\boldsymbol{y}$;

- if $\det D(\boldsymbol{y}^{-1} \circ \boldsymbol{x})(u_0, v_0) < 0$, replace $\boldsymbol{x}$ by $(\widetilde{U}, \widetilde{\boldsymbol{x}})$ given by $\widetilde{\boldsymbol{x}}(v, u) = \boldsymbol{x}(u, v)$, where $\widetilde{U} = \{(v, u) \in \mathbb{R}^2 \mid (u, v) \in U_{\boldsymbol{x}}\}$, and take $\mathfrak{O}$ to be the collection formed by $\widetilde{\boldsymbol{x}}$ and $\boldsymbol{y}$ only.

□

Corollary 3.1.32. *Surfaces of revolution and graphs of smooth functions defined in open subsets of the plane are orientable regular surfaces.*

To motivate one equivalence, which we'll give soon, with the definition of orientability, note that if $\boldsymbol{x}\colon U \to \mathbb{R}^3_\nu$ is an injective regular parametrized surface, then for each $(u, v) \in U$, the vector $\boldsymbol{x}_u(u, v) \times \boldsymbol{x}_v(u, v)$ is normal to $T_{\boldsymbol{x}(u,v)}\boldsymbol{x}(U)$. If the tangent planes to the parametrization are not lightlike, we have a well-defined map $N: x(U) \to \mathbb{R}^3_\nu$, which takes each point $x(U)$ to a unit normal vector, given by

$$N(x(u,v)) \doteq \frac{x_u(u,v) \times x_v(u,v)}{\|x_u(u,v) \times x_v(u,v)\|}.$$

Naturally, we would like to repeat this for arbitrary regular surfaces. Consider for now just the Euclidean inner product. If $M \subseteq \mathbb{R}^3$ is a regular surface, and (U, x) and $(\widetilde{U}, \widetilde{x})$ are two parametrizations for M with $x(U) \cap \widetilde{x}(\widetilde{U})$ non-empty and connected, we have that

$$\frac{x_u(u,v) \times x_v(u,v)}{\|x_u(u,v) \times x_v(u,v)\|} = \pm \frac{\widetilde{x}_{\widetilde{u}}(\widetilde{u},\widetilde{v}) \times \widetilde{x}_{\widetilde{v}}(\widetilde{u},\widetilde{v})}{\|\widetilde{x}_{\widetilde{u}}(\widetilde{u},\widetilde{v}) \times \widetilde{x}_{\widetilde{v}}(\widetilde{u},\widetilde{v})\|}$$

for each $(u,v) \in U$ and $(\widetilde{u},\widetilde{v}) \in \widetilde{U}$ such that $x(u,v) = \widetilde{x}(\widetilde{u},\widetilde{v})$, where the sign $\pm$ is the sign of the determinant of the derivative of the coordinate change. To summarize, if M is orientable, it is possible to choose parametrizations which make this "patching up" work, that is, that all the above signs are positive. This yields a *unit normal field* $N: M \to \mathbb{R}^3$ globally defined on the whole surface M, that is, a map N that associates to each point $p \in M$ a unit vector $N(p)$ normal to T_pM.

Since orientability is a notion which does not depend on the ambient product, we may indeed work only with the Euclidean inner product. For surfaces with non-degenerate tangent planes, the existence of a Euclidean unit normal field is equivalent to the existence of a Lorentzian one.

Theorem 3.1.33. *Let $M \subseteq \mathbb{R}^3_\nu$ be a regular surface. Then M is orientable if and only if there is a smooth Euclidean unit normal field $N: M \to \mathbb{R}^3_\nu$, defined on all of M.*

Proof: If M is orientable, the argument was sketched above: let $\mathfrak{O}$ be an orientation for M, and for each $p \in M$, take a parametrization (U, x) for M around p, which is in $\mathfrak{O}$. If $p = x(u_0, v_0)$, define

$$N(p) \doteq \frac{x_u(u_0,v_0) \times x_v(u_0,v_0)}{\|x_u(u_0,v_0) \times x_v(u_0,v_0)\|}.$$

This definition does not depend on the choice of parametrization in $\mathfrak{O}$, because all parametrizations there are pairwise compatible.

Conversely, assume that there is a smooth Euclidean unit normal field $N: M \to \mathbb{R}^3_\nu$ defined on all of M. Consider all the parametrizations (U, x) such that $x(U)$ is connected. Then, for each $(u,v) \in U$ we have that

$$N(x(u,v)) = \pm \frac{x_u(u,v) \times x_v(u,v)}{\|x_u(u,v) \times x_v(u,v)\|}.$$

If suffices to take $\mathfrak{O}$ as the collection of all the parametrizations of M with connected image for which the sign in the above formula is positive. □

Remark. If follows that if $M \subseteq \mathbb{R}^3_\nu$ is a regular surface and $p \in M$ is any point, there is an orientable open subset of M around p.

Corollary 3.1.34. *Let $\Omega \subseteq \mathbb{R}^3_\nu$ be open, $f: \Omega \to \mathbb{R}$ be smooth, and $a \in \mathbb{R}$ be a regular value for f. Then $M = f^{-1}(\{a\})$ is orientable.*

Proof: It suffices to note that $N: M \to \mathbb{R}^3_\nu$ given by

$$N(p) = \frac{\nabla f(p)}{\|\nabla f(p)\|_E}$$

is a smooth Euclidean unit normal field along M. □

Remark. A converse holds: if M is an orientable regular surface, then M is (globally) the inverse image of a regular value of a smooth function. For a proof in the case where M is compact, see [17, p. 130].

It is also interesting to analyze the relation between diffeomorphisms and orientations. Suppose that $M_1, M_2 \subseteq \mathbb{R}^3_\nu$ are regular surfaces, and that $f\colon M_1 \to M_2$ is a diffeomorphism. If M_1 is orientable, say, with an orientation $\mathfrak{O}_1$, then the collection $\mathfrak{O}_2 \doteq f_*\mathfrak{O}_1$ in M_2 consisting of the parametrizations $f \circ \boldsymbol{x}$, for $\boldsymbol{x}$ in $\mathfrak{O}_1$, is an orientation for M_2, called the *orientation induced by* f. Note that if two surfaces are diffeomorphic, then necessarily both are orientable, or both are non-orientable.

Definition 3.1.35. Let $M \subseteq \mathbb{R}^3_\nu$ be a regular surface equipped with an orientation $\mathfrak{O}$ and $f\colon M \to M$ be a diffeomorphism. We say that f *preserves orientation* if $f_*\mathfrak{O} = \mathfrak{O}$, and that it *reverses orientation* otherwise.

Proposition 3.1.36. *Let $M \subseteq \mathbb{R}^3_\nu$ be a connected and orientable surface, and $f\colon M \to M$ be a diffeomorphism. Then:*

(i) f preserving or reversing orientation depends only on f itself, and not on the orientation chosen for M a priori;

(ii) for each orientation $\mathfrak{O}$ of M we have that

$$\det D(\boldsymbol{y}^{-1} \circ f \circ \boldsymbol{x}) > 0 \quad \text{or} \quad \det D(\boldsymbol{y}^{-1} \circ f \circ \boldsymbol{x}) < 0,$$

for any parametrizations $\boldsymbol{x}$ and $\boldsymbol{y}$ in $\mathfrak{O}$. Moreover, f preserves orientation in the first case, and reverses it in the second.

Example 3.1.37. Consider the function $f\colon \mathbb{S}^2 \to \mathbb{S}^2$ given by $f(x,y,z) = (-x,y,z)$, and the Monge parametrization $\boldsymbol{x}\colon B_1(0) \to \mathbb{S}^2$, $\boldsymbol{x}(u,v) = \left(u, v, \sqrt{1-u^2-v^2}\right)$. We have that $\boldsymbol{x}^{-1} \circ f \circ \boldsymbol{x}(u,v) = (-u,v)$, whose derivative has negative determinant. We conclude that f inverts orientation.

From this point on, we will assume that all the surfaces we'll work with are orientable. To get an idea for the sort of surface we leaving out of future discussions, let's see a "non-example":

Example 3.1.38 (Möbius Strip)**.** The surface is constructed by considering a line segment of length ℓ, which is twisted while it rotates along a circle of radius $r > \ell$ in such a way that opposite ends of this line segment are identified when it returns to its original position in the circle.

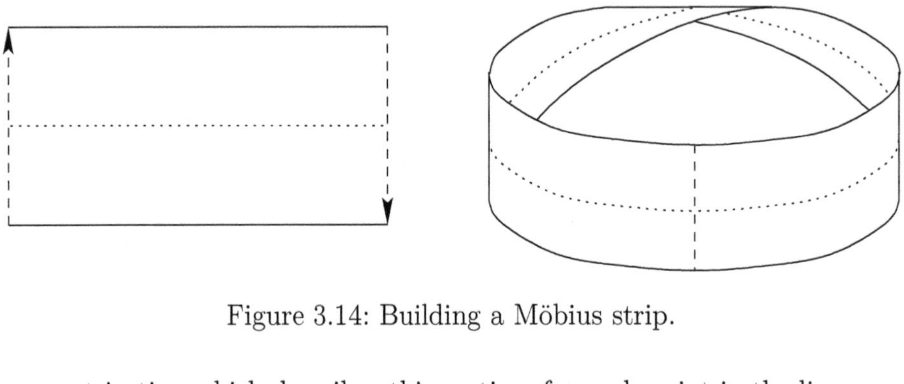

Figure 3.14: Building a Möbius strip.

A parametrization which describes this motion, for each point in the line segment, is

$$\boldsymbol{x}(u,v) = \boldsymbol{\alpha}(u) + (v - 1/2)\boldsymbol{\beta}(u),$$

where $\alpha(u) = 2(\cos u, \sin u, 0)$ and $\beta(u) = \cos(u/2)\alpha(u) + \sin(u/2)e_3$. In the curve β we use $u/2$ instead of u in the trigonometric functions to ensure that the line segment returns to its original position with the endpoints reversed. If we do not use $u/2$, the resulting surface would be orientable.

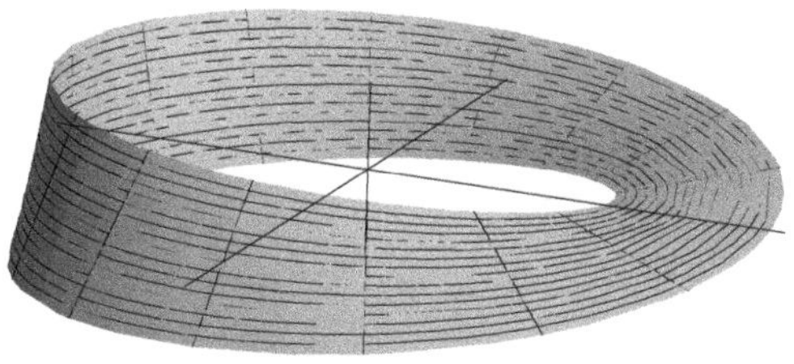

Figure 3.15: The image of x in $\mathbb{R}^3$.

Remark. By relaxing the definition of a regular surface, dropping the requirement that parametrizations are homeomorphisms onto its images (i.e., also allowing surfaces to have self-intersections), we may obtain more non-orientable surfaces, by mimicking the above construction. For instance, the *Klein bottle* arises from replacing the line segment used to construct a Möbius strip by a lemniscate:

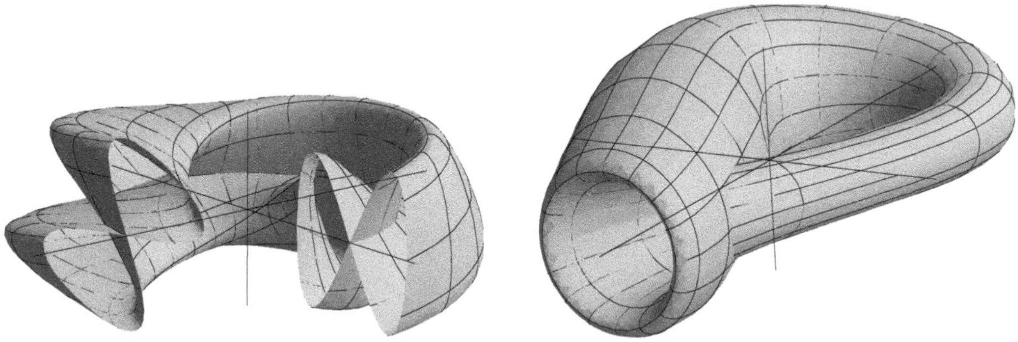

(a) The Klein bottle, according to the above construction.

(b) Another realization of the Klein bottle.

Figure 3.16: Realizations of the Klein bottle.

Both surfaces above are diffeomorphic and the second one justifies the name "bottle" in Klein bottle. For even more examples, see Chapter 11 in [27].

Exercises

Exercise 3.1.1 (Localization). Let $M \subseteq \mathbb{R}^3_\nu$ be a regular surface and $p \in M$. Show that we can always choose a parametrization x for M around p whose domain contains the origin in the plane, with $x(0,0) = p$.

Exercise† 3.1.2 (Stereographic Projection). Consider the *unit sphere*

$$\mathbb{S}^2 = \{(x,y,z) \in \mathbb{R}^3 \mid x^2+y^2+z^2=1\}.$$

Let $e_3 = (0,0,1)$ be the *north pole* of $\mathbb{S}^2$. For each $(u,v) \in \mathbb{R}^2$ the line in $\mathbb{R}^3$ joining $(u,v,0)$ to e_3 intercepts $\mathbb{S}^2 \setminus \{e_3\}$ in precisely one point $\mathsf{St}^{-1}(u,v)$. This determines a map $\mathsf{St}^{-1}\colon \mathbb{R}^2 \to \mathbb{S}^2 \setminus \{e_3\}$.

(a) Write an expression for St^{-1} and show that it is a regular parametrized surface.

(b) Write an expression for the inverse $\mathsf{St}\colon \mathbb{S}^2 \setminus \{e_3\} \to \mathbb{R}^2$.

Hint. Thinking geometrically and using a similar reasoning as in the previous item is easier than inverting the expression you found.

(c) Rewrite the expression for St^{-1} identifying $\mathbb{R}^2 \equiv \mathbb{C}$ via $(u,v) \mapsto z = u+iv$, in terms of $z, \bar{z}, \mathrm{Re}(z)$ and $\mathrm{Im}(z)$.

Exercise 3.1.3. Determine the values $c \in \mathbb{R}$ for which the set

$$M \doteq \{(x,y,z) \in \mathbb{R}^3_\nu \mid z(z+4) = 3xy + c\}$$

is a regular surface.

Exercise† 3.1.4. Prove Lemma 3.1.20 (p. 142).

Exercise 3.1.5.

(a) Let $\phi\colon \mathbb{R}^3_\nu \to \mathbb{R}^3_\nu$ be a diffeomorphism and $M \subseteq \mathbb{R}^3_\nu$ be a regular surface. Show that the image $\phi(M)$ is also a regular surface, and that $T_{\phi(p)}\phi(M) = \mathrm{d}\phi_p(T_pM)$ for all $p \in M$.

(b) Use the previous item to show that if $M = \mathbb{S}^2$, $\mathbb{S}^2_1$, or $\mathbb{H}^2$ and $f\colon M \to \mathbb{R}_{>0}$ is smooth, then $M(f) \doteq \{f(p)p \mid p \in M\}$ is a regular surface, diffeomorphic to M.

Exercise† 3.1.6.

(a) Let $M \subseteq \mathbb{R}^3_\nu$ be a regular surface. Show that

$$\mathrm{Diff}(M) \doteq \{f\colon M \to M \mid f \text{ is a diffeomorphism}\}$$

equipped with the operation of function composition is a group.

(b) Show that if $M_1, M_2 \subseteq \mathbb{R}^3_\nu$ are diffeomorphic regular surfaces, then we have that $\mathrm{Diff}(M_1) \cong \mathrm{Diff}(M_2)$.

Remark. In more general settings, we may consider actions of a subgroup $G \subseteq \mathrm{Diff}(M)$ on M and, under suitable conditions, obtain new "quotient surfaces" M/G (orbit spaces).

Exercise† 3.1.7. Show Proposition 3.1.15 (p. 140).

Hint. Use Proposition 3.1.14, analyzing the behavior of the function along curves in M.

Exercise 3.1.8. Let $f\colon \mathbb{R} \to \mathbb{R}$ be a smooth function, and consider the regular parametrized surface $\boldsymbol{x}\colon \mathbb{R}_{>0} \times \mathbb{R} \to \mathbb{R}^3_\nu$ given by $\boldsymbol{x}(u,v) = (u,v,uf(v/u))$. Show that all the tangent planes to this surface pass through the origin $\boldsymbol{0}$.

Exercise 3.1.9. Let $x\colon]0,2\pi[\times \mathbb{R} \to \mathbb{R}^3$ be a parametrized surface of the form $x(\theta,z) = (r(\theta,z)\cos\theta, r(\theta,z)\sin\theta, z)$, where r is a positive function. Show that x has *rotational symmetry*, that is, $\partial r/\partial\theta = 0$, if and only if all (Euclidean) normal lines to the image of x pass through the z-axis.

Exercise 3.1.10. Similarly to what was done above, let $x\colon \mathbb{R}^2 \to \mathbb{L}^3$ be a parametrized surface of the form $x(\varphi,x) = (x, \rho(\varphi,x)\cosh\varphi, \rho(\varphi,x)\sinh\varphi)$, where ρ is a positive function. Show that $\partial\rho/\partial\varphi = 0$ if and only if all (Lorentzian) normal lines to the image of x pass through the x-axis. Compare with the previous exercise.

Exercise 3.1.11 (Horocycles). Let $v \in \mathbb{L}^3$ be a future-directed lightlike vector and $c < 0$. The set $H_{v,c} \doteq \{x \in \mathbb{H}^2 \mid \langle x, v\rangle_L = c\}$ is called a *horocycle* of $\mathbb{H}^2$, based on v. Let $\alpha\colon I \to H_{v,c} \subseteq \mathbb{H}^2$ have unit speed.

(a) Show that

$$\alpha(s) = -\frac{s^2}{2c}v + sw_1 + w_2,$$

where w_1 and w_2 are unit and orthogonal vectors, with w_1 spacelike, w_2 timelike, w_1 orthogonal to v, and $\langle w_2, v\rangle_L = c$.

Hint. Write $\alpha''(s)$ as a combination of $\alpha(s)$, $\alpha'(s)$ and v, and also assume that $0 \in I$ to simplify. It is not necessary to parametrize $\mathbb{H}^2$ to solve this exercise.

(b) Conclude that α is semi-lightlike, with zero pseudo-torsion.

Exercise 3.1.12. Show that if $M \subseteq \mathbb{R}^3_\nu$ is a regular surface and S is open in M, then S is also a regular surface and $T_pS = T_pM$ for all $p \in S$. Moreover, the inclusion $\iota\colon S \hookrightarrow M$ is smooth.

Exercise† 3.1.13. Show Corollary 3.1.27 (p. 146).

Exercise 3.1.14. For each $p \in \mathbb{S}^2 \setminus \{(0,0,\pm 1)\}$, let $f(p)$ be the intersection of the ray passing through the origin and passing through p with the cylinder $\mathbb{S}^1 \times \mathbb{R}$. Show that $f : \mathbb{S}^2 \setminus \{(0,0,\pm 1)\} \to \mathbb{S}^1 \times \mathbb{R}$ thus defined is a diffeomorphism.

Exercise 3.1.15 (Lambert Projection for $\mathbb{S}^2_1$). For each $p \in \mathbb{S}^2_1$, consider $f(p) \in \mathbb{S}^1 \times \mathbb{R}$ the intersection of the horizontal ray starting at the z-axis passing through p, with the cylinder $\mathbb{S}^1 \times \mathbb{R}$. Show that the map $f\colon \mathbb{S}^2_1 \to \mathbb{S}^1 \times \mathbb{R}$ so defined is a diffeomorphism.

Exercise 3.1.16. Let $M \subseteq \mathbb{R}^3$ be a regular surface and $p_0 \notin M$. Show that the *central projection through* p_0, $f : M \to \mathbb{S}^2$ given by

$$f(p) \doteq \frac{p - p_0}{\|p - p_0\|_E}$$

is a local diffeomorphism if and only if $p - p_0 \notin T_pM$, for each $p \in M$.

Hint. Compute $\mathrm{d}f_p$ and observe what happens if you're able to evaluate $\mathrm{d}f_p(p - p_0)$.

Exercise 3.1.17 (Instructive challenge). Show that

$$M \doteq \{(x,y,z) \in \mathbb{R}^3 \mid \mathrm{e}^{x^2} + \mathrm{e}^{y^2} + \mathrm{e}^{z^2} = a\}$$

is a regular surface if $a > 3$, which is diffeomorphic to the sphere $\mathbb{S}^2$.

Hint. Verifying that M is a regular surface (not containing the origin) is a straightforward application of Theorem 3.1.7 (p. 135). To see that M is diffeomorphic to $\mathbb{S}^2$, follow the steps:

- for each $(x, y, z) \neq \mathbf{0}$, show that the (smooth) function $h\colon \mathbb{R} \to \mathbb{R}$ given by $h(t) = \mathrm{e}^{t^2x^2} + \mathrm{e}^{t^2y^2} + \mathrm{e}^{t^2z^2}$ is increasing and surjective onto the interval $[3, +\infty[$;

- conclude that the central projection through the origin $\phi\colon M \to \mathbb{S}^2$ is bijective. Use Exercise 3.1.16 above to conclude that ϕ is a local diffeomorphism (and hence global).

Exercise 3.1.18. Here we have a version of Exercise 2.1.21 (p. 77) in $\mathbb{R}^3$. Let $\boldsymbol{\alpha}\colon I \to \mathbb{R}^3$ be a regular parametrized curve and $\boldsymbol{x}\colon U \to \mathbb{R}^3$ be a regular parametrized surface. Suppose that $\boldsymbol{\alpha}$ and $\boldsymbol{x}$ *intersect transversally* at $\boldsymbol{p} = \boldsymbol{\alpha}(t_0) = \boldsymbol{x}(u_0, v_0)$, i.e., such that $\{\boldsymbol{\alpha}'(t_0), \boldsymbol{x}_u(u_0, v_0), \boldsymbol{x}_v(u_0, v_0)\}$ is linearly independent, where $t_0 \in I$ and $(u_0, v_0) \in U$. Let $\boldsymbol{v} \in \mathbb{R}^3$ be a unit vector and take an arbitrary $s \in \mathbb{R}$.

(a) Define $\boldsymbol{\alpha}^s\colon I \to \mathbb{R}^3$, by $\boldsymbol{\alpha}^s(t) = \boldsymbol{\alpha}(t) + s\boldsymbol{v}$. Show that for small enough s, the traces of $\boldsymbol{\alpha}^s$ and $\boldsymbol{x}$ intersect near $\boldsymbol{p}$.

(b) Define $\boldsymbol{x}^s\colon U \to \mathbb{R}^3$, by $\boldsymbol{x}^s(u, v) = \boldsymbol{x}(u, v) + s\boldsymbol{v}$. Show that for small enough s, the traces of $\boldsymbol{x}^s$ and $\boldsymbol{\alpha}$ intersect near $\boldsymbol{p}$.

Hint. Use the Implicit Function Theorem for a suitable function $F\colon \mathbb{R} \times I \times U \to \mathbb{R}^3$.

Exercise 3.1.19. Let $M \subseteq \mathbb{R}^3_\nu$ be a regular surface and I an open interval of the real line. Say that a function $F\colon M \times I \to \mathbb{R}^k$ is *smooth* if for each parametrization $(U, \boldsymbol{x})$ for M, the composition $F \circ (\boldsymbol{x} \times \mathrm{Id}_I)\colon U \times I \to \mathbb{R}^k$ is smooth as a map between Euclidean spaces. Show that:

(a) to verify whether such F is smooth, it suffices to check it for a collection of parametrizations whose images cover M;

(b) if $F, G\colon M \times I \to \mathbb{R}^k$ are smooth and $\lambda \in \mathbb{R}$, then $F + G, \lambda F$ and $\langle F, G\rangle$ are also smooth;

(c) given $\boldsymbol{p}_0 \in M$ and $t_0 \in I$, the maps $F_{t_0}\colon M \to \mathbb{R}^k$ and $\boldsymbol{\alpha}_{\boldsymbol{p}_0}\colon I \to \mathbb{R}^k$ given by $F_{t_0}(\boldsymbol{p}) \doteq F(\boldsymbol{p}, t_0)$ and $\boldsymbol{\alpha}_{\boldsymbol{p}_0}(t) \doteq F(\boldsymbol{p}_0, t)$ are smooth;

(d) the *differential* of F, $\mathrm{d}F_{(\boldsymbol{p},t)}\colon T_{\boldsymbol{p}}M \times \mathbb{R} \to \mathbb{R}^k$, given by

$$\mathrm{d}F_{(\boldsymbol{p},t)}(\boldsymbol{v}, a) \doteq \left.\frac{\mathrm{d}}{\mathrm{d}s}\right|_{s=0} F(\boldsymbol{\alpha}(s), t + sa),$$

where $\boldsymbol{\alpha}$ is any curve in M realizing $\boldsymbol{v}$, is well-defined (that is, it does not depend on the choice of $\boldsymbol{\alpha}$).

Remark. The ideas presented in this exercise naturally generalize for functions defined in the cartesian product of two (or more) regular surfaces. Moreover, one can put a third surface as the codomain (instead of some $\mathbb{R}^k$). Can you give statements of results similar to the ones presented above in this setting?

Exercise 3.1.20 (Inverse Function Theorem). Let $M \subseteq \mathbb{R}^3_\nu$ be a regular surface, I an interval in the real line, and $F\colon M \times I \to \mathbb{R}^3$ be a smooth function (as in the previous exercise). Show that if $(\boldsymbol{p}_0, t_0) \in M \times I$ is such that $\mathrm{d}F_{(\boldsymbol{p}_0,t_0)}$ is non-singular, there is an open subset V of M containing p_0, an open subset W of $\mathbb{R}^3$ containing $f(p_0, t_0)$, and $r > 0$ such that the restriction

$$F\big|_{V\times]t_0-r,t_0+r[}: V \times]t_0 - r, t_0 + r[\to W$$

is a diffeomorphism.

Exercise 3.1.21 (Implicit Function Theorem). Let $M_1, M_2, M_3 \subseteq \mathbb{R}^3_\nu$ be regular surfaces, $p_i \in M_i$ $(i = 1, 2, 3)$ and $f: M_1 \times M_2 \to M_3$ be a smooth function. Suppose that $f(p_1, p_2) = p_3$, and also that the *partial differential* $\mathrm{d}_2 f_{(p_1,p_2)}: T_{p_2}M_2 \to T_{p_3}M_3$ defined by $\mathrm{d}_2 f_{(p_1,p_2)}(w) \doteq \mathrm{d}f_{(p_1,p_2)}(\mathbf{0}, w)$ is non-singular. Show that there are neighborhoods $V_1 \subseteq M_1$ and $V_2 \subseteq M_2$ of p_1 and p_2, and a smooth function $\varphi: V_1 \to V_2$ such that $f(p, \varphi(p)) = p_3$ for each $p \in V_1$.

Hint. "Pull" everything down and apply the "old" Implicit Function Theorem.

Exercise 3.1.22. Let $M \subseteq \mathbb{R}^3_\nu$ be a regular surface, and suppose that M can be covered by two parametrizations (U_x, x) and (U_y, y), such that $W = x(U_x) \cap y(U_y)$ has two connected components. Call W_1 and W_2 the two connected components of the inverse image $x^{-1}(W)$. Show that if $\det D(y^{-1} \circ x)$ is positive on W_1 and negative on W_2, then M is non-orientable.

Exercise 3.1.23. Show that if $M_1, M_2 \subseteq \mathbb{R}^3_\nu$ are regular surfaces, M_2 is orientable, and $f: M_1 \to M_2$ is a local diffeomorphism, then M_1 is orientable.

Exercise 3.1.24. Show Proposition 3.1.36 (p. 150).

Exercise 3.1.25. Consider the *antipodal map* $A: \mathbb{R}^3_\nu \to \mathbb{R}^3_\nu$ given by $A(p) = -p$. One may consider the restrictions $A: M \to M$, for $M = \mathbb{S}^2$, $\mathbb{S}^2_1$ or $\mathbb{H}^2 \cup \mathbb{H}^2_-$. Investigate whether A preserves or inverts orientation. Does the answer depend on the choice of M we take here?

Exercise 3.1.26. Let $M \subseteq \mathbb{R}^3_\nu$ be a orientable regular surface, and $f: M \to M$ be a diffeomorphism. Investigate whether $f \circ f$ preserves or reverses orientation.

3.2 CAUSAL TYPE OF SURFACES, FIRST FUNDAMENTAL FORM

We finally begin the study of the actual geometry of regular surfaces, and now the ambient space ($\mathbb{R}^3$ or $\mathbb{L}^3$) from which the surface will get its geometry matters. The notion of causal character seen so far for vectors and subspaces of $\mathbb{L}^n$, and then generalized to curves, has a version for surfaces:

Definition 3.2.1 (Causal Character). Let $M \subseteq \mathbb{L}^3$ be a regular surface. We say that:

(i) M is *spacelike*, if for each $p \in M$, T_pM is a spacelike plane;

(ii) M is *timelike*, if for each $p \in M$, T_pM is a timelike plane;

(iii) M is *lightlike*, if for each $p \in M$, T_pM is a lightlike plane.

Remark.

- We will have a situation similar to what happened for curves: by continuity, if T_pM is spacelike or timelike for some $p \in M$, then T_qM will have the same causal type for each q in some neighborhood of p in M. This way, we may possibly restrict our attention to surfaces with constant causal character, when needed.
- Keeping consistency with the conventions adopted throughout the text so far, we will consider regular surfaces in $\mathbb{R}^3$ to be spacelike.
- If M has no points p for which T_pM is lightlike, we'll simply say that M is non-degenerate.
- If $x\colon U \to \mathbb{R}^3_\nu$ is an injective and regular parametrized surface, we will abuse terminology and give the causal character of the regular surface $x(U)$ to the map x itself. If x is not injective, we may define the causal character of x in a similar fashion to what was done above, but considering the tangent planes to x itself, $T_{(u,v)}x$.

Naturally, parametrizations are one of the main tools we have available to decide the causal type of a surface. We will register the next definition, which may prove convenient in the future to present in an easier way some expressions related to the geometry of a surface:

Definition 3.2.2 (Partial indicators). Let $M \subseteq \mathbb{R}^3_\nu$ be a non-degenerate regular surface and (U, x) be a parametrization for M. The *partial indicators* of x are defined as the indicators of the partial derivatives of x: $\epsilon_u \doteq \epsilon_{x_u(u,v)}$ and $\epsilon_v \doteq \epsilon_{x_v(u,v)}$.

Remark.

- This time, from the start, we'll allow these partial indicators to assume the value zero, in case one of the derivatives is lightlike.
- If we denote our coordinates by (s,t) instead of (u,v), the partial indicators will be ϵ_s and ϵ_t, for instance.

We know that the tangent planes to a regular surface are 2-dimensional vector subspaces of $\mathbb{R}^3_\nu$ and so, in $\mathbb{L}^3$, their causal type is determined by the causal type of their normal directions, whether they are Euclidean or Lorentzian. To summarize, we will register the:

Proposition 3.2.3. *Let $M \subseteq \mathbb{L}^3$ be a regular surface and (U, x) be a parametrization for M such that $x(u_0, v_0) = p \in M$. Then:*

(i) T_pM is spacelike if and only if $x_u(u_0, v_0) \times x_v(u_0, v_0)$ is timelike;

(ii) T_pM is timelike if and only if $x_u(u_0, v_0) \times x_v(u_0, v_0)$ is spacelike;

(iii) T_pM is lightlike if and only if $x_u(u_0, v_0) \times x_v(u_0, v_0)$ is lightlike.

With this in mind, we may particularize Proposition 3.1.9, seen previously, as follows:

Proposition 3.2.4. *Let $M \subseteq \mathbb{L}^3$ be a regular surface.*

(i) If M is spacelike or lightlike, then M is locally the graph of a smooth function whose domain is a subset of the plane $z = 0$.

(ii) If M is timelike, then M is locally the graph of a smooth function whose domain is a subset of the plane $x = 0$ or the plane $y = 0$.

Proof: Take any point $\boldsymbol{p} \in M$ and choose a parametrization $(U, \boldsymbol{x})$ for M with $\boldsymbol{x}(u_0, v_0) = \boldsymbol{p}$. From here on, the proof goes exactly like the proof of Proposition 3.1.9, with the only difference being that here we can use the extra information regarding the causal type of M to precisely pinpoint which of the submatrices of $D\boldsymbol{x}(u_0, v_0)$ will be non-singular, instead of making a generic assumption as before. Explicitly writing $\boldsymbol{x}(u,v) = (x(u,v), y(u,v), z(u,v))$ and noting that

$$\frac{\partial \boldsymbol{x}}{\partial u} \times_E \frac{\partial \boldsymbol{x}}{\partial v} = \left(\frac{\partial(y,z)}{\partial(u,v)}, -\frac{\partial(x,z)}{\partial(u,v)}, \frac{\partial(x,y)}{\partial(u,v)} \right),$$

we have that if M is spacelike (resp., lightlike), then the above vector is timelike (resp., lightlike) at (u_0, v_0), so that

$$\frac{\partial(x,y)}{\partial(u,v)}(u_0, v_0) \neq 0,$$

and so M admits a reparametrization around $\boldsymbol{p}$ of the form $(s, t, f(s,t))$ for some smooth function f. Similarly, if M is timelike, then this cross product in display is spacelike, whence

$$\frac{\partial(y,z)}{\partial(u,v)}(u_0, v_0) \neq 0 \quad \text{or} \quad \frac{\partial(x,z)}{\partial(u,v)}(u_0, v_0) \neq 0,$$

which would give us reparametrizations around $\boldsymbol{p}$ of the forms $(f(s,t), s, t)$ or $(s, f(s,t), t)$ for some smooth function f, respectively. □

In $\mathbb{L}^3$, the additional information given by the causal type of a surface may also impose strong restrictions on its topology:

Proposition 3.2.5. *There is no compact regular surface of constant causal type in $\mathbb{L}^3$.*

Proof: Consider the projection $\pi_1, \pi_3 : M \to \mathbb{R}$ given, respectively, by $\pi_1(x,y,z) = x$ and $\pi_3(x,y,z) = z$. Since M is compact and these functions are continuous, each one of them admits a maximum value in M, say, at $\boldsymbol{p}_1$ and $\boldsymbol{p}_2$, respectively. At these points, we have, for each $\boldsymbol{v} = (v_1, v_2, v_3) \in T_{\boldsymbol{p}_1}M$ and $\boldsymbol{w} = (w_1, w_2, w_3) \in T_{\boldsymbol{p}_2}M$, that

$$0 = \mathrm{d}(\pi_1)_{\boldsymbol{p}_1}(\boldsymbol{v}) = v_1 \quad \text{and} \quad 0 = \mathrm{d}(\pi_3)_{\boldsymbol{p}_2}(\boldsymbol{w}) = w_3,$$

whence $T_{\boldsymbol{p}_1}M = \boldsymbol{e}_1^{\perp}$ and $T_{\boldsymbol{p}_2}M = \boldsymbol{e}_3^{\perp}$ are tangent planes to M with distinct causal types. □

The next result allows us to determine the causal type of a surface without using parametrizations:

Theorem 3.2.6. *Let $\Omega \subseteq \mathbb{L}^3$ be open, $f : \Omega \to \mathbb{R}$ be a smooth function and $a \in \mathbb{R}$ be a regular value for f. If $M = f^{-1}(\{a\})$, then:*

(i) M is spacelike if and only if $\nabla f(\boldsymbol{p})$ is always timelike;

(ii) M is timelike if and only if $\nabla f(\boldsymbol{p})$ is always spacelike;

(iii) M is lightlike if and only if $\nabla f(\boldsymbol{p})$ is always lightlike.

Proof: We have previously seen that $T_{\boldsymbol{p}}M = \ker Df(\boldsymbol{p})$, for each $\boldsymbol{p} \in M$. Since the gradient $\nabla f(\boldsymbol{p})$ is the vector equivalent to $Df(\boldsymbol{p})$ under $\langle \cdot, \cdot \rangle_E$, we also have that the tangent plane at $\boldsymbol{p}$ is $T_{\boldsymbol{p}}M = (\nabla f(\boldsymbol{p}))^{\perp}$. □

Remark. Exercise 2.2.11 (p. 92) illustrates how the notion of "gradient" may depend on the ambient product chosen. Given $\Omega \subseteq \mathbb{R}^n_\nu$ open and $f\colon \Omega \to \mathbb{R}$ smooth, the *Euclidean gradient* of f is the vector $\nabla_E f(p)$ associated to $Df(p)$ via $\langle \cdot,\cdot\rangle_E$ by Riesz's Lemma. In the same way, the *Lorentzian gradient* of f is the vector $\nabla_L f(p)$ associated to $Df(p)$ in a similar fashion using $\langle \cdot,\cdot\rangle_L$ instead. The notation ∇_E is immediately discarded, as it coincides with the differential operator ∇ seen in basic Calculus courses. On the other hand, in the Lorentzian gradient we have a sign change only in the timelike component:

$$\nabla_L f(p) = \left(\frac{\partial f}{\partial x_1}(p), \cdots, \frac{\partial f}{\partial x_{n-1}}(p), -\frac{\partial f}{\partial x_n}(p)\right).$$

This same idea may be used to define the gradient of functions defined only along surfaces (or in even more general settings): see Exercise 3.2.6. The previous theorem remains valid using $\nabla_L f(p)$ instead of $\nabla f(p)$.

Example 3.2.7 (Causal type of spheres in $\mathbb{L}^3$)**.** We have seen that if $p \in \mathbb{L}^3$ and $r > 0$ are given, and $f\colon \mathbb{L}^3 \to \mathbb{R}$ is given by $f(q) = \langle q-p, q-p\rangle_L$, then we have that $Df(q) = 2\langle q-p,\cdot\rangle_L$. This says that $\nabla_L f(q) = 2(q-p)$, and thus if $q \in \mathbb{S}^2_1(p,r)$, then

$$\langle \nabla_L f(q), \nabla_L f(q)\rangle_L = 4r^2 > 0,$$

whence we conclude that $\mathbb{S}^2_1(p,r)$ is a timelike surface. Similarly, if $q \in \mathbb{H}^2_\pm(p,r)$, we have

$$\langle \nabla_L f(q), \nabla_L f(q)\rangle_L = -4r^2 < 0,$$

whence $\mathbb{H}^2_\pm(p,r)$ is a spacelike surface.

To study the geometry of the surface, we would have many ways to define "length" and "angle" over it. Let's use the previous existence of those notions in the ambient $\mathbb{R}^3_\nu$, motivating the following:

Definition 3.2.8 (First Fundamental Form)**.** Let $M \subseteq \mathbb{R}^3_\nu$ be a regular surface and $p \in M$. The *First Fundamental Form* of M at p is the bilinear map $\mathrm{I}_p\colon T_pM \times T_pM \to \mathbb{R}$ given by $\mathrm{I}_p(v,w) \doteq \langle v,w\rangle$.

Remark.

- We will abbreviate $\mathrm{I}_p(v,v)$ simply by $\mathrm{I}_p(v)$, let alone when we decide to omit the point p as well.
- In $\mathbb{L}^3$, I_p is also called the *Minkowski First Fundamental Form* of M at p.
- The First Fundamental Form is a particular case of a more general concept called a *pseudo-Riemannian metric*. When M is spacelike, the metric is called *Riemannian*, and when M is timelike the metric is called *Lorentzian*. We will briefly discuss this further in Chapter 4, ahead. For more details see, for instance, [54].

Definition 3.2.9 (Components of the First Form)**.** Let $M \subseteq \mathbb{R}^3_\nu$ be a regular surface and (U,x) be a parametrization for M. The *components of the First Fundamental Form relative to* x are defined by

$$E(u,v) \doteq \mathrm{I}_{x(u,v)}\left(\frac{\partial x}{\partial u}(u,v)\right),$$

$$F(u,v) \doteq \mathrm{I}_{x(u,v)}\left(\frac{\partial x}{\partial u}(u,v), \frac{\partial x}{\partial v}(u,v)\right) \quad \text{and}$$

$$G(u,v) \doteq \mathrm{I}_{x(u,v)}\left(\frac{\partial x}{\partial v}(u,v)\right).$$

Remark.

- It is also usual to write $g_{11} \doteq E$, $g_{12} = g_{21} \doteq F$ and $g_{22} \doteq G$, so that all the relevant information regarding the First Fundamental Form is encoded in the Gram matrix $(g_{ij}(u,v))_{1\leq i,j\leq 2}$ of $\mathrm{I}_{x(u,v)}$ relative to the basis $\mathcal{B}_x$. We will see shortly that this Gram matrix is non-singular precisely when M is non-degenerate, in which case we'll denote the inverse matrix by $(g^{ij})_{1\leq i,j\leq 2}$, with upper indices.

- A third way to represent the First Fundamental Form in coordinates is through *differential notation*:

$$\mathrm{d}s^2 = \sum_{i,j=1}^{2} g_{ij}\,\mathrm{d}u^i\,\mathrm{d}u^j = E(u,v)\,\mathrm{d}u^2 + 2F(u,v)\,\mathrm{d}u\,\mathrm{d}v + G(u,v)\,\mathrm{d}v^2,$$

  where we identify $u \leftrightarrow u^1$ and $v \leftrightarrow u^2$ (do not confuse this with exponents). We will repeat such identifications in the future without many comments, as well as use whichever way of representing the First Fundamental Form is more convenient in the moment.

We have seen in Chapter 1 how to use Sylvester's Criterion to determine the causal type of subspaces of $\mathbb{L}^n$. When dealing with surfaces, though, the next result makes our lives easier:

Proposition 3.2.10. *Let $\boldsymbol{x}\colon U \to \mathbb{L}^3$ be a regular parametrized surface. The causal type of $\boldsymbol{x}$ is decided by the sign of the determinant of its First Fundamental Form:*

(i) $\boldsymbol{x}$ is spacelike if and only if $\det\left((g_{ij})_{1\leq i,j\leq 2}\right) > 0$*;*

(ii) $\boldsymbol{x}$ is timelike if and only if $\det\left((g_{ij})_{1\leq i,j\leq 2}\right) < 0$*;*

(iii) $\boldsymbol{x}$ is lightlike if and only if $\det\left((g_{ij})_{1\leq i,j\leq 2}\right) = 0$.

Proof: This follows directly from Proposition 3.2.3 (p. 156) by using Lagrange's Identity in $\mathbb{L}^3$:

$$\langle \boldsymbol{x}_u \times_L \boldsymbol{x}_v, \boldsymbol{x}_u \times_L \boldsymbol{x}_v\rangle_L = -\begin{vmatrix} \langle \boldsymbol{x}_u, \boldsymbol{x}_u\rangle_L & \langle \boldsymbol{x}_u, \boldsymbol{x}_v\rangle_L \\ \langle \boldsymbol{x}_v, \boldsymbol{x}_u\rangle_L & \langle \boldsymbol{x}_v, \boldsymbol{x}_v\rangle_L \end{vmatrix} = -\det\left((g_{ij})_{1\leq i,j\leq 2}\right).$$

□

In principle, we could ask ourselves what is the importance of the First Fundamental Form of a regular surface, since it is nothing more than the restriction of the ambient product to each tangent plane. To some extent, the answer lies in the question itself: the First Fundamental Form allows us to study the geometry of the surface without making reference to the ambient space, once the restriction of $\langle\cdot,\cdot\rangle$ has been made. In other words, to study the intrinsic geometry of the surface it is not necessary to know how the ambient product acts on vectors which are not tangent. Let's see a few notions which become intrinsic:

Example 3.2.11 (Intrinsic concepts)**.** Let $M \subseteq \mathbb{R}^3_\nu$ be a regular surface, $\boldsymbol{p} \in M$, $(U, \boldsymbol{x})$ be a parametrization for M, and $\boldsymbol{\alpha}\colon I \to M$ be a curve in M.

(1) If M is spacelike, the angle between two vectors $\boldsymbol{u}, \boldsymbol{v} \in T_{\boldsymbol{p}}M \setminus \{\boldsymbol{0}\}$ is the number $\theta \in [0, 2\pi[$ determined by the relation

$$\cos\theta = \frac{\mathrm{I}_{\boldsymbol{p}}(\boldsymbol{u},\boldsymbol{v})}{\sqrt{\mathrm{I}_{\boldsymbol{p}}(\boldsymbol{u})}\sqrt{\mathrm{I}_{\boldsymbol{p}}(\boldsymbol{v})}}.$$

(2) If M is timelike in $\mathbb{L}^3$, the hyperbolic angle between two timelike vectors $\boldsymbol{u}, \boldsymbol{v} \in T_pM$, both future-directed or past-directed, is the number $\varphi \geq 0$ determined by the relation

$$\cosh \varphi = -\frac{\mathrm{I}_p(\boldsymbol{u}, \boldsymbol{v})}{\sqrt{-\mathrm{I}_p(\boldsymbol{u})}\sqrt{-\mathrm{I}_p(\boldsymbol{v})}}.$$

(3) The arclength of $\boldsymbol{\alpha}$ is expressed just as

$$L[\boldsymbol{\alpha}] = \int_I \sqrt{|\mathrm{I}_{\boldsymbol{\alpha}(t)}(\boldsymbol{\alpha}'(t))|}\, \mathrm{d}t.$$

This also justifies the differential notation adopted to express the First Fundamental Form in coordinates. Suppose that $M \subseteq \mathbb{R}^3$ and that the curve is written in coordinates as $\boldsymbol{\alpha}(t) = \boldsymbol{x}(u(t), v(t))$. If s is an arclength function for $\boldsymbol{\alpha}$, omitting points of evaluation, we have that

$$\mathrm{I}_{\boldsymbol{\alpha}(t)}(\boldsymbol{\alpha}'(t)) = \left(\frac{\mathrm{d}s}{\mathrm{d}t}\right)^2 = E\left(\frac{\mathrm{d}u}{\mathrm{d}t}\right)^2 + 2F\frac{\mathrm{d}u}{\mathrm{d}t}\frac{\mathrm{d}v}{\mathrm{d}t} + G\left(\frac{\mathrm{d}v}{\mathrm{d}t}\right)^2.$$

A more rigorous interpretation of the symbols $\mathrm{d}u$ and $\mathrm{d}v$ is the following: suppose that $\boldsymbol{x}(u_0, v_0) = \boldsymbol{p}$, and take $\boldsymbol{v} = a\boldsymbol{x}_u(u_0, v_0) + b\boldsymbol{x}_v(u_0, v_0) \in T_pM$. Regarding $\mathrm{d}u$ and $\mathrm{d}v$ as the basis for $T_p^*M \doteq (T_pM)^*$ (the so-called *cotangent plane* to M at $\boldsymbol{p}$) dual to $\boldsymbol{x}_u(u_0, v_0)$ and $\boldsymbol{x}_v(u_0, v_0)$, and denoting the coefficients of the First Fundamental Form at this point by E_0, F_0 G_0 only, we have:

$$\begin{aligned}
\mathrm{I}_p(\boldsymbol{v}) &= \mathrm{I}_p(a\boldsymbol{x}_u(u_0, v_0) + b\boldsymbol{x}_v(u_0, v_0)) \\
&= a^2\mathrm{I}_p(\boldsymbol{x}_u(u_0, v_0)) + 2ab\mathrm{I}_p(\boldsymbol{x}_u(u_0, v_0), \boldsymbol{x}_v(u_0, v_0)) + b^2\mathrm{I}_p(\boldsymbol{x}_v(u_0, v_0)) \\
&= E_0a^2 + 2F_0ab + G_0b^2 \\
&= E_0\,(\mathrm{d}u(\boldsymbol{v}))^2 + 2F_0\,\mathrm{d}u(\boldsymbol{v})\,\mathrm{d}v(\boldsymbol{v}) + G_0(\mathrm{d}v(\boldsymbol{w}))^2 \\
&= E_0\,\mathrm{d}u^2(\boldsymbol{v}) + 2F_0\,\mathrm{d}u(\boldsymbol{v})\,\mathrm{d}v(\boldsymbol{v}) + G_0\,\mathrm{d}v^2(\boldsymbol{v}) = \mathrm{d}s_p^2(\boldsymbol{v}).
\end{aligned}$$

(4) The energy of $\boldsymbol{\alpha}$ is expressed just as

$$E[\boldsymbol{\alpha}] = \frac{1}{2}\int_I \mathrm{I}_{\boldsymbol{\alpha}(t)}(\boldsymbol{\alpha}'(t))\, \mathrm{d}t.$$

This functional will have a crucial role in the study of geodesics that will be done in Section 3.6.

(5) If R is an open subset of M such that the closure $\overline{R}$ is compact and $R \subseteq \boldsymbol{x}(U)$, we define the *area* of R as

$$\mathcal{A}(R) \doteq \int_{\boldsymbol{x}^{-1}(R)} \sqrt{|\det(g_{ij}(u, v))|}\, \mathrm{d}u\, \mathrm{d}v.$$

In Exercise 3.2.7 we ask you to show that $\mathcal{A}(R)$ does not depend on the choice of parametrization $\boldsymbol{x}$.

We extend the last example above to regions not necessarily contained in the image of a single parametrization. For example, when R is covered by two parametrizations, we have the:

Definition 3.2.12. Let $M \subseteq \mathbb{R}^3_\nu$ be a regular surface and $(U_x, \boldsymbol{x})$ and $(U_y, \boldsymbol{y})$ be two parametrizations for M. If R is an open subset of M such that the closure $\overline{R}$ is compact and $R \subseteq \boldsymbol{x}(U_x) \cup \boldsymbol{y}(U_y)$, we define the *area* of R as

$$\mathscr{A}(R) \doteq \int_{\boldsymbol{x}^{-1}(R)\cap U_x} \sqrt{|\det(g_{ij}(u,v))|}\,\mathrm{d}u\,\mathrm{d}v + \int_{\boldsymbol{y}^{-1}(R)\cap U_y} \sqrt{|\det(\widetilde{g_{ij}}(\tilde{u},\tilde{v}))|}\,\mathrm{d}\tilde{u}\,\mathrm{d}\tilde{v} - \\ - \int_{\boldsymbol{x}^{-1}(R\cap \boldsymbol{x}(U_x)\cap \boldsymbol{y}(U_y))} \sqrt{|\det(g_{ij}(u,v))|}\,\mathrm{d}u\,\mathrm{d}v,$$

where $(g_{ij})_{1\leq i,j\leq 2}$ and $(\widetilde{g_{ij}})_{1\leq i,j\leq 2}$ denote the components of First Fundamental Form of $\boldsymbol{x}$ and $\boldsymbol{y}$, respectively.

Remark.

- It also follows from Exercise 3.2.7 that the third integral above may be replaced by

$$\int_{\boldsymbol{y}^{-1}(R\cap \boldsymbol{x}(U_x)\cap \boldsymbol{y}(U_y))} \sqrt{|\det(\widetilde{g_{ij}}(\tilde{u},\tilde{v}))|}\,\mathrm{d}\tilde{u}\,\mathrm{d}\tilde{v}.$$

- It is a known fact, whose proof is outside the scope of this text, that every regular surface in $\mathbb{R}^3_\nu$ may be covered by three or fewer parametrizations. For more details, see [6].

Example 3.2.13.

(1) Planes: we have seen that every plane is the image of some (injective) regular parametrized surface of the form $\boldsymbol{x}(u,v) = \boldsymbol{p} + u\boldsymbol{w}_1 + v\boldsymbol{w}_2$, where $\boldsymbol{p}, \boldsymbol{w}_1, \boldsymbol{w}_2 \in \mathbb{R}^3_\nu$ are given, and $\{\boldsymbol{w}_1, \boldsymbol{w}_2\}$ is linearly independent. If the plane is not lightlike, there is no loss of generality in assuming that $\boldsymbol{w}_1$ and $\boldsymbol{w}_2$ are orthogonal and both unit vectors. Observe that $\epsilon_u = \epsilon_{\boldsymbol{w}_1}$ and $\epsilon_v = \epsilon_{\boldsymbol{w}_2}$. Thus, we have that the First Fundamental Form of the plane in those coordinates is given by $\mathrm{d}s^2 = \epsilon_{\boldsymbol{w}_1}\mathrm{d}u^2 + \epsilon_{\boldsymbol{w}_2}\mathrm{d}v^2$.

In the pathological case where the plane is lightlike, if $\boldsymbol{w}_1$ is unit (spacelike) and $\boldsymbol{w}_2$ is lightlike and orthogonal to $\boldsymbol{w}_1$, we obtain just the degenerate $\mathrm{d}s^2 = \mathrm{d}u^2$.

(2) When expressing the First Fundamental Form of a surface in coordinates, one must pay attention to the domain of the chosen parametrization. Maps which are not regular may introduce singularities in the metric's coordinate expression which are artificial and unrelated to the geometry of the surface. For example, we have seen that $\mathbb{H}^2$ is a spacelike surface. Consider $\boldsymbol{x}\colon \mathbb{R} \times]0, 2\pi[\to \mathbb{H}^2$ given by

$$\boldsymbol{x}(u,v) = (\sinh u \cos v, \sinh u \sin v, \cosh u).$$

We have that

$$(g_{ij}(u,v))_{1\leq i,j\leq 2} = \begin{pmatrix} 1 & 0 \\ 0 & \sinh^2 u \end{pmatrix}, \quad \text{or} \quad \mathrm{d}s^2 = \mathrm{d}u^2 + \sinh^2 u\,\mathrm{d}v^2.$$

The Gram determinant is just $\sinh^2 u \geq 0$, which vanishes only for $u = 0$. This indeed shows that $\mathbb{H}^2$ is spacelike on the points for which $u \neq 0$, but it does not say that $\mathbb{H}^2$ is lightlike at $\boldsymbol{x}(0,v) = (0,0,1)$. This happens because the derivative $\boldsymbol{x}_v(0,v)$ is the zero vector, and so $\boldsymbol{x}$ is not a valid parametrization, unless the points of the form $(0,v)$ are removed from its domain.

This phenomenon is not particular to $\mathbb{L}^3$: for the sphere $\mathbb{S}^2$, we may take

$$y(u,v) = (\cos u \cos v, \cos u \sin v, \sin u),$$

and verify that in this case we have the same situation as above, with

$$(g_{ij}(u,v))_{1\le i,j\le 2} = \begin{pmatrix} 1 & 0 \\ 0 & \sin^2 u \end{pmatrix}, \quad \text{or} \quad \mathrm{d}s^2 = \mathrm{d}u^2 + \sin^2 u\, \mathrm{d}v^2.$$

(3) Graphs. If $f\colon U \to \mathbb{R}$ is smooth, we have seen that the Monge parametrization $x\colon U \to \mathbb{R}^3_\nu$ given by $x(u,v) = (u,v,f(u,v))$ is an injective and regular parametrized surface, so that $\mathrm{gr}(f) = x(U)$ is a surface regular. We have in $\mathbb{L}^3$ that

$$(g_{ij,L})_{1\le i,j\le 2} = \begin{pmatrix} 1-f_u^2 & -f_uf_v \\ -f_uf_v & 1-f_v^2 \end{pmatrix},$$

and thus $\det\big((g_{ij,L})_{1\le i,j\le 2}\big) = 1 - \|\nabla f\|_E^2$, where this Euclidean norm is (clearly) taken in $\mathbb{R}^2$. Then we conclude that

- $\mathrm{gr}(f)$ is spacelike if and only if $\|\nabla f\|_E < 1$;
- $\mathrm{gr}(f)$ is timelike if and only if $\|\nabla f\|_E > 1$;
- $\mathrm{gr}(f)$ is lightlike if and only if $\|\nabla f\|_E = 1$.

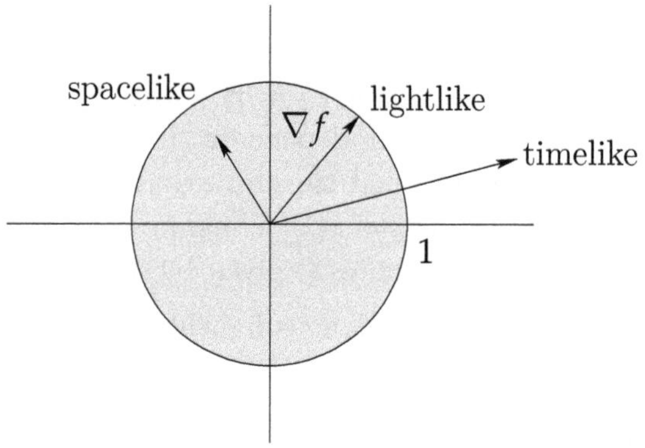

Figure 3.17: Illustrating the classification criterion for graphs over the plane $z = 0$.

The matrix of the First Fundamental Form induced by $\langle \cdot, \cdot \rangle_E$, i.e., regarding the graph as a surface in $\mathbb{R}^3$, is

$$(g_{ij,E})_{1\le i,j\le 2} = \begin{pmatrix} 1+f_u^2 & f_uf_v \\ f_uf_v & 1+f_v^2 \end{pmatrix}.$$

(4) The *cylinder* of radius $r > 0$

$$\mathbb{S}^1(r) \times \mathbb{R} = \{(x,y,z) \in \mathbb{R}^3_\nu \mid x^2 + y^2 = r^2\},$$

may be parametrized (excluding a meridian) by $x\colon \mathbb{R} \times]0,2\pi[\to \mathbb{R}^3_\nu$ given by $x(u,v) = (r\cos u, r\sin u, v)$. Its First Fundamental Form is given in coordinates by $\mathrm{d}s^2 = r^2\,\mathrm{d}u^2 + (-1)^\nu \mathrm{d}v^2$. We see from this that the cylinder is spacelike in $\mathbb{R}^3$ and timelike in $\mathbb{L}^3$.

Figure 3.18: The cylinder $\mathbb{S}^1 \times \mathbb{R}$.

(5) The lightcone $C_L(\mathbf{0})$ may be parametrized (excluding a light ray) by the map $\boldsymbol{x}\colon (\mathbb{R} \setminus \{\mathbf{0}\}) \times]0,2\pi[\to C_L(\mathbf{0})$ given by $\boldsymbol{x}(u,v) = (u\cos v, u\sin v, u)$. Its Minkowski First Fundamental Form is given by $\mathrm{d}s^2 = \mathrm{d}u^2$. That is, we see that $C_L(\mathbf{0})$ is a lightlike surface, as the name suggests.

In $\mathbb{R}^3$, its metric is given by $\mathrm{d}s^2 = \mathrm{d}u^2 + u^2\,\mathrm{d}v^2$.

(6) Surfaces of revolution: we may generalize the previous two examples. We have previously seen that if $\boldsymbol{\alpha}\colon I \to \mathbb{R}^3_\nu$ is an injective, regular, and smooth curve, of the form $\boldsymbol{\alpha}(u) = (f(u), 0, g(u))$ with $f(u) > 0$ for all $u \in I$, then $\boldsymbol{x}\colon I \times]0,2\pi[\to \mathbb{R}^3_\nu$ given by

$$\boldsymbol{x}(u,v) = (f(u)\cos v, f(u)\sin v, g(u))$$

is an injective and regular parametrized surface, whose image is the surface of revolution generated by $\boldsymbol{\alpha}$. Its First Fundamental Form is given by:

$$(g_{ij}(u,v))_{1\leq i,j\leq 2} = \begin{pmatrix} \langle \boldsymbol{\alpha}'(u), \boldsymbol{\alpha}'(u)\rangle & 0 \\ 0 & f(u)^2 \end{pmatrix},$$

no matter which is the ambient space considered. We then see that in $\mathbb{L}^3$, the causal type of $\boldsymbol{x}$ is the same one as $\boldsymbol{\alpha}$'s. When $\boldsymbol{\alpha}$ is not lightlike, it is usual to consider a unit speed reparametrization before generating the surface of revolution, and this gives us the metric

$$\mathrm{d}s^2 = \epsilon_{\boldsymbol{\alpha}}\,\mathrm{d}u^2 + f(u)^2\,\mathrm{d}v^2.$$

The relation with the causal type of $\boldsymbol{\alpha}$ is to be expected, in the following geometric sense: the vector $\boldsymbol{x}_v(u,v)$ (in fact, any tangent vector to a parallel of the surface) will always give us a spacelike direction, while $\boldsymbol{x}_u(u,v)$ is the image of $\boldsymbol{\alpha}'(u)$ under a rotation around the z-axis, which is a Lorentz transformation. In the same way, if we rotate a curve contained in the plane $x = 0$ or $y = 0$ around the x-axis or y-axis, respectively, the corresponding surface of revolution will have tangent planes of all possible causal types. Indeed, any tangent vector to a parallel of the surface will complete one full rotation around the revolution axis. The fact that Euclidean rotations around both the x and y-axes are not Lorentz transformations is further evidence for this.

However, *hyperbolic* rotations around both the x and y-axes are Lorentz transformations, which hints that in $\mathbb{L}^3$ we will have new types of surfaces of revolution. See Exercise 3.2.9.

Exercises

Exercise 3.2.1. Consider the cylinder $\mathbb{S}^1 \times \mathbb{R} = \{(x,y,z) \in \mathbb{R}^3_\nu \mid x^2 + y^2 = 1\}$.

(a) In $\mathbb{R}^3$, find all the curves in the cylinder which make a constant angle with all of its generating lines (vertical lines).

(b) In $\mathbb{L}^3$, find all timelike curves in the cylinder which make a constant hyperbolic angle with all of its generating lines. Is the result different from what you got in (a)?

Exercise 3.2.2. Fix $0 < \alpha_0 < \pi/2$ and consider $\boldsymbol{x}\colon \mathbb{R}_{>0} \times]0, 2\pi[\to \mathbb{L}^3$ given by

$$\boldsymbol{x}(u,v) = (u \cos v \tan \alpha_0, u \sin v \tan \alpha_0, u).$$

Show that $\boldsymbol{x}$ is an injective and regular parametrized surface, and discuss the causal type of $\boldsymbol{x}$ in terms of α_0. Make sketches of the image of $\boldsymbol{x}$ for $\alpha_0 = \pi/6, \pi/4$ and $\pi/3$.

Exercise 3.2.3 (Tangent surfaces - I). Let $\boldsymbol{\alpha}\colon I \to \mathbb{R}^3_\nu$ be an admissible curve whose curvature never vanishes. Consider its *tangent surface*, $\boldsymbol{x}\colon I \times \mathbb{R} \to \mathbb{R}^3_\nu$, given by

$$\boldsymbol{x}(t,v) = \boldsymbol{\alpha}(t) + v\boldsymbol{\alpha}'(t).$$

Show that $\boldsymbol{x}$ restricted to $U = \{(t,v) \in I \times \mathbb{R} \mid v \neq 0\}$ is a regular parametrized surface, discuss its causal type in terms of the causal type of the Frenet-Serret trihedron for $\boldsymbol{\alpha}$, and determine the tangent planes to $\boldsymbol{x}$ along its coordinate curves.

Exercise 3.2.4 (Helicoids). Let $a, b > 0$ and consider the helix $\boldsymbol{\alpha}\colon \mathbb{R} \to \mathbb{R}^3_\nu$ given by $\boldsymbol{\alpha}(t) = (a \cos t, a \sin t, bt)$.

(a) For each $t \in \mathbb{R}$, consider the line passing through $\boldsymbol{\alpha}(t)$ and orthogonally crossing the z-axis. Obtain a parametrized surface $\boldsymbol{x}\colon \mathbb{R}^2 \to \mathbb{R}^3_\nu$, regular and injective, whose image is the union of these lines: this surface is called a *helicoid* (we have seen in the text a particular case of this, with $a = b = 1$).

(b) In $\mathbb{L}^3$, discuss the causal type of $\boldsymbol{x}$ in terms of a and b and find the lightlike helix in the helicoid which divides it into two regions: a spacelike one and a timelike one.

Exercise 3.2.5 (Hyperbolic helicoids). Let's study now the Lorentzian analogue of the situation in the previous exercise. Let $a, b > 0$ and consider the helix $\boldsymbol{\alpha}\colon \mathbb{R} \to \mathbb{L}^3$ given by $\boldsymbol{\alpha}(t) = (bt, a \cosh t, a \sinh t)$.

(a) For each $t \in \mathbb{R}$, consider the line passing through $\boldsymbol{\alpha}(t)$ and orthogonally crossing the x-axis. Obtain a parametrized surface $\boldsymbol{x}\colon \mathbb{R}^2 \to \mathbb{L}^3$, regular and injective, whose image is the union of these lines.

(b) Discuss the causal type of $\boldsymbol{x}$ in terms of a and b and find the lightlike helix in the image $\boldsymbol{x}(\mathbb{R}^2)$ which divides it into two regions, like in the previous exercise.

Exercise† 3.2.6 (Surface gradient).

(a) Let $M \subseteq \mathbb{R}^3_\nu$ be a non-degenerate regular surface and $f\colon M \to \mathbb{R}$ be a smooth function. For each $\boldsymbol{p} \in M$, since $\mathrm{d}f_{\boldsymbol{p}}$ is a linear functional defined in $T_{\boldsymbol{p}}M$, the non-degeneracy of M allows us to apply Riesz's Lemma to obtain a tangent vector $\operatorname{grad} f(\boldsymbol{p}) \in T_{\boldsymbol{p}}M$ satisfying

$$\mathrm{d}f_{\boldsymbol{p}}(\boldsymbol{v}) = \langle \operatorname{grad} f(\boldsymbol{p}), \boldsymbol{v}\rangle, \quad \text{for all } \boldsymbol{v} \in T_{\boldsymbol{p}}M.$$

If $(U, \boldsymbol{x})$ is a parametrization for M, show that

$$(\operatorname{grad} f) \circ \boldsymbol{x} = \frac{f_u G - f_v F}{EG - F^2}\frac{\partial \boldsymbol{x}}{\partial u} + \frac{f_v E - f_u F}{EG - F^2}\frac{\partial \boldsymbol{x}}{\partial v},$$

where f_u and f_v stand for the partial derivatives of the composition $f \circ \boldsymbol{x}$. In particular, this shows that $\operatorname{grad} f\colon M \to \mathbb{R}^3_\nu$ is smooth.

(b) Use item (a) to verify that in $\mathbb{R}^2$ we have $\operatorname{grad} f(x,y) = \nabla f(x,y)$, and in $\mathbb{L}^2$ we have $\operatorname{grad} f(x,y) = \nabla_L f(x,y)$.

(c) Use item (a) to deduce an expression for the gradient of a function defined on $\mathbb{R}^2_\nu$ in polar and Rindler coordinates (see Exercise 2.2.14, p. 94). More precisely, considering $\boldsymbol{x}(r,\theta) = (r\cos\theta, r\sin\theta)$ in $\mathbb{R}^2 \setminus \{\boldsymbol{0}\}$ and $\boldsymbol{y}(\rho,\varphi) = (\rho\cosh\varphi, \rho\sinh\varphi)$ in the Rindler wedge of $\mathbb{L}^2$, show that

$$(\operatorname{grad} f) \circ \boldsymbol{x} = \frac{\partial f}{\partial r}\boldsymbol{e}_r + \frac{1}{r}\frac{\partial f}{\partial \theta}\boldsymbol{e}_\theta \text{ and}$$
$$(\operatorname{grad} f) \circ \boldsymbol{y} = \frac{\partial f}{\partial \rho}\boldsymbol{e}_\rho - \frac{1}{\rho}\frac{\partial f}{\partial \varphi}\boldsymbol{e}_\varphi,$$

where $\boldsymbol{e}_r$, $\boldsymbol{e}_\theta$, $\boldsymbol{e}_\rho$ and $\boldsymbol{e}_\varphi$ denote the unit vectors in the direction of the respective derivatives of $\boldsymbol{x}$ and $\boldsymbol{y}$.

Hint. If you want to, you may regard $\mathbb{R}^2$ and $\mathbb{L}^2$ as coordinate planes inside $\mathbb{L}^3$. Verify that $\mathrm{d}x^2 + \mathrm{d}y^2 = \mathrm{d}r^2 + r^2\,\mathrm{d}\theta^2$ and $\mathrm{d}x^2 - \mathrm{d}y^2 = \mathrm{d}\rho^2 - \rho^2\,\mathrm{d}\varphi^2$.

Exercise 3.2.7. Let $M \subset \mathbb{R}^3_\nu$ be a regular surface, and $(U_{\boldsymbol{x}}, \boldsymbol{x})$ and $(U_{\boldsymbol{y}}, \boldsymbol{y})$ be parametrizations for M. Suppose that R is an open subset of M for which $\overline{R}$ is compact, $R \subseteq \boldsymbol{x}(U_{\boldsymbol{x}}) \cap \boldsymbol{y}(U_{\boldsymbol{y}})$. Show that the area of R does not depend on the choice of parametrization, that is:

$$\int_{\boldsymbol{x}^{-1}(R)} \sqrt{|\det(g_{ij}(u,v))|}\,\mathrm{d}u\,\mathrm{d}v = \int_{\boldsymbol{y}^{-1}(R)} \sqrt{|\det(\widetilde{g_{ij}}(\tilde{u},\tilde{v}))|}\,\mathrm{d}\tilde{u}\,\mathrm{d}\tilde{v},$$

where $(g_{ij})_{1\le i,j\le 2}$ and $(\widetilde{g_{ij}})_{1\le i,j\le 2}$ denote the components of the First Fundamental Forms of $\boldsymbol{x}$ and $\boldsymbol{y}$, respectively.

Exercise 3.2.8 (More graphs). Let $f\colon U \to \mathbb{R}$ be a smooth function and $\boldsymbol{x}\colon U \to \mathbb{L}^3$ be a parametrized surface, given by $\boldsymbol{x}(u,v) = (u, f(u,v), v)$ or $(f(u,v), u, v)$. Show that $\boldsymbol{x}$ is:

- spacelike if and only if $\langle \mathfrak{f}(\nabla f), \mathfrak{f}(\nabla f)\rangle_L > 1$;
- timelike if and only if $\langle \mathfrak{f}(\nabla f), \mathfrak{f}(\nabla f)\rangle_L < 1$;
- lightlike if and only if $\langle \mathfrak{f}(\nabla f), \mathfrak{f}(\nabla f)\rangle_L = 1$,

where the "flip" operator $\mathfrak{f}\colon \mathbb{L}^2 \to \mathbb{L}^2$ is given by $\mathfrak{f}(x,y) = (y,x)$. Make a drawing similar to Figure 3.17 (p. 162) to illustrate this criterion.

Hint. Switching the first two coordinates in $\mathbb{L}^3$ is a Poincaré transformation, so you only need to do one of the cases.

Exercise 3.2.9 (Surfaces of hyperbolic revolutions – I)**.** Let $\boldsymbol{\alpha}\colon I \to \mathbb{L}^3$ be a regular, smooth and injective curve, of the form $\boldsymbol{\alpha}(u) = (f(u), 0, g(u))$, with $g(u) > 0$ for each $u \in I$. Consider the surface generated by the hyperbolic rotation of $\boldsymbol{\alpha}$ around the x-axis, parametrized by the map $\boldsymbol{x}\colon I \times \mathbb{R} \to \boldsymbol{x}(I \times \mathbb{R}) \subseteq \mathbb{L}^3$ given by

$$\boldsymbol{x}(u,v) = (f(u), g(u)\sinh v, g(u)\cosh v).$$

Show that $\boldsymbol{x}$ is an injective and regular parametrized surface whose causal type is the same causal type of $\boldsymbol{\alpha}$, and compute its Minkowski First Fundamental Form. What does the metric look like (in differential notation) when $\boldsymbol{\alpha}$ has unit speed?

Remark.

- Following the usual terminology, we will say that the coordinate curves $u = \text{cte.}$ are *parallels* and the curves $v = \text{cte.}$ are *meridians*. The parametrization $\boldsymbol{x}$ above omits one meridian of the surface.
- One may also consider generating curves in other planes, and apply suitable hyperbolic rotations. How many possibilities do we have?

Exercise† 3.2.10. Let $\boldsymbol{x}\colon U \to \mathbb{R}^3_\nu$ be a non-degenerate regular parametrized surface. Define

$$\boldsymbol{N}(u,v) \doteq \frac{\boldsymbol{x}_u(u,v) \times \boldsymbol{x}_v(u,v)}{\|\boldsymbol{x}_u(u,v) \times \boldsymbol{x}_v(u,v)\|}$$

and use the indices E and L to distinguish between ambient spaces.

(a) Show that $\boldsymbol{N}_L(u,v) = \sqrt{\left|\dfrac{\det(g_{ij,E}(u,v))}{\det(g_{ij,L}(u,v))}\right|}\,\mathrm{Id}_{2,1}(\boldsymbol{N}_E(u,v))$.

(b) If θ is the (Euclidean) angle between $\boldsymbol{N}_E(u,v)$ and the plane $z = 0$, show that $|\det(g_{ij,L})| = |\cos 2\theta| \det(g_{ij,E})$. In particular, we obtain the inequality $|\det(g_{ij,L})| \leq \det(g_{ij,E})$.

Exercise 3.2.11. Let $f\colon U \subseteq \mathbb{R}^2 \to \mathbb{R}^3_\nu$ be a smooth function, where U is connected, and consider the graph

$$\mathrm{gr}(f) = \{(u, v, f(u,v)) \in \mathbb{R}^3_\nu \mid (u,v) \in U\}.$$

(a) Show that $\pi\colon \mathrm{gr}(f) \to U$ given by $\pi(x, y, f(x,y)) = (x,y)$ is a diffeomorphism.

(b) Show that in $\mathbb{R}^3$, π decreases areas: if $R \subseteq \mathrm{gr}(f)$ is open, $\mathscr{A}(R) \geq \mathscr{A}(\pi(R))$. Also verify that the equality holds if and only if R is contained in a horizontal plane.

(c) Show that in $\mathbb{L}^3$, if $\mathrm{gr}(f)$ is spacelike, then π increases areas: if $R \subseteq \mathrm{gr}(f)$ is open, $\mathscr{A}(R) \leq \mathscr{A}(\pi(R))$. Verify again that equality holds if and only if R is contained in a horizontal plane.

(d) Give counter-examples when $\mathrm{gr}(f)$ is timelike in $\mathbb{L}^3$.

Exercise 3.2.12. Consider an isosceles triangle T in the plane, with base $b > 0$ and height $h > 0$, say, conveniently positioned inside $\mathbb{L}^3$ with its vertices at $(-b/2, 0, 0)$, $(b/2, 0, 0)$ and $(0, h, 0)$. We know that the area of such a triangle is $\mathcal{A}(T) = bh/2$. Let $\theta \in [0, 2\pi]$ be a fixed angle and consider

$$R = \begin{pmatrix} 1 & 0 & 0 \\ 0 & \cos\theta & -\sin\theta \\ 0 & \sin\theta & \cos\theta \end{pmatrix}.$$

We have that R is an orthogonal map (and thus preserves Euclidean areas), but it is not a Lorentz transformation. Show that if $T_\theta \doteq R(T)$ is the slanted triangle, the Lorentzian area of T_θ is

$$\mathcal{A}(T_\theta) = \frac{bh\sqrt{|\cos 2\theta|}}{2}.$$

When is this area minimal?

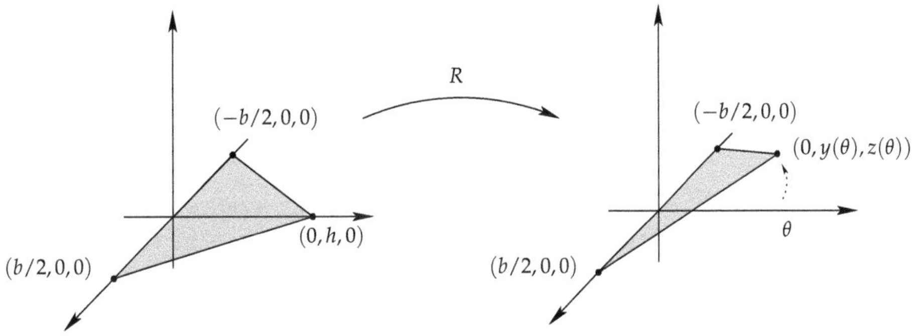

Figure 3.19: Illustrating the triangle T_θ.

Exercise 3.2.13 (Girard's Formula). Recall that a *great circle* in $\mathbb{S}^2$ is the intersection of $\mathbb{S}^2$ with a plane passing through the origin of $\mathbb{R}^3$. A *spherical triangle* is the region in $\mathbb{S}^2$ bounded by three arcs of great circles. The goal of this exercise is to show *Girard's Formula*, which gives the area of a spherical triangle in terms of its interior angles (more precisely, the angles formed between the tangent vectors to the great circles at the vertices of the triangle).

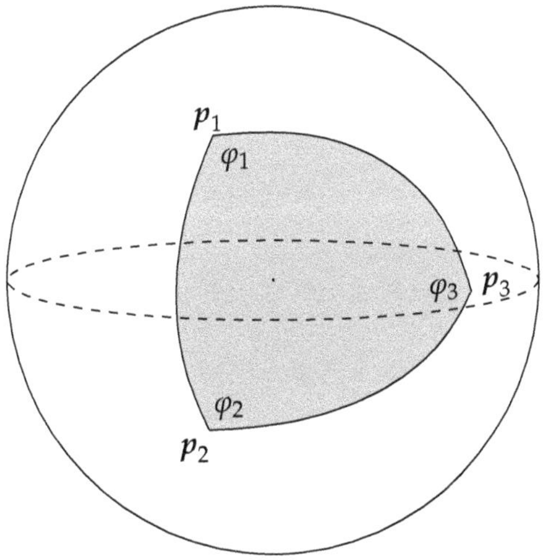

Figure 3.20: A spherical triangle in $\mathbb{S}^2$.

(a) Show that $\mathscr{A}(\mathbb{S}^2) = 4\pi$, and that the area of the fuse bounded by $v = 0$ and $v = v_0$ equals $2v_0$, where $(u, v) \in]-\pi/2, \pi/2[\times]0, 2\pi[$ are spherical coordinates for $\mathbb{S}^2$.

Remark. By the symmetry of $\mathbb{S}^2$, this shows that the area of any fuse of amplitude v_0 is $2v_0$. This may be formalized using the notion of isometry, to be seen soon.

(b) Let $T \subseteq \mathbb{S}^2$ be a spherical triangle whose vertices are $\boldsymbol{p}_1, \boldsymbol{p}_2, \boldsymbol{p}_3 \in \mathbb{S}^2$, with interior angles φ_1, φ_2 and φ_3, respectively (as in the previous figure). Show that

$$\mathscr{A}(T) = \varphi_1 + \varphi_2 + \varphi_3 - \pi.$$

Hint. Use a combinatorics argument: for $1 \leq i < j \leq 3$, let C_{ij} be the great circle in $\mathbb{S}^2$ passing through $\boldsymbol{p}_i$ and $\boldsymbol{p}_j$. Given two such great circles, two fuses of amplitudes φ_k are defined, where k is the index of the common point of the two great circles considered. If Δ_k is the union of these two fuses, argue that

$$\mathscr{A}(\Delta_1) + \mathscr{A}(\Delta_2) + \mathscr{A}(\Delta_3) = \mathscr{A}(\mathbb{S}^2) + 2(\mathscr{A}(T) + \mathscr{A}(\widetilde{T})),$$

where $\widetilde{T}$ is the spherical triangle with vertices $-\boldsymbol{p}_1$, $-\boldsymbol{p}_2$ and $-\boldsymbol{p}_3$. Use item (a) and conclude it.

Remark.

- In *spherical geometry*, the sum of the interior angles of a spherical triangle is always larger than π. Moreover, "similar" spherical triangles are, in fact, "congruent";
- Considering a "hyperbolic triangle" T in $\mathbb{H}^2$ defined in the same fashion as above, and keeping the same notation, it holds that $\mathscr{A}(T) = \pi - (\varphi_1 + \varphi_2 + \varphi_3)$ (Lambert's Formula).

(c) In general, a n-sided *spherical polygon* P, is the region in $\mathbb{S}^2$ bounded by n arcs of great circles. Show that

$$\mathscr{A}(P) = \sum_{i=1}^{n} \varphi_i - (n-2)\pi,$$

where $\varphi_1, \ldots, \varphi_n$ are the interior angles of P.

Hint. Divide P into spherical triangles and use (b).

Remark. Girard's Formula is generalized to some spacelike surfaces other than the sphere. Such generalization is known as the *Gauss-Bonnet Theorem*. For more details see, for example, [17].

Exercise 3.2.14.

(a) For each $r > 1$, show that the area of the "disk"

$$D_r \doteq \{(x, y, z) \in \mathbb{H}^2 \mid z < r\}$$

is $\mathscr{A}(D_r) = 2\pi(r-1)$.

(b) For each $r > 0$, show that the area of the "strip"

$$S_r \doteq \{(x, y, z) \in \mathbb{S}^2_1 \mid |z| < r\}$$

is $\mathscr{A}(S_r) = 4\pi r$.

Hint. Use parametrizations of revolution. Those will omit a single meridian, which has no area.

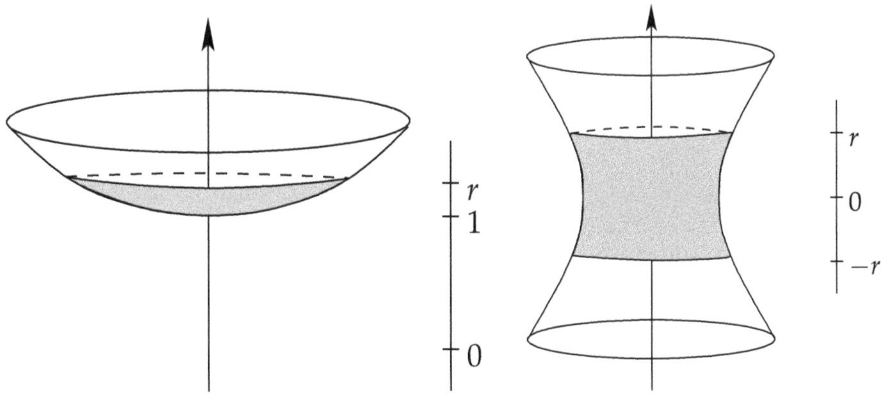

(a) The region D_r.　　(b) The region S_r.

Figure 3.21: "Disks" in $\mathbb{H}^2$ and $\mathbb{S}^2_1$.

Remark. Extending the definition of area to unbounded regions in a surface, possibly allowing improper integrals, we may conclude that $\mathscr{A}(\mathbb{H}^2) = \mathscr{A}(\mathbb{S}^2_1) = +\infty$.

3.2.1 Isometries between surfaces

We will conclude this section by introducing a class of smooth functions between surfaces which is extremely important for everything we will do in the text from here on:

Definition 3.2.14 (Isometry)**.** Let $M_1, M_2 \subseteq \mathbb{R}^3_\nu$ be regular surfaces.

(i) A diffeomorphism $\phi\colon M_1 \to M_2$ is called an *isometry* if given any $p \in M_1$ and vectors $v_1, v_2 \in T_pM_1$, we have that

$$\mathrm{I}_p(v_1, v_2) = \mathrm{I}_{\phi(p)}\big(\mathrm{d}\phi_p(v_1), \mathrm{d}\phi_p(v_2)\big).$$

(ii) A smooth map $\phi\colon M_1 \to M_2$ is called a *local isometry* if for every $p \in M_1$ there is an open subset U of M_1 containing p such that the restriction $\phi|_U\colon U \to \phi(U)$ is an isometry.

(iii) We say that M_1 and M_2 are *isometric* (resp. *locally isometric*) if there is an isometry (resp. local isometry) between M_1 and M_2.

Remark.

- Above, we're using I to denote the First Fundamental Form of M_1 on the left side of $\mathrm{I}_p(v_1, v_2) = \mathrm{I}_{\phi(p)}\big(\mathrm{d}\phi_p(v_1), \mathrm{d}\phi_p(v_2)\big)$, and also to denote the First Fundamental Form of M_2 on the right side.
- We could also ask ourselves if a surface in $\mathbb{R}^3$ is isometric to another surface in $\mathbb{L}^3$. In this setup, we consider the First Fundamental Forms in the above definition induced by the ambient spaces where each surface lies.

Example 3.2.15. Let $M_1, M_2 \subseteq \mathbb{R}^3_\nu$ be regular surfaces *in the same ambient space* such that $M_2 = F(M_1)$, for some $F \in \mathrm{E}_\nu(3, \mathbb{R})$. Then M_1 and M_2 are isometric. Indeed, writing $F = T_{\boldsymbol{a}} \circ A$ (with $A \in \mathrm{O}_\nu(3, \mathbb{R})$ and $\boldsymbol{a} \in \mathbb{R}^3$) we will have, for each $\boldsymbol{p} \in M_1$, that $\mathrm{d}F_{\boldsymbol{p}} = DF(\boldsymbol{p})\big|_{T_{\boldsymbol{p}}M_1} = A\big|_{T_{\boldsymbol{p}}M_1}$ and thus

$$\mathrm{I}_{F(\boldsymbol{p})}(\mathrm{d}F_{\boldsymbol{p}}(\boldsymbol{v}), \mathrm{d}F_{\boldsymbol{p}}(\boldsymbol{w})) = \mathrm{I}_{F(\boldsymbol{p})}(A\boldsymbol{v}, A\boldsymbol{w}) = \langle A\boldsymbol{v}, A\boldsymbol{w}\rangle = \langle \boldsymbol{v}, \boldsymbol{w}\rangle = \mathrm{I}_{\boldsymbol{p}}(\boldsymbol{v}, \boldsymbol{w}),$$

for all $\boldsymbol{v}, \boldsymbol{w} \in T_{\boldsymbol{p}}M_1$. Since F is a diffeomorphism, F is an isometry. In this situation, we say that M_1 and M_2 are *congruent*. That is, congruent surfaces are isometric. We will soon see that isometric surfaces are not always congruent, that is, there are isometric surfaces which "do not look like each other".

With this concept in hand, we see that the theory of surfaces is, to some extent, parallel to the theory of curves, focusing on the following question: when are two regular surfaces congruent? Seeking the answer, in the next sections we will see how to define certain geometric invariants of a surface, aiming towards a "Fundamental Theorem of Surfaces", similar to the Fundamental Theorem of Curves.

Example 3.2.16. We see from the definition that two isometric surfaces must necessarily have the same causal type. For example, the cylinder $\mathbb{S}^1(r) \times \mathbb{R}$ is not isometric to itself when considered inside both ambients $\mathbb{R}^3$ and $\mathbb{L}^3$. In particular, the identity map itself is not an isometry. Thus, when dealing with isometries, it is essential to know clearly which are the First Fundamental Forms being considered.

Before we present a few examples, it will be convenient to register some relations between isometries and parametrizations. The next result follows immediately from the definition of isometry:

Proposition 3.2.17. *Let $M_1, M_2 \subseteq \mathbb{R}^3_\nu$ be regular surfaces and $\phi\colon M_1 \to M_2$ be a local isometry. If $(U, \boldsymbol{x})$ is a parametrization for M_1 and $(U, \phi \circ \boldsymbol{x})$ is the corresponding parametrization for M_2, we have that*

$$g_{ij}(u, v) = \tilde{g}_{ij}(u, v),$$

for all $(u, v) \in U$ and $1 \leq i, j \leq 2$, where the $\tilde{g}_{ij}$ denote the coefficients of the First Fundamental Form of M_2 relative to $\phi \circ \boldsymbol{x}$.

Remark. Or, in another suggestive notation:

$$E(u, v) = \tilde{E}(u, v), \quad F(u, v) = \tilde{F}(u, v) \quad \text{and} \quad G(u, v) = \tilde{G}(u, v).$$

As expected, isometries also preserve lengths and areas:

Proposition 3.2.18. *Let $M_1, M_2 \subseteq \mathbb{R}^3_\nu$ be regular surfaces and $\phi\colon M_1 \to M_2$ an isometry. Then:*

(i) if $\boldsymbol{\alpha}\colon I \to M_1$ is a curve in M_1, then $\phi \circ \boldsymbol{\alpha}$ is a curve in M_2 with the same causal type as $\boldsymbol{\alpha}$, satisfying $L[\boldsymbol{\alpha}] = L[\phi \circ \boldsymbol{\alpha}]$ and $E[\boldsymbol{\alpha}] = E[\phi \circ \boldsymbol{\alpha}]$;

(ii) if R is an open subset of M_1 with $\overline{R}$ compact, then $\mathcal{A}(R) = \mathcal{A}(\phi(R))$.

Proof:

(i) We have that

$$\begin{aligned} L[\phi\circ\alpha] &= \int_I \sqrt{|\mathrm{I}_{\phi\circ\alpha(t)}((\phi\circ\alpha)'(t))|}\,\mathrm{d}t \\ &= \int_I \sqrt{|\mathrm{I}_{\phi\circ\alpha(t)}(\mathrm{d}\phi_{\alpha(t)}(\alpha'(t)))|}\,\mathrm{d}t \\ &= \int_I \sqrt{|\mathrm{I}_{\alpha(t)}(\alpha'(t))|}\,\mathrm{d}t = L[\alpha], \end{aligned}$$

and similarly for the energy.

(ii) Suppose without loss of generality that $R \subseteq \boldsymbol{x}(U)$ for some parametrization $(U, \boldsymbol{x})$ of M. We have that $\phi(R) \subseteq \phi(\boldsymbol{x}(U))$. Since $\phi\circ\boldsymbol{x}$ is a parametrization for M_2, $\phi(R)$ is open in M, and $\overline{\phi(R)}$ is compact (since ϕ is, in particular, a diffeomorphism), we have:

$$\begin{aligned} \mathcal{A}(\phi(R)) &= \int_{(\phi\circ\boldsymbol{x})^{-1}(\phi(R))} \sqrt{|\det \widetilde{g_{ij}}(u,v)|}\,\mathrm{d}u\,\mathrm{d}v \\ &= \int_{\boldsymbol{x}^{-1}(R)} \sqrt{|\det\ g_{ij}(u,v)|}\,\mathrm{d}u\,\mathrm{d}v = \mathcal{A}(R). \end{aligned}$$

□

It is also very convenient to relate in a more general way the coordinate expression of an isometry with the First Fundamental Forms of the given surfaces:

Proposition 3.2.19. *Let $M_1, M_2 \subseteq \mathbb{R}^3_\nu$ be regular surfaces and $\phi\colon M_1 \to M_2$ an isometry. Suppose that:*

- *$(U, \boldsymbol{x})$ and $(\widetilde{U}, \widetilde{\boldsymbol{x}})$ are parametrizations for M_1 and M_2 such that $\phi(\boldsymbol{x}(U)) \subseteq \widetilde{\boldsymbol{x}}(\widetilde{U})$;*
- *for any $(u,v) \in U$ and $(\widetilde{u}, \widetilde{v}) \doteq \widetilde{\boldsymbol{x}}^{-1}\circ\phi\circ\boldsymbol{x}(u,v)$, the matrix representing the differential of ϕ, relative to the bases associated to the parametrizations (in the correct points) is denoted by $A \doteq \left[\mathrm{d}\phi_{\boldsymbol{x}(u,v)}\right]_{\mathcal{B}_{\boldsymbol{x}},\mathcal{B}_{\widetilde{\boldsymbol{x}}}} = (a^i{}_j(u,v))_{1\le i,j\le 2}$;*
- *$G \doteq (g_{ij}(u,v))_{1\le i,j\le 2}$ and $\widetilde{G} \doteq (\widetilde{g}_{ij}(\widetilde{u},\widetilde{v}))_{1\le i,j\le 2}$ are the Gram matrices of the First Fundamental Forms of $\boldsymbol{x}$ and $\widetilde{\boldsymbol{x}}$.*

Then $G = A^\top \widetilde{G} A$.

Proof: It suffices to compute each $g_{ij}(u,v)$ by using all the given assumptions. Identify $u \leftrightarrow u^1$ and $v \leftrightarrow u^2$, and similarly for the coordinates in $\widetilde{\boldsymbol{x}}$. We have:

$$\begin{aligned} g_{ij}(u^1,u^2) &= \mathrm{I}_{\boldsymbol{x}(u^1,u^2)}\left(\frac{\partial \boldsymbol{x}}{\partial u^i}(u^1,u^2), \frac{\partial \boldsymbol{x}}{\partial u^j}(u^1,u^2)\right) \\ &= \mathrm{I}_{\phi\circ\boldsymbol{x}(u^1,u^2)}\left(\mathrm{d}\phi_{\boldsymbol{x}(u^1,u^2)}\left(\frac{\partial \boldsymbol{x}}{\partial u^i}(u^1,u^2)\right), \mathrm{d}\phi_{\boldsymbol{x}(u^1,u^2)}\left(\frac{\partial \boldsymbol{x}}{\partial u^j}(u^1,u^2)\right)\right) \\ &= \mathrm{I}_{\phi\circ\boldsymbol{x}(u^1,u^2)}\left(\sum_{k=1}^2 a^k{}_i(u^1,u^2)\frac{\partial \widetilde{\boldsymbol{x}}}{\partial \widetilde{u}^k}(\widetilde{u}^1,\widetilde{u}^2), \sum_{\ell=1}^2 a^\ell{}_j(u^1,u^2)\frac{\partial \widetilde{\boldsymbol{x}}}{\partial \widetilde{u}^\ell}(\widetilde{u}^1,\widetilde{u}^2)\right) \\ &= \sum_{k,\ell=1}^2 a^k{}_i(u^1,u^2)a^\ell{}_j(u^1,u^2)\mathrm{I}_{\phi\circ\boldsymbol{x}(u^1,u^2)}\left(\frac{\partial \widetilde{\boldsymbol{x}}}{\partial \widetilde{u}^k}(\widetilde{u}^1,\widetilde{u}^2), \frac{\partial \widetilde{\boldsymbol{x}}}{\partial \widetilde{u}^\ell}(\widetilde{u}^1,\widetilde{u}^2)\right) \\ &= \sum_{k,\ell=1}^2 a^k{}_i(u^1,u^2)a^\ell{}_j(u^1,u^2)\widetilde{g}_{k\ell}(\widetilde{u}^1,\widetilde{u}^2). \end{aligned}$$

This last expression is precisely the entry in position (i,j) of the matrix $A^\top \widetilde{G} A$. □

Remark. Compare the above result with Lemma 1.2.6 (p. 6).

For non-degenerate surfaces, we have the following auxiliary result:

Proposition 3.2.20. *Let $M_1, M_2 \subseteq \mathbb{R}^3_\nu$ be two non-degenerate regular surfaces, and $\phi\colon M_1 \to M_2$ be a smooth map. If $\mathrm{d}\phi_p$ preserves the First Fundamental Forms for each $p \in M_1$, then ϕ is automatically a local isometry.*

Proof: Take $p \in M_1$ and an orthonormal basis (v_1, v_2) for T_pM_1. By assumption, $(\mathrm{d}\phi_p(v_1), \mathrm{d}\phi_p(v_2))$ is an orthonormal subset of $T_{\phi(p)}M_2$ and, thus, is linearly independent. Hence $\mathrm{d}\phi_p$ is surjective, and since the dimension of the tangent planes is the same, it follows that $\mathrm{d}\phi_p$ is non-singular. As p was arbitrary, the Inverse Function Theorem says that ϕ is a local diffeomorphism, as wanted. □

The same calculations done in the proof of Proposition 3.2.19 and the above result also give us that:

Proposition 3.2.21. *Let $x\colon U \to \mathbb{R}^3_\nu$ and $\widetilde{x}\colon \widetilde{U} \to \mathbb{R}^3_\nu$ be injective and regular parametrized surfaces, and $\varphi\colon U \to \widetilde{U}$ be a diffeomorphism. Suppose that for all $(u,v) \in U$, if $G = (g_{ij}(u,v))_{1\le i,j\le 2}$, $\widetilde{G} = (\widetilde{g}_{ij}(\varphi(u,v)))_{1\le i,j\le 2}$ and $A = D\varphi(u,v)$, we have that*

$$G = A^\top \widetilde{G} A.$$

Then $\phi \doteq \widetilde{x} \circ \varphi \circ x^{-1}\colon x(U) \to \widetilde{x}(\widetilde{U})$ is an isometry between $x(U)$ and $\widetilde{x}(\widetilde{U})$.

Let's see how to use those results in practice:

Example 3.2.22.

(1) Lambert's cylindrical projection, $f : \mathbb{S}^2 \setminus \{(0,0,\pm 1)\} \to \mathbb{S}^1 \times \mathbb{R}$, seen in Example 3.1.21 (p. 143) is not an isometry. Indeed, note that the First Fundamental Form of x is $\mathrm{d}u^2 + \cos^2 u\,\mathrm{d}v^2$ and the one for $\widetilde{x}$ is $\mathrm{d}\widetilde{u}^2 + \mathrm{d}\widetilde{v}^2$. Since $f(x(u,v)) = \widetilde{x}(\sin u, v)$, we let $\widetilde{u} = \sin u$ and $\widetilde{v} = v$, whence $\mathrm{d}\widetilde{u} = \cos u\,\mathrm{d}u$ and $\mathrm{d}\widetilde{v} = \mathrm{d}v$. But

$$\mathrm{d}\widetilde{u}^2 + \mathrm{d}\widetilde{v}^2 = \cos^2 u\,\mathrm{d}u^2 + \mathrm{d}v^2 \neq \mathrm{d}u^2 + \cos^2 u\,\mathrm{d}v^2.$$

Despite this, note that f locally preserves areas, since

$$\det\begin{pmatrix} \cos^2 u & 0 \\ 0 & 1 \end{pmatrix} = \det\begin{pmatrix} 1 & 0 \\ 0 & \cos^2 u \end{pmatrix}.$$

(2) The plane $\mathbb{R}^2$ is locally isometric to the cylinder $\mathbb{S}^1 \times \mathbb{R}$ seen in $\mathbb{R}^3$. The local isometry is $F\colon \mathbb{R}^2 \to \mathbb{S}^1 \times \mathbb{R}$ given by $F(u,v) = (\cos v, \sin v, u)$. Considering the identity map as a parametrization for $\mathbb{R}^2$ along with the usual parametrization of revolution for the cylinder, $\widetilde{x}(\widetilde{u},\widetilde{v}) = (\cos\widetilde{v}, \sin\widetilde{v}, \widetilde{u})$, we see that the local expression for F is precisely the identity (to wit, $F(\mathrm{id}_{\mathbb{R}^2}(u,v)) = \widetilde{x}(u,v)$), whose derivative is non-singular, giving that F is a local diffeomorphism (but not global, as it is not injective). That F is a local isometry then follows from the First Fundamental Form for the given cylinder parametrization being $\mathrm{d}\widetilde{u}^2 + \mathrm{d}\widetilde{v}^2 = \mathrm{d}u^2 + \mathrm{d}v^2$.

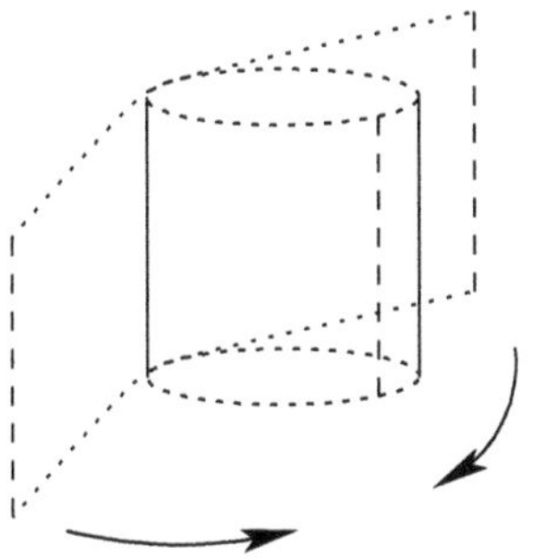

Figure 3.22: Local isometry between the plane and the cylinder.

There is no global isometry between these surfaces because there is not even a diffeomorphism between them. The proof of this fact is beyond the scope of this text, but may be found in [42].

(3) The surface of revolution generated by the *catenary* $\boldsymbol{\alpha}\colon \mathbb{R} \to \mathbb{R}^3$ (given by $\boldsymbol{\alpha}(u) = (\cosh u, 0, u)$) is called the *catenoid*, and it may be parametrized (excluding one meridian, as usual) by the map $\boldsymbol{x} : \mathbb{R} \times]0, 2\pi[\to \boldsymbol{x}(\mathbb{R} \times]0, 2\pi[) \subseteq \mathbb{R}^3$ given by $\boldsymbol{x}(u, v) = (\cosh u \cos v, \cosh u \sin v, u)$. The First Fundamental Form, in differential notation, is $\cosh^2 u(\mathrm{d}u^2 + \mathrm{d}v^2)$.

Also consider again the helicoid, parametrized by $\widetilde{\boldsymbol{x}} :]0, 2\pi[\times \mathbb{R} \to \widetilde{\boldsymbol{x}}(]0, 2\pi[\times \mathbb{R})$, given by $\widetilde{\boldsymbol{x}}(\widetilde{u}, \widetilde{v}) = (\widetilde{v} \cos \widetilde{u}, \widetilde{v} \sin \widetilde{u}, \widetilde{u})$, with First Fundamental Form in differential notation is $(1 + \widetilde{v}^2)\mathrm{d}\widetilde{u}^2 + \mathrm{d}\widetilde{v}^2$.

Define $F : \boldsymbol{x}(\mathbb{R} \times]0, 2\pi[) \to \widetilde{\boldsymbol{x}}(]0, 2\pi[\times \mathbb{R})$ by $F(\boldsymbol{x}(u, v)) = \widetilde{\boldsymbol{x}}(u, \sinh v)$. Let's see that F is an isometry. To wit, the map $(u, v) \mapsto (v, \sinh u)$ is a diffeomorphism, and so F is as well. To verify that F preserves First Fundamental Forms, let $\widetilde{u} = v$ and $\widetilde{v} = \sinh u$, so that $\mathrm{d}\widetilde{u} = \mathrm{d}v$ and $\mathrm{d}\widetilde{v} = \cosh u \,\mathrm{d}u$. Hence,

$$(1 + \widetilde{v}^2)\mathrm{d}\widetilde{u}^2 + \mathrm{d}\widetilde{v}^2 = (1 + \sinh^2 u)\mathrm{d}v^2 + \cosh^2 u \,\mathrm{d}u^2 = \cosh^2 u(\mathrm{d}u^2 + \mathrm{d}v^2).$$

Observe that F maps, respectively, meridians and parallels of the catenoid into lines and helices in the helicoid, as Figure 3.23 shows:

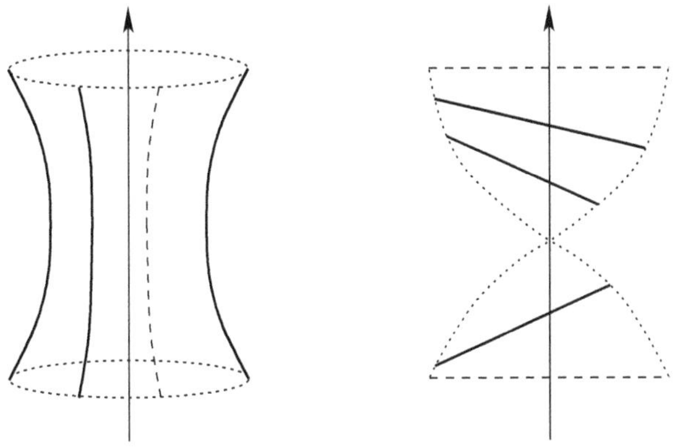

Figure 3.23: Isometry between parts of the catenoid and helicoid.

We have seen that even though isometries preserve areas, there are functions which preserve areas but are not isometries (e.g., Lambert's projection). We will see next that, while preserving areas is not enough to characterize isometries, preserving lengths will do the trick:

Proposition 3.2.23. *Let $M_1, M_2 \subseteq \mathbb{R}^3_\nu$ be two regular surfaces and $\phi\colon M_1 \to M_2$ be a (local) diffeomorphism whose differential preserves the causal type of tangent vectors. If, for every curve $\boldsymbol{\alpha}\colon I \to M_1$ in M_1 we have that $L[\phi \circ \boldsymbol{\alpha}] = L[\boldsymbol{\alpha}]$, then ϕ is a (local) isometry.*

Proof: Fix $\boldsymbol{p} \in M_1$ and $\boldsymbol{v} \in T_{\boldsymbol{p}}M_1$. Let $\boldsymbol{\alpha}\colon\,]-\epsilon,\epsilon[\, \to M_1$ be a curve in M_1 realizing $\boldsymbol{v}$. The assumption says that

$$\int_0^t \|\boldsymbol{\alpha}'(u)\|\,du = \int_0^t \|(\phi \circ \boldsymbol{\alpha})'(u)\|\,du = \int_0^t \|d\phi_{\boldsymbol{\alpha}(u)}(\boldsymbol{\alpha}'(u))\|\,du,$$

for all $t \in\,]-\epsilon,\epsilon[$. Differentiating both sides with respect to t, and evaluating at $t = 0$, we obtain $\|\boldsymbol{v}\| = \|d\phi_{\boldsymbol{p}}(\boldsymbol{v})\|$. Since $d\phi_{\boldsymbol{p}}$ preserves causal types, it follows from this that $\mathrm{I}_{\boldsymbol{p}}(\boldsymbol{v}) = \mathrm{I}_{\phi(\boldsymbol{p})}(d\phi_{\boldsymbol{p}}(\boldsymbol{v}))$. Polarizing, we conclude that

$$\mathrm{I}_{\boldsymbol{p}}(\boldsymbol{v}_1, \boldsymbol{v}_2) = \mathrm{I}_{\phi(\boldsymbol{p})}(d\phi_{\boldsymbol{p}}(\boldsymbol{v}_1), d\phi_{\boldsymbol{p}}(\boldsymbol{v}_2))$$

for all $\boldsymbol{v}_1, \boldsymbol{v}_2 \in T_{\boldsymbol{p}}M_1$. But since $\boldsymbol{p}$ was arbitrary and ϕ is a (local) diffeomorphism, we conclude that ϕ is a (local) isometry. □

Remark.

- The hypothesis of preservation of causal type is automatically satisfied when the ambient space considered is just $\mathbb{R}^3$, but it is crucial in the general case. The "flip" operator $\mathfrak{f}\colon \mathbb{L}^2 \to \mathbb{L}^2$ given by $\mathfrak{f}(x,y) = (y,x)$, which we have seen in a few exercises so far, is a witness for that.
- A similar result holds replacing arclength by energy. See Exercise 3.2.20.

Exercises

Exercise† 3.2.15. Show Proposition 3.2.17 (p. 170).

Exercise 3.2.16.

(a) Let $M \subseteq \mathbb{R}^3_\nu$ be a regular surface. Show that

$$\mathrm{Iso}(M) \doteq \{\phi\colon M \to M \mid \phi \text{ is an isometry}\}$$

is a subgroup of $\mathrm{Diff}(M)$ (see Exercise 3.1.6, p. 152). Is it a normal subgroup?

(b) Show that if $M_1, M_2 \subseteq \mathbb{R}^3_\nu$ are isometric regular surfaces, then $\mathrm{Iso}(M_1) \cong \mathrm{Iso}(M_2)$.

Remark.

- We have then defined an action of $\mathrm{Iso}(M)$ on M by $\phi \cdot \boldsymbol{p} \doteq \phi(\boldsymbol{p})$.
- Intuitively, the "larger" is $\mathrm{Iso}(M)$, the simpler is the geometry of M, since we have many more "symmetries". For example, we have already determined in Chapter 1 the groups $\mathrm{Iso}(\mathbb{R}^2)$ and $\mathrm{Iso}(\mathbb{L}^2)$.

Exercise 3.2.17 (Isometries of $\mathbb{S}^2$).

(a) Show that $\mathrm{Iso}(\mathbb{S}^2) = \left\{ C|_{\mathbb{S}^2} \mid C \in \mathrm{O}(3,\mathbb{R}) \right\}$.

Hint. If $\phi \in \mathrm{Iso}(\mathbb{S}^2)$, define $C\colon \mathbb{R}^3 \to \mathbb{R}^3$ by

$$C(\boldsymbol{p}) = \begin{cases} \|\boldsymbol{p}\|\,\phi\left(\dfrac{\boldsymbol{p}}{\|\boldsymbol{p}\|}\right), & \text{if } \boldsymbol{p} \neq \boldsymbol{0} \\ \boldsymbol{0}, & \text{if } \boldsymbol{p} = \boldsymbol{0}. \end{cases}$$

(b) Compute the stabilizers[3] of $(1,0,0)$ and $(0,0,1)$ in $\mathrm{Iso}(\mathbb{S}^2)$.

Exercise 3.2.18 (Isometries of $\mathbb{H}^2$).

(a) Show that $\mathrm{Iso}(\mathbb{H}^2) = \left\{ \Lambda|_{\mathbb{H}^2} \mid \Lambda \in \mathrm{O}_1^{\uparrow}(3,\mathbb{R}) \right\}$.

Hint. To avoid issues with lightlike vectors in the analogue of the construction suggested in the hint of Exercise 3.2.17 above, we may directly explore the vector space structure of $\mathbb{L}^3$, as follows: take $\boldsymbol{p}_1, \boldsymbol{p}_2 \in \mathbb{H}^2$ such that $\{\boldsymbol{p}_1, \boldsymbol{p}_2, \boldsymbol{e}_3\}$ is linearly independent and linearly extend ϕ to $\Lambda\colon \mathbb{L}^3 \to \mathbb{L}^3$ from those points.

Note that

$$\phi(\mathbb{H}^2) = \mathbb{H}^2, \quad \Lambda|_{T_{\boldsymbol{e}_3}\mathbb{H}^2} = d\phi_{\boldsymbol{e}_3} \quad \text{and} \quad d\phi_{\boldsymbol{e}_3}(\boldsymbol{e}_i) \in T_{\phi(\boldsymbol{e}_3)}\mathbb{H}^2 = \phi(\boldsymbol{e}_3)^{\perp},$$

for $1 \leq i \leq 2$. Use these facts to show that $\langle \Lambda \boldsymbol{e}_i, \Lambda \boldsymbol{e}_j \rangle_L = \eta_{ij}$, for $1 \leq i,j \leq 3$.

Observe that by construction, Λ is orthochronous. Moreover Λ does not depend on the choice of $\boldsymbol{p}_1$ and $\boldsymbol{p}_2$ on the given conditions.

(b) Compute the stabilizer of $(0,0,1)$ in $\mathrm{Iso}(\mathbb{H}^2)$.

Exercise 3.2.19 (Isometries of $\mathbb{S}_1^2$).

(a) Show that $\mathrm{Iso}(\mathbb{S}_1^2) = \left\{ \Lambda|_{\mathbb{S}_1^2} \mid \Lambda \in \mathrm{O}_1(3,\mathbb{R}) \right\}$.

Hint. Adapt what was done in item (a) of Exercise 3.2.18 above, using $\boldsymbol{e}_1$ instead of $\boldsymbol{e}_3$.

(b) Compute the stabilizer of $(1,0,0)$ in $\mathrm{Iso}(\mathbb{S}_1^2)$.

Exercise 3.2.20. Let $M_1, M_2 \subseteq \mathbb{R}^3_\nu$ be non-degenerate regular surfaces and consider a diffeomorphism $\phi\colon M_1 \to M_2$. If ϕ is energy-preserving, i.e., for every curve $\boldsymbol{\alpha}\colon I \to M_1$ we have that $E[\phi \circ \boldsymbol{\alpha}] = E[\boldsymbol{\alpha}]$, then ϕ is an isometry.

Exercise† 3.2.21. Show Proposition 3.2.21 (p. 172).

Exercise 3.2.22. Consider usual parametrizations for the cone and the plane, $\boldsymbol{x}\colon \mathbb{R}_{>0} \times]0,2\pi[\to \mathbb{R}^3$ and $\tilde{\boldsymbol{x}}\colon \mathbb{R}_{>0} \times]0,2\pi[\to \mathbb{R}^2$, given by

$$\boldsymbol{x}(u,v) = (u\cos v, u\sin v, u) \quad \text{and} \quad \tilde{\boldsymbol{x}}(\tilde{u},\tilde{v}) = (\tilde{u}\cos\tilde{v}, \tilde{u}\sin\tilde{v}),$$

respectively. Show that the map $F\colon \boldsymbol{x}(\mathbb{R}_{>0} \times]0,2\pi[) \to \tilde{\boldsymbol{x}}(\mathbb{R}_{>0} \times]0,2\pi[)$ given by $F(\boldsymbol{x}(u,v)) = \tilde{\boldsymbol{x}}(u\sqrt{2}, v/\sqrt{2})$ is an isometry onto its image, and determine it.

[3]Recall that if G is a group acting (on the left) on a set X, the stabilizer of an element $x \in X$ is $G_x \doteq \{g \in G \mid g \cdot x = x\}$.

Exercise 3.2.23. Consider the cylinder

$$\mathbb{S}^1 \times \mathbb{R} = \{(x,y,z) \in \mathbb{R}^3_\nu \mid x^2 + y^2 = 1\}.$$

Exhibit an isometry $f\colon \mathbb{S}^1 \times \mathbb{R} \to \mathbb{S}^1 \times \mathbb{R}$ with exactly two fixed points.

Hint. There is an isometry that works no matter in which ambient space we consider the cylinder.

Exercise 3.2.24.

(a) Let $\boldsymbol{\alpha}\colon \mathbb{R} \to \mathbb{R}^3$ be a unit speed injective parametrized curve, of the particular form $\boldsymbol{\alpha}(u) = (x(u), y(u), 0)$. Define $\boldsymbol{x}\colon \mathbb{R}^2 \to \mathbb{R}^3$ by $\boldsymbol{x}(u,v) = (x(u), y(u), v)$. Show that $\boldsymbol{x}$ is an injective and regular parametrized surface, and that $\boldsymbol{x}(\mathbb{R}^2)$ and $\mathbb{R}^2$ are isometric.

(b) Suppose now that $\boldsymbol{\alpha}$ is seen in $\mathbb{L}^3$, has unit speed, and has the particular form $\boldsymbol{\alpha}(u) = (0, y(u), z(u))$. This time, define $\boldsymbol{x}\colon \mathbb{R}^2 \to \mathbb{L}^3$ by $\boldsymbol{x}(u,v) = (v, y(u), z(u))$ instead. Show that $\boldsymbol{x}(\mathbb{R}^2)$ is isometric to $\mathbb{R}^2$ if $\boldsymbol{\alpha}$ is spacelike, and isometric to $\mathbb{L}^2$ if $\boldsymbol{\alpha}$ is timelike.

Exercise 3.2.25. Let $U \subseteq \mathbb{R}^2$ be open and connected, $f, g\colon U \to \mathbb{R}$ be two smooth functions, and consider $\phi\colon \operatorname{gr}(f) \to \operatorname{gr}(g)$ given by $\phi(u,v,f(u,v)) = (u,v,g(u,v))$.

(a) Show that if both graphs are seen inside the same ambient space and ϕ is an isometry, then $f(u,v) = \pm g(u,v) + c$, for some constant $c \in \mathbb{R}$, for all $(u,v) \in U$.

(b) Show that if the graphs are seen in different ambient spaces, i.e., if $\operatorname{gr}(f) \subseteq \mathbb{R}^3$ and $\operatorname{gr}(g) \subseteq \mathbb{L}^3$, and ϕ is an isometry, then both f and g are constant.

Remark. That is, item (b) says that the "direct projection" is an isometry between graphs in different ambient spaces if and only if both graphs are actually horizontal spacelike planes (and hence isometric to $\mathbb{R}^2$). A priori, we could have some other isometry between such graphs. Can you think of a concrete example? We already know that this is impossible if $\|\nabla g\|_E \geq 1$.

Exercise 3.2.26 (Tangent Surfaces – II).

(a) Let $\boldsymbol{\alpha}_1, \boldsymbol{\alpha}_2\colon I \to \mathbb{R}^3_\nu$ be two unit speed admissible curves, and consider their tangent surfaces (as done in Exercise 3.2.3, p. 164), $\boldsymbol{x}_1, \boldsymbol{x}_2\colon I \times \mathbb{R} \to \mathbb{R}^3_\nu$ given by

$$\boldsymbol{x}_i(s,v) = \boldsymbol{\alpha}_i(s) + v\boldsymbol{\alpha}_i'(s), \quad i = 1,2.$$

Fix $(s_0, v_0) \in I \times \mathbb{R}$ with $v_0 \neq 0$ and take a neighborhood V of (s_0, v_0) for which $\boldsymbol{x}_1(V)$ and $\boldsymbol{x}_2(V)$ are both regular surfaces. Suppose that $\kappa_{\boldsymbol{\alpha}_1}(s) = \kappa_{\boldsymbol{\alpha}_2}(s) \neq 0$ for all s and that the causal type of the Frenet-Serret trihedrons for both curves is always the same. Show that the composition $\boldsymbol{x}_1 \circ \boldsymbol{x}_2^{-1}\colon \boldsymbol{x}_2(V) \to \boldsymbol{x}_1(V)$ is an isometry.

(b) Show that if $\boldsymbol{\alpha}\colon I \to \mathbb{R}^3_\nu$ is an admissible curve with non-zero curvature, and $\boldsymbol{x}\colon I \times \mathbb{R} \to \mathbb{R}^3_\nu$ is its tangent surface, then for every $(t_0, v_0) \in I \times \mathbb{R}$ with $v_0 \neq 0$, there is a neighborhood V of (t_0, v_0) such that $\boldsymbol{x}(V)$ is a regular surface isometric to an open subset of $\mathbb{R}^2$ or $\mathbb{L}^2$, depending on the causal type of the Frenet-Serret trihedron for $\boldsymbol{\alpha}$.

Hint. Suppose without loss of generality that $\boldsymbol{\alpha}$ has unit speed, and combine the Fundamental Theorem of Curves with item (a) above.

Exercise 3.2.27. In Chapter 1, we mentioned that it was usual in the literature to consider the Lorentzian product defined with a negative sign in the first term instead of the last one, as we have done. Let $\mathbb{L}^2_{(+,-)} \doteq \mathbb{L}^2$, and $\mathbb{L}^2_{(-,+)}$ stand for the plane $\mathbb{R}^2$ equipped with the scalar product

$$\langle (u_1, u_2), (v_1, v_2) \rangle_{(-,+)} \doteq -u_1v_1 + u_2v_2$$

or, in other words, equipped with the First Fundamental Form $-\mathrm{d}u^2 + \mathrm{d}v^2$. Justify the force and ubiquity of the "flip" operator $\mathfrak{f}\colon \mathbb{L}^2_{(+,-)} \to \mathbb{L}^2_{(+,-)}$ given by $\mathfrak{f}(x,y) = (y,x)$, by showing that it is, in this setting, an isometry.

Exercise 3.2.28 (Conformal mappings). Let $M_1, M_2 \subseteq \mathbb{R}^3_\nu$ be regular surfaces. A (local) diffeomorphism $\psi\colon M_1 \to M_2$ is called *(locally) conformal* if there is a smooth function $\lambda\colon M_1 \to \mathbb{R}_{>0}$ such that, given $\boldsymbol{p} \in M_1$ and $\boldsymbol{v}, \boldsymbol{w} \in T_{\boldsymbol{p}}M_1$, the relation

$$\mathrm{I}_{\psi(\boldsymbol{p})}\big(\mathrm{d}\psi_{\boldsymbol{p}}(\boldsymbol{v}), \mathrm{d}\psi_{\boldsymbol{p}}(\boldsymbol{w})\big) = \lambda(\boldsymbol{p})\mathrm{I}_{\boldsymbol{p}}(\boldsymbol{v}, \boldsymbol{w})$$

holds. The function λ is called the *conformality coefficient of* ψ.

(a) Show that a conformal mapping must preserve the causal types of surfaces, angles between spacelike tangent vectors, and hyperbolic angles between tangent timelike vectors with the same time direction (in timelike surfaces).

(b) Show that if a diffeomorphism between spacelike surfaces preserves angles between tangent vectors, then it is actually a conformal mapping.

(c) Show that the stereographic projection seen in Exercise 3.1.2 (p. 152) is a conformal mapping.

Hint. You may regard $\mathbb{R}^2$ inside $\mathbb{R}^3$ as the coordinate plane $z = 0$, as usual.

Exercise 3.2.29. Let $f\colon \mathbb{R}^2 \to \mathbb{R}^2$ be smooth and given by $f(x,y) = (u(x,y), v(x,y))$.

(a) Suppose that the functions u and v satisfy the Cauchy-Riemann equations $u_x = v_y$ and $u_y = -v_x$. If

$$Q \doteq \{(x,y) \in \mathbb{R}^2 \mid u_x(x,y)^2 + u_y(x,y)^2 \neq 0\},$$

show that f is a locally conformal mapping from Q into $\mathbb{R}^2$.

(b) To obtain a result similar to item (a) in $\mathbb{L}^2$, we will again resort to the analogies of $\mathbb{C}$ with the set of split-complex numbers

$$\mathbb{C}' = \{x + \mathrm{h}y \mid x, y \in \mathbb{R} \text{ and } \mathrm{h}^2 = 1\},$$

informally presented in Exercise 2.2.16 (p. 95). Suppose that the functions u and v satisfy the *revised Cauchy-Riemann equations* $u_x = v_y$ and $u_y = v_x$ (which motivate a notion of "split-holomorphicity", as we will see in Chapter 4).

Adopting the notation given in Exercise 3.2.27 above, and setting

$$Q_+ \doteq \{(x,y) \in \mathbb{L}^2_{(+,-)} \mid u_x(x,y)^2 - u_y(x,y)^2 > 0\} \text{ and}$$
$$Q_- \doteq \{(x,y) \in \mathbb{L}^2_{(-,+)} \mid u_x(x,y)^2 - u_y(x,y)^2 < 0\},$$

show that f is a locally conformal map from Q_+ into $\mathbb{L}^2_{(+,-)}$, and from Q_- into $\mathbb{L}^2_{(+,-)}$.

Hint. Do like we did for isometries: write $\mathrm{d}u = u_x\,\mathrm{d}x + u_y\,\mathrm{d}y$, similarly for $\mathrm{d}v$, and compute $\mathrm{d}u^2 \pm \mathrm{d}v^2$ according to each ambient space.

3.3 SECOND FUNDAMENTAL FORM AND CURVATURES

We have previously seen that the orientability of a regular surface $M \subseteq \mathbb{R}^3_\nu$ is equivalent to the existence of a unit normal vector field, $\boldsymbol{N}: M \to \mathbb{R}^3_\nu$, smooth and defined on all of M. At this stage, we are ready to present a few avatars of the notion of "curvature" for surfaces.

The motivation is simple: we wish to know how the surface M bends near a given point $\boldsymbol{p} \in M$. Near enough to said point, it would be reasonable to turn our attention to the "linearization" of M at $\boldsymbol{p}$, namely, the tangent plane $T_{\boldsymbol{p}}M$. But knowing how $T_{\boldsymbol{p}}M$ changes with $\boldsymbol{p}$ is clearly equivalent to knowing how $\boldsymbol{N}(\boldsymbol{p})$ changes with $\boldsymbol{p}$. This indicates the crucial role that the map $\mathbf{d}\boldsymbol{N}_{\boldsymbol{p}}$ will play in the definition of curvature. We start to formalize such ideas now:

Definition 3.3.1 (Gauss map). Let $M \subseteq \mathbb{R}^3_\nu$ be a non-degenerate regular surface. A *Gauss (normal) map* for M is a smooth field $\boldsymbol{N}: M \to \mathbb{R}^3_\nu$ of unit vectors normal to M, that is, $\boldsymbol{N}(\boldsymbol{p}) \in (T_{\boldsymbol{p}}M)^\perp$, for all $\boldsymbol{p} \in M$. We denote by ϵ_M the indicator of the normal direction to M, namely, $\epsilon_M = 1$ if M is timelike, and $\epsilon_M = -1$ if M is spacelike.

Remark. In general, we can precisely say what the codomain of the Gauss map is:

- if $M \subseteq \mathbb{R}^3$, we have $\boldsymbol{N}: M \to \mathbb{S}^2$;
- if $M \subseteq \mathbb{L}^3$ is spacelike, we have $\boldsymbol{N}: M \to \mathbb{H}^2_\pm$;
- if $M \subseteq \mathbb{L}^3$ is timelike, we have $\boldsymbol{N}: M \to \mathbb{S}^2_1$.

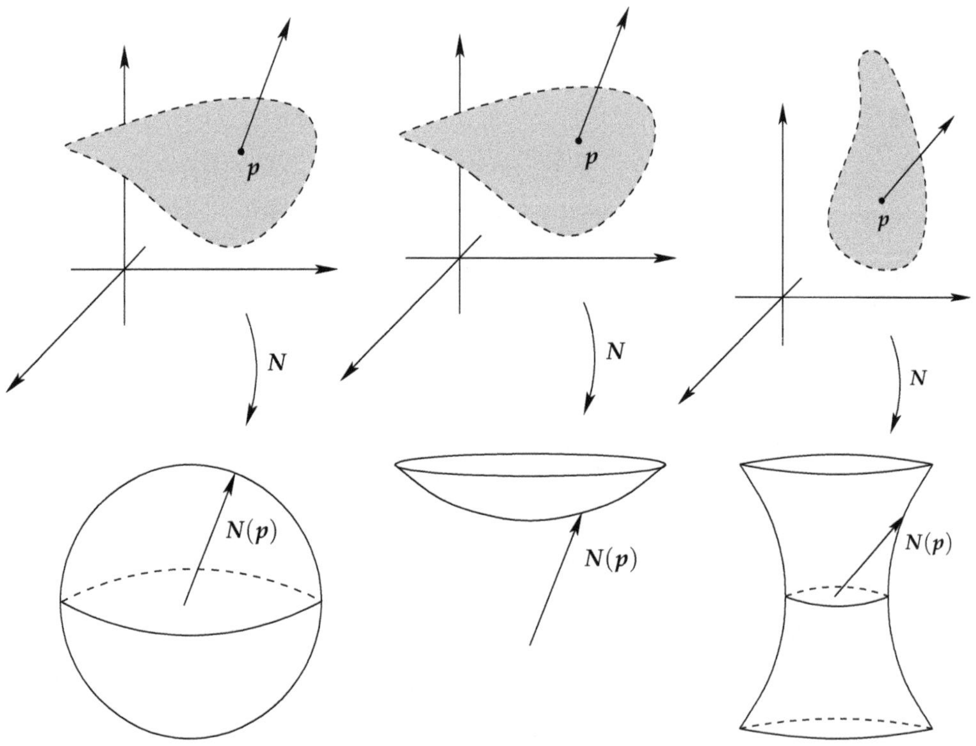

(a) In $\mathbb{R}^3$. (b) In $\mathbb{L}^3$, $\epsilon_M = -1$. (c) In $\mathbb{L}^3$, $\epsilon_M = 1$.

Figure 3.24: The Gauss map of a surface.

We note that in $\mathbb{R}^3$, T_pM and $T_{N(p)}\mathbb{S}^2$ are the same vector space, namely, the orthogonal complement of the line which has $N(p)$ as direction. The same remark holds in $\mathbb{L}^3$, with $\mathbb{S}^2_1$ or $\mathbb{H}^2_\pm$ instead of $\mathbb{S}^2$. With this, we may regard the differential of the Gauss map as a linear operator in T_pM.

Definition 3.3.2 (Weingarten Map). Let $M \subseteq \mathbb{R}^3_\nu$ be non-degenerate regular surface, and N be a Gauss map for M. The *Weingarten map* of M at p is the differential $-\mathrm{d}N_p \colon T_pM \to T_pM$.

Remark. The Weingarten map is also known as the *shape operator* of M. A naive justification for the negative sign in the above definition, seemingly artificial, is that it is meant to reduce the quantity of negative signs in future computations.

Proposition 3.3.3. *Let $M \subseteq \mathbb{R}^3_\nu$ be a non-degenerate regular surface and N be a Gauss map for M. Then, for each $p \in M$, the Weingarten map $-\mathrm{d}N_p$ is self-adjoint relative to the First Fundamental Form I_p.*

Proof: Let (U, x) be a parametrization for M around $p = x(u_0, v_0)$. By definition of differential, we have:

$$\frac{\partial(N \circ x)}{\partial u}(u,v) = \mathrm{d}N_{x(u,v)}\left(\frac{\partial x}{\partial u}(u,v)\right) \quad \text{and} \quad \frac{\partial(N \circ x)}{\partial v}(u,v) = \mathrm{d}N_{x(u,v)}\left(\frac{\partial x}{\partial v}(u,v)\right),$$

for all $(u, v) \in U$. We also know that:

$$\left\langle \frac{\partial x}{\partial u}(u,v), N(x(u,v))\right\rangle = 0 = \left\langle \frac{\partial x}{\partial v}(u,v), N(x(u,v))\right\rangle$$

for all $(u, v) \in U$. Then, differentiating the first relation with respect to v and the second one with respect to u we obtain:

$$\left\langle \frac{\partial^2 x}{\partial v \partial u}(u,v), N(x(u,v))\right\rangle + \left\langle \frac{\partial x}{\partial u}(u,v), \mathrm{d}N_{x(u,v)}\left(\frac{\partial x}{\partial v}(u,v)\right)\right\rangle = 0$$
$$\left\langle \frac{\partial^2 x}{\partial u \partial v}(u,v), N(x(u,v))\right\rangle + \left\langle \frac{\partial x}{\partial v}(u,v), \mathrm{d}N_{x(u,v)}\left(\frac{\partial x}{\partial u}(u,v)\right)\right\rangle = 0,$$

whence it follows that:

$$\left\langle \frac{\partial x}{\partial u}(u,v), \mathrm{d}N_{x(u,v)}\left(\frac{\partial x}{\partial v}(u,v)\right)\right\rangle = \left\langle \frac{\partial x}{\partial v}(u,v), \mathrm{d}N_{x(u,v)}\left(\frac{\partial x}{\partial u}(u,v)\right)\right\rangle.$$

Since $x_u(u_0, v_0)$ and $x_v(u_0, v_0)$ span T_pM, it follows from linearity, evaluating everything at (u_0, v_0), that

$$\langle -\mathrm{d}N_p(v), w\rangle = \langle v, -\mathrm{d}N_p(w)\rangle,$$

for all $v, w \in T_pM$, as wanted. ☐

If M is a spacelike surface, then the First Fundamental Form is positive-definite, and so $-\mathrm{d}N_p$ is diagonalizable, by the Real Spectral Theorem. If M is timelike, we cannot guarantee that this will happen.

Definition 3.3.4. Let $M \subseteq \mathbb{R}^3_\nu$ be a non-degenerate regular surface and N be a Gauss map for M. Given $p \in M$, if the Weingarten map $-\mathrm{d}N_p$ is diagonalizable, its two eigenvalues $\kappa_1(p)$ and $\kappa_2(p)$ are called the *principal curvatures* of M at p. The associated (orthogonal) eigenvectors are called the *principal directions* of M at p.

Remark. In the usual development of the theory done when considering only surfaces in $\mathbb{R}^3$, it is usual to define the curvatures of M from the principal curvatures defined above, which will always exist. In our case, we need an alternative approach that includes timelike surfaces in $\mathbb{L}^3$ as well.

Definition 3.3.5 (Second Fundamental Form). Let $M \subseteq \mathbb{R}^3_\nu$ be a non-degenerate regular surface and $\boldsymbol{N}$ be a Gauss map for M. The *Second Fundamental Form* of M at $\boldsymbol{p}$ (associated to $\boldsymbol{N}$) is the symmetric bilinear map $\mathrm{II}_{\boldsymbol{p}}\colon T_{\boldsymbol{p}}M \times T_{\boldsymbol{p}}M \to (T_{\boldsymbol{p}}M)^\perp$ defined by the relation

$$\langle \mathrm{II}_{\boldsymbol{p}}(\boldsymbol{v},\boldsymbol{w}), \boldsymbol{N}(\boldsymbol{p})\rangle = \langle -\mathrm{d}\boldsymbol{N}_{\boldsymbol{p}}(\boldsymbol{v}), \boldsymbol{w}\rangle,$$

for all $\boldsymbol{v},\boldsymbol{w} \in T_{\boldsymbol{p}}M$.

Remark.

- As done for $\mathrm{I}_{\boldsymbol{p}}$, we will abbreviate $\mathrm{II}_{\boldsymbol{p}}(\boldsymbol{v},\boldsymbol{v})$ to $\mathrm{II}_{\boldsymbol{p}}(\boldsymbol{v})$ only. Occasionally it will be convenient to omit the $\boldsymbol{p}$ as well.
- In $\mathbb{L}^3$, $\mathrm{II}_{\boldsymbol{p}}$ is also called the *Minkowski Second Fundamental Form* of M at $\boldsymbol{p}$.

Definition 3.3.6 (Components of the Second Form). Let $M \subseteq \mathbb{R}^3_\nu$ be a regular surface, $\boldsymbol{N}$ be a Gauss map for M, and $(U,\boldsymbol{x})$ be a parametrization for M. The *components of the Second Fundamental Form relative to* $\boldsymbol{x}$ are defined by

$$\begin{aligned}
e(u,v) &\doteq \langle \boldsymbol{x}_{uu}(u,v), \boldsymbol{N}(\boldsymbol{x}(u,v))\rangle,\\
f(u,v) &\doteq \langle \boldsymbol{x}_{uv}(u,v), \boldsymbol{N}(\boldsymbol{x}(u,v))\rangle \text{ and}\\
g(u,v) &\doteq \langle \boldsymbol{x}_{vv}(u,v), \boldsymbol{N}(\boldsymbol{x}(u,v))\rangle.
\end{aligned}$$

Remark.

- The above definition is justified by noting that, identifying indices $u \leftrightarrow 1$ and $v \leftrightarrow 2$, we have

$$\begin{aligned}
\langle \mathrm{II}_{\boldsymbol{x}(u,v)}(\boldsymbol{x}_i(u,v), \boldsymbol{x}_j(u,v)), \boldsymbol{N}(\boldsymbol{x}(u,v))\rangle &= \langle -\mathrm{d}\boldsymbol{N}_{\boldsymbol{x}(u,v)}(\boldsymbol{x}_i(u,v)), \boldsymbol{x}_j(u,v)\rangle\\
&= \langle \boldsymbol{x}_{ij}(u,v), \boldsymbol{N}(\boldsymbol{x}(u,v))\rangle.
\end{aligned}$$

- It is also usual to write $\ell \equiv h_{11} = e$, $m \equiv h_{12} = h_{21} = f$ and $n = h_{22} = g$, which allows us to gather all the necessary information about $\mathrm{II}_{\boldsymbol{x}(u,v)}$ in the matrix $(h_{ij}(u,v))_{1\leq i,j\leq 2}$.
- It follows from the above considerations that

$$\begin{aligned}
\mathrm{II}_{\boldsymbol{x}(u,v)}(\boldsymbol{x}_u(u,v)) &= \epsilon_M e(u,v)\boldsymbol{N}(\boldsymbol{x}(u,v)),\\
\mathrm{II}_{\boldsymbol{x}(u,v)}(\boldsymbol{x}_u(u,v), \boldsymbol{x}_v(u,v)) &= \epsilon_M f(u,v)\boldsymbol{N}(\boldsymbol{x}(u,v)) \text{ and}\\
\mathrm{II}_{\boldsymbol{x}(u,v)}(\boldsymbol{x}_v(u,v)) &= \epsilon_M g(u,v)\boldsymbol{N}(\boldsymbol{x}(u,v)).
\end{aligned}$$

Lemma 3.3.7. *Let $M \subseteq \mathbb{R}^3_\nu$ be a non-degenerate regular surface, $\boldsymbol{N}$ be a Gauss map for M and $(U,\boldsymbol{x})$ be a parametrization for M. Identifying the indices $u \leftrightarrow 1$ and $v \leftrightarrow 2$ and omitting all points of evaluation, we have that if $(h^i{}_j)_{1\leq i,j\leq 2} \doteq [-\mathrm{d}\boldsymbol{N}]_{\mathcal{B}_{\boldsymbol{x}}}$, then we have that $h^i{}_j = \sum_{k=1}^2 g^{ik}h_{kj}$.*

Proof: By definition of the matrix of a linear transformation, for each $j = 1,2$ we have that $-\mathrm{d}\boldsymbol{N}(\boldsymbol{x}_j) = \sum_{i=1}^{2} h^i{}_j \boldsymbol{x}_i$. Taking products with $\boldsymbol{x}_k$ on both sides, we have that $\langle -\mathrm{d}\boldsymbol{N}(\boldsymbol{x}_j), \boldsymbol{x}_k \rangle = \sum_{i=1}^{2} h^i{}_j g_{ik}$. It follows that $h_{jk} = \langle \boldsymbol{N} \circ \boldsymbol{x}, \boldsymbol{x}_{jk} \rangle = \sum_{i=1}^{2} h^i{}_j g_{ik}$. Since M is non-degenerate, we have the inverse matrix $(g^{ij})_{1 \leq i,j \leq 2}$. Hence, multiplying both sides by $g^{k\ell}$, summing over k and renaming $\ell \to i$, we obtain precisely $h^i{}_j = \sum_{k=1}^{2} g^{ik} h_{kj}$, as wanted. □

With this basic language, we may return to our initial idea: looking at the Weingarten map of M at $\boldsymbol{p}$. We know, from Linear Algebra, that the trace and the determinant of a linear operator are invariant under change of basis. The Second Fundamental Form is a vector-valued bilinear form so that, a priori, we wouldn't have its trace and determinant available. Precisely to avoid this hindrance, we have seen in Lemmas 1.6.7 and 1.6.8 (p. 57) in Chapter 1 how to define the trace and determinant of a bilinear form relative to the ambient scalar product. But since this "metric determinant" was defined only for scalar-valued bilinear forms, we consider instead $\widetilde{\mathrm{II}}_{\boldsymbol{p}}(\boldsymbol{v}, \boldsymbol{w}) \doteq \langle \mathrm{II}_{\boldsymbol{p}}(\boldsymbol{v}, \boldsymbol{w}), \boldsymbol{N}(\boldsymbol{p}) \rangle$.

Thus we may write the:

Definition 3.3.8 (Mean and Gaussian curvatures). Let $M \subseteq \mathbb{R}^3_\nu$ be a non-degenerate regular surface, and $\boldsymbol{N}$ be a Gauss map for M. The *mean curvature vector* and the *Gaussian curvature* of M at $\boldsymbol{p}$ are defined by:

$$\boldsymbol{H}(\boldsymbol{p}) \doteq \frac{1}{2}\,\mathrm{tr}_{\mathrm{I}_{\boldsymbol{p}}}(\mathrm{II}_{\boldsymbol{p}}) = \frac{1}{2}(\epsilon_{\boldsymbol{v}_1}\mathrm{II}_{\boldsymbol{p}}(\boldsymbol{v}_1) + \epsilon_{\boldsymbol{v}_2}\mathrm{II}_{\boldsymbol{p}}(\boldsymbol{v}_2)) \text{ and}$$
$$K(\boldsymbol{p}) \doteq (-1)^\nu \mathrm{det}_{\mathrm{I}_{\boldsymbol{p}}} \widetilde{\mathrm{II}}_{\boldsymbol{p}} = (-1)^\nu \det\left((\widetilde{\mathrm{II}}_{\boldsymbol{p}}(\boldsymbol{v}_i, \boldsymbol{v}_j))_{1 \leq i,j \leq 2}\right),$$

where $\{\boldsymbol{v}_1, \boldsymbol{v}_2\}$ is any orthonormal basis for $T_{\boldsymbol{p}}M$. Moreover, the *mean curvature* of M at $\boldsymbol{p}$ is the number $H(\boldsymbol{p})$ determined by the relation $\boldsymbol{H}(\boldsymbol{p}) = H(\boldsymbol{p})\boldsymbol{N}(\boldsymbol{p})$.

Remark.

- Note that replacing $\boldsymbol{N}$ by $-\boldsymbol{N}$, the sign of the mean curvature is reversed, but not the sign of the Gaussian curvature, since the matrix whose determinant is computed has even order.
- We recall that the presence of indicators in the definition of the mean curvature is indeed natural: without them, the quantity to be defined is not invariant under a change of orthonormal basis.
- The coefficient $(-1)^\nu$ in the definition of K, in turn, has the purpose of recovering the information about the causal type of the surface which is lost in $\mathbb{L}^3$ (but not in $\mathbb{R}^3$), when considering the scalar $\widetilde{\mathrm{II}}$ instead of II.

Naturally, we need to know how to express those new objects in terms of coordinates. For that end, we need the following technical lemma:

Lemma 3.3.9. *Let $M \subseteq \mathbb{R}^3_\nu$ be a non-degenerate regular surface, $\boldsymbol{N}$ be a Gauss map for M, and $(U, \boldsymbol{x})$ a parametrization for M. Then, omitting points of evaluation, we define*

$$\boldsymbol{E}_1 \doteq \frac{\boldsymbol{x}_u}{\|\boldsymbol{x}_u\|} \quad \textit{and} \quad \boldsymbol{E}_2 \doteq \frac{\boldsymbol{x}_v - (F/E)\boldsymbol{x}_u}{\|\boldsymbol{x}_v - (F/E)\boldsymbol{x}_u\|}.$$

Then, we have that $\{\boldsymbol{E}_1(u,v), \boldsymbol{E}_2(u,v)\}$ *is an orthonormal basis for* $T_{\boldsymbol{x}(u,v)}M$ *satisfying:*

$$\begin{aligned}
\mathrm{II}(\boldsymbol{E}_1) &= \frac{\epsilon_1\epsilon_M e}{E}\boldsymbol{N}\circ\boldsymbol{x}, \\
\mathrm{II}(\boldsymbol{E}_1,\boldsymbol{E}_2) &= \frac{\epsilon_M(Ef-Fe)}{E\sqrt{|EG-F^2|}}\boldsymbol{N}\circ\boldsymbol{x} \\
\mathrm{II}(\boldsymbol{E}_2) &= \frac{\epsilon_2\epsilon_M\left(Eg-2Ff+F^2e/E\right)}{EG-F^2}\boldsymbol{N}\circ\boldsymbol{x},
\end{aligned}$$

where ϵ_1 *and* ϵ_2 *stand for the indicators of* $\boldsymbol{E}_1$ *and* $\boldsymbol{E}_2$.

Proof: That $\{\boldsymbol{E}_1(u,v), \boldsymbol{E}_2(u,v)\}$ is an orthonormal basis for $T_{\boldsymbol{x}(u,v)}M$ is nothing more than a direct consequence of the Gram-Schmidt process. Initially, we have:

$$\mathrm{II}(\boldsymbol{E}_1) = \frac{1}{\|\boldsymbol{x}_u\|^2}\mathrm{II}(\boldsymbol{x}_u) = \frac{1}{\epsilon_1 E}\epsilon_M e\boldsymbol{N}\circ\boldsymbol{x} = \frac{\epsilon_1\epsilon_M e}{E}\boldsymbol{N}\circ\boldsymbol{x}.$$

Next, noting that:

$$\left\langle \boldsymbol{x}_v - \frac{F}{E}\boldsymbol{x}_u, \boldsymbol{x}_v - \frac{F}{E}\boldsymbol{x}_u \right\rangle = G - \frac{2F^2}{E} + \frac{F^2}{E^2}E = \frac{EG-F^2}{E},$$

we have that:

$$\mathrm{II}\left(\boldsymbol{E}_1,\boldsymbol{E}_2\right) = \frac{\mathrm{II}\left(\boldsymbol{x}_u, \boldsymbol{x}_v - \frac{F}{E}\boldsymbol{x}_u\right)}{\sqrt{|EG-F^2|}} = \frac{\epsilon_M(f-\frac{F}{E}e)}{\sqrt{|EG-F^2|}}\boldsymbol{N}\circ\boldsymbol{x} = \frac{\epsilon_M(Ef-Fe)}{E\sqrt{|EG-F^2|}}\boldsymbol{N}\circ\boldsymbol{x},$$

and lastly, noting that $\epsilon_1\epsilon_2\epsilon_M = (-1)^\nu$, we have:

$$\begin{aligned}
\mathrm{II}(\boldsymbol{E}_2) &= \frac{\mathrm{II}\left(\boldsymbol{x}_v - \frac{F}{E}\boldsymbol{x}_u\right)}{\left\|\boldsymbol{x}_v - \frac{F}{E}\boldsymbol{x}_u\right\|^2} = \frac{|E|}{|EG-F^2|}\mathrm{II}\left(\boldsymbol{x}_v - \frac{F}{E}\boldsymbol{x}_u\right) \\
&= \frac{\epsilon_1 E}{(-1)^\nu\epsilon_M(EG-F^2)}\epsilon_M\left(g - \frac{2F}{E}f + \frac{F^2}{E^2}e\right)\boldsymbol{N}\circ\boldsymbol{x} \\
&= \frac{\epsilon_2\epsilon_M\left(Eg-2Ff+F^2e/E\right)}{EG-F^2}\boldsymbol{N}\circ\boldsymbol{x}.
\end{aligned}$$

□

Proposition 3.3.10 (Local curvature expressions). *Let* $M \subseteq \mathbb{R}^3_\nu$ *be a non-degenerate regular surface,* $\boldsymbol{N}$ *be a Gauss map for* M*, and* $(U, \boldsymbol{x})$ *a parametrization for* M *compatible with* $\boldsymbol{N}$*. Then we have:*

$$\begin{aligned}
H\circ\boldsymbol{x} &= \frac{\epsilon_M}{2}\mathrm{tr}(-\mathrm{d}\boldsymbol{N}) = \frac{\epsilon_M}{2}\frac{Eg-2Ff+Ge}{EG-F^2} \quad \text{and} \\
K\circ\boldsymbol{x} &= \epsilon_M\det(-\mathrm{d}\boldsymbol{N}) = \epsilon_M\frac{eg-f^2}{EG-F^2}.
\end{aligned}$$

Remark. $\boldsymbol{N}$ and $\boldsymbol{x}$ are *compatible* if $\boldsymbol{N}\circ\boldsymbol{x} = \dfrac{\boldsymbol{x}_u\times\boldsymbol{x}_v}{\|\boldsymbol{x}_u\times\boldsymbol{x}_v\|}$.

Proof: Let $\boldsymbol{E}_1$ and $\boldsymbol{E}_2$ be as in Lemma 3.3.9 above, and omit points of evaluation. Let's deal with the mean curvature first. We have that:

$$\begin{aligned} \boldsymbol{H}\circ\boldsymbol{x} &= \frac{1}{2}\left(\epsilon_1 \mathrm{I\!I}\left(\boldsymbol{E}_1\right)+\epsilon_2 \mathrm{I\!I}\left(\boldsymbol{E}_2\right)\right) \\ &= \frac{1}{2}\left(\frac{\epsilon_M e}{E}+\frac{\epsilon_M(Eg-2Ff+F^2e/E)}{EG-F^2}\right)\boldsymbol{N}\circ\boldsymbol{x} \\ &= \frac{\epsilon_M}{2}\frac{Eg-2Ff+Ge}{EG-F^2}\boldsymbol{N}\circ\boldsymbol{x}. \end{aligned}$$

It follows from the definition that:

$$H\circ\boldsymbol{x} = \frac{\epsilon_M}{2}\frac{eG-2fF+Eg}{EG-F^2}.$$

To check the relation between $H\circ\boldsymbol{x}$ and $\mathbf{tr}(-\mathbf{d}\boldsymbol{N})$, we now use Lemma 3.3.7:

$$\begin{aligned} \mathbf{tr}(-\mathbf{d}\boldsymbol{N}) &= h^1_1+h^2_2 \\ &= g^{11}h_{11}+g^{12}h_{21}+g^{21}h_{12}+g^{22}h_{22} \\ &= \frac{G}{EG-F^2}e-\frac{F}{EG-F^2}f-\frac{F}{EG-F^2}f+\frac{E}{EG-F^2}g \\ &= \frac{eG-2Ff+Eg}{EG-F^2}. \end{aligned}$$

For the Gaussian curvature, in turn, we have:

$$\begin{aligned} K\circ\boldsymbol{x} &= (-1)^\nu\left(\widetilde{\mathrm{I\!I}}(\boldsymbol{E}_1)\widetilde{\mathrm{I\!I}}(\boldsymbol{E}_2)-\widetilde{\mathrm{I\!I}}(\boldsymbol{E}_1,\boldsymbol{E}_2)^2\right) \\ &= (-1)^\nu\left(\left(\frac{\epsilon_1}{E}e\right)\left(\frac{\epsilon_2(Eg-2Ff+F^2e/E)}{(EG-F^2)}\right)-\left(\frac{(Ef-Fe)}{E\sqrt{|EG-F^2|}}\right)^2\right) \\ &= \frac{(-1)^\nu\epsilon_1\epsilon_2}{EG-F^2}\left(\frac{Eeg-2Fef+F^2e^2/E}{E}-\frac{E^2f^2-2EFef+F^2e^2}{E^2}\right) \\ &= \frac{\epsilon_M}{EG-F^2}\left(\frac{E^2eg-E^2f^2}{E^2}\right) \\ &= \epsilon_M\frac{eg-f^2}{EG-F^2}. \end{aligned}$$

The relation between $K\circ\boldsymbol{x}$ and $\mathbf{det}(-\mathbf{d}\boldsymbol{N})$ also follows from Lemma 3.3.7, which essentially says that the matrix of $-\mathbf{d}\boldsymbol{N}_{\boldsymbol{p}}$ is the product of the inverse matrix of $\mathrm{I}_{\boldsymbol{p}}$ with the matrix of $\widetilde{\mathrm{I\!I}}_{\boldsymbol{p}}$, from where it follows that

$$\mathbf{det}(-\mathbf{d}\boldsymbol{N}) = \frac{\det\left((h_{ij})_{1\le i,j\le 2}\right)}{\det\left((g_{ij})_{1\le i,j\le 2}\right)} = \frac{eg-f^2}{EG-F^2},$$

as wanted. □

Example 3.3.11.

(1) Consider a plane $\Pi\subseteq\mathbb{R}^3_\nu$, non-degenerate, passing through a certain point $\boldsymbol{p}_0\in\mathbb{R}^3_\nu$ with a unit vector $\boldsymbol{n}\in\mathbb{R}^3_\nu$ giving the normal direction. A Gauss map in this case is simply $\boldsymbol{N}\colon\Pi\to\mathbb{R}^3_\nu$ given by $\boldsymbol{N}(\boldsymbol{p})=\boldsymbol{n}$. This way $\mathbf{d}\boldsymbol{N}_{\boldsymbol{p}}$ is the zero operator for all $\boldsymbol{p}\in\Pi$, whence we conclude that $K=H\equiv 0$.

Note that in this case the Weingarten map is trivially diagonalizable, both principal curvatures vanish, and every direction is principal.

(2) For the "spheres" $\mathbb{S}^2(r)$, $\mathbb{S}^2_1(r)$ and $\mathbb{H}^2(r)$ with radius $r > 0$, a Gauss map is simply N given by $N(p) = p/r$. Directly, we have that $\mathrm{d}N_p(v) = v/r$, for every vector v tangent at p, and thus

$$-\mathrm{d}N_p = -\frac{1}{r}\mathrm{id}_{T_p\mathbb{S}^2(r)} \implies \det(-\mathrm{d}N_p) = \frac{1}{r^2} \quad \text{and} \quad \frac{1}{2}\mathrm{tr}(-\mathrm{d}N_p) = -\frac{1}{r}.$$

Hence, both $\mathbb{S}^2(r)$ and $\mathbb{S}^2_1(r)$ have constant and positive Gaussian curvature $1/r^2$, and also constant mean curvature, equal to $-1/r$. The hyperbolic plane $\mathbb{H}^2(r)$, in turn, has constant and negative Gaussian curvature $-1/r^2$, and mean curvature equal to $1/r$.

In these cases, the Weingarten map is again diagonalizable, with all the directions being principal.

(3) Considering now a straight cylinder $\mathbb{S}^1(r) \times \mathbb{R}$ with radius $r > 0$, and the projection $\pi\colon \mathbb{R}^3_\nu \to \mathbb{R}^3_\nu$ onto the first two components, we have that a Gauss map is $N\colon \mathbb{S}^1(r) \times \mathbb{R} \to \mathbb{S}^2_\nu$, given by $N(p) = \pi(p)/r$. It follows from this that the derivative is given by $\mathrm{d}N_p(v) = \pi(v)/r$ for every $v \in T_p(\mathbb{S}^1(r) \times \mathbb{R})$, since N is the composition of the restrictions of linear maps. Fixed $p \in \mathbb{S}^1(r) \times \mathbb{R}$, we may consider the orthonormal basis of $T_p(\mathbb{S}^1(r) \times \mathbb{R})$ formed by the vector u_1 tangent to the cylinder and horizontal (take any of the two possible vectors here), and the vector $u_2 = (0,0,1)$.

Relative to the basis $\mathcal{B} = (u_1, u_2)$, we have that

$$[-\mathrm{d}N_p]_{\mathcal{B}} = \begin{pmatrix} -1/r & 0 \\ 0 & 0 \end{pmatrix} \implies K(p) = 0 \quad \text{and} \quad H(p) = -\frac{1}{2r},$$

independently of the ambient space considered. This in particular illustrates that it is possible, when considering different ambient spaces, that surfaces which are not congruent might have the same curvatures. Another way to obtain the same conclusions is by doing coordinate computations, considering the parametrization $x\colon]0,2\pi[\times \mathbb{R} \to \mathbb{R}^3_\nu$ given by $x(u,v) = (r\cos u, r\sin u, v)$, and computing all the g_{ij} and h_{ij}.

Note that even in $\mathbb{L}^3$, with $\mathbb{S}^1(r) \times \mathbb{R}$ being timelike, the Weingarten map is diagonalizable. The principal vectors are precisely u_1 and u_2 chosen above.

(4) Let $f : U \subseteq \mathbb{R}^2 \to \mathbb{R}$ be a smooth function, and consider the usual Monge parametrization for its graph: $x\colon U \to \mathrm{gr}(f)$ given by $x(u,v) = (u,v,f(u,v))$. Suppose that the graph of f is non-degenerate. Denoting the curvature with indices according to the ambient space, abbreviating the partial derivatives of f and omitting points of evaluation, we have that

$$K_E \circ x = \frac{f_{uu}f_{vv} - f_{uv}^2}{(1+f_u^2+f_v^2)^2} \quad \text{and} \quad H_E \circ x = \frac{f_{uu}(1+f_v^2) - 2f_uf_vf_{uv} + f_{vv}(1+f_u^2)}{2\,(1+f_u^2+f_v^2)^{3/2}}$$

in $\mathbb{R}^3$, and

$$K_L \circ x = \frac{f_{uv}^2 - f_{uu}f_{vv}}{(-1+f_u^2+f_v^2)^2}, \quad H_L \circ x = \frac{f_{uu}(-1+f_v^2) - 2f_uf_vf_{uv} + f_{vv}(-1+f_u^2)}{2|-1+f_u^2+f_v^2|^{3/2}}$$

in $\mathbb{L}^3$. We ask you to verify this in Exercise 3.3.4.

(5) Suppose that $\boldsymbol{\alpha}: I \to \mathbb{R}^3_\nu$ is a smooth, regular, injective and non-degenerate curve of the form $\boldsymbol{\alpha}(u) = (f(u), 0, g(u))$, for certain smooth functions f and g such that $f(u) > 0$ for all $u \in I$. This way, we may consider the revolution surface around the z-axis generated by $\boldsymbol{\alpha}$, which will also be non-degenerate. Considering the usual parametrization $\boldsymbol{x}: I \times]0, 2\pi[\to \boldsymbol{x}(I \times]0, 2\pi[) \subseteq \mathbb{R}^3_\nu$ given by $\boldsymbol{x}(u,v) = (f(u)\cos v, f(u)\sin v, g(u))$, as in the previous example, we have that

$$K_E \circ \boldsymbol{x} = \frac{g'}{f}\frac{(-f''g' + f'g'')}{\langle \boldsymbol{\alpha}', \boldsymbol{\alpha}' \rangle_E^2} \quad \text{and} \quad H_E \circ \boldsymbol{x} = \frac{g'(\langle \boldsymbol{\alpha}', \boldsymbol{\alpha}' \rangle_E - ff'') + ff'g''}{2f \langle \boldsymbol{\alpha}', \boldsymbol{\alpha}' \rangle_E^{3/2}}$$

for $\mathbb{R}^3$ and

$$K_L \circ \boldsymbol{x} = \frac{g'}{f}\frac{(f''g' - f'g'')}{\langle \boldsymbol{\alpha}', \boldsymbol{\alpha}' \rangle_L^2} \quad \text{and} \quad H_L \circ \boldsymbol{x} = \frac{g'(-\langle \boldsymbol{\alpha}', \boldsymbol{\alpha}' \rangle_L + ff'') - ff'g''}{2f\,|\langle \boldsymbol{\alpha}', \boldsymbol{\alpha}' \rangle_L|^{3/2}}$$

for $\mathbb{L}^3$. The verification of those formulas is left to Exercise 3.3.5. Note that these expressions undergo a great simplification when $\boldsymbol{\alpha}$ has unit speed.

(6) Consider again the helicoid, image of the regular and injective parametrized surface $\boldsymbol{x}: \mathbb{R} \times \mathbb{R} \to \mathbb{R}^3_\nu$ given by $\boldsymbol{x}(u,v) = (u\cos v, u\sin v, v)$. For $u \neq \pm 1$, $\boldsymbol{x}$ is non-degenerate and:

$$(g_{ij})_{1 \leq i,j \leq 2} = \begin{pmatrix} 1 & 0 \\ 0 & (-1)^\nu + u^2 \end{pmatrix} \quad \text{and} \quad (h_{ij})_{1 \leq i,j \leq 2} = \begin{pmatrix} 0 & \frac{-1}{\sqrt{|(-1)^\nu + u^2|}} \\ \frac{-1}{\sqrt{|(-1)^\nu + u^2|}} & 0 \end{pmatrix}$$

and, thus,

$$K(\boldsymbol{x}(u,v)) = \frac{(-1)^{\nu+1}}{|(-1)^\nu + u^2|^2} \quad \text{and} \quad H(\boldsymbol{x}(u,v)) = 0.$$

In possession of the relations of H and K with the trace and determinant of the Weingarten map, previously seen, Lemma 1.6.9 (p. 58) from Chapter 1 gives us the:

Proposition 3.3.12. *Let $M \subseteq \mathbb{R}^3_\nu$ be a non-degenerate regular surface, with orientation given by the unit normal field $\mathbf{N}$. For each $\boldsymbol{p} \in M$, we have:*

$$\begin{aligned} \mathrm{d}\mathbf{N}_{\boldsymbol{p}}(\boldsymbol{v}) \times \mathrm{d}\mathbf{N}_{\boldsymbol{p}}(\boldsymbol{w}) &= \epsilon_M K(\boldsymbol{p})\, \boldsymbol{v} \times \boldsymbol{w} \\ \mathrm{d}\mathbf{N}_{\boldsymbol{p}}(\boldsymbol{v}) \times \boldsymbol{w} + \boldsymbol{v} \times \mathrm{d}\mathbf{N}_{\boldsymbol{p}}(\boldsymbol{w}) &= -2\epsilon_M H(\boldsymbol{p})\, \boldsymbol{v} \times \boldsymbol{w}, \end{aligned}$$

for all linearly independent $\boldsymbol{v}, \boldsymbol{w} \in T_{\boldsymbol{p}}M$.

Remark. For an interesting application of this last proposition, see Exercise 3.3.7.

At this point, we have enough tools to raise the following natural question: are the mean and Gaussian curvatures invariant under congruences, really deserving the name of "curvatures"? The affirmative answer to this first question is now easy to obtain:

Proposition 3.3.13. *Let $M_1, M_2 \subseteq \mathbb{R}^3_\nu$ be non-degenerate regular surfaces such that there is $F \in \mathrm{E}_\nu(3, \mathbb{R})$ with $M_2 = F(M_1)$. Then, if K_1, K_2, H_1 and H_2 denote the Gaussian and mean curvatures of M_1 and M_2, we have the relations $K_1(\boldsymbol{p}) = K_2(F(\boldsymbol{p}))$ and $H_1(\boldsymbol{p}) = H_2(F(\boldsymbol{p}))$, for all $\boldsymbol{p} \in M_1$.*

Remark. The equality between the mean curvatures only holds indeed without the absolute value, once a convenient choice of a Gauss map for M_2 has been made, "compatible" with F, in a sense to be made precise in the following proof.

Proof: Suppose that $F \in \mathrm{E}_\nu(3,\mathbb{R})$ is written as $F = T_{\boldsymbol{a}} \circ A$, with $A \in \mathrm{O}_\nu(3,\mathbb{R})$ and $\boldsymbol{a} \in \mathbb{R}^3_\nu$. If $\boldsymbol{N}_1$ is a Gauss map for M_1, then $\boldsymbol{N}_2 \doteq A \circ \boldsymbol{N}_1 \circ F^{-1}$ is a Gauss map for M_2. More precisely, if $\boldsymbol{N}_1(\boldsymbol{p})$ is normal to M_1 at $\boldsymbol{p}$, then $A(\boldsymbol{N}_1(\boldsymbol{p}))$ is normal to M_2 at $F(\boldsymbol{p})$. Take linearly independent vectors $\boldsymbol{v}, \boldsymbol{w} \in T_{\boldsymbol{p}}M_1$. Then $A\boldsymbol{v}, A\boldsymbol{w} \in T_{\boldsymbol{p}}M_2$ are also linearly independent, and Proposition 3.3.12 now gives that:

$$d(\boldsymbol{N}_1)_{\boldsymbol{p}}(\boldsymbol{v}) \times d(\boldsymbol{N}_1)_{\boldsymbol{p}}(\boldsymbol{w}) = \epsilon_{M_1} K_1(\boldsymbol{p})\boldsymbol{v} \times \boldsymbol{w}, \quad \text{and}$$
$$d(\boldsymbol{N}_2)_{F(\boldsymbol{p})}(A\boldsymbol{v}) \times d(\boldsymbol{N}_2)_{F(\boldsymbol{p})}(A\boldsymbol{w}) = \epsilon_{M_2} K_2(F(\boldsymbol{p}))A\boldsymbol{v} \times A\boldsymbol{w}.$$

Noting that $d(\boldsymbol{N}_2)_{F(\boldsymbol{p})} = A \circ d(\boldsymbol{N}_1)_{\boldsymbol{p}} \circ A^{-1}$, the second equation in display reduces to

$$A(d(\boldsymbol{N}_1)_{\boldsymbol{p}}(\boldsymbol{v})) \times A(d(\boldsymbol{N}_1)_{\boldsymbol{p}}(\boldsymbol{w})) = \epsilon_{M_2} K_2(F(\boldsymbol{p}))A\boldsymbol{v} \times A\boldsymbol{w}.$$

Now, directly using Lemma 1.6.9 (p. 58) and canceling $\det A$ on both sides, it follows that

$$d(\boldsymbol{N}_1)_{\boldsymbol{p}}(\boldsymbol{v}) \times d(\boldsymbol{N}_1)_{\boldsymbol{p}}(\boldsymbol{w}) = \epsilon_{M_2} K_2(F(\boldsymbol{p}))\boldsymbol{v} \times \boldsymbol{w}.$$

Since congruent surfaces have the same causal type, we have that $\epsilon_{M_1} = \epsilon_{M_2}$, so that by direct comparison we obtain $K_1(\boldsymbol{p}) = K_2(F(\boldsymbol{p}))$, as wanted.

The reasoning for the mean curvature is similar: Proposition 3.3.12 gives us two more relations to be compared:

$$d(\boldsymbol{N}_1)_{\boldsymbol{p}}(\boldsymbol{v}) \times \boldsymbol{w} + \boldsymbol{v} \times d(\boldsymbol{N}_1)_{\boldsymbol{p}}(\boldsymbol{w}) = -2\epsilon_{M_1} H_1(\boldsymbol{p})\boldsymbol{v} \times \boldsymbol{w}, \quad \text{and}$$
$$d(\boldsymbol{N}_2)_{F(\boldsymbol{p})}(A\boldsymbol{v}) \times A\boldsymbol{w} + A\boldsymbol{v} \times d(\boldsymbol{N}_2)_{F(\boldsymbol{p})}(A\boldsymbol{w}) = -2\epsilon_{M_2} H_2(F(\boldsymbol{p}))A\boldsymbol{v} \times A\boldsymbol{w}.$$

The same remarks done for the Gaussian curvature simplify the second expression above to

$$d(\boldsymbol{N}_1)_{\boldsymbol{p}}(\boldsymbol{v}) \times \boldsymbol{w} + \boldsymbol{v} \times d(\boldsymbol{N}_1)_{\boldsymbol{p}}(\boldsymbol{w}) = -2\epsilon_{M_2} H_2(F(\boldsymbol{p}))\boldsymbol{v} \times \boldsymbol{w},$$

and M_1 and M_2 having the same causal type again allows us to conclude, by comparing, that $H_1(\boldsymbol{p}) = H_2(F(\boldsymbol{p}))$. □

The above result raises a slightly subtler question: are the mean and Gaussian curvatures invariant under local isometries? We may focus our attention on local isometries instead of necessarily global ones, since the values of the curvatures at a given point are inherently local quantities, with coordinate expressions. This new question is fundamentally distinct than the previous one, since isometries between surfaces need not be restrictions of rigid motions defined on the ambient space. This observation justifies the usual terminology used in geometry: objects invariant under isometries (local or global) are called *intrinsic* to the surface.

We have previously seen, though, that the mean curvature is not intrinsic to the surface (this might have been hinted at by the sign ambiguity in its definition): the plane and the cylinder are locally isometric, but the plane has zero mean curvature, while the cylinder does not.

It remains to understand what happens with the Gaussian curvature. The answer is registered in one of the most beautiful theorems in all of Mathematics, established by Gauss himself in 1827:

Theorem 3.3.14 (Theorema Egregium). *Let $M_1, M_2 \subseteq \mathbb{R}^3_\nu$ be non-degenerate regular surfaces. If $\phi\colon M_1 \to M_2$ is a local isometry, and K_1 and K_2 denote the Gaussian curvatures of M_1 and M_2, respectively, then $K_1(\boldsymbol{p}) = K_2(\phi(\boldsymbol{p}))$, for all $\boldsymbol{p} \in M_1$. In other words,* the Gaussian curvature of a surface is intrinsic to it.

A more geometric interpretation: inhabitants of a surface M are able to determine the Gaussian curvature of M by only measuring angles, distances and ratios in M, without any reference to the "outside world", the ambient space $\mathbb{R}^3_\nu$.

The proof of this theorem will be presented on a more opportune moment ahead, but its idea basically consists of expressing the Gaussian curvature in terms of the First Fundamental Form only, but not the Second. As local isometries preserve the First Fundamental Form of a surface, they will also preserve any object which depends only on it. Namely, if $(U, \boldsymbol{x})$ is a parametrization for a non-degenerate regular surface M, and $\boldsymbol{x}$ is *orthogonal* (that is, it satisfies $F = 0$), it is possible to show that

$$K \circ \boldsymbol{x} = \frac{-1}{\sqrt{|EG|}} \left(\epsilon_u \left(\frac{(\sqrt{|G|})_u}{\sqrt{|E|}} \right)_u + \epsilon_v \left(\frac{(\sqrt{|E|})_v}{\sqrt{|G|}} \right)_v \right).$$

As expected, the formula when $F \neq 0$ is much more complicated and its practical usefulness is questionable.

Gauss' Theorema Egregium is one of the most powerful tools we have to decide when any given surfaces are not isometric. See an example of this idea in Exercise 3.3.9.

Exercises

Exercise† 3.3.1 (Alternative expressions for K and $\boldsymbol{H}$). Let $M \subseteq \mathbb{R}^3_\nu$ be a non-degenerate regular surface.

(a) Show that if $\boldsymbol{p} \in M$ and $\{\boldsymbol{v}, \boldsymbol{w}\}$ is any basis for $T_{\boldsymbol{p}}M$, then *Gauss' equation*

$$K(\boldsymbol{p}) = \frac{\langle \mathrm{II}_{\boldsymbol{p}}(\boldsymbol{v}), \mathrm{II}_{\boldsymbol{p}}(\boldsymbol{w})\rangle - \langle \mathrm{II}_{\boldsymbol{p}}(\boldsymbol{v},\boldsymbol{w}), \mathrm{II}_{\boldsymbol{p}}(\boldsymbol{w},\boldsymbol{v})\rangle}{\langle \boldsymbol{v},\boldsymbol{v}\rangle\langle \boldsymbol{w},\boldsymbol{w}\rangle - \langle \boldsymbol{v},\boldsymbol{w}\rangle\langle \boldsymbol{w},\boldsymbol{v}\rangle}$$

holds.

(b) Show that if $(U, \boldsymbol{x})$ is any parametrization for M, then

$$\boldsymbol{H} \circ \boldsymbol{x} = \frac{1}{2} \sum_{i,j=1}^{2} g^{ij} \mathrm{II}(\boldsymbol{x}_i, \boldsymbol{x}_j),$$

where we identify $u \leftrightarrow 1$ and $v \leftrightarrow 2$.

Exercise 3.3.2. Let $M \subseteq \mathbb{R}^3_\nu$ be a non-degenerate regular surface, $\boldsymbol{N}$ be a Gauss map for M, and $\boldsymbol{p} \in M$ be any point. The *Third Fundamental Form* of M at $\boldsymbol{p}$ is the map $\mathrm{III}_{\boldsymbol{p}} \colon T_{\boldsymbol{p}}M \times T_{\boldsymbol{p}}M \to \mathbb{R}$ given by

$$\mathrm{III}_{\boldsymbol{p}}(\boldsymbol{v}, \boldsymbol{w}) \doteq \langle \mathrm{d}\boldsymbol{N}_{\boldsymbol{p}}(\boldsymbol{v}), \mathrm{d}\boldsymbol{N}_{\boldsymbol{p}}(\boldsymbol{w})\rangle.$$

Show that the relation

$$\mathrm{III}_{\boldsymbol{p}} - 2\epsilon_M H(\boldsymbol{p})\widetilde{\mathrm{II}}_{\boldsymbol{p}} + \epsilon_M K(\boldsymbol{p})\mathrm{I}_{\boldsymbol{p}} = 0$$

holds, for all $\boldsymbol{p} \in M$. Thus III gives no new geometric information about M.

Hint. Cayley-Hamilton.

Exercise 3.3.3. Consider the Monge parametrization $x\colon \mathbb{R}^2 \to x(\mathbb{R}^2) \subseteq \mathbb{R}^3_\nu$ given by $x(u,v) = (u,v,uv)$.

(a) In $\mathbb{R}^3$, show that $K(x(u,v)) < 0$ for all $(u,v) \in \mathbb{R}^2$, that $K(x(u,v))$ depends only on the distance between $x(u,v)$ and the z-axis, and that $K(x(u,v)) \to 0$ when such distance goes to $+\infty$.

(b) In $\mathbb{L}^3$, show that $K(x(u,v)) < 0$ for all $(u,v) \in \mathbb{R}^2$ wherever x is spacelike, and also that $K(x(u,v)) > 0$ for all $(u,v) \in \mathbb{R}^2$ wherever x is timelike. In this latter case, $K(x(u,v))$ also depends only on the Euclidean distance between $x(u,v)$ and the z-axis, and $K(x(u,v)) \to 0$ when such distance goes to $+\infty$.

Exercise 3.3.4 (Graphs)**.** Let $f\colon U \to \mathbb{R}$ be a smooth function. Assuming that its graph is non-degenerate, and considering the usual Monge parametrization x, show that its curvatures are given by

$$K_E \circ x = \frac{f_{uu}f_{vv} - f_{uv}^2}{(1+f_u^2+f_v^2)^2} \quad \text{and} \quad H_E \circ x = \frac{f_{uu}(1+f_v^2) - 2f_uf_vf_{uv} + f_{vv}(1+f_u^2)}{2\,(1+f_u^2+f_v^2)^{3/2}}$$

in $\mathbb{R}^3$, and:

$$K_L \circ x = \frac{f_{uv}^2 - f_{uu}f_{vv}}{(-1+f_u^2+f_v^2)^2} \quad \text{and} \quad H_L \circ x = \frac{f_{uu}(-1+f_v^2) - 2f_uf_vf_{uv} + f_{vv}(-1+f_u^2)}{2\,|-1+f_u^2+f_v^2|^{3/2}}$$

in $\mathbb{L}^3$.

Exercise 3.3.5 (Surfaces of Revolution)**.** Let $\alpha\colon I \to \mathbb{R}^3_\nu$ be a smooth, regular, injective and non-degenerate curve of the form $\alpha(u) = (f(u), 0, g(u))$ for certain functions f and g, with $f(u) > 0$ for all $u \in I$. Assuming that α is not lightlike, consider the usual revolution parametrization $x\colon I \times]0,2\pi[\to x(I \times]0,2\pi[) \subseteq \mathbb{R}^3_\nu$ given by $x(u,v) = (f(u)\cos v, f(u)\sin v, g(u))$. Compute the coefficients of the Second Fundamental Form of x and show that

$$K_E \circ x = \frac{g'}{f}\frac{(-f''g' + f'g'')}{\langle \alpha', \alpha' \rangle_E^2} \quad \text{and} \quad H_E \circ x = \frac{g'(\langle \alpha', \alpha' \rangle_E - ff'') + ff'g''}{2f\,\langle \alpha', \alpha' \rangle_E^{3/2}}$$

in $\mathbb{R}^3$ and

$$K_L \circ x = \frac{g'}{f}\frac{(f''g' - f'g'')}{\langle \alpha', \alpha' \rangle_L^2} \quad \text{and} \quad H_L \circ x = \frac{g'(-\langle \alpha', \alpha' \rangle_L + ff'') - ff'g''}{2f\,|\langle \alpha', \alpha' \rangle_L|^{3/2}}$$

in $\mathbb{L}^3$. Note that if α has unit speed, we have that:

$$K_E \circ x = -\frac{f''}{f} \quad \text{and} \quad K_L \circ x = -\epsilon_\alpha \frac{f''}{f},$$

only.

Exercise 3.3.6 (Surfaces of hyperbolic revolutions – II)**.** Consider again a curve smooth, regular and injective curve $\alpha\colon I \to \mathbb{L}^3$ of the form $\alpha(u) = (f(u), 0, g(u))$, with $g(u) > 0$, and the associated surface of hyperbolic revolution around the x-axis, $x\colon I \times \mathbb{R} \to \mathbb{L}^3$ given by:

$$x(u,v) = (f(u), g(u)\sinh v, g(u)\cosh v),$$

as in Exercise 3.2.9 (p. 166), where we had asked you to show that the First Fundamental

Form for this parametrization is given by $ds^2 = \langle \boldsymbol{\alpha}'(u), \boldsymbol{\alpha}'(u)\rangle_L du^2 + g(u)^2 dv^2$. Suppose, in addition, that $\boldsymbol{\alpha}$ is not lightlike. Compute the coefficients of the Minkowski Second Fundamental Form of $\boldsymbol{x}$ and show that

$$K \circ \boldsymbol{x} = \frac{f'}{g}\frac{(f''g' - f'g'')}{\langle \boldsymbol{\alpha}', \boldsymbol{\alpha}'\rangle_L^2} \quad \text{and} \quad H \circ \boldsymbol{x} = \frac{-f'(\langle \boldsymbol{\alpha}', \boldsymbol{\alpha}'\rangle_L + gg'') + gg'f''}{2g\,|\langle \boldsymbol{\alpha}', \boldsymbol{\alpha}'\rangle_L|^{3/2}}.$$

Note that if $\boldsymbol{\alpha}$ has unit speed, then we simply have $K \circ \boldsymbol{x} = -\epsilon_{\boldsymbol{\alpha}} g''/g$, in a similar fashion to what happened for usual revolution surfaces in $\mathbb{R}^3$.

Exercise 3.3.7 (Parallel Surfaces). Let $\boldsymbol{x}\colon U \to \mathbb{R}^3_\nu$ be an injective regular parametrized surface. Define $\boldsymbol{y}\colon U \to \mathbb{R}^3_\nu$ by

$$\boldsymbol{y}(u,v) \doteq \boldsymbol{x}(u,v) + a\boldsymbol{N}(\boldsymbol{x}(u,v)),$$

where $a \in \mathbb{R}$ is fixed. For a small enough, $\boldsymbol{y}$ is also regular and has the same causal type as $\boldsymbol{x}$ (by continuity).

(a) Show that

$$\boldsymbol{y}_u \times \boldsymbol{y}_v = (1 - 2\epsilon_M aH + \epsilon_M a^2 K)\boldsymbol{x}_u \times \boldsymbol{x}_v,$$

where $H \equiv H \circ \boldsymbol{x}$ and $K \equiv K \circ \boldsymbol{x}$ stand for the mean and Gaussian curvatures of $\boldsymbol{x}$.

(b) Show that the mean and Gaussian curvatures of $\boldsymbol{y}$, $H_a \equiv H_a \circ \boldsymbol{y}$ and $K_a \equiv K_a \circ \boldsymbol{y}$, are respectively given by

$$H_a = \frac{H - aK}{1 - 2a\epsilon_M H + a^2\epsilon_M K} \quad \text{and} \quad K_a = \frac{K}{1 - 2a\epsilon_M H + a^2\epsilon_M K}.$$

Hint. If $\boldsymbol{N}_1 = \boldsymbol{N}$ is a Gauss map for $\boldsymbol{x}$ and $\boldsymbol{N}_2$ is a Gauss map for $\boldsymbol{y}$, it follows from the previous item that $d\boldsymbol{N}_1(\boldsymbol{x}_u) = d\boldsymbol{N}_2(\boldsymbol{y}_u)$, and similarly for derivatives with respect to v. Why?

(c) Show that if the mean curvature of $\boldsymbol{x}$ is a constant c and $\boldsymbol{y}$ is $a = \epsilon_M/(2c)$ far from $\boldsymbol{x}$, then the Gaussian curvature of $\boldsymbol{y}$ is constant and equals $4\epsilon_M c^2$.

Exercise 3.3.8. Let $M \subseteq \mathbb{R}^3_\nu$ be a non-degenerate regular surface, $\boldsymbol{N}$ be a Gauss map for M, and $\lambda > 0$. Consider the homothetic image $\lambda M = \{\lambda \boldsymbol{p} \in \mathbb{R}^3_\nu \mid \boldsymbol{p} \in M\}$, which we know to be a regular surface. Show that λM is also non-degenerate, and that its mean and Gaussian curvatures, H_λ and K_λ, are given by

$$H_\lambda(\lambda \boldsymbol{p}) = \frac{H(\boldsymbol{p})}{\lambda} \quad \text{and} \quad K_\lambda(\lambda \boldsymbol{p}) = \frac{K(\boldsymbol{p})}{\lambda^2},$$

where H and K denote the mean and Gaussian curvatures of M.

Exercise 3.3.9. Give curves $\boldsymbol{\alpha}, \widetilde{\boldsymbol{\alpha}}\colon I \to \mathbb{R}^3$ whose images are contained in the plane $y = 0$, which are congruent under an element of $\mathrm{E}_\nu(2,\mathbb{R})$ (acting only on the place $y = 0$), such that their associated surfaces of revolution are not isometric.

Hint. Look for surfaces of revolution whose Gaussian curvatures have opposite signs — Theorema Egregium ensures that this will be a legitimate counter-example.

3.4 THE DIAGONALIZATION PROBLEM

We have previously seen that if $M \subseteq \mathbb{R}^3_\nu$ is a non-degenerate regular surface which is spacelike, and $\boldsymbol{N}$ is a Gauss map for M, it follows from the Spectral Theorem that the Weingarten map $-\mathrm{d}\boldsymbol{N}_p$ is always diagonalizable. Our goal now is to give conditions ensuring the possibility of diagonalization in the general case. We start recording the:

Proposition 3.4.1. *Let $M \subseteq \mathbb{R}^3_\nu$ be a non-degenerate regular surface with diagonalizable Weingarten map. Then it holds that*

$$H(\boldsymbol{p}) = \epsilon_M \frac{\kappa_1(\boldsymbol{p}) + \kappa_2(\boldsymbol{p})}{2} \quad \text{and} \quad K(\boldsymbol{p}) = \epsilon_M \kappa_1(\boldsymbol{p})\kappa_2(\boldsymbol{p}).$$

Remark. In $\mathbb{R}^3$ it is usual to *define* the mean and Gaussian curvatures by the above expressions. The expression for H also justifies the name "mean curvature".

A particular class of points for which the Weingarten map is diagonalizable are the so-called umbilic points:

Definition 3.4.2. Let $M \subseteq \mathbb{R}^3_\nu$ be a non-degenerate regular surface, and $\boldsymbol{p} \in M$. We'll say that $\boldsymbol{p}$ is *umbilic* if there is $\lambda(\boldsymbol{p}) \in \mathbb{R}$ such that

$$\widetilde{\mathrm{II}}_{\boldsymbol{p}}(\boldsymbol{v}, \boldsymbol{w}) = \lambda(\boldsymbol{p})\langle \boldsymbol{v}, \boldsymbol{w} \rangle,$$

for all $\boldsymbol{v}, \boldsymbol{w} \in T_{\boldsymbol{p}}M$. We'll also say that M is *totally umbilic* if all its points are umbilic.

Informally, a point is umbilic if the fundamental forms of M at said point are "linearly dependent".

Proposition 3.4.3. *Let $M \subseteq \mathbb{R}^3_\nu$ be a non-degenerate regular surface, and $\boldsymbol{p} \in M$ be an umbilic point. Then the Weingarten map at $\boldsymbol{p}$ is a multiple of the identity map.*

Proof: Just take $\boldsymbol{v}, \boldsymbol{w} \in T_{\boldsymbol{p}}M$ and compute

$$\langle \lambda(\boldsymbol{p})\boldsymbol{v}, \boldsymbol{w} \rangle = \widetilde{\mathrm{II}}_{\boldsymbol{p}}(\boldsymbol{v}, \boldsymbol{w}) = \langle -\mathrm{d}\boldsymbol{N}_{\boldsymbol{p}}(\boldsymbol{v}), \boldsymbol{w} \rangle.$$

The result follows from non-degeneracy of $\langle \cdot, \cdot \rangle$ restricted to $T_{\boldsymbol{p}}M$. □

The computations done in previous examples show that non-degenerate planes and the "spheres" are totally umbilic surfaces (in their respective ambient spaces). The next result states that this short list of examples is complete:

Theorem 3.4.4 (Characterization of totally umbilic surfaces)**.** *Let $M \subseteq \mathbb{R}^3_\nu$ be a connected and totally umbilic non-degenerate regular surface. Then M is contained in a plane of $\mathbb{R}^3_\nu$, or there are $\boldsymbol{c} \in \mathbb{R}^3_\nu$ and $r > 0$ such that:*

(i) if $M \subseteq \mathbb{R}^3$, then $M \subseteq \mathbb{S}^2(\boldsymbol{c}, r)$;

(ii) if $M \subseteq \mathbb{L}^3$ is spacelike, then $M \subseteq \mathbb{H}^2(\boldsymbol{c}, r)$ or $M \subseteq \mathbb{H}^2_-(\boldsymbol{c}, r)$;

(iii) if $M \subseteq \mathbb{L}^3$ is timelike, then $M \subseteq \mathbb{S}^2_1(\boldsymbol{c}, r)$.

Remark. In item *(ii)* above, what decides between $\mathbb{H}^2(\boldsymbol{c}, r)$ and $\mathbb{H}^2_-(\boldsymbol{c}, r)$ is the causal type of $\boldsymbol{p} - \boldsymbol{c}$ for one (and hence all) $\boldsymbol{p} \in M$, by continuity. The time orientation of the Gauss map is not relevant at all here (why?).

Proof: Since M is totally umbilic, for every $p \in M$ there is $\lambda(p) \in \mathbb{R}$ such that $-\mathrm{d}N_p = \lambda(p)\,\mathrm{id}_{T_pM}$, and thus we have defined a map $\lambda\colon M \to \mathbb{R}$, which is automatically smooth by the previous expression. Consider an arbitrary parametrization (U, x), with U connected. Identifying $N \equiv N \circ x$ and similarly for λ, we have that:

$$\begin{cases} -N_u(u,v) = -\mathrm{d}N_{x(u,v)}(x_u(u,v)) = \lambda(u,v)x_u(u,v) \\ -N_v(u,v) = -\mathrm{d}N_{x(u,v)}(x_v(u,v)) = \lambda(u,v)x_v(u,v), \end{cases}$$

for all $(u,v) \in U$. Differentiating the first equality with respect to v and the second with respect to u, we obtain:

$$\begin{cases} -N_{uv}(u,v) = \lambda_v(u,v)x_u(u,v) + \lambda(u,v)x_{uv}(u,v) \\ -N_{vu}(u,v) = \lambda_u(u,v)x_v(u,v) + \lambda(u,v)x_{vu}(u,v). \end{cases}$$

Since second order partial derivatives commute, and $x_u(u,v)$ and $x_v(u,v)$ are linearly independent, it follows that:

$$\lambda_v(u,v)x_u(u,v) - \lambda_u(u,v)x_v(u,v) = 0 \implies \lambda_u \equiv \lambda_v \equiv 0 \text{ em } U,$$

and by connectedness of U, we have that $\lambda \circ x$ is constant. As M is covered by parametrizations with connected domains, it follows that λ itself is locally constant. Now from the connectedness of M, we conclude that the function λ is constant.

- If $\lambda = 0$, $\mathrm{d}N_p = 0$ for all $p \in M$, so that the Gauss map $N(p) = N_0$ is constant, and then M is contained in some plane in $\mathbb{R}^3_\nu$ with N_0 giving the normal direction.
- If $\lambda \neq 0$, define $c\colon M \to \mathbb{R}^3_\nu$ by

  $$c(p) \doteq p + \frac{1}{\lambda}N(p).$$

  Let's see that c is constant. To wit, we have that

  $$\mathrm{d}c_p = \mathrm{id}_{T_pM} + \frac{1}{\lambda}\mathrm{d}N_p = \mathrm{id}_{T_pM} + \frac{1}{\lambda}\big(-\lambda\,\mathrm{id}_{T_pM}\big) = 0$$

  and M is connected. With this, note that

  $$\langle p - c, p - c\rangle = \frac{\epsilon_M}{\lambda^2},$$

  and the conclusion follows.

□

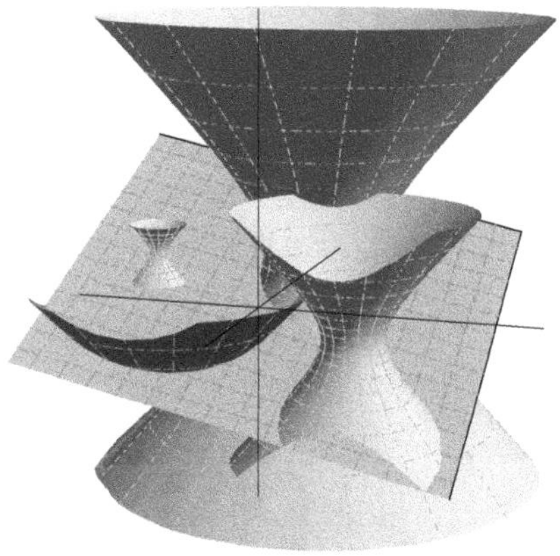

Figure 3.25: The totally umbilic surfaces in $\mathbb{L}^3$.

Back to the discussion of the possibility of diagonalization of the Weingarten map, we have the:

Proposition 3.4.5. *Let $M \subseteq \mathbb{L}^3$ be a non-degenerate regular surface, and $\boldsymbol{p} \in M$ such that the Weingarten map at $\boldsymbol{p}$ is diagonalizable. Then $H(\boldsymbol{p})^2 - \epsilon_M K(\boldsymbol{p}) \geq 0$, with equality holding if and only if $\boldsymbol{p}$ is umbilic.*

Proof: Directly, we have:

$$\begin{aligned}
0 \leq \left(\frac{\kappa_1(\boldsymbol{p}) - \kappa_2(\boldsymbol{p})}{2}\right)^2 &= \frac{\kappa_1(\boldsymbol{p})^2 - 2\kappa_1(\boldsymbol{p})\kappa_2(\boldsymbol{p}) + \kappa_2(\boldsymbol{p})^2}{4} \\
&= \frac{\kappa_1(\boldsymbol{p})^2 + 2\kappa_1(\boldsymbol{p})\kappa_2(\boldsymbol{p}) + \kappa_2(\boldsymbol{p})^2}{4} - \kappa_1(\boldsymbol{p})\kappa_2(\boldsymbol{p}) \\
&= \left(\frac{\kappa_1(\boldsymbol{p}) + \kappa_2(\boldsymbol{p})}{2}\right)^2 - \kappa_1(\boldsymbol{p})\kappa_2(\boldsymbol{p}) \\
&= (\epsilon_M H(\boldsymbol{p}))^2 - \epsilon_M K(\boldsymbol{p}) = H(\boldsymbol{p})^2 - \epsilon_M K(\boldsymbol{p}).
\end{aligned}$$

Equality holds if and only if $\kappa_1(\boldsymbol{p}) = \kappa_2(\boldsymbol{p})$, that is to say, if $\boldsymbol{p}$ is umbilic. □

The above proposition gives a necessary (but not sufficient) condition for $-\mathrm{d}\boldsymbol{N}_{\boldsymbol{p}}$ to be diagonalizable, and emphasizes the importance of the quantity $H(\boldsymbol{p})^2 - \epsilon_M K(\boldsymbol{p})$, which will be explored in the proof of the:

Theorem 3.4.6 (Diagonalization in $\mathbb{L}^3$). *Let $M \subseteq \mathbb{L}^3$ be a non-degenerate regular surface, $\boldsymbol{N}$ be a Gauss map for M, and $\boldsymbol{p} \in M$. Then:*

(i) if $H(\boldsymbol{p})^2 - \epsilon_M K(\boldsymbol{p}) > 0$, $-\mathrm{d}\boldsymbol{N}_{\boldsymbol{p}}$ is diagonalizable;

(ii) if $H(\boldsymbol{p})^2 - \epsilon_M K(\boldsymbol{p}) < 0$, $-\mathrm{d}\boldsymbol{N}_{\boldsymbol{p}}$ is not diagonalizable;

(iii) if $H(\boldsymbol{p})^2 - \epsilon_M K(\boldsymbol{p}) = 0$ and M is spacelike, then $\boldsymbol{p}$ is umbilic, and hence $-\mathrm{d}\boldsymbol{N}_{\boldsymbol{p}}$ is diagonalizable.

Remark. With the above notation, if $H(\boldsymbol{p})^2 - \epsilon_M K(\boldsymbol{p}) = 0$ and M is timelike, the criterion is inconclusive and the Weingarten map may or may not be diagonalizable. In the following, we will see examples illustrating both situations.

Proof: Consider the characteristic polynomial $c(t)$ of $-\mathrm{d}\boldsymbol{N}_{\boldsymbol{p}}$, given by

$$c(t) = t^2 - \mathrm{tr}(-\mathrm{d}\boldsymbol{N}_{\boldsymbol{p}})\, t + \det(-\mathrm{d}\boldsymbol{N}_{\boldsymbol{p}}) = t^2 - 2\epsilon_M H(\boldsymbol{p})t + \epsilon_M K(\boldsymbol{p}),$$

whose discriminant is:

$$(-2\epsilon_M H(\boldsymbol{p}))^2 - 4(\epsilon_M K(\boldsymbol{p})) = 4(H(\boldsymbol{p})^2 - \epsilon_M K(\boldsymbol{p})).$$

- If $H(\boldsymbol{p})^2 - \epsilon_M K(\boldsymbol{p}) > 0$, then $c(t)$ has two distinct roots, which are the eigenvalues of $-\mathrm{d}\boldsymbol{N}_{\boldsymbol{p}}$, which then admits two linearly independent eigenvectors (hence diagonalizable).

- If $H(\boldsymbol{p})^2 - \epsilon_M K(\boldsymbol{p}) < 0$, $c(t)$ does not have any real roots. Thus $-\mathrm{d}\boldsymbol{N}_{\boldsymbol{p}}$ has no real eigenvalues, and hence it is not diagonalizable.

- Now assume that $H(p)^2 - \epsilon_M K(p) = 0$ and that M is spacelike, that is, that $K(p) = -H(p)^2$. From the expression given for the discriminant of $c(t)$, it follows that $-H(p)$ is an eigenvalue of $-\mathrm{d}N_p$. So, there is a unit (spacelike) vector $u_1 \in T_pM$ such that $\mathrm{d}N_p(u_1) = H(p)u_1$. Consider then an orthogonal basis $\mathcal{B} \doteq (u_1, u_2)$ of T_pM. Then:

$$[\mathrm{d}N_p]_{\mathcal{B}} = \begin{pmatrix} H(p) & a \\ 0 & b \end{pmatrix}, \quad \text{where } \mathrm{d}N_p(u_2) = au_1 + bu_2.$$

It suffices to check that $a = 0$ and $b = H(p)$ to conclude the proof. Applying $\langle \cdot, u_1 \rangle_L$, we have:

$$a = \langle \mathrm{d}N_p(u_2), u_1 \rangle_L = \langle u_2, \mathrm{d}N_p(u_1) \rangle_L = \langle u_2, H(p)u_1 \rangle_L = H(p)\langle u_2, u_1 \rangle_L = 0.$$

On the other hand:

$$-H(p)^2 = K(p) = -\det(-\mathrm{d}N_p) = -\det(\mathrm{d}N_p) = -H(p)b,$$

so that $H(p)b = H(p)^2$. If $H(p) = 0$, then $\mathrm{d}N_p$ is the zero map (hence diagonalizable). If $H(p) \neq 0$, we obtain $b = H(p)$, as wanted. Note that in this case p is umbilic.

□

Observe that in the above proof, we do not have any control over the causal type of the eigenvector u_1 in the last described situation when M is timelike. If u_1 were lightlike, we could not consider the basis $\mathcal{B}$ to proceed with the argument. With this in mind, we extend the above result, with the due adaptations:

Corollary 3.4.7. *Let $M \subseteq \mathbb{L}^3$ be a timelike regular surface and $p \in M$ be a point with $H(p)^2 - K(p) = 0$. If $-\mathrm{d}N_p$ has no lightlike eigenvectors, then it is diagonalizable and p is umbilic, with both principal curvatures equal to $-H(p)$.*

Now, let's see the promised examples illustrating the situation where $H(p)^2 = K(p)$ and the Weingarten map may or may not be diagonalizable.

Example 3.4.8.

(1) We have previously seen that for the de Sitter space $\mathbb{S}^2_1$, we had $-\mathrm{d}N_p = -\mathrm{id}_{T_p(\mathbb{S}^2_1)}$, hence diagonalizable, with $K = 1$ and $H = -1$, so that $H^2 - K = 0$.

(2) Consider a lightlike curve $\alpha\colon I \to \mathbb{L}^3$, with arc-photon parametrization. Recall that $T_\alpha(\phi) = \alpha'(\phi)$, $N_\alpha(\phi) = \alpha''(\phi)$ and that if $(T_\alpha(\phi), N_\alpha(\phi))$ is positive, then $B_\alpha(\phi)$ is the unique lightlike vector orthogonal to $N_\alpha(\phi)$ with $\langle T_\alpha(\phi), B_\alpha(\phi) \rangle_L = -1$, thus making the basis $(T_\alpha(\phi), N_\alpha(\phi), B_\alpha(\phi))$ positive as well. Moreover, we have Cartan's equations

$$\begin{pmatrix} T'_\alpha(\phi) \\ N'_\alpha(\phi) \\ B'_\alpha(\phi) \end{pmatrix} = \begin{pmatrix} 0 & 1 & 0 \\ \tau_\alpha(\phi) & 0 & 1 \\ 0 & \tau_\alpha(\phi) & 0 \end{pmatrix} \begin{pmatrix} T_\alpha(\phi) \\ N_\alpha(\phi) \\ B_\alpha(\phi) \end{pmatrix},$$

where $\tau_\alpha(\phi)$ is the pseudo-torsion of α.

Define the B-*scroll* associated to α, $x\colon I \times \mathbb{R} \to \mathbb{L}^3$ by

$$x(\phi, t) \doteq \alpha(\phi) + tB_\alpha(\phi).$$

Restricting the domain of $\boldsymbol{x}$ enough, we may assume that its image is a regular surface. Let, for each ϕ:

$$D(\phi) \doteq \det\left(\boldsymbol{T}_\alpha(\phi), \boldsymbol{N}_\alpha(\phi), \boldsymbol{B}_\alpha(\phi)\right) > 0.$$

Computing the derivatives

$$\boldsymbol{x}_\phi(\phi,t) = \boldsymbol{T}_\alpha(\phi) + t\tau_\alpha(\phi)\boldsymbol{N}_\alpha(\phi) \quad \text{and} \quad \boldsymbol{x}_t(\phi,t) = \boldsymbol{B}_\alpha(\phi),$$

we immediately have that

$$(g_{ij}(\phi,t))_{1\leq i,j\leq 2} = \begin{pmatrix} t^2\tau_\alpha(\phi)^2 & -1 \\ -1 & 0 \end{pmatrix},$$

whence $\boldsymbol{x}$ is timelike. Furthermore, note that $|\det((g_{ij}(\phi,t))_{1\leq i,j\leq 2})| = 1$, and then we directly obtain

$$\boldsymbol{N}(\boldsymbol{x}(\phi,t)) = \boldsymbol{T}_\alpha(\phi) \times_L \boldsymbol{B}_\alpha(\phi) + t\tau_\alpha(\phi)\boldsymbol{N}_\alpha(\phi) \times_L \boldsymbol{B}_\alpha(\phi).$$

Computing the second order derivatives

$$\begin{aligned}
\boldsymbol{x}_{\phi\phi}(\phi,t) &= t\tau_\alpha(\phi)^2\boldsymbol{T}_\alpha(\phi) + (1 + t\tau_\alpha'(\phi))\boldsymbol{N}_\alpha(\phi) + t\tau_\alpha(\phi)\boldsymbol{B}_\alpha(\phi),\\
\boldsymbol{x}_{\phi t}(\phi,t) &= \tau_\alpha(\phi)\boldsymbol{N}_\alpha(\phi) \quad \text{and}\\
\boldsymbol{x}_{tt}(\phi,t) &= \boldsymbol{0},
\end{aligned}$$

we obtain the Second Fundamental Form

$$(h_{ij}(\phi,t))_{1\leq i,j\leq 2} = \begin{pmatrix} (-1 - t\tau_\alpha'(\phi) + t^2\tau_\alpha(\phi)^3)D(\phi) & -\tau_\alpha(\phi)D(\phi) \\ -\tau_\alpha(\phi)D(\phi) & 0 \end{pmatrix}.$$

Thus, we have

$$K(\boldsymbol{x}(\phi,t)) = \tau_\alpha(\phi)^2 D(\phi)^2 \quad \text{and} \quad H(\boldsymbol{x}(\phi,t)) = \tau_\alpha(\phi)D(\phi).$$

Hence, we know that at each point $\boldsymbol{x}(\phi,t)$, $-\mathrm{d}\boldsymbol{N}_{\boldsymbol{x}(\phi,t)}$ has only one eigenvalue, namely, $\tau_\alpha(\phi)D(\phi)$. Then, it suffices to see that there are points for which the associated eigenspace has dimension 1, so that the Weingarten maps at such points are not diagonalizable. Using Lemma 3.3.7 (p. 180), we have:

$$\left[-\mathrm{d}\boldsymbol{N}_{\boldsymbol{x}(\phi,t)}\right]_{\mathcal{B}_{\boldsymbol{x}}} = D(\phi)\begin{pmatrix} \tau_\alpha(\phi) & 0 \\ 1 + t\tau_\alpha(\phi) & \tau_\alpha(\phi) \end{pmatrix}.$$

The kernel of the matrix

$$\begin{pmatrix} 0 & 0 \\ 1 + t\tau_\alpha(\phi) & 0 \end{pmatrix}$$

clearly has dimension 1 whenever $1 + t\tau_\alpha(\phi) \neq 0$ (for instance, along the curve $\boldsymbol{\alpha}$ itself, when $t = 0$).

3.4.1 Interpretations for curvatures

At this point we are ready to present some geometric interpretations for the mean and Gaussian curvatures of a non-degenerate regular surface in $\mathbb{R}^3_\nu$. Aiming towards this end, we need a special parametrization explicitly emphasizing the Second Fundamental Form of the surface. The existence of one such parametrization is given in the:

Theorem 3.4.9 (Inertial coordinates). *Let $M \subseteq \mathbb{R}^3_\nu$ be a non-degenerate regular surface, $\boldsymbol{N}$ be a Gauss map for M, $\boldsymbol{p} \in M$, and $(\boldsymbol{w}_1, \boldsymbol{w}_2)$ an orthonormal basis for $T_{\boldsymbol{p}}M$ such that $(\boldsymbol{w}_1, \boldsymbol{w}_2, \boldsymbol{N}(\boldsymbol{p}))$ is a positive basis for $\mathbb{R}^3_\nu$. Then there is a parametrization $(U, \boldsymbol{x})$ around $\boldsymbol{p}$ satisfying the following conditions:*

(i) $(0,0) \in U$ *and* $\boldsymbol{x}(0,0) = \boldsymbol{p}$;

(ii) $g_{ij}(0,0) = \epsilon_{\boldsymbol{w}_i}\delta_{ij}$, *for* $1 \leq i,j \leq 2$;

(iii) $(\partial g_{ij}/\partial u^k)(0,0) = 0$, *for* $1 \leq i,j,k \leq 2$;

(iv) up to second order:

$$\boldsymbol{x}(u^1,u^2) - \boldsymbol{p} = u^1\boldsymbol{w}_1 + u^2\boldsymbol{w}_2 + \frac{\epsilon_M}{2}\sum_{i,j=1}^{2} h_{ij}(0,0)u^iu^j\,\boldsymbol{N}(\boldsymbol{p}) + \boldsymbol{R}(u^1,u^2),$$

where $\boldsymbol{R}(u^1,u^2)/\|(u^1,u^2)\|_E^2 \to \boldsymbol{0}$ *if* $(u^1,u^2) \to (0,0)$.

We will say that a parametrization satisfying these four properties is inertial.

Remark. Condition *(iv)* above is equivalent to

$$\boldsymbol{x}(u^1,u^2) - \boldsymbol{p} = u^1\boldsymbol{w}_1 + u^2\boldsymbol{w}_2 + \frac{1}{2}\sum_{i,j=1}^{2} u^iu^j\mathrm{I\!I}_{\boldsymbol{p}}(\boldsymbol{w}_i,\boldsymbol{w}_j) + \boldsymbol{R}(u^1,u^2),$$

with the same $\boldsymbol{R}(u^1,u^2)$ of the original statement.

Proof: The strategy is to start with an arbitrary parametrization around $\boldsymbol{p}$ and perform successive changes of coordinates until the end result satisfies all the needed requirements. That said, take an initial parametrization $(U_1, \overline{\boldsymbol{x}})$ around $\boldsymbol{p}$. By Exercise 3.1 (p. 151), we may assume from the start that $(0,0) \in U_1$ and that $\overline{\boldsymbol{x}}(0,0) = \boldsymbol{p}$.

Now, let $A\colon \mathbb{R}^2 \to \mathbb{R}^2$ be the linear isomorphism defined by $A\boldsymbol{e}_i = \boldsymbol{v}_i$, where the vectors $\boldsymbol{v}_1, \boldsymbol{v}_2 \in \mathbb{R}^2$ are such that $D\overline{\boldsymbol{x}}(0,0)(\boldsymbol{v}_i) = \boldsymbol{w}_i$, for $1 \leq i \leq 2$. Let $U_2 \doteq A^{-1}(U_1)$ and define $\widetilde{\boldsymbol{x}} \doteq \overline{\boldsymbol{x}} \circ A\colon U_2 \to \widetilde{\boldsymbol{x}}(U_2) = \overline{\boldsymbol{x}}(U_1) \subseteq M$. Clearly we still have $\widetilde{\boldsymbol{x}}(0,0) = \boldsymbol{p}$, but now we have in addition that

$$\begin{aligned}\frac{\partial \widetilde{\boldsymbol{x}}}{\partial u^i}(0,0) &= D\widetilde{\boldsymbol{x}}(0,0)(\boldsymbol{e}_i) = D\overline{\boldsymbol{x}}(A(0,0)) \circ DA(0,0)(\boldsymbol{e}_i)\\ &= D\overline{\boldsymbol{x}}(0,0)(A\boldsymbol{e}_i) = D\overline{\boldsymbol{x}}(0,0)(\boldsymbol{v}_i) = \boldsymbol{w}_i.\end{aligned}$$

We need one last reparametrization. Consider the second order Taylor expansion, centered at $(0,0)$:

$$\widetilde{\boldsymbol{x}}(u^1,u^2) - \boldsymbol{p} = u^1\boldsymbol{w}_1 + u^2\boldsymbol{w}_2 + \frac{1}{2}\sum_{i,j=1}^{2} \frac{\partial^2 \widetilde{\boldsymbol{x}}}{\partial u^i \partial u^j}(0,0)u^iu^j + \widetilde{\boldsymbol{R}}(u^1,u^2),$$

where $\widetilde{\boldsymbol{R}}(u^1,u^2)/\|(u^1,u^2)\|_E^2 \to 0$ if $(u^1,u^2) \to (0,0)$. But, applying orthonormal expansion for these second order partial derivatives, we obtain:

$$\begin{aligned}\widetilde{\boldsymbol{x}}(u^1,u^2) - \boldsymbol{p} = u^1\boldsymbol{w}_1 + u^2\boldsymbol{w}_2 + \frac{\epsilon_M}{2}\sum_{i,j=1}^{2} u^iu^j\left\langle \frac{\partial^2 \widetilde{\boldsymbol{x}}}{\partial u^i \partial u^j}(0,0), \boldsymbol{N}(\boldsymbol{p})\right\rangle \boldsymbol{N}(\boldsymbol{p}) +\\ + \frac{1}{2}\sum_{i,j,k=1}^{2} \epsilon_{\boldsymbol{w}_k}u^iu^j\left\langle \frac{\partial^2 \widetilde{\boldsymbol{x}}}{\partial u^i \partial u^j}(0,0), \boldsymbol{w}_k\right\rangle \boldsymbol{w}_k + \widetilde{\boldsymbol{R}}(u^1,u^2).\end{aligned}$$

Our goal, then, is to perform a change of coordinates to eliminate the triple summation above. For such end, consider $\psi\colon U_2 \to \mathbb{R}^2$ given (in matrix notation, for convenience) by

$$\psi(u^1,u^2) \doteq \begin{pmatrix} u^1 + \dfrac{\epsilon_{w_1}}{2}\displaystyle\sum_{i,j=1}^{2} u^i u^j \left\langle \dfrac{\partial^2 \widetilde{x}}{\partial u^i \partial u^j}(0,0), w_1 \right\rangle \\ u^2 + \dfrac{\epsilon_{w_2}}{2}\displaystyle\sum_{i,j=1}^{2} u^i u^j \left\langle \dfrac{\partial^2 \widetilde{x}}{\partial u^i \partial u^j}(0,0), w_2 \right\rangle \end{pmatrix}.$$

To simplify notation in what follows, abbreviate

$$\widetilde{u}^k \doteq u^k + \frac{\epsilon_{w_k}}{2}\sum_{i,j=1}^{2} u^i u^j \left\langle \frac{\partial^2 \widetilde{x}}{\partial u^i \partial u^j}(0,0), w_k \right\rangle.$$

Clearly we have $\psi(0,0) = (0,0)$, and also $D\psi(0,0) = \mathrm{Id}_2$, whence the Inverse Function Theorem provides open subsets $U_2' \subseteq U_2$ and U of $\mathbb{R}^2$ around $(0,0)$ such that $\psi\colon U_2' \to U$ is a diffeomorphism. Define, finally, $x \doteq \widetilde{x}\big|_{U_2'} \circ \psi^{-1}\colon U \to x(U) = \widetilde{x}(U_2') \subseteq M$. Let's verify that x now satisfies all the requirements. To begin with, we have

$$x(0,0) = \widetilde{x}(\psi^{-1}(0,0)) = \widetilde{x}(0,0) = p,$$

which shows condition (i). Moreover:

$$\begin{aligned} \frac{\partial x}{\partial \widetilde{u}^i}(0,0) &= Dx(0,0)(e_i) = D(\widetilde{x} \circ \psi^{-1})(0,0)(e_i) \\ &= D\widetilde{x}(\psi^{-1}(0,0)) \circ D\psi^{-1}(0,0)(e_i) = D\widetilde{x}(0,0) \circ \mathrm{Id}_2(e_i) \\ &= D\widetilde{x}(0,0)(e_i) = w_i, \end{aligned}$$

whence $g_{ij}(0,0) = \epsilon_{w_i}\delta_{ij}$ and we conclude (ii). Factoring out w_1 and w_2 in the Taylor formula for $\widetilde{x}$, and using the definition of the coefficients $\widetilde{h}_{ij}$ of the Second Fundamental Form of $\widetilde{x}$, we have:

$$\begin{aligned} x(\widetilde{u}^1,\widetilde{u}^2) - p &= \widetilde{x}(\psi^{-1}(\widetilde{u}^1,\widetilde{u}^2)) - p \\ &= \widetilde{x}(u^1,u^2) - p \\ &= \sum_{k=1}^{2}\left(u^k + \frac{\epsilon_{w_k}}{2}\sum_{i,j=1}^{2} u^i u^j \left\langle \frac{\partial^2 \widetilde{x}}{\partial u^i \partial u^j}(0,0), w_k \right\rangle\right) w_k + \\ &\quad + \frac{\epsilon_M}{2}\sum_{i,j=1}^{2} u^i u^j \left\langle \frac{\partial^2 \widetilde{x}}{\partial u^i \partial u^j}(0,0), N(p) \right\rangle N(p) + \widetilde{R}(u^1,u^2) \\ &= \widetilde{u}^1 w_1 + \widetilde{u}^2 w_2 + \frac{\epsilon_M}{2}\sum_{i,j=1}^{2} u^i u^j \widetilde{h}_{ij}(0,0)\, N(p) + \widetilde{R}(u^1,u^2). \end{aligned}$$

Since $\widetilde{u}^k = u^k + R_k(u^1,u^2)$ with $R_k(u^1,u^2)/\|(u^1,u^2)\|_E^2 \to 0$ as $(u^1,u^2) \to (0,0)$, we may group all the terms that go to zero fast and conclude that

$$x(\widetilde{u}^1,\widetilde{u}^2) - p = \widetilde{u}^1 w_1 + \widetilde{u}^2 w_2 + \frac{\epsilon_M}{2}\sum_{i,j=1}^{2} \widetilde{u}^i \widetilde{u}^j \widetilde{h}_{ij}(0,0)\, N(p) + R(\widetilde{u}^1,\widetilde{u}^2),$$

where $R(\widetilde{u}^1,\widetilde{u}^2)/\|(\widetilde{u}^1,\widetilde{u}^2)\|_E^2 \to \mathbf{0}$ if $(\widetilde{u}^1,\widetilde{u}^2) \to (0,0)$. Since the partial derivatives up to second order of x and $\widetilde{x}$ agree at $(0,0)$, we have that

$$h_{ij}(0,0) = \widetilde{\mathrm{II}}_p(w_i,w_j) = \widetilde{h}_{ij}(0,0),$$

so that we conclude (iv). To conclude (iii), we differentiate (iv) and observe that in general

$$g_{ij}(\widetilde{u}^1,\widetilde{u}^2) = \epsilon_{w_i}\delta_{ij} + R_{ij}(u^1,u^2)$$

holds, where $R_{ij}(\widetilde{u}^1,\widetilde{u}^2)/\|(\widetilde{u}^1,\widetilde{u}^2)\|_E \to 0$ if $(\widetilde{u}^1,\widetilde{u}^2) \to (0,0)$, whence it follows that all the possible first order partial derivatives of the g_{ij} evaluated at $(0,0)$ vanish. Renaming the parameters $(\widetilde{u}^1,\widetilde{u}^2) \to (u^1,u^2)$, we even obtain the conditions with the same notation as in the original statement. □

This theorem combined with Proposition 3.1.29 (p. 147) says that given a point $\boldsymbol{p}$ in a non-degenerate regular surface $M \subseteq \mathbb{R}^3_\nu$, we may regard M locally as a graph over $T_{\boldsymbol{p}}M$, discarding terms of order 3 and higher, of the quadratic form

$$\frac{\epsilon_M}{2}\sum_{i,j=1}^{2} h_{ij}(0,0)u^i u^j.$$

Then, of course, the behavior of the surface, near $\boldsymbol{p}$, will be controlled by the matrix $(h_{ij}(0,0))_{1\leq i,j\leq 2}$ or, in other words, by the scalar version of the Second Fundamental Form $\widetilde{\mathrm{II}}_{\boldsymbol{p}}$. With this in mind, we just need some last piece of terminology to start discussing geometric interpretations for the Gaussian curvature:

Definition 3.4.10. Let $M \subseteq \mathbb{R}^3_\nu$ be a non-degenerate regular surface and $\boldsymbol{p} \in M$. We'll say that $\boldsymbol{p}$ is:

(i) *elliptic* if $\widetilde{\mathrm{II}}_{\boldsymbol{p}}$ is definite;

(ii) *hyperbolic* if $\widetilde{\mathrm{II}}_{\boldsymbol{p}}$ is indefinite;

(iii) *parabolic* if $\widetilde{\mathrm{II}}_{\boldsymbol{p}} \neq 0$ is degenerate;

(iv) *planar* if $\widetilde{\mathrm{II}}_{\boldsymbol{p}} = 0$.

If $(U,\boldsymbol{x})$ is a parametrization, we see from the coordinate expression for $K\circ \boldsymbol{x}$ in terms of g_{ij} and h_{ij} that $K(\boldsymbol{x}(u,v))$ and $\det\left((h_{ij}(u,v))_{1\leq i,j\leq 2}\right)$ have the same sign in $\mathbb{R}^3$, and opposite signs in $\mathbb{L}^3$. The conclusion we make from this observation, by applying Sylvester's Criterion for $(h_{ij}(u,v))_{1\leq i,j\leq 2}$, is then recorded in:

Proposition 3.4.11. *Let $M \subseteq \mathbb{R}^3_\nu$ be a non-degenerate regular surface and $\boldsymbol{p} \in M$. Then:*

(i) in $\mathbb{R}^3$, if $K(\boldsymbol{p}) > 0$ (resp. < 0), $\boldsymbol{p}$ is elliptic (resp. hyperbolic);

(ii) in $\mathbb{L}^3$, if $K(\boldsymbol{p}) > 0$ (resp. < 0), $\boldsymbol{p}$ is hyperbolic (resp. elliptic);

(iii) if $K(\boldsymbol{p}) = 0$ and $H(\boldsymbol{p}) \neq 0$, $\boldsymbol{p}$ is parabolic;

(iv) if $K(\boldsymbol{p}) = H(\boldsymbol{p}) = 0$, $\boldsymbol{p}$ é planar.

The details of the proof are left for Exercise 3.4.5.

Example 3.4.12.

(1) If $\Pi \subseteq \mathbb{R}^3_\nu$ is a non-degenerate plane, all the points of Π are planar (surprise?).

(2) All the points of $\mathbb{S}^2$ and $\mathbb{H}^2$ are elliptic, while the points of $\mathbb{S}^2_1$ are hyperbolic.

(3) All the points of $\mathbb{S}^1 \times \mathbb{R}$ are parabolic.

(4) All the points of the $\boldsymbol{B}$-scroll over a lightlike curve in $\mathbb{L}^3$ (seen in Example 3.4.8 above, p. 193) are hyperbolic.

(5) All the points in the helicoid seen in Example 3.3.11 (p. 183), where it is non-degenerate, are hyperbolic.

We have the following situations which justify the terminology given in Definition 3.4.10 above:

(I) if $\boldsymbol{p}$ is elliptic, $\widetilde{\mathrm{II}}_{\boldsymbol{p}}$ is positive or negative-definite and then M is approximated by an elliptic paraboloid:

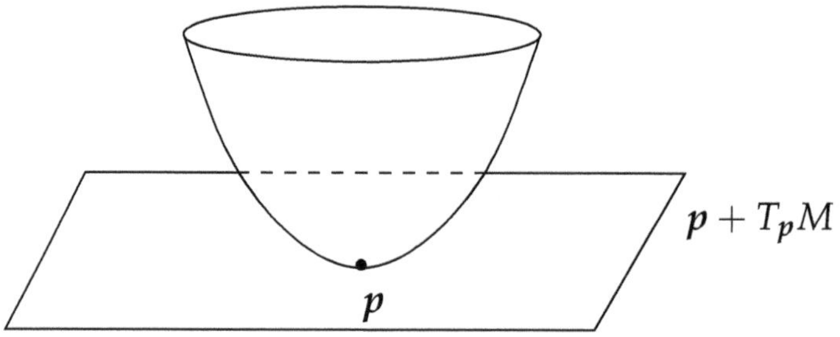

(II) if $\boldsymbol{p}$ is hyperbolic, $\widetilde{\mathrm{II}}_{\boldsymbol{p}}$ is indefinite and then M is approximated by a hyperbolic paraboloid:

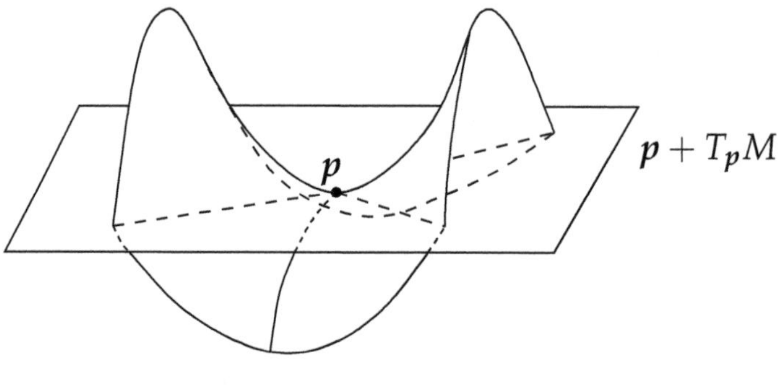

(III) if $\boldsymbol{p}$ is parabolic, $\widetilde{\mathrm{II}}_{\boldsymbol{p}} \neq 0$ is degenerate and then M is approximated by a "parabolic cylinder":

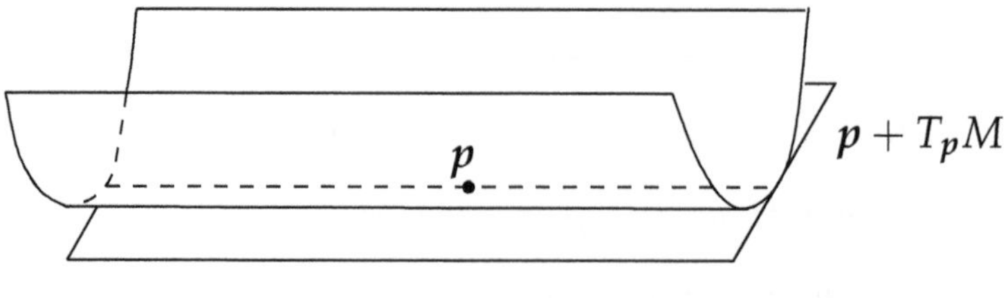

(IV) if $\boldsymbol{p}$ is planar, $\widetilde{\mathrm{II}}_{\boldsymbol{p}} = 0$ and M coincides with $T_{\boldsymbol{p}}M$ up to terms of order 3.

In terms of the sign of the Gaussian curvature only, we also have the following interpretation, which follows from the discussion after the proof of Theorem 3.4.9:

Proposition 3.4.13. *Let $M \subseteq \mathbb{R}^3$ be a regular surface and $\boldsymbol{p} \in M$. Then:*

(i) if $K(\boldsymbol{p}) > 0$, there is an open neighborhood of $\boldsymbol{p}$ in M entirely contained in one of the half-spaces determined by $T_{\boldsymbol{p}}M$;

(ii) if $K(\boldsymbol{p}) < 0$, then every open neighborhood of $\boldsymbol{p}$ in M has points in both half-spaces determined by $T_{\boldsymbol{p}}M$.

The same conclusion holds if $M \subseteq \mathbb{L}^3$ is non-degenerate, reversing the sign of $K(\boldsymbol{p})$.

Remark. A priori, to compute the Gaussian curvature of a non-degenerate regular surface at a certain point, it is not necessary for the Gauss map to be defined in all of the surface, but only on a neighborhood of this point. That is, the orientability assumption made initially could have been dropped for this local analysis. However, it is a consequence of the above proposition that every regular surface M in $\mathbb{R}^3$ (resp. non-degenerate in $\mathbb{L}^3$) with strictly positive (resp. negative) Gaussian curvature in all points is automatically orientable. It suffices to consider, for each $\boldsymbol{p} \in M$, the neighborhood $W_{\boldsymbol{p}}$ of $\boldsymbol{p}$ in M given by the above proposition, and take the unit normal vector $\boldsymbol{N}(\boldsymbol{p})$ pointing towards the half-space determined by $T_{\boldsymbol{p}}M$ which contains $W_{\boldsymbol{p}}$. This defines an orientation for M.

This last proposition also allows us to visually verify that the Gaussian curvatures of $\mathbb{S}^2$ and $\mathbb{S}^2_1$ are positive, while the curvature of $\mathbb{H}^2$ is negative. Unfortunately, this criterion is not decisive for parabolic and planar points (see Exercise 3.4.6).

We may also relate the sign of $K(\boldsymbol{p}) \neq 0$ with $\mathrm{d}\boldsymbol{N}_{\boldsymbol{p}}$ preserving or reversing orientation in $T_{\boldsymbol{p}}M$, simply by recalling that $K(\boldsymbol{p}) = \epsilon_M \det(\mathrm{d}\boldsymbol{N}_p)$, as follows:

- in $\mathbb{R}^3$: $K(\boldsymbol{p}) > 0$ if and only if $\mathrm{d}\boldsymbol{N}_p$ preserves orientation;
- in $\mathbb{L}^3$: if M is timelike, $K(\boldsymbol{p}) > 0$ if and only if $\mathrm{d}\boldsymbol{N}_p$ preserves orientation;
- in $\mathbb{L}^3$: if M is spacelike, $K(\boldsymbol{p}) < 0$ if and only if $\mathrm{d}\boldsymbol{N}_p$ preserves orientation.

To conclude the interpretations for the Gaussian curvature for now, we have the:

Proposition 3.4.14. *Let $M \subseteq \mathbb{R}^3_\nu$ be a non-degenerate regular surface, $\boldsymbol{N}$ a Gauss map for M, and $\boldsymbol{p} \in M$. Then*

$$|K(\boldsymbol{p})| = \lim_{\substack{\mathscr{A}(R)\to 0 \\ \boldsymbol{p}\in R}} \frac{\mathscr{A}(\boldsymbol{N}(R))}{\mathscr{A}(R)}.$$

Remark.

- The above limit may be formally understood by thinking of a classical ϵ-δ definition, as follows: given $\epsilon > 0$, there is $\delta > 0$ such that, if R is a region in M containing $\boldsymbol{p}$ with $\mathscr{A}(R) < \delta$, then $\left|\frac{\mathscr{A}(\boldsymbol{N}(R))}{\mathscr{A}(R)} - |K(\boldsymbol{p})|\right| < \epsilon$.
- In other words, K may be seen as an infinitesimal ratio of areas of images under the Gauss map over the original areas.

Proof: Since $\mathcal{A}(R) \to 0$, we may evaluate the limit along regions all contained inside the image of a fixed parametrization $(U, \boldsymbol{x})$ around $\boldsymbol{p}$. We have that

$$\begin{aligned}
\lim_{\substack{\mathcal{A}(R)\to 0\\ \boldsymbol{p}\in R}} \frac{\mathcal{A}(\boldsymbol{N}(R))}{\mathcal{A}(R)} &= \lim_{\substack{\mathcal{A}(R)\to 0\\ \boldsymbol{p}\in R}} \frac{1}{\mathcal{A}(R)} \int_{(\boldsymbol{N}\circ\boldsymbol{x})^{-1}(\boldsymbol{N}(R))} \|(\boldsymbol{N}\circ\boldsymbol{x})_u(u,v) \times (\boldsymbol{N}\circ\boldsymbol{x})_v(u,v)\| \,\mathrm{d}u\,\mathrm{d}v \\
&\overset{(1)}{=} \lim_{\substack{\mathcal{A}(R)\to 0\\ \boldsymbol{p}\in R}} \frac{1}{\mathcal{A}(R)} \int_{\boldsymbol{x}^{-1}(R)} |K(\boldsymbol{x}(u,v))| \, \|\boldsymbol{x}_u(u,v) \times \boldsymbol{x}_v(u,v)\| \,\mathrm{d}u\,\mathrm{d}v \\
&\overset{(2)}{=} \lim_{\substack{\mathcal{A}(R)\to 0\\ \boldsymbol{p}\in R}} \frac{|K(\boldsymbol{x}(u_R, v_R))|\mathcal{A}(R)}{\mathcal{A}(R)} \\
&\overset{(3)}{=} |K(\boldsymbol{p})|,
\end{aligned}$$

where in (1) we use Proposition 3.3.12 (p. 185), in (2) the Mean Value Theorem for integrals and, in (3), that $(u_R, v_R) \in \boldsymbol{x}^{-1}(R)$, $\mathcal{A}(R) \to 0$ and $\boldsymbol{p} \in R$ together imply that $\boldsymbol{x}(u_R, v_R) \to \boldsymbol{p}$. □

The Gaussian curvature is not the only curvature that may be interpreted in terms of areas, as seen above. One of the most important interpretations for the mean curvature comes from *the first variation of area* for a non-degenerate parametrized regular surface.

Definition 3.4.15. Let $\boldsymbol{x}\colon \overline{U} \to \boldsymbol{x}(\overline{U}) \subseteq \mathbb{R}^3_\nu$ be a smooth mapping such that $\boldsymbol{x}\big|_U$ is a non-degenerate regular parametrized surface. An *admissible variation* of $\boldsymbol{x}$, for t in the interval $]-\epsilon, \epsilon[$, is a mapping $\boldsymbol{x}^t\colon \overline{U} \to \boldsymbol{x}^t(\overline{U}) \subseteq \mathbb{R}^3_\nu$, given by

$$\boldsymbol{x}^t(u,v) \doteq \boldsymbol{x}(u,v) + t\boldsymbol{V}(u,v),$$

where $\boldsymbol{V} : \overline{U} \to \mathbb{R}^3_\nu$ is a smooth map that vanishes on the boundary $\partial U = \overline{U} \setminus \mathring{U}$.

Remark.

- We may assume, in the above definition, that $\epsilon > 0$ is small enough so that each $\boldsymbol{x}^t\big|_U$ is also regular and with the same causal type as $\boldsymbol{x}\big|_U$.
- In the above definition, we ask those mappings to be defined in the closure $\overline{U}$, because we'll need to analyze the curve $\boldsymbol{x}(\partial U)$ in $\mathbb{R}^3_\nu$.

Definition 3.4.16. Let $\boldsymbol{x}^t\colon \overline{U} \to \boldsymbol{x}^t(\overline{U}) \subseteq \mathbb{R}^3_\nu$ be an admissible variation of a smooth mapping $\boldsymbol{x}\colon \overline{U} \to \boldsymbol{x}(\overline{U}) \subseteq \mathbb{R}^3_\nu$. The *area functional* associated to this variation is

$$\mathcal{A}(t) \doteq \int_U \|\boldsymbol{x}^t_u(u,v) \times \boldsymbol{x}^t_v(u,v)\| \,\mathrm{d}u\,\mathrm{d}v.$$

Theorem 3.4.17 (First Variation of Area). *Let $U \subseteq \mathbb{R}^2$ be a bounded open set with regular boundary, $\boldsymbol{x} : \overline{U} \to \boldsymbol{x}(\overline{U}) \subseteq \mathbb{R}^3_\nu$ a smooth map such that $\boldsymbol{x}\big|_U$ is a non-degenerate and injective regular parametrized surface. Then $\boldsymbol{x}\big|_U$ has zero mean curvature if and only if $\mathcal{A}'(0) = 0$, for every admissible variation $\boldsymbol{x}^t$ of $\boldsymbol{x}$.*

Proof: Initially, we compute $\mathcal{A}'(0)$ for an arbitrary variation $\boldsymbol{x}^t$. We will omit the points of evaluation (u, v) to simplify the notation. Also set $\epsilon_M \doteq \epsilon_{\boldsymbol{x}(U)}$. Since $\overline{U}$ is compact and all the relevant functions here are smooth, we may differentiate under the integral sign to obtain

$$\mathcal{A}'(0) = \int_U \frac{\mathrm{d}}{\mathrm{d}t}\bigg|_{t=0} \|\boldsymbol{x}^t_u \times \boldsymbol{x}^t_v\| \,\mathrm{d}u\,\mathrm{d}v.$$

By Proposition 2.1.13 (p. 70), we have

$$\mathscr{A}'(0) = \int_U \frac{\epsilon_M \left\langle (\boldsymbol{x}_u^t \times \boldsymbol{x}_v^t)'\big|_{t=0}, \boldsymbol{x}_u \times \boldsymbol{x}_v \right\rangle}{\|\boldsymbol{x}_u \times \boldsymbol{x}_v\|} \, du \, dv.$$

Directly, we see that

$$(\boldsymbol{x}_u^t \times \boldsymbol{x}_v^t)'\big|_{t=0} = \boldsymbol{V}_u \times \boldsymbol{x}_v + \boldsymbol{x}_u \times \boldsymbol{V}_v,$$

whence

$$\mathscr{A}'(0) = \int_U \epsilon_M \langle \boldsymbol{V}_u \times \boldsymbol{x}_v + \boldsymbol{x}_u \times \boldsymbol{V}_v, \boldsymbol{N} \circ \boldsymbol{x} \rangle \, du \, dv.$$

The strategy now is to apply the Green-Stokes theorem (which is possible since U has regular boundary), considering $P \doteq \langle \boldsymbol{N} \circ \boldsymbol{x}, \boldsymbol{V} \times \boldsymbol{x}_u \rangle$ and $Q \doteq \langle \boldsymbol{N} \circ \boldsymbol{x}, \boldsymbol{V} \times \boldsymbol{x}_v \rangle$. We have

$$\begin{aligned} \frac{\partial Q}{\partial u} - \frac{\partial P}{\partial v} &= \langle \boldsymbol{N} \circ \boldsymbol{x}, \boldsymbol{V}_u \times \boldsymbol{x}_v + \boldsymbol{x}_u \times \boldsymbol{V}_v \rangle - \langle \boldsymbol{V}, (\boldsymbol{N} \circ \boldsymbol{x})_u \times \boldsymbol{x}_v + \boldsymbol{x}_u \times (\boldsymbol{N} \circ \boldsymbol{x})_v \rangle \\ &= \langle \boldsymbol{N} \circ \boldsymbol{x}, \boldsymbol{V}_u \times \boldsymbol{x}_v + \boldsymbol{x}_u \times \boldsymbol{V}_v \rangle + 2\epsilon_M \langle \boldsymbol{V}, (H \circ \boldsymbol{x}) \boldsymbol{x}_u \times \boldsymbol{x}_v \rangle, \end{aligned}$$

by Proposition 3.3.12 (p. 185). So, Stokes' formula

$$\int_U \frac{\partial Q}{\partial u} - \frac{\partial P}{\partial v} \, du \, dv = \oint_{\partial U} P \, du + Q \, dv$$

boils down to

$$\begin{aligned} \epsilon_M \mathscr{A}'(0) + 2\epsilon_M \int_U \langle \boldsymbol{V}, (H \circ \boldsymbol{x}) \boldsymbol{x}_u \times \boldsymbol{x}_v \rangle \, du \, dv &= \\ = \oint_{\partial U} \langle \boldsymbol{N} \circ \boldsymbol{x}, \boldsymbol{V} \times \boldsymbol{x}_u \rangle \, du + \langle \boldsymbol{N} \circ \boldsymbol{x}, \boldsymbol{V} \times \boldsymbol{x}_v \rangle \, dv &= 0, \end{aligned}$$

since $\boldsymbol{V}\big|_{\partial U} = \boldsymbol{0}$. Setting $dA \doteq \|\boldsymbol{x}_u \times \boldsymbol{x}_v\| \, du \, dv$, we finally obtain

$$\mathscr{A}'(0) = -2 \int_U \langle \boldsymbol{V}, \boldsymbol{H} \circ \boldsymbol{x} \rangle \, dA.$$

If $\boldsymbol{H} \circ \boldsymbol{x} = \boldsymbol{0}$, then $\mathscr{A}'(0) = 0$ for every admissible variation. On the other hand, if $\boldsymbol{H} \circ \boldsymbol{x} \neq \boldsymbol{0}$ (and it is necessarily always spacelike or timelike, since $\boldsymbol{x}$ is non-degenerate), it is possible to construct an admissible variation of $\boldsymbol{x}$ for which $\mathscr{A}'(0) \neq 0$: considering $\boldsymbol{V} = \boldsymbol{H} \circ \boldsymbol{x}$ does not yield an admissible variation since $\boldsymbol{H} \circ \boldsymbol{x}$ does not necessarily vanish on ∂U (a priori it might not even be defined there). To remedy this, it suffices to multiply $\boldsymbol{H} \circ \boldsymbol{x}$ by a smooth function whose support is contained in U. □

It is not an easy task to compute an explicit expression for $\mathscr{A}''(0)$ which, in more general settings, may be given in terms of certain geometric objects which are outside the scope of this text (see [3], for instance). From such expression, one may conclude that:

- in $\mathbb{R}^3$, if $H = 0$ then $\boldsymbol{x}$ is a local minimum of the area functional;
- in $\mathbb{L}^3$, if $H = 0$ and $\boldsymbol{x}$ is spacelike, then $\boldsymbol{x}$ is a local maximum of the area functional;
- in $\mathbb{L}^3$, if $H = 0$ and $\boldsymbol{x}$ is timelike, then $\boldsymbol{x}$ is a local minimum of the area functional.

This motivates the:

Definition 3.4.18. Let $M \subseteq \mathbb{R}^3_\nu$ be a non-degenerate regular surface. If its mean curvature vanishes, we say that the surface is *critical*. In particular, M is *minimal* if $M \subseteq \mathbb{R}^3$ or if it is timelike, and *maximal* if $M \subseteq \mathbb{L}^3$ is spacelike.

The Gaussian and mean curvatures impose restrictions on the topology of the surface. One example of this is given in the:

Proposition 3.4.19. *Let $M \subseteq \mathbb{R}^3$ be a compact regular surface. Then there is a point $\boldsymbol{p}_0 \in M$ such that $K(\boldsymbol{p}_0) > 0$.*

Proof: Consider the (smooth) function $f\colon M \to \mathbb{R}$ given by $f(\boldsymbol{p}) = \langle \boldsymbol{p}, \boldsymbol{p} \rangle$. Since M is compact, f has a global maximum at some point $\boldsymbol{p}_0 \in M$. If $r \doteq \|\boldsymbol{p}_0\| > 0$, note that $\mathrm{d}f_{\boldsymbol{p}_0} = 2\langle \boldsymbol{p}_0, \cdot \rangle = 0$ tells us that $T_{\boldsymbol{p}_0}M = T_{\boldsymbol{p}_0}\mathbb{S}^2(r)$. Moreover, if $\boldsymbol{v} \in T_{\boldsymbol{p}_0}M$ is any tangent vector realized by a curve $\boldsymbol{\alpha}\colon \,]-\epsilon, \epsilon[\, \to M$, and $\boldsymbol{N}(\boldsymbol{p}_0)$ denotes the unit normal to M (and also to $\mathbb{S}^2(r)$) at $\boldsymbol{p}_0$, we have that

$$\left.\frac{\mathrm{d}^2}{\mathrm{d}t^2}\right|_{t=0} f(\boldsymbol{\alpha}(t)) \leq 0 \implies 2\left(\langle \boldsymbol{\alpha}''(0), \boldsymbol{p}_0 \rangle + \langle \boldsymbol{v}, \boldsymbol{v} \rangle\right) \leq 0,$$

which may be reorganized as:

$$\langle \boldsymbol{\alpha}''(0), r\boldsymbol{N}(\boldsymbol{p}_0) \rangle \leq -\langle \boldsymbol{v}, \boldsymbol{v} \rangle \implies r\langle \boldsymbol{v}, -\mathrm{d}\boldsymbol{N}_{\boldsymbol{p}_0}(\boldsymbol{v}) \rangle \leq -\langle \boldsymbol{v}, \boldsymbol{v} \rangle.$$

This in particular holds for the unit principal directions $\boldsymbol{v}_1, \boldsymbol{v}_2 \in T_{\boldsymbol{p}_0}M$, satisfying $-\mathrm{d}\boldsymbol{N}_{\boldsymbol{p}_0}(\boldsymbol{v}_i) = \kappa_i(\boldsymbol{p})\boldsymbol{v}_i$ for $1 \leq i \leq 2$, whence

$$r\kappa_i(\boldsymbol{p}) \leq -1 \implies \kappa_i(\boldsymbol{p}) \leq \frac{-1}{r},$$

for $1 \leq i \leq 2$. Thus

$$K(\boldsymbol{p}_0) = \kappa_1(\boldsymbol{p}_0)\kappa_2(\boldsymbol{p}_0) \geq \frac{1}{r^2} > 0,$$

as wanted. □

Corollary 3.4.20. *There is no minimal and compact surface in $\mathbb{R}^3$.*

Proof: It suffices to note that in $\mathbb{R}^3$, we always have that $H(\boldsymbol{p})^2 - K(\boldsymbol{p}) \geq 0$. If $H = 0$ then $K(\boldsymbol{p}) \leq 0$ for all $\boldsymbol{p} \in M$. □

Suppose now that the surface $M \subseteq \mathbb{R}^3_\nu$ is the graph of some smooth function $f : U \subseteq \mathbb{R}^2 \to \mathbb{R}$. By taking a Monge parametrization, for instance, we see that according to Exercise 3.3.4, M is critical if and only if

$$f_{uu}((-1)^\nu + f_v^2) - 2f_u f_v f_{uv} + f_{vv}((-1)^\nu + f_u^2) = 0,$$

which is a quasi-linear Partial Differential Equation (PDE). Clearly all affine (linear) functions are solutions to this equation, i.e., planes are the simplest examples for critical surfaces.

If the domain of the function f is the whole plane, we say that f is *entire* (and that its graph is an *entire graph*). Demanding an entire graph to be critical is very restrictive, as the following theorem (whose proof uses techniques beyond the scope of this text) shows.

Theorem 3.4.21 (Calabi-Bernstein). *Let $M \subseteq \mathbb{R}^3_\nu$ be the graph of an entire smooth function $f : \mathbb{R}^2 \to \mathbb{R}$, satisfying the critical surfaces equation above. If $M \subseteq \mathbb{R}^3$ or if M is spacelike in $\mathbb{L}^3$, then f is affine and M is an entire plane.*

Example 3.4.22. In general, solving PDEs such as the above can be very complicated. It is natural, then, to look for solutions having certain types of symmetries. For example, suppose that f depends smoothly only on the distance between the point (u, v) and the origin, that is, that f has the form $f(u, v) = g(\sqrt{u^2 + v^2})$ for some single-variable smooth function g. Letting $r \doteq \sqrt{u^2 + v^2}$ and omitting points of evaluation, we compute the derivatives $f_u = ug'(r)/r$ and $f_v = vg'(r)/r$, as well as the second order derivatives:

$$f_{uu} = \frac{u^2}{r^2}g''(r) + \frac{v^2}{r^3}g'(r), \quad f_{uv} = \frac{uv}{r^2}g''(r) - \frac{uv}{r^3}g'(r) \text{ and } f_{vv} = \frac{v^2}{r^2}g''(r) + \frac{u^2}{r^3}g'(r).$$

A direct substitution leads us to the Ordinary Differential Equation

$$g''(r) + \frac{g'(r)(1 + (-1)^\nu g'(r)^2)}{r} = 0.$$

After order reduction, this is a *Bernoulli equation*, which can be solved by a well-known algorithm: set $h \doteq g'$ and rewrite the equation as

$$h'(r) + \frac{h(r)}{r} + (-1)^\nu \frac{h(r)^3}{r} = 0 \implies h(r)^{-3}h'(r) + \frac{h(r)^{-2}}{r} + \frac{(-1)^\nu}{r} = 0.$$

Let $w \doteq h^{-2}$, so that $w' = -2h^{-3}h'$, and so we may further rewrite the equation as $rw'(r) - 2w(r) = 2(-1)^\nu$. Multiplying both sides by r^{-3} and integrating, we finally obtain

$$w(r) = (-1)^{\nu+1} + cr^2,$$

for some constant $c \in \mathbb{R}$.

Now we'll start discussing the situation in each ambient space, in terms of the sign of the constant c:

- In $\mathbb{R}^3$, the relation $w = h^{-2} > 0$ does not allow the possibility that $c \leq 0$. Hence $c > 0$ we proceed to solve the equation. We have

  $$h(r) = \pm\frac{1}{\sqrt{cr^2 - 1}} \implies g(r) = \pm\frac{\log\left(\sqrt{c}\sqrt{cr^2 - 1} + cr\right)}{\sqrt{c}} + b,$$

  for some $b \in \mathbb{R}$, where r ranges over some adequate domain. We then obtain a family of catenoids.

- In $\mathbb{L}^3$, we indeed have two situations according to the sign of the constant c. Recall that $\|\nabla f(u, v)\|_E = |g'(r)|$ controls the causal type of M. With this in mind, we see from $h(r)^{-2} = 1 + cr^2$ that $c > 0$ if and only if M is spacelike, and $c < 0$ if and only if M is timelike (the case $c = 0$ corresponds to lightlike M and it is outside the scope of this discussion). If $c > 0$, we have

  $$h(r) = \pm\frac{1}{\sqrt{1 + (\sqrt{c}r)^2}} \implies g(r) = \pm\frac{1}{\sqrt{c}} \operatorname{arcsinh}\left(\sqrt{c}r\right) + b$$

  and if $c < 0$, it follows that

  $$h(r) = \pm\frac{1}{\sqrt{1 - (\sqrt{-c}r)^2}} \implies g(r) = \pm\frac{1}{\sqrt{-c}} \arcsin\left(\sqrt{-c}r\right) + b,$$

  for certain $b \in \mathbb{R}$.

For the surfaces obtained above to be graphs over the plane $z = 0$, the domains of the solutions must be, respectively, $\mathbb{R}_{>0}$ and $\left]0, 1/\sqrt{-c}\right[$.

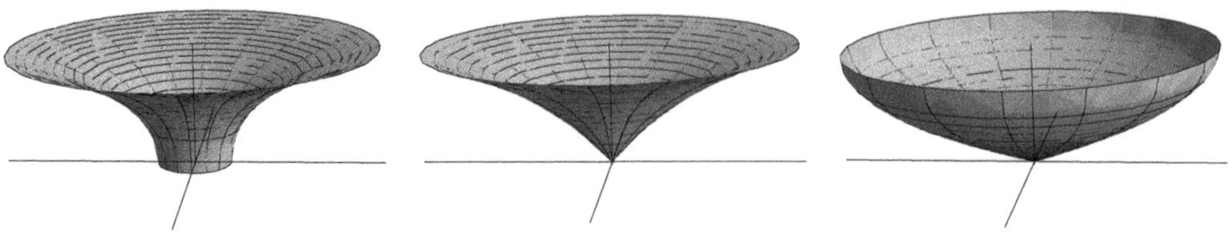

Figure 3.26: Catenoids in $\mathbb{R}^3$, $\mathbb{L}^3$ (spacelike) and $\mathbb{L}^3$ (timelike).

The above calculations in particular show that the assumption of f being entire is essential in the Calabi-Bernstein theorem: the surfaces, in $\mathbb{R}^3$ and spacelike in $\mathbb{L}^3$, found above, are graphs of non-affine functions defined in proper open subsets of $\mathbb{R}^2$. Moreover, for spacelike surfaces in $\mathbb{L}^3$, we have that $g(0) = b$ and $g'(0) = \pm 1$, that is, their graphs are asymptotic to the lightcone centered at $(0, 0, b)$.

Finally we observe that, in $\mathbb{L}^3$, the solutions obtained are odd functions, and so may have their domains extended to $\mathbb{R}$ and $\left]-1/\sqrt{-c}, 1/\sqrt{-c}\right[$, respectively, with the planes of the form $z = \dfrac{k\pi}{2\sqrt{-c}} + b$ (with integer k) are symmetry planes for the surface in the second case.

Another example of a situation where it is interesting to look for solutions of the critical surface PDE with certain symmetries is seen in Exercise 3.4.3, where we ask you to investigate the critical *translation surfaces* in $\mathbb{R}^3_\nu$.

Exercises

Exercise 3.4.1. Let $M_1, M_2 \subseteq \mathbb{R}^3_\nu$ be non-degenerate regular surfaces, and $F \in \mathrm{E}_\nu(3, \mathbb{R})$ be such that $F(M_1) = M_2$. Show that if the Weingarten map of M_1 is diagonalizable at $\boldsymbol{p}$, then the Weingarten map of M_2 is diagonalizable at $F(\boldsymbol{p})$.

Exercise 3.4.2 (A factory of critical surfaces)**.** Let $\boldsymbol{\alpha}\colon I \to \mathbb{R}^3_\nu$ and $\boldsymbol{\beta}\colon J \to \mathbb{R}^3_\nu$ be two regular curves with $\{\boldsymbol{\alpha}'(u), \boldsymbol{\beta}'(v)\}$ linearly independent for all possibilities of u and v. Define $\boldsymbol{x}\colon I \times J \to \mathbb{L}^3$ by setting $\boldsymbol{x}(u, v) = \boldsymbol{\alpha}(u) + \boldsymbol{\beta}(v)$.

(a) Show that $\boldsymbol{x}$ is a parametrized regular surface.

(b) Show that the tangent planes along a fixed coordinate curve are all parallel to a single line.

(c) In $\mathbb{L}^3$, show that if $\boldsymbol{\alpha}$ and $\boldsymbol{\beta}$ are lightlike, then $\boldsymbol{x}$ is timelike with zero mean curvature.

Remark. Actually, the converse holds: every timelike surface with zero mean curvature is locally the sum of lightlike curves, i.e., around every point there is a parametrization as above. Furthermore, if the Gaussian curvature of M also vanishes, we have that $\langle \boldsymbol{\alpha}'(u), \boldsymbol{\beta}'(v) \rangle_L = c \neq 0$ for all $(u, v) \in I \times J$. See [13] for more details.

Exercise 3.4.3 (Translation surfaces). Consider the particular Monge parametrization $x\colon \mathbb{R}^2 \to x(\mathbb{R}^2) \subseteq \mathbb{R}^3_\nu$ given by $x(u,v) = (u, v, h(u) + \ell(v))$, where h and ℓ are certain smooth functions.

(a) In $\mathbb{R}^3$, show that x is minimal if and only if

$$\frac{h''(u)}{1+h'(u)^2} = -\frac{\ell''(v)}{1+\ell'(v)^2} = a$$

for some constant $a \in \mathbb{R}$. Solve the differential equations and conclude that all the minimal surfaces of this type are given by

$$h(u) = \frac{1}{a}\log\cos(au+b_1) + c_1 \quad \text{and} \quad \ell(v) = \frac{1}{a}\log\cos(-av+b_2) + c_2,$$

when $a \neq 0$, and $h(u) = b_1 u + c_1$ and $\ell(v) = b_2 v + c_2$ if $a = 0$, for suitable constants $b_1, b_2, c_1, c_2 \in \mathbb{R}$.

(b) In $\mathbb{L}^3$, assuming that x is non-degenerate, show that x is critical if and only if

$$\frac{h''(u)}{1-h'(u)^2} = -\frac{\ell''(v)}{1-\ell'(v)^2} = a$$

for some constant $a \in \mathbb{R}$. Solve the differential equations and conclude that all the critical surfaces of this type are given by

$$h(u) = \frac{1}{a}\log\cosh(au+b_1) + c_1 \quad \text{and} \quad \ell(v) = \frac{1}{a}\log\cosh(-av+b_2) + c_2,$$

when $a \neq 0$, and $h(u) = b_1 u + c_1$ and $\ell(v) = b_2 v + c_2$ if $a = 0$, for suitable constants $b_1, b_2, c_1, c_2 \in \mathbb{R}$.

Hint. Recall that $\operatorname{arctanh} x = (1/2)\log((1+x)/(1-x))$ for all $x \in \mathbb{R}$.

Exercise† 3.4.4. Show Corollary 3.4.7 (p. 193).

Exercise† 3.4.5. Show Proposition 3.4.11 (p. 197).

Exercise 3.4.6. Consider the *Sherlock hat*, image of the regular parametrized surface $x\colon \left]-1,1\right[\times \left]0,2\pi\right[\to \mathbb{R}^3$ given by

$$x(u,v) = \big((1-u^3)\cos v, u, (1-u^3)\sin v + 1\big).$$

Show that given a parabolic point $x(0,v_0)$ in the image of x, there are points $x(u,v)$ in both half-spaces determined by $T_{(0,v_0)}x$.

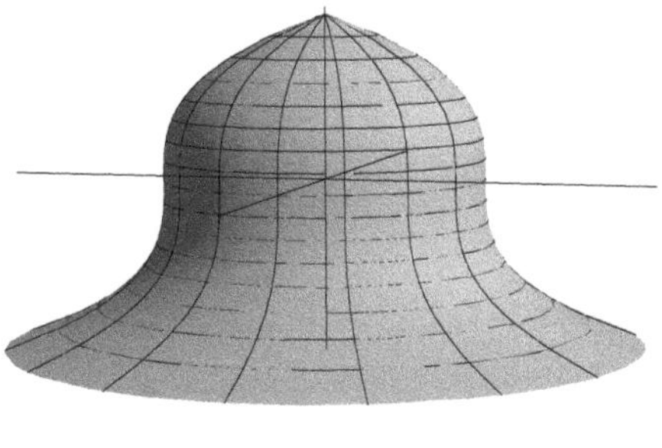

Figure 3.27: "Elementary, my dear Watson!"

Exercise 3.4.7. Let $M \subseteq \mathbb{R}^3_\nu$ be a non-degenerate regular surface and $(U, \boldsymbol{x})$ be a parametrization for M.

(a) If $\boldsymbol{N}_E$ and $\boldsymbol{N}_L$ stand for the Euclidean and Lorentzian unit normal fields along M, respectively, and $\boldsymbol{N}_E$ and $\boldsymbol{N}_L$ have their last components with opposite signs, then show that $h_{ij,E} = h_{ij,L}$, where $h_{ij,E}$ and $h_{ij,L}$ denote the components of the Euclidean and Minkowski Second Fundamental Forms of $\boldsymbol{x}$, respectively.

(b) Conclude that the classification of points in a surface as elliptic, hyperbolic, parabolic or planar is independent of the ambient space where the surface is. Also conclude that if K_E and K_L denote the Gaussian curvatures of M equipped with the metrics from $\mathbb{R}^3$ and $\mathbb{L}^3$, respectively, then $K_E(\boldsymbol{p})K_L(\boldsymbol{p}) \leq 0$, for every $\boldsymbol{p} \in M$.

Exercise 3.4.8. Consider the torus $\mathbb{T}^2$ in $\mathbb{R}^3$ generated by the revolution of the unit circle centered at $(0,2,0)$ and contained in the plane $x = 0$, around the z-axis. Compute the Gaussian curvature of $\mathbb{T}^2$ and conclude that $\mathbb{T}^2$ has elliptic, hyperbolic and parabolic points. Sketch a picture.

Exercise 3.4.9. Let $M \subseteq \mathbb{R}^3_\nu$ be a non-degenerate regular surface and $\boldsymbol{N}$ be a Gauss map for M.

(a) Show that if $\boldsymbol{\alpha}\colon I \to M$ is a regular curve in M such that $\boldsymbol{N}$ is constant along $\boldsymbol{\alpha}$, then all the points in $\boldsymbol{\alpha}$ are parabolic or planar.

(b) Show that if $\boldsymbol{\alpha}, \boldsymbol{\beta}\colon I \to M$ are two regular curves in M, $t_0 \in I$ is such that $\boldsymbol{p} \doteq \boldsymbol{\alpha}(t_0) = \boldsymbol{\beta}(t_0)$ and $\{\boldsymbol{\alpha}'(t_0), \boldsymbol{\beta}'(t_0)\}$ are linearly independent, and $\boldsymbol{N}$ is constant along $\boldsymbol{\alpha}$ and $\boldsymbol{\beta}$, then $\boldsymbol{p}$ is planar.

Exercise 3.4.10. Let $M \subseteq \mathbb{R}^3_\nu$ be the graph of a smooth function $f\colon U \to \mathbb{R}$. Suppose that f depends smoothly only on the angle formed between the x-axis and the point (u,v) in the domain U, that is, that $f(u,v) = g(\arctan(v/u))$ for a certain single-variable smooth function g.

If $r = \sqrt{u^2+v^2}$ and $\theta = \arctan(v/u)$, verify that $f_u = -vg'(\theta)/r^2$, $f_v = ug'(\theta)/r^2$,

$$\begin{aligned} f_{uu} &= \frac{v^2}{r^4}g''(\theta) + \frac{2uv}{r^4}g'(\theta), \\ f_{uv} &= -\frac{u^2}{r^4}g'(\theta) + \frac{v^2}{r^4}g'(\theta) - \frac{uv}{r^4}g''(\theta) \quad \text{and} \\ f_{vv} &= \frac{u^2}{r^4}g''(\theta) - \frac{2uv}{r^4}g'(\theta), \end{aligned}$$

and use this to show that the critical surfaces equation boils down to $g''(\theta) = 0$, in both ambient spaces. Determine, as done in Example 3.4.22 (p. 203), all the surfaces with this property.

3.5 CURVES IN A SURFACE

In Chapter 2, we studied the theory of curves in space. At this point we may start investigating the relation between curves whose images are contained in a non-degenerate regular surface $M \subseteq \mathbb{R}^3_\nu$ and the geometry of M itself.

Naturally, the first step is to look for a relation between the curvature of curves in M with the curvatures of M itself. We start building this bridge by using the First and Second Fundamental Forms of M to introduce yet another notion of curvature. For each $\boldsymbol{p} \in M$, let $T_p^\dagger M \doteq \{\boldsymbol{v} \in T_pM \mid \langle \boldsymbol{v}, \boldsymbol{v}\rangle \neq 0\}$. We have the:

Definition 3.5.1 (Normal curvature). Let $M \subseteq \mathbb{R}^3_\nu$ be a non-degenerate regular surface and $\boldsymbol{N}$ be a Gauss map for M. Given $\boldsymbol{p} \in M$, the *normal curvature* of M at $\boldsymbol{p}$ is the function $\kappa_n \colon T_p^\dagger M \to \mathbb{R}$ given by

$$\kappa_n(\boldsymbol{v}) \doteq \frac{\widetilde{\mathrm{II}}_p(\boldsymbol{v})}{\mathrm{I}_p(\boldsymbol{v})}.$$

Moreover, we say that a vector $\boldsymbol{v} \in T_p^\dagger M$ is *asymptotic* if $\kappa_n(\boldsymbol{v}) = 0$.

Remark.

- Note that for all $\boldsymbol{v} \in T_p^\dagger M$ and all non-zero $\lambda \in \mathbb{R}$, we have that $\kappa_n(\lambda \boldsymbol{v}) = \kappa_n(\boldsymbol{v})$, so that we may restrict ourselves to unit vectors and talk about the normal curvature at $\boldsymbol{p}$ *along a general tangent direction* (non-zero and non-degenerate) to M.
- To justify the name "asymptotic vector" see Exercise 3.5.8.

For a geometric interpretation, consider a unit speed admissible curve $\boldsymbol{\alpha} \colon I \to M$ in M. Using the first Frenet-Serret equation for $\boldsymbol{\alpha}$ (that is, the definition of curvature of $\boldsymbol{\alpha}$), we have that

$$\kappa_n(\boldsymbol{\alpha}'(s)) = \frac{\langle -\mathrm{d}\boldsymbol{N}_{\boldsymbol{\alpha}(s)}(\boldsymbol{\alpha}'(s)), \boldsymbol{\alpha}'(s)\rangle}{\langle \boldsymbol{\alpha}'(s), \boldsymbol{\alpha}'(s)\rangle} = \epsilon_{\boldsymbol{\alpha}}\langle \boldsymbol{N}(\boldsymbol{\alpha}(s)), \boldsymbol{\alpha}''(s)\rangle = \epsilon_{\boldsymbol{\alpha}}\kappa_{\boldsymbol{\alpha}}(s)\langle \boldsymbol{N}(\boldsymbol{\alpha}(s)), \boldsymbol{N}_{\boldsymbol{\alpha}}(s)\rangle.$$

This way, if $M \subseteq \mathbb{R}^3$, for each $s \in I$ the angle $\theta(s) \in [0, 2\pi[$ between $\boldsymbol{N}(\boldsymbol{\alpha}(s))$ and $\boldsymbol{N}_{\boldsymbol{\alpha}}(s)$ is well-defined, and the relation $\kappa_n(\boldsymbol{\alpha}'(s)) = \kappa_{\boldsymbol{\alpha}}(s)\cos\theta(s)$ holds. If $s_0 \in I$ is such that $\theta(s_0) = 0$ or π, then we have that $|\kappa_n(\boldsymbol{\alpha}'(s_0))| = |\kappa_{\boldsymbol{\alpha}}(s_0)|$.

In general, we say that a *normal section* of M at $\boldsymbol{p}$ along a non-zero vector $\boldsymbol{v} \in T_pM$ is the intersection of M with the plane passing through $\boldsymbol{p}$ and spanned by $\boldsymbol{N}(\boldsymbol{p})$ and $\boldsymbol{v}$. That is, the above calculation tells us, in $\mathbb{R}^3$, that if $\boldsymbol{v}$ is a unit vector, then the normal curvature of M at $\boldsymbol{p}$ along $\boldsymbol{v}$ and the curvature of the normal section of M at $\boldsymbol{p}$ along $\boldsymbol{v}$ are equal in absolute value, justifying the terminology adopted.

If $M \subseteq \mathbb{L}^3$ is spacelike, then $\boldsymbol{N}(\boldsymbol{\alpha}(s))$ is a timelike vector. If $\boldsymbol{N}_{\boldsymbol{\alpha}}(s)$ is also timelike, with the same time direction as $\boldsymbol{N}(\boldsymbol{\alpha}(s))$, we may consider the hyperbolic angle $\varphi(s)$ between $\boldsymbol{N}(\boldsymbol{\alpha}(s))$ and $\boldsymbol{N}_{\boldsymbol{\alpha}}(s)$, and the relation $\kappa_n(\boldsymbol{\alpha}'(s)) = -\kappa_{\boldsymbol{\alpha}}(s)\cosh\varphi(s)$ holds. Note that given $s_0 \in I$, we have that $\varphi(s_0) = 0$ if and only if $\kappa_n(\boldsymbol{\alpha}'(s_0)) = -\kappa_{\boldsymbol{\alpha}}(s_0)$, in which case not only $\boldsymbol{\alpha}$ works as a sort of normal section, but the vectors $\boldsymbol{N}(\boldsymbol{\alpha}(s))$ and $\boldsymbol{N}_{\boldsymbol{\alpha}}(s)$ are actually equal.

Lastly, if $M \subseteq \mathbb{L}^3$ is timelike, then the causal type of $\boldsymbol{\alpha}$ may be anything, and the interpretation to be given depends on the causal type of the Frenet-Serret trihedron of $\boldsymbol{\alpha}$. For instance, if $\boldsymbol{N}_{\boldsymbol{\alpha}}(s)$ is spacelike and such that the plane it spans with $\boldsymbol{N}(\boldsymbol{\alpha}(s))$ is also spacelike, the discussion goes exactly like what was done in $\mathbb{R}^3$.

Our conclusion from this analysis is the:

Theorem 3.5.2 (Meusnier). *Let $M \subseteq \mathbb{R}^3_\nu$ be a non-degenerate regular surface. Then the curves in M which, at a point $\boldsymbol{p} \in M$, have the same non-lightlike tangent line all have the same normal curvature at $\boldsymbol{p}$.*

It is to be expected, when the Weingarten map of M is diagonalizable, that nice relations between the normal curvature and principal curvatures and direction will appear.

Proposition 3.5.3 (Euler Formulas). *Let $M \subseteq \mathbb{R}^3_\nu$ be a non-degenerate regular surface, $\boldsymbol{p} \in M$, and $\boldsymbol{N}$ be a Gauss map for M. Suppose that $-\mathrm{d}\boldsymbol{N}_{\boldsymbol{p}}$ is diagonalizable and consider $\boldsymbol{u}_1, \boldsymbol{u}_2 \in T_{\boldsymbol{p}}M$ the unit principal vectors associated to the principal curvatures $\kappa_1(\boldsymbol{p})$ and $\kappa_2(\boldsymbol{p})$, with $\boldsymbol{u}_1$ spacelike. Then:*

(i) if M is spacelike, every unit vector in $T_{\boldsymbol{p}}M$ may be uniquely written in the form $\boldsymbol{v}(\theta) = \cos\theta\, \boldsymbol{u}_1 + \sin\theta\, \boldsymbol{u}_2$ for some $\theta \in [0, 2\pi[$ and

$$\kappa_n(\theta) \equiv \kappa_n(\boldsymbol{v}(\theta)) = \kappa_1(\boldsymbol{p})\cos^2\theta + \kappa_2(\boldsymbol{p})\sin^2\theta$$

holds;

(ii) if M is timelike, every unit spacelike vector in $T_{\boldsymbol{p}}M$ may be uniquely written in the form $\boldsymbol{v}(\varphi) = \pm\cosh\varphi\, \boldsymbol{u}_1 + \sinh\varphi\, \boldsymbol{u}_2$ and every unit timelike vector has the unique form $\boldsymbol{w}(\varphi) = \sinh\varphi\, \boldsymbol{u}_1 \pm \cosh\varphi\, \boldsymbol{u}_2$ for some $\varphi \in \mathbb{R}$, where $\pm$ indicates exactly on which branch of the "unit hyperbola" the vector is. Then

$$\kappa_n^{\text{s.l.}}(\varphi) \doteq \kappa_n(\boldsymbol{v}(\varphi)) = \kappa_1(\boldsymbol{p})\cosh^2\varphi - \kappa_2(\boldsymbol{p})\sinh^2\varphi, \text{ and}$$
$$\kappa_n^{\text{t.l.}}(\varphi) \doteq \kappa_n(\boldsymbol{w}(\varphi)) = \kappa_2(\boldsymbol{p})\cosh^2\varphi - \kappa_1(\boldsymbol{p})\sinh^2\varphi$$

hold.

In particular, if $\boldsymbol{p}$ is umbilic, the normal curvature of M at $\boldsymbol{p}$ is constant.

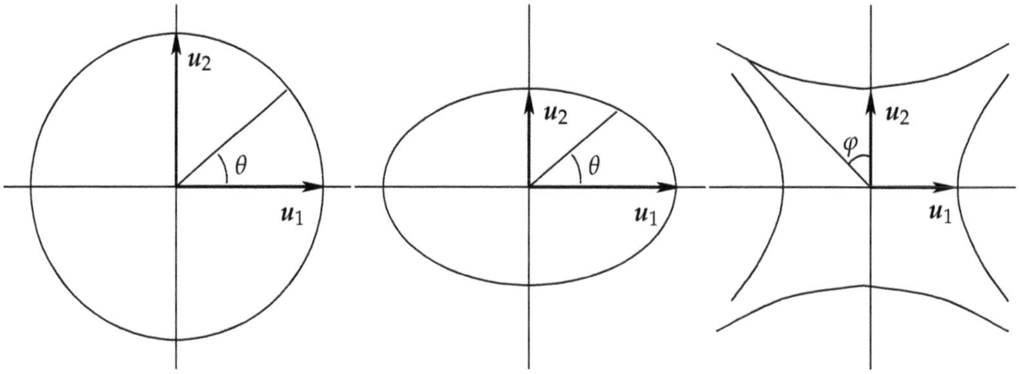

(a) In $\mathbb{R}^3$. (b) In $\mathbb{L}^3$, $\epsilon_M = -1$. (c) In $\mathbb{L}^3$, $\epsilon_M = 1$.

Figure 3.28: Unit vectors in $T_{\boldsymbol{p}}M$.

Proof:

(i) If M is spacelike, we have

$$\begin{aligned}\kappa_n(\theta) &= \langle -\mathrm{d}\boldsymbol{N}_{\boldsymbol{p}}(\cos\theta\, \boldsymbol{u}_1 + \sin\theta\, \boldsymbol{u}_2), \cos\theta\, \boldsymbol{u}_1 + \sin\theta\, \boldsymbol{u}_2\rangle \\ &= \langle \kappa_1(\boldsymbol{p})\cos\theta\, \boldsymbol{u}_1 + \kappa_2(\boldsymbol{p})\sin\theta\, \boldsymbol{u}_2, \cos\theta\, \boldsymbol{u}_1 + \sin\theta\, \boldsymbol{u}_2\rangle \\ &= \kappa_1(\boldsymbol{p})\cos^2\theta + \kappa_2(\boldsymbol{p})\sin^2\theta.\end{aligned}$$

(ii) If M is timelike, let's do the first case:

$$\begin{aligned}\kappa_n^{\text{s.l.}}(\varphi) &= \langle -\mathrm{d}\boldsymbol{N}_{\boldsymbol{p}}(\pm\cosh\varphi\,\boldsymbol{u}_1+\sinh\varphi\,\boldsymbol{u}_2), \pm\cosh\varphi\,\boldsymbol{u}_1+\sinh\varphi\,\boldsymbol{u}_2\rangle \\ &= \langle \pm\kappa_1(\boldsymbol{p})\cosh\varphi\,\boldsymbol{u}_1+\kappa_2(\boldsymbol{p})\sinh\varphi\,\boldsymbol{u}_2, \pm\cosh\varphi\,\boldsymbol{u}_1+\sinh\varphi\,\boldsymbol{u}_2\rangle \\ &= \kappa_1(\boldsymbol{p})\cosh^2\varphi-\kappa_2(\boldsymbol{p})\sinh^2\varphi.\end{aligned}$$

The computation of $\kappa_n^{\text{t.l.}}(\varphi)$ is similar, recalling that since $\boldsymbol{w}(\varphi)$ is timelike, the denominator in the definition of normal curvature is now -1.

□

Corollary 3.5.4. *Let $M \subseteq \mathbb{R}^3_\nu$ be a non-degenerate regular surface, $\boldsymbol{p} \in M$, and $\boldsymbol{N}$ a Gauss map for M. If $-\mathrm{d}\boldsymbol{N}_{\boldsymbol{p}}$ is diagonalizable, the critical points for the normal curvature of M at $\boldsymbol{p}$ are precisely the principal directions of M at $\boldsymbol{p}$. And the normal curvatures along these directions at $\boldsymbol{p}$ are precisely the principal curvatures of M at $\boldsymbol{p}$.*

Proof: Keep all notation from Proposition 3.5.3 above. Suppose that $\boldsymbol{p}$ is not umbilic, otherwise there is nothing to prove.

If M is spacelike, we have $\kappa_n(\theta) = \kappa_1(\boldsymbol{p})\cos^2\theta + \kappa_2(\boldsymbol{p})\sin^2\theta$. To find the critical points of κ_n, note that

$$\kappa_n'(\theta) = (-\kappa_1(\boldsymbol{p}) + \kappa_2(\boldsymbol{p}))\sin 2\theta$$

vanishes if and only if $\theta = 0$, $\pi/2$, π or $3\pi/2$, and observe that

$$\kappa_n(0) = \kappa_n(\pi) = \kappa_1(\boldsymbol{p}) \quad \text{and} \quad \kappa_n(\pi/2) = \kappa_n(3\pi/2) = \kappa_2(\boldsymbol{p}).$$

If M is timelike, let's first make our search among spacelike vectors in the unit hyperbola, using $\kappa_n^{\text{s.l.}}(\varphi) = \kappa_1(\boldsymbol{p})\cosh^2\varphi - \kappa_2(\boldsymbol{p})\sinh^2\varphi$. We have that

$$(\kappa_n^{\text{s.l.}})'(\varphi) = (\kappa_1(\boldsymbol{p}) - \kappa_2(\boldsymbol{p}))\sinh 2\varphi$$

vanishes if and only if $\varphi = 0$, and $\kappa_n^{\text{s.l.}}(0) = \kappa_1(\boldsymbol{p})$. The search among timelike vectors using the formula for $\kappa_n^{\text{t.l.}}(\varphi)$ is similar, and one obtains $\kappa_n^{\text{t.l.}}(0) = \kappa_2(\boldsymbol{p})$. □

Remark. In the above proof, the nature of the critical points is determined by the sign of the difference $\kappa_1(\boldsymbol{p}) - \kappa_2(\boldsymbol{p})$, and no critical point is an inflection point if $\boldsymbol{p}$ is not umbilic. Usually, the convention in $\mathbb{R}^3$ is to assume that $\kappa_1(\boldsymbol{p}) \leq \kappa_2(\boldsymbol{p})$, by renaming the curvatures if necessary, but this is not possible in $\mathbb{L}^3$ since the principal vectors may have distinct causal types.

The Euler formulas seen above may be used to give further interpretations of the mean curvature as an (either discrete or continuous) average of the normal curvatures, even when M is timelike and the "unit circle" in the tangent plane is no longer compact. See Exercises 3.5.1 and 3.5.2.

We now introduce the first special type of curve in a surface, whose existence is intimately related to the diagonalization problem for the Weingarten map:

Definition 3.5.5 (Lines of curvature). Let $M \subseteq \mathbb{R}^3_\nu$ be a non-degenerate regular surface and $\boldsymbol{N}$ be a Gauss map for M. We'll say that a curve $\boldsymbol{\alpha}\colon I \to M$ is a *line of curvature* of M if $\boldsymbol{\alpha}'(t)$ is an eigenvector of $-\mathrm{d}\boldsymbol{N}_{\boldsymbol{\alpha}(t)}$, for all $t \in I$.

It follows from what we have seen so far about the Weingarten map that given a non-umbilic point $\boldsymbol{p} \in M$, there are at most two lines of curvature of M passing through $\boldsymbol{p}$ and, when M is spacelike, there are always exactly two. We also have a very concrete criterion to determine whether a given curve is or not a line of curvature, in terms of their coordinates relative to a parametrization of the surface:

Proposition 3.5.6. *Let $M \subseteq \mathbb{R}^3_\nu$ be a non-degenerate regular surface, $\mathbf{N}$ be a Gauss map for M, and $(U,\boldsymbol{x})$ be a parametrization for M. If $\boldsymbol{\alpha}\colon I \to M$ is given in coordinates by $\boldsymbol{\alpha}(t) = \boldsymbol{x}(u(t),v(t))$, then $\boldsymbol{\alpha}$ is a line of curvature of M if and only if:*

$$\begin{vmatrix} (v')^2 & -u'v' & (u')^2 \\ E & F & G \\ e & f & g \end{vmatrix} = 0, \quad \textit{para todo } t \in I,$$

where u',v' are evaluated in t, and the coefficients of the Fundamental Forms are evaluated in $(u(t),v(t))$.

Remark.

- Clearly there is a pointwise version of the above result: if $\boldsymbol{p} = \boldsymbol{x}(u_0,v_0)$ and $\boldsymbol{v} = a\boldsymbol{x}_u(u_0,v_0) + b\boldsymbol{x}_v(u_0,v_0)$, then $\boldsymbol{v}$ is a principal direction if and only if

$$\begin{vmatrix} b^2 & -ab & a^2 \\ E & F & G \\ e & f & g \end{vmatrix} = 0,$$

  where the coefficients of the Fundamental Forms are evaluated in (u_0,v_0).

- From here onwards, it will be convenient to denote derivatives taken with respect to t with dots, as usual in Physics.

Proof: Let's identify $u \leftrightarrow u^1$ and $v \leftrightarrow u^2$, as usual. Omitting points of application, the chain rule gives us that $\boldsymbol{\alpha}' = \sum_{i=1}^2 \dot{u}^i\boldsymbol{x}_i$. If $\boldsymbol{\alpha}$ is a line of curvature, there is a smooth $\lambda\colon I \to \mathbb{R}$ with $-\mathrm{d}\mathbf{N}(\boldsymbol{\alpha}') = \lambda\boldsymbol{\alpha}'$. We have that

$$-\mathrm{d}\mathbf{N}\left(\sum_{i=1}^2 \dot{u}^i\boldsymbol{x}_i\right) = \sum_{i=1}^2 \dot{u}^i(-\mathrm{d}\mathbf{N}(\boldsymbol{x}_i)) = \sum_{i,j=1}^2 \dot{u}^i h^j{}_i \boldsymbol{x}_j,$$

whence it follows (from linear independence) that

$$\sum_{i=1}^2 \dot{u}^i h^j{}_i = \lambda \dot{u}^j,$$

for $1 \le j \le 2$. Using Lemma 3.3.7, this equality reads as

$$\sum_{i,k=1}^2 \dot{u}^i h_{ik} g^{kj} = \lambda \dot{u}^j.$$

Back to index-free notation, this last expression is nothing more than the two following equations together:

$$\begin{aligned} \lambda u' &= \left(\frac{eG - fF}{EG - F^2}\right)u' + \left(\frac{fG - gF}{EG - F^2}\right)v' \\ \lambda v' &= \left(\frac{-eF + fE}{EG - F^2}\right)u' + \left(\frac{-fF + gE}{EG - F^2}\right)v'. \end{aligned}$$

Multiplying the first by $v'(EG - F^2)$, the second one by $u'(EG - F^2)$, and equating them, we have:

$$(eG - fF)u'v' + (fG - gF)(v')^2 = (-eF + fE)(u')^2 + (-fF + gE)u'v',$$

which simplifies as:

$$(fG - gF)(v')^2 + (eG - gE)u'v' + (eF - fE)(u')^2 = 0.$$

This expression is precisely the Laplace expansion of the determinant given in the statement, through the first row. From these calculations, it is easy to see that the converse holds. To wit, all the steps are reversible since M is non-degenerate and thus $EG - F^2 \neq 0$. □

Remark. If the Weingarten map is diagonalizable, we know that the two principal directions are orthogonal at each point. Thus, the equation above gives us that a necessary and sufficient condition for the principal directions to be the coordinate curves of some parametrization $\boldsymbol{x}$ is that both F and f vanish. In this case, $\boldsymbol{x}$ is called a *principal parametrization*. Moreover, it follows from the general theory of Ordinary Differential Equations that if $\boldsymbol{p} \in M$ is not umbilic, there is a principal parametrization for M around $\boldsymbol{p}$.

Before we present examples, we'll introduce the second special type of curve in a surface, more closely related to the normal curvature:

Definition 3.5.7 (Asymptotic lines)**.** Let $M \subseteq \mathbb{R}^3_\nu$ be a non-degenerate regular surface. A non-lightlike regular curve $\boldsymbol{\alpha}\colon I \to M$ is called an *asymptotic line* of M if $\boldsymbol{\alpha}'(t)$ is an asymptotic vector for all $t \in I$.

The first thing to investigate are conditions for the existence of asymptotic lines in a surface. Before that, the existence of asymptotic vectors in a given tangent plane $T_{\boldsymbol{p}}M$ is characterized by the following:

Proposition 3.5.8. *Let $M \subseteq \mathbb{R}^3_\nu$ be a non-degenerate regular surface, $\mathbf{N}$ be a Gauss map for M, and $\boldsymbol{p} \in M$. Suppose that $-\mathrm{d}\mathbf{N}_{\boldsymbol{p}}$ is diagonalizable. Then there are asymptotic vectors in $T^{\dagger}_{\boldsymbol{p}}M$ if and only if $(-1)^\nu K(\boldsymbol{p}) \leq 0$.*

Remark. In particular, given an asymptotic line $\boldsymbol{\alpha}\colon I \to M$, if the Weingarten map of M is diagonalizable along $\boldsymbol{\alpha}$, it holds that $(-1)^\nu K(\boldsymbol{\alpha}(t)) \leq 0$ for all $t \in I$.

Proof: Observe that $(-1)^\nu K(\boldsymbol{p}) \leq 0$ if and only if $\kappa_1(\boldsymbol{p})$ and $\kappa_2(\boldsymbol{p})$ have opposite signs when M is spacelike and equal signs when M is timelike. We then have the following cases to discuss:

(i) If M is spacelike, the Euler formula is $\kappa_n(\theta) = \kappa_1(\boldsymbol{p})\cos^2\theta + \kappa_2(\boldsymbol{p})\sin^2\theta$. Suppose that $(-1)^\nu K(\boldsymbol{p}) \leq 0$. If one of the principal curvatures vanishes, the corresponding principal direction is asymptotic. If both of the principal curvatures are non-zero, the equation $\kappa_n(\theta) = 0$ has solutions

$$\theta = \pm \arctan\sqrt{-\frac{\kappa_1(\boldsymbol{p})}{\kappa_2(\boldsymbol{p})}}.$$

Conversely, suppose that there is $\theta_0 \in [0, \pi[$ with $\kappa_n(\theta_0) = 0$. If $\theta_0 = 0$ or $\theta_0 = \pi/2$, then $\kappa_1(\boldsymbol{p})$ or $\kappa_2(\boldsymbol{p})$ vanishes. Otherwise

$$\frac{\kappa_1(\boldsymbol{p})}{\kappa_2(\boldsymbol{p})} = -\tan^2\theta_0 < 0.$$

(ii) If M is timelike, it suffices to look for spacelike directions, for which the Euler formula is $\kappa_n^{\text{s.l.}}(\varphi) = \kappa_1(\boldsymbol{p})\cosh^2\varphi - \kappa_2(\boldsymbol{p})\sinh^2\varphi$, as the situation for timelike directions is completely similar. Suppose that $K(\boldsymbol{p}) \geq 0$. If $\kappa_2(\boldsymbol{p}) = 0$ it follows that $\kappa_1(\boldsymbol{p}) = 0$ as well, so that all directions are asymptotic. Otherwise, the equation $\kappa_n^{\text{s.l.}}(\varphi) = 0$ has solution

$$\varphi = \pm \operatorname{arctanh} \sqrt{\frac{\kappa_1(\boldsymbol{p})}{\kappa_2(\boldsymbol{p})}}.$$

Conversely, suppose that there is $\varphi_0 \in \mathbb{R}$ such that $\kappa_n^{\text{s.l.}}(\varphi_0) = 0$. If $\varphi_0 = 0$, then $\kappa_1(\boldsymbol{p}) = 0$ and thus $K(\boldsymbol{p}) \geq 0$. Otherwise, both the principal curvatures vanish, or neither of them do. If both vanish, the situation is trivial. If neither vanishes, we have

$$\frac{\kappa_1(\boldsymbol{p})}{\kappa_2(\boldsymbol{p})} = \tanh^2\varphi_0 > 0.$$

□

As we have done for lines of curvature, let's establish a criterion for identifying asymptotic lines in terms of their coordinates relative to a parametrization of the surface:

Proposition 3.5.9. *Let $M \subseteq \mathbb{R}^3_\nu$ be a non-degenerate regular surface and $(U, \boldsymbol{x})$ be a parametrization for M. If $\boldsymbol{\alpha}\colon I \to M$ is a regular curve given in coordinates by $\boldsymbol{\alpha}(t) = \boldsymbol{x}(u(t), v(t))$, then $\boldsymbol{\alpha}$ is an asymptotic line if and only if:*

$$e(u')^2 + 2fu'v' + g(v')^2 = 0, \quad \text{for all } t \in I,$$

where the coefficients of the Second Fundamental Form are evaluated in $(u(t), v(t))$, and u', v' are evaluated in t.

Proof: As in the proof of Proposition 3.5.6 above, we have that $\boldsymbol{\alpha}' = \sum_{i=1}^2 \dot{u}^i \boldsymbol{x}_i$, and $\boldsymbol{\alpha}$ is an asymptotic line if and only if $\langle -\mathrm{d}\boldsymbol{N}(\boldsymbol{\alpha}'), \boldsymbol{\alpha}'\rangle = 0$. However:

$$\begin{aligned}
\langle -\mathrm{d}\boldsymbol{N}(\boldsymbol{\alpha}'), \boldsymbol{\alpha}'\rangle &= \left\langle -\mathrm{d}\boldsymbol{N}\left(\sum_{i=1}^2 \dot{u}^i \boldsymbol{x}_i\right), \sum_{j=1}^2 \dot{u}^j \boldsymbol{x}_j \right\rangle = \sum_{i=1}^2\sum_{j=1}^2 \dot{u}^i\dot{u}^j \langle -\mathrm{d}\boldsymbol{N}(\boldsymbol{x}_i), \boldsymbol{x}_j\rangle \\
&= \sum_{i,j=1}^2 \dot{u}^i\dot{u}^j \left\langle \sum_{k=1}^2 h^k_i \boldsymbol{x}_k, \boldsymbol{x}_j \right\rangle = \sum_{i,j,k=1}^2 \dot{u}^i\dot{u}^j h^k_i g_{kj} \\
&= \sum_{i,j=1}^2 \dot{u}^i\dot{u}^j h_{ij} = e(u')^2 + 2fu'v' + g(v')^2.
\end{aligned}$$

□

Remark. Note that a necessary and sufficient condition for the coordinate curves of a parametrization $\boldsymbol{x}$ around a hyperbolic point $\boldsymbol{p}$ (which satisfies $(-1)^\nu K(\boldsymbol{p}) \leq 0$) to be asymptotic lines is that both e and g vanish. To wit, if $u' = 0$, the equation boils down to $g(v')^2 = 0$, whence $g = 0$. Similarly, $v' = 0$ implies that $e = 0$. On the other hand, if $e = g = 0$, the equation becomes $fu'v' = 0$, which is trivially satisfied by the coordinate curves. In this case $\boldsymbol{x}$ is called an *asymptotic parametrization*.

At last, the long-awaited examples:

Example 3.5.10.

(1) Let's consider again the cylinder with radius $r > 0$, $\mathbb{S}^1(r) \times \mathbb{R}$, parametrized by $\boldsymbol{x}\colon\ ]0, 2\pi[\times \mathbb{R} \to \mathbb{S}^1(r) \times \mathbb{R} \subseteq \mathbb{R}^3_\nu$, given by $\boldsymbol{x}(u,v) = (r\cos u, r\sin u, v)$. Its lines of curvature are characterized by:

$$\begin{vmatrix} (v')^2 & -u'v' & (u')^2 \\ r^2 & 0 & (-1)^\nu \\ -r & 0 & 0 \end{vmatrix} = 0.$$

Expanding through the third row, one easily sees that the equation boils down to $u'v' = 0$, whence $u' = 0$ or $v' = 0$. So, in the first case, we have that $u \equiv u_0$ and $\boldsymbol{\alpha}(t) = (r\cos u_0, r\sin u_0, v(t))$ is a vertical line in the cylinder, while in the second case we have that $v \equiv v_0$ and $\boldsymbol{\alpha}(t) = (r\cos u(t), r\sin u(t), v_0)$, and $\boldsymbol{\alpha}$ is a horizontal circle in the cylinder.

The equation for asymptotic lines, in turn, is just $-r(u')^2 = 0$, whence we conclude that the asymptotic lines in the cylinder are vertical lines. Observe that the conclusion was the same for both ambient spaces.

(2) Consider the surface of revolution generated by an injective and unit speed curve $\boldsymbol{\alpha}\colon I \to \mathbb{R}^3_\nu$, of the form $\boldsymbol{\alpha}(u) = (f(u), 0, g(u))$, with $f(u) > 0$ for all $u \in I$, and the usual revolution parametrization $\boldsymbol{x}\colon I \times]0, 2\pi[\to \mathbb{R}^3_\nu$, given by $\boldsymbol{x}(u,v) = (f(u)\cos v, f(u)\sin v, g(u))$. The equation of the lines of curvature, in both ambient spaces, is:

$$\begin{vmatrix} (v')^2 & -u'v' & (u')^2 \\ \epsilon_\alpha & 0 & f^2 \\ -f''g' + f'g'' & 0 & fg' \end{vmatrix} = 0 \iff u'v'(\epsilon_\alpha fg' + f^2 f''g' - f^2 f'g'') = 0.$$

There are at most two lines of curvature passing through each non-umbilic point. Clearly the coordinate curves are solutions to this equation and, thus, the meridians and parallels are the only lines of curvature passing through non-umbilic points of a surface of revolution. On the other hand, if $\boldsymbol{x}(u_0, v_0)$ is umbilic, then

$$\epsilon_\alpha f(u_0)g'(u_0) + f(u_0)^2 f''(u_0)g'(u_0) - f(u_0)^2 f'(u_0)g''(u_0) = 0.$$

The equation for the asymptotic lines is, in turn:

$$(-f''g' + f'g'')(u')^2 + (fg')(v')^2 = 0,$$

which is better studied in particular cases.

(3) Consider the graph of a smooth function $f\colon U \subseteq \mathbb{R}^2 \to \mathbb{R}$, with the usual Monge parametrization $\boldsymbol{x}\colon U \to \mathbb{R}^3_\nu$, $\boldsymbol{x}(u,v) = (u, v, f(u,v))$. The equation for the lines of curvature is:

$$\begin{vmatrix} (v')^2 & -u'v' & (u')^2 \\ 1 + (-1)^\nu f_u^2 & (-1)^\nu f_u f_v & 1 + (-1)^\nu f_v^2 \\ f_{uu} & f_{uv} & f_{vv} \end{vmatrix} = 0.$$

Expanding this determinant gives us a non-linear ordinary differential equation, in general unmanageable. We can, however, study simpler cases, such as when f depends only on u or v.

Suppose that $f_u = 0$. With this we also have $f_{uu} = f_{uv} = 0$ and the equation boils down to $f_{vv}u'v' = 0$, in both ambient spaces. If $f_{vv} = 0$, all the curves will be lines of curvature (to wit, the graph of f is a plane). And if $f_{vv} \neq 0$, we have $u'v' = 0$, whence u' or v' is zero, and the coordinate curves are the lines of curvature we're looking for.

The equation for the asymptotic lines, in both ambient spaces, is:

$$f_{uu}(u')^2 + 2f_{uv}u'v' + f_{vv}(v')^2 = 0.$$

Again, looking at the case where $f_u = 0$, we have that the equation boils down to $f_{vv}(v')^2 = 0$. If $f_{vv} = 0$, all the curves are asymptotic lines, and if $f_{vv} \neq 0$, it follows that the coordinate curve $v \equiv v_0$ is an asymptotic line.

(4) Consider again the helicoid with the parametrization $\boldsymbol{x}\colon \mathbb{R} \times \mathbb{R}_{\geq 0} \to \mathbb{R}^3_\nu$ given by $\boldsymbol{x}(u, v) = (v \cos u, v \sin u, u)$. The equation for the lines of curvature is:

$$\begin{vmatrix} (v')^2 & -u'v' & (u')^2 \\ (-1)^\nu + v^2 & 0 & 1 \\ 0 & 1 & 0 \end{vmatrix} = 0.$$

In $\mathbb{R}^3$, this becomes $(v')^2 - (u')^2(1 + v^2) = 0$, and then we see that $v' \neq 0$ (or else we would have $u' \equiv 0$ as well, and the curve described by u and v would not be regular). Hence we may write:

$$\frac{du}{dv} = \pm\frac{1}{\sqrt{1+v^2}},$$

using v as implicit parameter (more precisely, we invert $v = v(t)$ and use the chain rule). Integrating with respect to v, we obtain $u(v) = \pm \ln |v + \sqrt{1+v^2}| + c$, where $c \in \mathbb{R}$ is a constant to be determined by prescribing a point in the surface on which the curve will pass through.

We have to be careful when repeating this in $\mathbb{L}^3$. To wit, in the strip $0 < v < 1$ the helicoid is timelike and its Weingarten map is not diagonalizable. The equation for the lines of curvature in this case is $(v')^2 - (u')^2(-1 + v^2) = 0$, and a similar argument to the previous one (assuming $v \neq 1$ only) gives us that $v' \neq 0$, allowing us to use v as a parameter again. The equation:

$$\left(\frac{du}{dv}\right)^2 = \frac{1}{-1+v^2}$$

is incompatible on $0 < v < 1$. We'll suppose then that $v > 1$, so that:

$$\frac{du}{dv} = \pm\frac{1}{\sqrt{-1+v^2}} \implies u(v) = \pm \ln |v + \sqrt{-1+v^2}| + c,$$

as before.

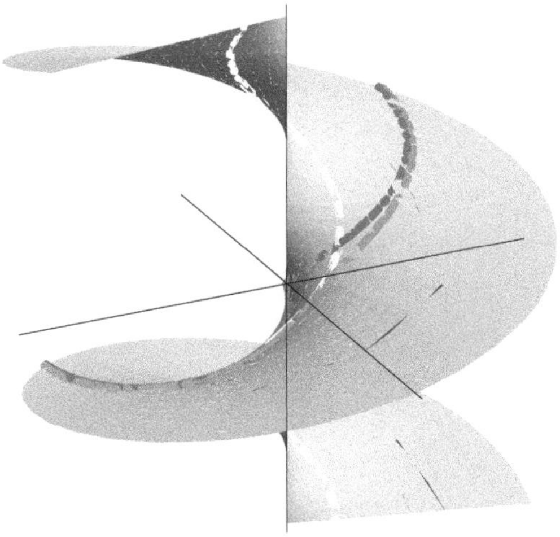

Figure 3.29: The lines of curvature in a helicoid—the inner one in $\mathbb{R}^3$ and the outer one in $\mathbb{L}^3$. The middle one is the horizon of causal type change in the surface.

Observe that the Lorentzian line of curvature does not enter the timelike region of the surface, as expected.

The analysis of the asymptotic lines is immediate, once the equation becomes just $u'v' = 0$, whence we conclude that the asymptotic lines are the coordinate curves, in both ambient spaces.

Example 3.5.11 ("*Funnel surface*")**.** In the last example above, we have seen that the parametrization given for the helicoid is asymptotic. In general, this is not the case, but when we do have asymptotic directions, we may integrate the equation for the asymptotic lines to obtain an asymptotic parametrization.

We will illustrate this procedure by producing an asymptotic parametrization to the graph M, seen in $\mathbb{R}^3$, of the radial function $h\colon \mathbb{R}^2 \setminus \{(0,0)\} \to \mathbb{R}$, given by $h(x,y) = \log\sqrt{x^2+y^2}$. Instead of starting with the usual Monge parametrization, consider instead $\boldsymbol{x}\colon \mathbb{R}_{>0} \times]0,2\pi[\to M$ defined by $\boldsymbol{x}(r,\theta) = (r\cos\theta, r\sin\theta, \log r)$. For this parametrization, we obtain

$$e(r,\theta) = \frac{-1}{r\sqrt{1+r^2}}, \quad f = 0 \quad \text{and} \quad g(r,\theta) = \frac{r}{\sqrt{1+r^2}},$$

whence the equation for the asymptotic lines boils down to

$$-\frac{r'(t)^2}{r(t)} + r(t)\theta'(t)^2 = 0.$$

This leads us to two equations, $\theta'(t) = \pm r'(t)/r(t)$, whence

$$\theta(t) + 2u = \log r(t) \quad \text{and} \quad \theta(t) + 2v = -\log r(t),$$

for certain constants of integration u and v, which will be the new parameters. Solving for u and v in terms of r and θ, we obtain a new parametrization

$$\boldsymbol{y}(u,v) \doteq \boldsymbol{x}(\mathrm{e}^{u-v}, -u-v) = \left(\mathrm{e}^{u-v}\cos(u+v), -\mathrm{e}^{u-v}\sin(u+v), u-v\right),$$

which is asymptotic. You may verify this in Exercise 3.5.10.

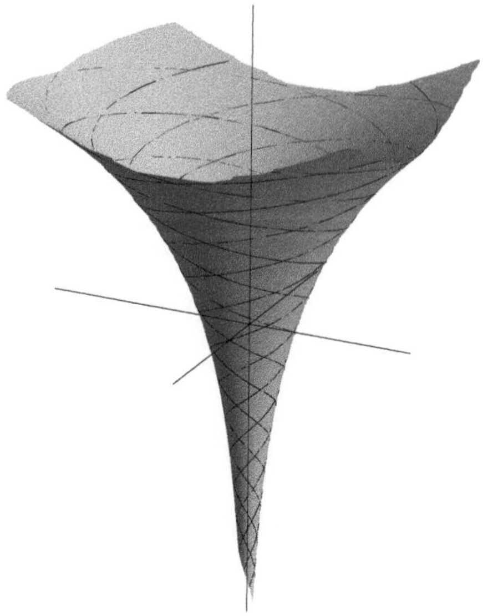

Figure 3.30: The funnel surface.

To geometrically understand the whole situation, it is interesting to take the point of view of an inhabitant of M: more precisely, if $\boldsymbol{\alpha}\colon I \to M$ is regular, we know that the velocity vector $\boldsymbol{\alpha}'(t)$ is always tangent to M at $\boldsymbol{\alpha}(t)$, but in general we do not know anything about the acceleration vector $\boldsymbol{\alpha}''(t)$. Since M is non-degenerate, we always have an orthogonal direct sum decomposition $\mathbb{R}^3_\nu = T_{\boldsymbol{\alpha}(t)}M \oplus (T_{\boldsymbol{\alpha}(t)}M)^\perp$, which allows us to write

$$\boldsymbol{\alpha}''(t) = \boldsymbol{\alpha}''(t)_{\tan} + \boldsymbol{\alpha}''(t)_{\rm nor} \doteq \frac{D\boldsymbol{\alpha}'}{dt}(t) + \mathrm{II}_{\boldsymbol{\alpha}(t)}(\boldsymbol{\alpha}'(t)),$$

where $(D\boldsymbol{\alpha}'/dt)(t) \in T_{\boldsymbol{\alpha}(t)}M$ is the so-called *covariant derivative* of $\boldsymbol{\alpha}'$ at t, being exactly the part of the acceleration of $\boldsymbol{\alpha}$ which is "detected" by an inhabitant of M.

This way, asymptotic lines in M are the curves for which an inhabitant of M feels the *full* acceleration of motion. To wit, $\kappa_n(\boldsymbol{\alpha}'(t)) = 0$ directly implies that $\mathrm{II}_{\boldsymbol{\alpha}(t)}(\boldsymbol{\alpha}'(t)) = \mathbf{0}$, so that $\boldsymbol{\alpha}''(t) = (D\boldsymbol{\alpha}'/dt)(t)$ is tangent to M.

Remark. We will see in the next section the dual notion, of a *geodesic*: the curves for which an inhabitant of M does not feel *any* acceleration, that is, curves satisfying $D\boldsymbol{\alpha}'/dt = \mathbf{0}$. In a general surface, geodesics play the role that straight lines play in a plane, and they are of fundamental importance in Differential Geometry. In particular, the above discussion tells us that if a line is contained in a surface, it is simultaneously asymptotic and geodesic (in fact, the converse also holds).

Exercises

Exercise 3.5.1 (Averages). Let $M \subseteq \mathbb{R}^3_\nu$ be a spacelike surface and $\boldsymbol{p} \in M$. Adopting the notation from Proposition 3.5.3 (p. 208), show that:

(a) $H(\boldsymbol{p}) = (-1)^\nu \dfrac{\kappa_n(\theta) + \kappa_n(\theta + \pi/2)}{2}$, for each $\theta \in [0, 2\pi[$;

(b) $H(\boldsymbol{p}) = \dfrac{(-1)^\nu}{m} \sum_{k=1}^{m} \kappa_n(\theta_k)$, where $m \geq 2$ and $\theta_k \doteq 2k\pi/m$;

Hint. First show that $\sum_{k=1}^{m} \cos(2\theta_k) = 0$. Use complex numbers and look at a certain geometric progression.

(c) $H(\boldsymbol{p}) = \dfrac{(-1)^\nu}{2\pi} \displaystyle\int_0^{2\pi} \kappa_n(\theta)\, \mathrm{d}\theta$.

Exercise 3.5.2 (More averages). Even though the functions $\cosh$ and $\sinh$ are not periodic, it is possible to obtain similar results to the ones in the previous exercise. Let $M \subseteq \mathbb{L}^3$ be a timelike surface and $\boldsymbol{p} \in M$. Assuming that $-\mathrm{d}\boldsymbol{N}_{\boldsymbol{p}}$ is diagonalizable and adopting the notation from Proposition 3.5.3, show that:

(a) $H(\boldsymbol{p}) = -\dfrac{1}{m} \sum_{k=1}^{m} \kappa_n^{\text{s.l.}}(\varphi_k) + \kappa_n^{\text{t.l.}}(\varphi_k)$, where $m \geq 2$ and $\varphi_1, \ldots, \varphi_m \in \mathbb{R}$ are arbitrary parameters;

(b) $H(\boldsymbol{p}) = -\dfrac{1}{2\varphi} \displaystyle\int_0^{\varphi} \kappa_n^{\text{s.l.}}(\tau) + \kappa_n^{\text{t.l.}}(\tau)\, \mathrm{d}\tau$, for all $\varphi > 0$.

Exercise 3.5.3. Show that if $M \subseteq \mathbb{R}^3_\nu$ is a spacelike surface and $\boldsymbol{p} \in M$ is hyperbolic, then the principal directions at $\boldsymbol{p}$ bisect the asymptotic ones. What does the situation look like if M is timelike?

Exercise 3.5.4. Show that the meridians of a torus in $\mathbb{R}^3$ are lines of curvature.

Remark. You may consider, for concreteness, the torus from Exercise 3.4.8 (p. 206), to be more precise.

Exercise 3.5.5. Consider the paraboloid $M \subseteq \mathbb{R}^3$, graph of the smooth function $f\colon \mathbb{R}^2 \to \mathbb{R}$, $f(u,v) = u^2 + v^2$. Show that the origin $\mathbf{0}$ is an umbilic point of M with the following property: for every non-zero $\boldsymbol{v} \in T_{\mathbf{0}}M$, there is a line of curvature $\boldsymbol{\alpha}\colon I \to M$ with $\boldsymbol{\alpha}(0) = \mathbf{0}$ and $\boldsymbol{\alpha}'(0) = \boldsymbol{v}$.

Exercise 3.5.6. Let $M_1, M_2 \subseteq \mathbb{R}^3_\nu$ be non-degenerate regular surfaces, $\boldsymbol{N}_1$ and $\boldsymbol{N}_2$ be Gauss maps for M_1 and M_2, and $\boldsymbol{\alpha}\colon I \to M_1 \cap M_2$ be a curve in both surfaces. Show that if $\langle \boldsymbol{N}_1(\boldsymbol{\alpha}(t)), \boldsymbol{N}_2(\boldsymbol{\alpha}(t)) \rangle \neq \pm 1$ independent of t, then $\boldsymbol{\alpha}$ is a line of curvature for one of the surfaces if and only if it is for the other one as well.

Hint. By symmetry, it suffices to show that if $\boldsymbol{\alpha}$ is a line of curvature for M_1, then it is also one for M_2. Write $\mathrm{d}(\boldsymbol{N}_2)_{\boldsymbol{\alpha}(t)}(\boldsymbol{\alpha}'(t))$ as a linear combination of $\boldsymbol{\alpha}'(t)$, $\boldsymbol{N}_1(\boldsymbol{\alpha}(t))$ and $\boldsymbol{N}_2(\boldsymbol{\alpha}(t))$ (the assumptions ensure that this is possible).

Remark. This happens, for instance, if the surfaces are in $\mathbb{R}^3$ and the normal directions form a constant angle along $\boldsymbol{\alpha}$. Being more precise, the exercise treats the situation where these normal directions are not parallel (because if this is the case, the result trivially holds).

Exercise 3.5.7. Let $M \subseteq \mathbb{R}^3_\nu$ be a non-degenerate regular surface with non-vanishing Gaussian curvature, $\boldsymbol{N}$ be a Gauss map for M, and $(U, \boldsymbol{x})$ a principal parametrization for M. Show that $\boldsymbol{y}\colon U \to \mathbb{R}^3_\nu$ defined by $\boldsymbol{y}(u,v) \doteq \boldsymbol{N}(\boldsymbol{x}(u,v))$ is a regular parametrized surface whose First Fundamental Form coefficients are given by $E^{(\boldsymbol{y})} = \kappa_1^2 E^{(\boldsymbol{x})}$, $F^{(\boldsymbol{y})} = 0$ and $G^{(\boldsymbol{y})} = \kappa_2^2 G^{(\boldsymbol{x})}$, where $\kappa_i \equiv \kappa_i \circ \boldsymbol{x}$ are the principal curvatures of M.

Exercise† 3.5.8 (Dupin Indicatrix). Let $M \subseteq \mathbb{R}^3_\nu$ be a non-degenerate regular surface and $(U, \boldsymbol{x})$ be a parametrization for M around $\boldsymbol{p} \doteq \boldsymbol{x}(u_0, v_0)$. Assume that $\boldsymbol{p}$ is non-planar and, when $(-1)^\nu K(\boldsymbol{p}) \geq 0$, choose the normal $\boldsymbol{N}$ in such a way that $h_{11}(u_0, v_0), h_{22}(u_0, v_0) \geq 0$.

(a) We define the *Dupin indicatrix* of M at $\boldsymbol{p}$ as the conic in $T_{\boldsymbol{p}}M$ given by the equation $\widetilde{\mathrm{II}}_{\boldsymbol{p}}(\boldsymbol{v}) = 1$. Show that if $\boldsymbol{p}$ is an elliptic (resp. hyperbolic, parabolic) point, then the Dupin indicatrix of M at $\boldsymbol{p}$ is an ellipse (resp. hyperbola, pair of parallel lines).

(b) Show that if $\boldsymbol{p}$ is hyperbolic, then the asymptotes of the Dupin indicatrix of M at $\boldsymbol{p}$ are given by the equation $\widetilde{\mathrm{II}}_{\boldsymbol{p}}(\boldsymbol{v}) = 0$, that is, the collection of asymptotic vectors of M at $\boldsymbol{p}$.

Exercise 3.5.9. Determine, in both ambient spaces, the lines of curvature and the asymptotic lines for the graph of the function $h\colon \mathbb{R}^2 \to \mathbb{R}$, $h(x, y) = xy$.

Exercise 3.5.10. Verify that the parametrization $\boldsymbol{y}$ obtained in Example 3.5.11 (p. 215) is indeed asymptotic.

Exercise 3.5.11. Let $h\colon \mathbb{R}_{\geq 0} \to \mathbb{R}$ be a smooth function. Show that the equation for the asymptotic lines of a:

(a) polar parametrization $\boldsymbol{x}\colon U \to \mathbb{R}^3_\nu$ of the form $\boldsymbol{x}(r, \theta) = (r\cos\theta, r\sin\theta, h(r))$, is

$$h''(r(t))r'(t)^2 + h'(r(t))r(t)\theta'(t)^2 = 0,$$

in both ambient spaces;

(b) Rindler parametrization $\boldsymbol{x}\colon U \to \mathbb{L}^3$ of the form $\boldsymbol{x}(\rho, \varphi) = (h(\rho), \rho\cosh\varphi, \rho\sinh\varphi)$, is

$$h''(\rho(t))\rho'(t)^2 - h'(\rho(t))\rho(t)\varphi'(t)^2 = 0.$$

Exercise 3.5.12. Repeat the strategy adopted in Example 3.5.11 (p. 215) to deduce that

$$\boldsymbol{y}(u, v) = \left(\mathrm{e}^{(u-v)/\sqrt{1-2\alpha}}\cos(u+v), -\mathrm{e}^{(u-v)/\sqrt{1-2\alpha}}\sin(u+v), \mathrm{e}^{2\alpha(u-v)/\sqrt{1-2\alpha}}\right)$$

is an asymptotic parametrization for the graph, in $\mathbb{R}^3$, of the smooth function $h\colon \mathbb{R}^2 \to \mathbb{R}$ given by $h(x, y) = (x^2 + y^2)^\alpha$, where $0 \neq \alpha < 1/2$.

Hint. Start with a polar parametrization and let the constants of integration obtained from the differential equation be denoted by $2u$ and $2v$ instead of u and v, to simplify the notation later.

Exercise 3.5.13 (Beltrami-Enneper). Let $M \subseteq \mathbb{R}^3_\nu$ be a non-degenerate regular surface and $\boldsymbol{\alpha}\colon I \to M$ be an admissible unit speed asymptotic. Show that the Gaussian curvature along $\boldsymbol{\alpha}$ is given by $K(\boldsymbol{\alpha}(s)) = (-1)^{\nu+1}\tau_{\boldsymbol{\alpha}}(s)^2$ for all $s \in I$.

Hint. If $\boldsymbol{\alpha}$ is asymptotic, then $\boldsymbol{\alpha}''(s)$ is always tangent to M and thus $T_{\boldsymbol{\alpha}(s)}M$ is always spanned by $\boldsymbol{T}_{\boldsymbol{\alpha}}(s)$ and $\boldsymbol{N}_{\boldsymbol{\alpha}}(s)$. Moreover, up to sign $\boldsymbol{B}_{\boldsymbol{\alpha}}(s)$ is a unit normal to M along $\boldsymbol{\alpha}$. Compute $\tau_{\boldsymbol{\alpha}}$ in terms of $\widetilde{\mathrm{II}}_{\boldsymbol{p}}$ and $K(\boldsymbol{\alpha}(s))$ by using the Gauss equation given in Exercise 3.3.1 (p. 187).

3.6 GEODESICS, VARIATIONAL METHODS AND ENERGY

Following the conclusion from the discussion in the end of the previous section, we'll start registering a few definitions:

Definition 3.6.1. Let $M \subseteq \mathbb{R}^3_\nu$ be a non-degenerate regular surface and $\boldsymbol{\alpha}\colon I \to M$ be a regular curve. A *vector field along* $\boldsymbol{\alpha}$ is a smooth map $\boldsymbol{V}\colon I \to \mathbb{R}^3_\nu$ such that for all $t \in I$, $\boldsymbol{V}(t) \in T_{\boldsymbol{\alpha}(t)}M$.

We're interested in knowing how vector fields change along $\boldsymbol{\alpha}$, but from the point of view of an inhabitant of M. This motivates the:

Definition 3.6.2. Let $M \subseteq \mathbb{R}^3_\nu$ be a non-degenerate regular surface, $\boldsymbol{\alpha}\colon I \to M$ be a regular curve and $\boldsymbol{V}\colon I \to \mathbb{R}^3_\nu$ be a vector field along $\boldsymbol{\alpha}$. The *covariant derivative of* $\boldsymbol{V}$ *along* $\boldsymbol{\alpha}$ is defined by

$$\frac{D\boldsymbol{V}}{dt}(t) \doteq \text{proj}_{T_{\boldsymbol{\alpha}(t)}M}\boldsymbol{V}'(t) \equiv \boldsymbol{V}'(t)_{\tan}.$$

We'll also say that $\boldsymbol{V}$ is *parallel* if $D\boldsymbol{V}/dt = \boldsymbol{0}$.

Remark.

- In this definition it is crucial that M is non-degenerate, or else we cannot consider projections in the tangent planes.
- Some basic properties of the operator D/dt are listed in Exercise 3.6.1.

Example 3.6.3.

(1) Our simplest example of a regular surface is a plane. Let $\Pi = \{\boldsymbol{p} \in \mathbb{R}^3_\nu \mid \langle \boldsymbol{p}, \boldsymbol{n}\rangle = c\}$ be a non-degenerate plane. For each $\boldsymbol{p} \in \Pi$, we have that the tangent plane is $T_{\boldsymbol{p}}\Pi = \{\boldsymbol{v} \in \mathbb{R}^3_\nu \mid \langle \boldsymbol{v}, \boldsymbol{n}\rangle = 0\}$. If $\boldsymbol{\alpha}\colon I \to \Pi$ is any regular curve and $\boldsymbol{V}\colon I \to \mathbb{R}^3_\nu$ is a vector field along $\boldsymbol{\alpha}$, differentiating the relation $\langle \boldsymbol{V}(t), \boldsymbol{n}\rangle = 0$, we obtain the relation $\langle \boldsymbol{V}'(t), \boldsymbol{n}\rangle = 0$, whence $\boldsymbol{V}'(t) \in T_{\boldsymbol{\alpha}(t)}\Pi$ for all $t \in I$. In other words, we see that in this case

$$\frac{D\boldsymbol{V}}{dt}(t) = \boldsymbol{V}'(t),$$

so that the covariant derivative may be seen as a generalization of the usual derivative, that now takes into account the geometry of the surface on where the base curve lies.

(2) Now let M be $\mathbb{S}^2$, $\mathbb{S}^2_1$ or $\mathbb{H}^2$, on the adequate ambient spaces. Let $\boldsymbol{\alpha}\colon I \to M$ be a regular curve and $\boldsymbol{V}\colon I \to \mathbb{R}^3_\nu$ be a vector field along $\boldsymbol{\alpha}$. Since a Gauss map for M is the position vector field itself, we have that

$$\begin{aligned}\frac{D\boldsymbol{V}}{dt}(t) &= \boldsymbol{V}'(t) - \frac{\langle \boldsymbol{V}'(t), \boldsymbol{\alpha}(t)\rangle}{\langle \boldsymbol{\alpha}(t), \boldsymbol{\alpha}(t)\rangle}\boldsymbol{\alpha}(t) \\ &= \boldsymbol{V}'(t) - \epsilon_M\langle \boldsymbol{V}'(t), \boldsymbol{\alpha}(t)\rangle\boldsymbol{\alpha}(t) \\ &= \boldsymbol{V}'(t) + \epsilon_M\langle \boldsymbol{V}(t), \boldsymbol{\alpha}'(t)\rangle\boldsymbol{\alpha}(t),\end{aligned}$$

where the last equality follows from the fact that $\langle \boldsymbol{V}(t), \boldsymbol{\alpha}(t)\rangle = 0$ holds for all $t \in I$. In particular, if $\boldsymbol{\alpha}$ has unit speed we conclude that

$$\frac{D\boldsymbol{\alpha}'}{ds}(s) = \boldsymbol{\alpha}''(s) + \epsilon_M\epsilon_{\boldsymbol{\alpha}}\boldsymbol{\alpha}(s).$$

In the most general case, when M is a non-degenerate regular surface, we have a correction factor given precisely by the Second Fundamental Form:

$$V'(t) = \frac{DV}{dt}(t) + \mathrm{II}_{\alpha(t)}\big(\alpha'(t), V(t)\big).$$

The main example of a vector field we should keep in mind for now, already mentioned in item (2) above, is the *velocity field* of α, $\alpha' \colon I \to \mathbb{R}^3_\nu$ itself.

Definition 3.6.4 (Geodesics)**.** Let $M \subseteq \mathbb{R}^3_\nu$ be a non-degenerate regular surface. We'll say that a curve $\alpha \colon I \to M$ is a *geodesic* if $D\alpha'/dt = \mathbf{0}$ or, equivalently, if $\alpha''(t)$ is normal to $T_{\alpha(t)}M$, for all $t \in I$.

A basic fact about geodesics that must be immediately registered is the:

Proposition 3.6.5. *Let $M \subseteq \mathbb{R}^3_\nu$ be a non-degenerate regular surface, and $\alpha \colon I \to M$ be a geodesic. Then α has constant speed.*

Proof: It suffices to note that since $\alpha''(t)$ is always normal to M and $\alpha'(t)$ is always tangent, then

$$\frac{d}{dt}\langle \alpha'(t), \alpha'(t)\rangle = 2\langle \alpha''(t), \alpha'(t)\rangle = 0.$$

□

Remark. In particular, we have that geodesics have constant causal character. Moreover, we'll say that a geodesic is *normalized* if it has unit speed.

Example 3.6.6. It follows from item (1) in Example 3.6.3 above that the geodesics in any non-degenerate plane Π are precisely the lines contained in it. To wit, every geodesic α satisfies $\alpha''(t) = (D\alpha'/dt)(t) = \mathbf{0}$.

Example 3.6.7. Consider the cylinder $\mathbb{S}^1(r) \times \mathbb{R} \subseteq \mathbb{R}^3_\nu$, with the usual parametrization $\mathbf{x} \colon \,]0, 2\pi[\, \times \mathbb{R} \to \mathbb{S}^1(r) \times \mathbb{R}$ given by $\mathbf{x}(u,v) = (r\cos u, r\sin u, v)$. Let $\alpha \colon I \to \mathbb{S}^1(r) \times \mathbb{R}$ be a curve given in coordinates as

$$\alpha(t) = \mathbf{x}(u(t), v(t)) = (r\cos u(t), r\sin u(t), v(t)).$$

Differentiating twice, we obtain

$$\begin{aligned} \alpha''(t) &= \Big(-ru''(t)\sin u(t) - ru'(t)^2\cos u(t), ru''(t)\cos u(t) - ru'(t)^2 \sin u(t), v''(t)\Big) \\ &= \Big(-ru''(t)\sin u(t), ru''(t)\cos u(t), v''(t)\Big) - ru'(t)^2\Big(\cos u(t), \sin u(t), 0\Big). \end{aligned}$$

Noting that a Gauss map for the cylinder, in both ambient spaces, is given by $N(\mathbf{x}(u,v)) = (\cos u, \sin u, 0)$, we conclude that

$$\frac{D\alpha'}{dt}(t) = (-ru''(t)\sin u(t), ru''(t)\cos u(t), v''(t)).$$

This way, we see that α is a geodesic if and only if $u''(t) = v''(t) = 0$, that is, if $u(t) = at + b$ and $v(t) = ct + d$. So the geodesics of the cylinder are given by

$$\alpha(t) = (r\cos(at+b), r\sin(at+b), ct+d),$$

for certain constants $a, b, c, d \in \mathbb{R}$. In particular, note that the geodesics are the same, no matter the ambient space where the cylinder is. There are four possibilities:

- if $a = c = 0$, the curve collapses to a point;
- if $a \neq 0$ and $c = 0$, we have horizontal circles;
- if $a = 0$ and $c \neq 0$, we have vertical lines;
- if $a \neq 0$ and $c \neq 0$, we have helices.

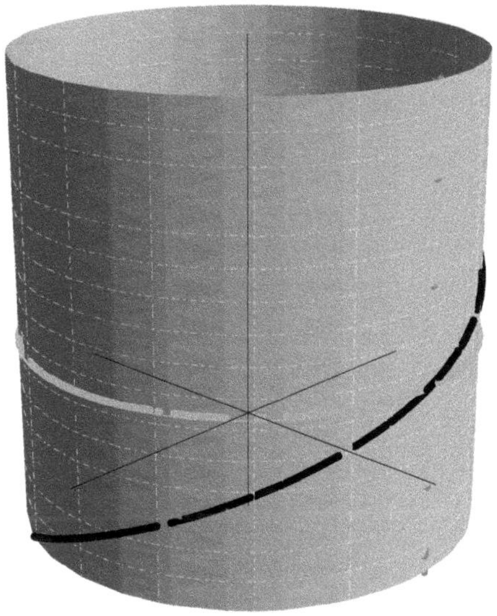

Figure 3.31: Geodesics in the cylinder.

We'll leave more examples for later, when we have more powerful variational techniques at our disposal. Such techniques are also useful and interesting on their own. Moreover, we will see a much larger class of examples when we discuss abstract metrics.

Occasionally, the following notion is also useful:

Definition 3.6.8 (Pre-geodesics). Let $M \subseteq \mathbb{R}^3_\nu$ be a non-degenerate regular surface. We'll say that a curve $\boldsymbol{\alpha}\colon I \to M$ is a *pre-geodesic* if $\boldsymbol{\alpha}$ admits a reparametrization which is a geodesic.

In other words, a pre-geodesic is nothing more than a geodesic, up to reparametrization. Pre-geodesics are characterized in the following:

Proposition 3.6.9. *Let $M \subseteq \mathbb{R}^3_\nu$ be a non-degenerate regular surface, and $\boldsymbol{\alpha}\colon I \to M$ be a regular curve. Then $\boldsymbol{\alpha}$ is a pre-geodesic if and only if $(D\boldsymbol{\alpha}'/\mathrm{d}t)(t)$ and $\boldsymbol{\alpha}'(t)$ are collinear, for all $t \in I$.*

Proof: Suppose initially that $\boldsymbol{\alpha}$ is a pre-geodesic and let $h\colon J \to I$ be a diffeomorphism such that the curve $\boldsymbol{\beta} \doteq \boldsymbol{\alpha} \circ h\colon J \to M$ is a geodesic. Differentiating twice the relation $\boldsymbol{\beta}(s) = \boldsymbol{\alpha}(h(s))$ with respect to s and setting $t = h(s)$ we obtain

$$\boldsymbol{\beta}''(s) = h'(h^{-1}(t))^2\boldsymbol{\alpha}''(t) + h''(h^{-1}(t))\boldsymbol{\alpha}'(t).$$

Projecting the above in the tangent plane $T_{\boldsymbol{\beta}(s)}M = T_{\boldsymbol{\alpha}(t)}M$, we have that

$$\mathbf{0} = \frac{D\boldsymbol{\beta}'}{\mathrm{d}s}(s) = h'(h^{-1}(t))^2\frac{D\boldsymbol{\alpha}'}{\mathrm{d}t}(t) + h''(h^{-1}(t))\boldsymbol{\alpha}'(t),$$

whence

$$\frac{D\boldsymbol{\alpha}'}{dt}(t) = -\frac{h''(h^{-1}(t))}{h'(h^{-1}(t))^2}\boldsymbol{\alpha}'(t)$$

are proportional for all $t \in I$, as wanted.

Conversely, suppose that we have $(D\boldsymbol{\alpha}'/dt)(t) = f(t)\boldsymbol{\alpha}'(t)$, for some smooth function $f\colon I \to \mathbb{R}$. Let's see what condition the diffeomorphism h must satisfy for $\boldsymbol{\alpha} \circ h$ to be a geodesic, and use such condition to try and define it. Well, the computation done above shows that

$$\big(h'(s)^2 f(h(s)) + h''(s)\big)\boldsymbol{\alpha}'(t) = \mathbf{0}$$

must hold for all s in the domain of h, which by regularity of $\boldsymbol{\alpha}$ implies that

$$h''(s) + f\,(h(s))\,h'(s)^2 = 0$$

for all s. We then look for a solution of this differential equation. Define $F(t) = \int f(t)\,dt$ and $g(t) = \int e^{F(t)}\,dt$. Since $g'(t) = e^{F(t)} > 0$, g is a positive diffeomorphism of I onto $J \doteq g(I)$, and so we may consider $h \doteq g^{-1}\colon J \to I$. Let's show that h is a solution for the latter ODE (which ensures that $\boldsymbol{\alpha} \circ h$ is a geodesic). Set $t = h(s)$ and $s = g(t)$, as before. Differentiating $h(g(t)) = t$ with respect to t twice gives us the relations

$$h'(s) = \frac{1}{g'(t)} \quad \text{and} \quad h''(s) = -\frac{g''(t)}{g'(t)^3}.$$

Moreover, $g''(t) = f(t)e^{F(t)}$, and with this we have:

$$\begin{aligned} h''(s) + f(h(s))h'(s)^2 &= -\frac{g''(t)}{g'(t)^3} + \frac{f(t)}{g'(t)^2} \\ &= \frac{1}{g'(t)^3}(-g''(t) + f(t)g'(t)) \\ &= \frac{1}{g'(t)^3}(-f(t)e^{F(t)} + f(t)e^{F(t)}) = 0, \end{aligned}$$

as wanted. □

Remark. It also follows from this that pre-geodesics with non-zero constant speed are, in fact, geodesics. To wit, differentiating the relation $\langle \boldsymbol{\alpha}'(t), \boldsymbol{\alpha}'(t)\rangle = c \neq 0$ gives us that $\langle \boldsymbol{\alpha}''(t), \boldsymbol{\alpha}'(t)\rangle = 0$. Discarding the normal component of $\boldsymbol{\alpha}''(t)$ we obtain

$$0 = \left\langle \frac{D\boldsymbol{\alpha}'}{dt}(t), \boldsymbol{\alpha}'(t)\right\rangle = cf(t) \implies f(t) = 0 \implies \frac{D\boldsymbol{\alpha}'}{dt} = \mathbf{0}.$$

On the other hand, what we'll do in what follows motivates a relatively simple proof that every lightlike curve is a pre-geodesic (see Exercise 3.6.10).

3.6.1 Darboux-Ribaucour frame

In Chapter 2, our main tool in the study of curves was the Frenet-Serret trihedron: a frame for $\mathbb{R}^3_\nu$ adapted to the curve. To simultaneously study lines of curvature, asymptotic lines and geodesics, we'll introduce a new trihedron, adapted not only to the curve itself, but also to the surface where the curve lies.

Let $M \subseteq \mathbb{R}^3_\nu$ be a non-degenerate regular surface, $\mathbf{N}$ be a Gauss map for M, and $\boldsymbol{\alpha}\colon I \to M$ be a unit speed admissible curve. Motivated by the Frenet-Serret trihedron, but this time requiring the unit normal field to M along $\boldsymbol{\alpha}$ to play the role of the binormal field $\boldsymbol{B}_{\boldsymbol{\alpha}}$, we look for a third element $\boldsymbol{V}_{\boldsymbol{\alpha}}$ such that $\big(\boldsymbol{T}_{\boldsymbol{\alpha}}(s), \boldsymbol{V}_{\boldsymbol{\alpha}}(s), \boldsymbol{N}(\boldsymbol{\alpha}(s))\big)$ is a positive orthonormal basis for $\mathbb{R}^3_\nu$, for all $s \in I$.

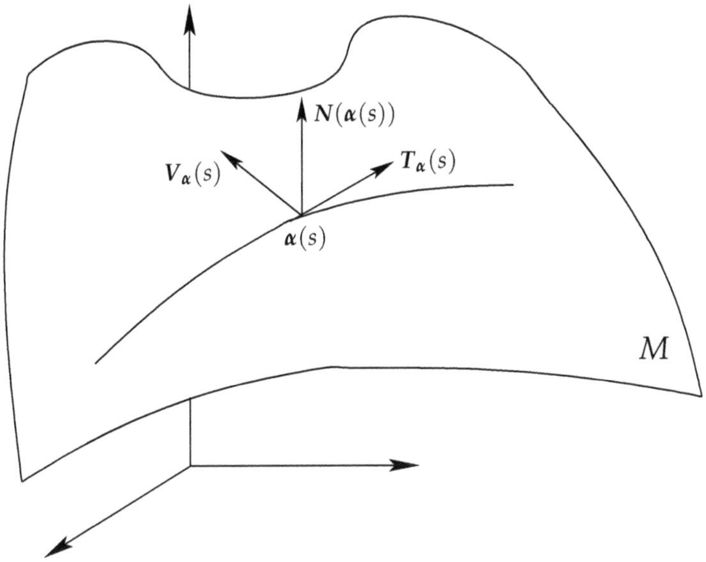

Figure 3.32: Constructing a frame adapted to M.

The orientability analysis done in Section 1.6 tells us that

$$\boldsymbol{V}_{\boldsymbol{\alpha}}(s) \doteq (-1)^{\nu+1}\epsilon_{\boldsymbol{\alpha}}\epsilon_M \boldsymbol{T}_{\boldsymbol{\alpha}}(s) \times \boldsymbol{N}(\boldsymbol{\alpha}(s))$$

is the vector we are looking for.

Definition 3.6.10 (Darboux-Ribaucour Trihedron)**.** Let $M \subseteq \mathbb{R}^3_\nu$ be a non-degenerate regular surface, $\boldsymbol{N}$ be a Gauss map for M, and $\boldsymbol{\alpha}\colon I \to M$ be a unit speed admissible curve. The *Darboux-Ribaucour frame* of $\boldsymbol{\alpha}$ at $s \in I$ is the positive orthonormal basis $\big(\boldsymbol{T}_{\boldsymbol{\alpha}}(s), \boldsymbol{V}_{\boldsymbol{\alpha}}(s), \boldsymbol{N}(\boldsymbol{\alpha}(s))\big)$ constructed above.

The next step, naturally, is to express the derivatives of those vectors as combinations of the vectors themselves, via orthonormal expansions. As before, some coefficients cannot be expressed in terms of objects already known, and so they will be baptized:

Definition 3.6.11 (Geodesic curvature and torsion)**.** Let $M \subseteq \mathbb{R}^3_\nu$ be a non-degenerate regular surface, $\boldsymbol{N}$ be a Gauss map for M, and $\boldsymbol{\alpha}\colon I \to M$ be a unit speed admissible curve.

(i) The *geodesic curvature* of $\boldsymbol{\alpha}$ at s is the component of $\boldsymbol{\alpha}''(s) = \boldsymbol{T}'_{\boldsymbol{\alpha}}(s)$ in the direction of $\boldsymbol{V}_{\boldsymbol{\alpha}}(s)$, and it is denoted by $\kappa_{g,\boldsymbol{\alpha}}(s)$.

(ii) The *geodesic torsion* of $\boldsymbol{\alpha}$ at s is the component of $\boldsymbol{V}'_{\boldsymbol{\alpha}}(s)$ in the direction of $\boldsymbol{N}(\boldsymbol{\alpha}(s))$, and it is denoted by $\tau_{g,\boldsymbol{\alpha}}(s)$.

Remark.

- We may express $\kappa_{g,\boldsymbol{\alpha}}$ and $\tau_{g,\boldsymbol{\alpha}}$ in terms of projections:

$$\frac{D\boldsymbol{\alpha}'}{\mathrm{d}s}(s) = \kappa_{g,\boldsymbol{\alpha}}(s)\boldsymbol{V}_{\boldsymbol{\alpha}}(s) \text{ and } \mathrm{proj}_{\boldsymbol{N}(\boldsymbol{\alpha}(s))}\boldsymbol{V}'_{\boldsymbol{\alpha}}(s) = \tau_{g,\boldsymbol{\alpha}}(s)\boldsymbol{N}(\boldsymbol{\alpha}(s)),$$

where in the first expression $\boldsymbol{T}'_{\alpha}(s) = \boldsymbol{\alpha}''(s)$ could be replaced by the covariant derivative of $\boldsymbol{\alpha}'$, since $\boldsymbol{V}_{\alpha}(s)$ is always tangent to M. In particular, we'll have that

$$\kappa_{g,\alpha}(s)^2 = (-1)^{\nu}\epsilon_{\alpha}\epsilon_M \left\langle \frac{D\boldsymbol{\alpha}'}{ds}(s), \frac{D\boldsymbol{\alpha}'}{ds}(s) \right\rangle.$$

- Since $\boldsymbol{\alpha}$ has constant speed, $\boldsymbol{T}'_{\alpha}(s)$ has no component in the direction of $\boldsymbol{T}_{\alpha}(s)$, so that the geodesic curvature effectively measures how much $\boldsymbol{T}'_{\alpha}(s)$ deviates from being normal to M and, hence, how much $\boldsymbol{\alpha}$ deviates from being a geodesic.
- The definition of geodesic torsion is meant to mimic what we have for torsion, when studying the Frenet-Serret trihedron in Chapter 2, with $\boldsymbol{V}_{\alpha}$ and $\boldsymbol{N}\circ\boldsymbol{\alpha}$ playing the roles of $\boldsymbol{N}_{\alpha}$ and $\boldsymbol{B}_{\alpha}$, respectively.

We are then ready to state and establish the:

Theorem 3.6.12 (Darboux equations). *Let $M \subseteq \mathbb{R}^3_{\nu}$ be a non-degenerate regular surface, $\boldsymbol{N}$ be a Gauss map for M, and $\boldsymbol{\alpha}\colon I \to M$ be a unit speed admissible curve. We have that:*

$$\begin{pmatrix} \boldsymbol{T}'_{\alpha}(s) \\ \boldsymbol{V}'_{\alpha}(s) \\ (\boldsymbol{N}\circ\boldsymbol{\alpha})'(s) \end{pmatrix} = \begin{pmatrix} 0 & \kappa_{g,\alpha}(s) & \epsilon_{\alpha}\epsilon_M\kappa_n(s) \\ (-1)^{\nu+1}\epsilon_M\kappa_{g,\alpha}(s) & 0 & \tau_{g,\alpha}(s) \\ -\kappa_n(s) & (-1)^{\nu+1}\epsilon_{\alpha}\tau_{g,\alpha}(s) & 0 \end{pmatrix} \begin{pmatrix} \boldsymbol{T}_{\alpha}(s) \\ \boldsymbol{V}_{\alpha}(s) \\ \boldsymbol{N}(\boldsymbol{\alpha}(s)) \end{pmatrix},$$

where we abbreviate the normal curvature of M at $\boldsymbol{\alpha}(s)$ by $\kappa_n(\boldsymbol{\alpha}'(s)) \equiv \kappa_n(s)$.

Proof: We'll repeatedly use orthonormal expansion. Since all the vectors in the Darboux-Ribaucour trihedron are unit vectors, it follows that the diagonal of the matrix of coefficients indeed consists of only zeros. The component of $\boldsymbol{T}'_{\alpha}(s)$ in the direction of $\boldsymbol{N}(\boldsymbol{\alpha}(s))$ is

$$\begin{aligned} \epsilon_M\langle \boldsymbol{T}'_{\alpha}(s), \boldsymbol{N}(\boldsymbol{\alpha}(s))\rangle &= \epsilon_M\langle \boldsymbol{T}_{\alpha}(s), -(\boldsymbol{N}\circ\boldsymbol{\alpha})'(s)\rangle \\ &= \epsilon_M\langle \boldsymbol{T}_{\alpha}(s), -d\boldsymbol{N}_{\boldsymbol{\alpha}(s)}(\boldsymbol{T}_{\alpha}(s))\rangle \\ &= \epsilon_{\alpha}\epsilon_M\kappa_n(s), \end{aligned}$$

and thus we have obtained the first equation. For the second equation, it only remains to compute the component of $\boldsymbol{V}'_{\alpha}(s)$ in the direction of $\boldsymbol{T}_{\alpha}(s)$, observing that by definition of geodesic curvature, we have that $\kappa_{g,\alpha}(s) = (-1)^{\nu}\epsilon_{\alpha}\epsilon_M\langle \boldsymbol{T}'_{\alpha}(s), \boldsymbol{V}_{\alpha}(s)\rangle$. We then have that

$$\epsilon_{\alpha}\langle \boldsymbol{V}'_{\alpha}(s), \boldsymbol{T}_{\alpha}(s)\rangle = -\epsilon_{\alpha}\langle \boldsymbol{V}_{\alpha}(s), \boldsymbol{T}'_{\alpha}(s)\rangle = (-1)^{\nu+1}\epsilon_M\kappa_{g,\alpha}(s),$$

and so we have obtained the second equation.

For the last equation, note that from the definition of geodesic torsion we have $\tau_{g,\alpha}(s) = \epsilon_M\langle \boldsymbol{V}'_{\alpha}(s), \boldsymbol{N}(\boldsymbol{\alpha}(s))\rangle$. So, the component of $(\boldsymbol{N}\circ\boldsymbol{\alpha})'(s)$ in the direction of $\boldsymbol{T}_{\alpha}(s)$ is

$$\epsilon_{\alpha}\langle (\boldsymbol{N}\circ\boldsymbol{\alpha})'(s), \boldsymbol{T}_{\alpha}(s)\rangle = -\epsilon_{\alpha}\langle -d\boldsymbol{N}_{\boldsymbol{\alpha}(s)}(\boldsymbol{T}_{\alpha}(s)), \boldsymbol{T}_{\alpha}(s)\rangle = -\kappa_n(s),$$

while the component in the direction of $\boldsymbol{V}_{\alpha}(s)$ is

$$\begin{aligned} (-1)^{\nu}\epsilon_{\alpha}\epsilon_M\langle (\boldsymbol{N}\circ\boldsymbol{\alpha})'(s), \boldsymbol{V}_{\alpha}(s)\rangle &= (-1)^{\nu+1}\epsilon_{\alpha}\epsilon_M\langle \boldsymbol{N}(\boldsymbol{\alpha}(s)), \boldsymbol{V}'_{\alpha}(s)\rangle \\ &= (-1)^{\nu+1}\epsilon_{\alpha}\tau_{g,\alpha}(s). \end{aligned}$$

□

With these equations in place, we may decompose the curvature of $\boldsymbol{\alpha}$, seen as a curve in $\mathbb{R}^3_\nu$, into a component tangent to M and into a component normal to M. In other words, the geodesic curvature is the part of the curvature of $\boldsymbol{\alpha}$ detected by an inhabitant of M. Or yet: in M, a geodesic $\boldsymbol{\alpha}$ is "straight", and so the curvature of $\boldsymbol{\alpha}$ in $\mathbb{R}^3_\nu$ is forced by the curvature of M in $\mathbb{R}^3_\nu$. These more qualitative interpretations are formalized in the:

Corollary 3.6.13 (Pythagoras). *Let $M \subseteq \mathbb{R}^3_\nu$ be a non-degenerate regular surface, $\boldsymbol{N}$ be a Gauss map for M, and $\boldsymbol{\alpha}\colon I \to M$ be a unit speed admissible curve. Then*

$$\eta_{\boldsymbol{\alpha}}\kappa_{\boldsymbol{\alpha}}(s)^2 = (-1)^\nu \epsilon_{\boldsymbol{\alpha}}\epsilon_M \kappa_{g,\boldsymbol{\alpha}}(s)^2 + \epsilon_M \kappa_n(s)^2,$$

for all $s \in I$.

Proof: It suffices to apply $\langle\cdot,\cdot\rangle$ in both sides of the first Darboux equation, bearing in mind that $\boldsymbol{T}'_{\boldsymbol{\alpha}}(s) = \kappa_{\boldsymbol{\alpha}}(s)\boldsymbol{N}_{\boldsymbol{\alpha}}(s)$ and that $\kappa_n(s) = \epsilon_{\boldsymbol{\alpha}}\kappa_{\boldsymbol{\alpha}}(s)\langle \boldsymbol{N}(\boldsymbol{\alpha}(s)), \boldsymbol{N}_{\boldsymbol{\alpha}}(s)\rangle$. ☐

Remark. In particular note that, in $\mathbb{R}^3$, the above formula boils down to $\kappa_{\boldsymbol{\alpha}}^2 = \kappa_{g,\boldsymbol{\alpha}}^2 + \kappa_n^2$. The usefulness of this corollary is to provide a practical method for computing $\kappa_{g,\boldsymbol{\alpha}}$, when necessary.

And lastly, the geodesic torsion "completes the toolbox" provided by the Darboux-Ribaucour trihedron:

Theorem 3.6.14. *Let $M \subseteq \mathbb{R}^3_\nu$ be a non-degenerate regular surface, and $\boldsymbol{\alpha}\colon I \to M$ be a unit speed admissible curve. Then:*

$$\begin{cases} \boldsymbol{\alpha} \text{ is a geodesic} \iff \kappa_{g,\boldsymbol{\alpha}} \equiv 0; \\ \boldsymbol{\alpha} \text{ is an asymptotic line} \iff \kappa_n \equiv 0; \\ \boldsymbol{\alpha} \text{ is a line of curvature} \iff \tau_{g,\boldsymbol{\alpha}} \equiv 0. \end{cases}$$

Remark. It is not complicated to redo the construction of the Darboux-Ribaucour trihedron for curves not necessarily having unit speed. The Pythagorean relation still holds without modifications, but "geodesic" has to be replaced by "pre-geodesic" in the criterion above.

The proof of this last theorem is direct and we ask you to carry it out in Exercise 3.6.5.

Exercises

Exercise 3.6.1. Let $M \subseteq \mathbb{R}^3_\nu$ be a non-degenerate regular surface, $\boldsymbol{\alpha}\colon I \to M$ be a regular curve, $\boldsymbol{V}, \boldsymbol{W}\colon I \to \mathbb{R}^3_\nu$ be vector fields along $\boldsymbol{\alpha}$, and $f\colon I \to \mathbb{R}$ be a smooth function. Show that for all $t \in I$ the following hold:

(a) $\dfrac{D(\boldsymbol{V}+\boldsymbol{W})}{dt}(t) = \dfrac{D\boldsymbol{V}}{dt}(t) + \dfrac{D\boldsymbol{W}}{dt}(t)$;

(b) $\dfrac{D(f\boldsymbol{V})}{dt}(t) = f'(t)\boldsymbol{V}(t) + f(t)\dfrac{D\boldsymbol{V}}{dt}(t)$.

Exercise 3.6.2. Let $M \subseteq \mathbb{R}^3_\nu$ be a non-degenerate regular surface, $\boldsymbol{\alpha}\colon I \to M$ be a regular curve, and $\boldsymbol{V}, \boldsymbol{W}\colon I \to \mathbb{R}^3_\nu$ be vector fields along $\boldsymbol{\alpha}$. Show that if $\boldsymbol{V}$ and $\boldsymbol{W}$ are both parallel vector fields along $\boldsymbol{\alpha}$, then $\langle \boldsymbol{V}(t), \boldsymbol{W}(t)\rangle$ is constant. Conclude that parallel fields have constant speed and causal character.

Remark. In particular, this shows that if M is spacelike, parallel vector fields always make a constant angle along $\boldsymbol{\alpha}$.

Exercise 3.6.3. Let $M \subseteq \mathbb{R}^3_\nu$ be a non-degenerate regular surface, and $\boldsymbol{\alpha}\colon I \to M$ be a (regular) geodesic. Show that if $h\colon J \to I$ is a diffeomorphism such that $\boldsymbol{\beta} \doteq \boldsymbol{\alpha} \circ h$ is also a geodesic, then h has the form $h(s) = as + b$, for certain $a, b \in \mathbb{R}$, $a \neq 0$.

Exercise† **3.6.4.** Let $M \subseteq \mathbb{R}^3_\nu$ be a non-degenerate regular surface, $\boldsymbol{\alpha}\colon I \to M$ be a unit speed admissible curve, and $\boldsymbol{V}_{\boldsymbol{\alpha}}\colon I \to \mathbb{R}^3_\nu$ defined by

$$\boldsymbol{V}_{\boldsymbol{\alpha}}(s) = (-1)^{\nu+1}\epsilon_{\boldsymbol{\alpha}}\epsilon_M \boldsymbol{T}_{\boldsymbol{\alpha}}(s) \times \boldsymbol{N}(\boldsymbol{\alpha}(s)).$$

Verify that $\big(\boldsymbol{T}_{\boldsymbol{\alpha}}(s), \boldsymbol{V}_{\boldsymbol{\alpha}}(s), \boldsymbol{N}(\boldsymbol{\alpha}(s))\big)$ is a positive basis for $\mathbb{R}^3_\nu$ for all possibilities of causal types for $\boldsymbol{\alpha}$ and M.

Exercise 3.6.5. Prove Theorem 3.6.14 (p. 225).

Exercise 3.6.6. Let $M \subseteq \mathbb{R}^3_\nu$ be a non-degenerate regular surface, $\boldsymbol{N}$ be a Gauss map for M, $\boldsymbol{\alpha}\colon I \to M$ be a unit speed admissible curve, and $\boldsymbol{p} \doteq \boldsymbol{\alpha}(s_0) \in M$. Show that if $\boldsymbol{\beta}\colon I \to M$ given by

$$\boldsymbol{\beta}(s) \doteq \boldsymbol{\alpha}(s) - \operatorname{proj}_{\boldsymbol{N}(\boldsymbol{p})}(\boldsymbol{\alpha}(s) - \boldsymbol{p})$$

is the orthogonal projection of $\boldsymbol{\alpha}$ onto the affine tangent plane $\boldsymbol{p} + T_{\boldsymbol{p}}M$, then we have that $|\kappa_{g,\boldsymbol{\alpha}}(s_0)| = \kappa_{\boldsymbol{\beta}}(s_0)$.

Exercise 3.6.7. Let $M \subseteq \mathbb{R}^3_\nu$ be a non-degenerate regular surface, $\boldsymbol{N}$ be a Gauss map for M, and $\boldsymbol{\alpha}\colon I \to M$ be a unit speed admissible curve. Show that $\boldsymbol{\alpha}$ is an asymptotic line for M if and only if for each $s \in I$ the osculating plane to $\boldsymbol{\alpha}$ at s is precisely $T_{\boldsymbol{\alpha}(s)}M$.

Exercise 3.6.8. Let $M \subseteq \mathbb{R}^3_\nu$ be a non-degenerate regular surface, $\boldsymbol{N}$ be a Gauss map for M, and $\boldsymbol{\alpha}\colon I \to M$ be a unit speed admissible curve. Show that:

(a) $\boldsymbol{\alpha}$ is simultaneously a geodesic and a line of curvature if and only the image of $\boldsymbol{\alpha}$ is contained in a plane normal to M along $\boldsymbol{\alpha}$;

(b) $\boldsymbol{\alpha}$ is simultaneously a line of curvature and an asymptotic line if and only if the image of $\boldsymbol{\alpha}$ is contained in a plane tangent to M along $\boldsymbol{\alpha}$.

Exercise 3.6.9. Let $M \subseteq \mathbb{R}^3_\nu$ be a non-degenerate regular surface and $\boldsymbol{\alpha}\colon I \to M$ be a unit speed admissible curve.

(a) Show that if $\boldsymbol{\alpha}$ is a plane geodesic which is not a straight line, then $\boldsymbol{\alpha}$ is a line of curvature.

Hint. Fix a Gauss map $\boldsymbol{N}$ for M, consider the Darboux frame for $\boldsymbol{\alpha}$, and show that the normal $\boldsymbol{v}$ to the plane containing the image of $\boldsymbol{\alpha}$ satisfies $\boldsymbol{v} = \lambda(s)\boldsymbol{V}_{\boldsymbol{\alpha}}(s)$ for all $s \in I$, where $\lambda\colon I \to \mathbb{R}$ is a non-vanishing smooth function.

(b) Give an example of a line of curvature which is a plane curve but not a geodesic.

Exercise 3.6.10. Let $M \subseteq \mathbb{L}^3$ be a timelike surface and $\boldsymbol{\alpha}\colon I \to M$ be a lightlike curve. Show that $\boldsymbol{\alpha}$ is a pre-geodesic.

Hint. For all $t \in I$, the vector $\mathbf{N}(\boldsymbol{\alpha}(t))$ is spacelike, and so you may write $(D\boldsymbol{\alpha}'/\mathrm{d}t)(t)$ as a linear combination of $\boldsymbol{\alpha}'(t)$ and $\mathbf{N}(\boldsymbol{\alpha}(t)) \times_L \boldsymbol{\alpha}'(t)$.

Exercise 3.6.11. Let $M \subseteq \mathbb{R}^3_\nu$ be a non-degenerate regular surface and $(U, \boldsymbol{x})$ be a parametrization for M. Show that:

(a) The coordinate curve $\boldsymbol{x}(u, v_0)$, with v_0 constant, is a geodesic if and only if $E_u(u, v_0) = 0$ and $E_v(u, v_0) = 2F_u(u, v_0)$ for all u in its domain.

(b) The coordinate curve $\boldsymbol{x}(u_0, v)$, with u_0 constant, is a geodesic if and only if $G_v(u_0, v) = 0$ and $G_u(u_0, v) = 2F_v(u_0, v)$ for all v in its domain.

Hint. Compute the products $\langle \boldsymbol{x}_{uu}, \boldsymbol{x}_u \rangle$, etc., in a convenient way.

Exercise 3.6.12. Let $f\colon U \subseteq \mathbb{R}^2 \to \mathbb{R}$ be a smooth function with the symmetry $f(u, v) = f(u, -v)$ for all $(u, v) \in U$ (of course, we also assume that U is symmetric about the u-axis) and consider the usual Monge parametrization $\boldsymbol{x}\colon U \to \mathrm{gr}(f) \subseteq \mathbb{R}^3_\nu$, $\boldsymbol{x}(u, v) = (u, v, f(u, v))$. Show that the coordinate curve $v = 0$ is a pre-geodesic.

Exercise 3.6.13 (Horocycles, again). Let $\boldsymbol{v} \in \mathbb{L}^3$ be a future-directed lightlike vector and $c < 0$. We have seen in Exercise 3.1.11 (p. 153) that the horocycle $H_{\boldsymbol{v},c} \subseteq \mathbb{H}^2$ admits a parametrization $\boldsymbol{\alpha}\colon \mathbb{R} \to H_{\boldsymbol{v},c}$ for the form

$$\boldsymbol{\alpha}(s) = -\frac{s^2}{2c}\boldsymbol{v} + s\boldsymbol{w}_1 + \boldsymbol{w}_2,$$

where $\boldsymbol{w}_1$ and $\boldsymbol{w}_2$ are orthogonal unit vectors, with $\boldsymbol{w}_1$ spacelike, $\boldsymbol{w}_2$ timelike, $\boldsymbol{w}_1$ orthogonal to $\boldsymbol{v}$ and $\langle \boldsymbol{w}_2, \boldsymbol{v} \rangle_L = c$.

(a) Show that the geodesic curvature of $\boldsymbol{\alpha}$ is constant, $\kappa_{g,\boldsymbol{\alpha}} = 1$.

(b) The following converse holds: if $\boldsymbol{\alpha}\colon I \to \mathbb{H}^2 \subseteq \mathbb{L}^3$ is a unit speed curve with constant geodesic curvature equal to 1, then the image of $\boldsymbol{\alpha}$ is contained in some horocycle $H_{\boldsymbol{v},c}$.

Hint. Show that in these conditions, $\boldsymbol{\alpha}$ is semi-lightlike. By Theorem 2.3.32 (p. 123), the image of $\boldsymbol{\alpha}$ is contained in an (affine) lightlike plane. Take an orthogonal basis $\{\boldsymbol{v}, \boldsymbol{w}_1\}$ for this plane and proceed from there.

Exercise 3.6.14. Compute the geodesic curvature of a circle of "latitude" $u = u_0$ in $\mathbb{S}^2$ and $\mathbb{S}^2_1$ (use the usual parametrizations of revolution).

Exercise 3.6.15. Find all the curves in $\mathbb{S}^2$ with constant geodesic curvature.

Exercise 3.6.16. Let $M \subseteq \mathbb{R}^3_\nu$ be a non-degenerate regular surface, $\mathbf{N}$ be a Gauss map for M, $\boldsymbol{p} \in M$ such that $-\mathrm{d}\mathbf{N}_{\boldsymbol{p}}$ is diagonalizable, and $\boldsymbol{\alpha}\colon I \to M$ be a unit speed curve such that $\boldsymbol{\alpha}(0) = \boldsymbol{p}$. Assume that the principal vectors $\boldsymbol{u}_1, \boldsymbol{u}_2 \in T_{\boldsymbol{p}}M$ are such that $\boldsymbol{u}_1$ is spacelike and $(\boldsymbol{u}_1, \boldsymbol{u}_2, \mathbf{N}(\boldsymbol{p}))$ is a positive basis for $\mathbb{R}^3_\nu$.

(a) If M is spacelike, show that

$$\tau_{g,\boldsymbol{\alpha}}(0) = \frac{(-1)^\nu}{2}(\kappa_2(\boldsymbol{p}) - \kappa_1(\boldsymbol{p}))\sin 2\theta,$$

where θ is the (oriented) angle between the principal vector $\boldsymbol{u}_1$ and $\boldsymbol{\alpha}'(0)$.

(b) If M is timelike, show that

$$\tau_{g,\alpha}(0) = \pm\frac{1}{2}(\kappa_2(p) - \kappa_1(p))\sinh 2\varphi,$$

no matter whether the velocity vector $\alpha'(0) = \pm\cosh\varphi\, u_1 + \sinh\varphi\, u_2$ is spacelike or $\alpha'(0) = \sinh\varphi\, u_1 \pm \cosh\varphi\, u_2$ is timelike.

That is, the conclusion here is that the geodesic torsion of a curve in a given instant depends only on the position of the curve and on its velocity vector in this instant.

Hint. Recall the proof of Euler's Formula, and the diagrams for cross products (p. 56, Chapter 1) may be useful.

3.6.2 Christoffel symbols

We concluded the previous subsection by discussing the Darboux-Ribaucour trihedron, yet another moving frame associated to a curve. But, if $M \subseteq \mathbb{R}^3_\nu$ is a non-degenerate regular surface, N is a Gauss map for M, and (U, x) is a parametrization for M, we have an "obvious" trihedron: the basis for $T_{x(u,v)}M$ associated to the parametrization x gives us, for each $(u,v) \in U$, a basis $\{x_u(u,v), x_v(u,v), N(x(u,v))\}$ for $\mathbb{R}^3_\nu$. With this, we may express any vector as a combination of x_u, x_v and $N \equiv N \circ x$, at adequate points. In particular, the second order derivatives of the parametrization x itself. Omitting points of application, and identifying $u \leftrightarrow 1$, $v \leftrightarrow 2$, we write:

$$x_{ij} = \sum_{k=1}^{2} \Gamma_{ij}^k x_k + \epsilon_M h_{ij} N, \quad 1 \leq i,j \leq 2,$$

where the h_{ij} are precisely the coefficients of the Second Fundamental Form of the parametrization x.

Definition 3.6.15 (Christoffel Symbols)**.** Keeping the above notation, the functions $\Gamma_{ij}^k \colon U \to \mathbb{R}$, $1 \leq i,j,k \leq 2$, are called the *Christoffel symbols* of (U, x).

A priori, the Christoffel symbols depend on the second order derivatives of the parametrization, which are not necessarily tangent to the surface. The next result allows us to express them only in terms of the First Fundamental Form:

Proposition 3.6.16. *If $M \subseteq \mathbb{R}^3_\nu$ is a non-degenerate regular surface and (U, x) is a parametrization for M, the Christoffel symbols of x are given by:*

$$\Gamma_{ij}^k = \sum_{r=1}^{2} \frac{1}{2} g^{kr}\left(\frac{\partial g_{ik}}{\partial u^j} + \frac{\partial g_{jk}}{\partial u^i} - \frac{\partial g_{ij}}{\partial u^r}\right).$$

Proof: Omitting points of application, by definition of the Christoffel symbols Γ_{ij}^k we have that $x_{ij} = \sum_{k=1}^2 \Gamma_{ij}^k x_k + \epsilon_M h_{ij} N$. Applying $\langle \cdot, x_r \rangle$ to both sides of this relation, we have:

$$\langle x_{ij}, x_r \rangle = \sum_{k=1}^{2} \Gamma_{ij}^k g_{kr}.$$

Multiplying both sides by $g^{r\ell}$ and summing over r, we get:

$$\sum_{r=1}^{2} g^{r\ell}\langle \boldsymbol{x}_{ij}, \boldsymbol{x}_r\rangle = \sum_{k,r=1}^{2} \Gamma_{ij}^k g_{kr} g^{r\ell} = \sum_{k=1}^{2} \Gamma_{ij}^k \delta_k^\ell = \Gamma_{ij}^\ell,$$

which after renaming indices reads as $\Gamma_{ij}^k = \sum_{r=1}^2 g^{kr}\langle \boldsymbol{x}_{ij}, \boldsymbol{x}_r\rangle$. It remains to find out what this product is. Using that mixed partial derivatives commute and the product rule, we have:

$$\begin{aligned}
\langle \boldsymbol{x}_{ij}, \boldsymbol{x}_r\rangle &= \frac{\partial}{\partial u^i}\langle \boldsymbol{x}_j, \boldsymbol{x}_r\rangle - \langle \boldsymbol{x}_j, \boldsymbol{x}_{ri}\rangle \\
&= \frac{\partial g_{jr}}{\partial u^i} - \frac{\partial}{\partial u^r}\langle \boldsymbol{x}_j, \boldsymbol{x}_i\rangle + \langle \boldsymbol{x}_{jr}, \boldsymbol{x}_i\rangle \\
&= \frac{\partial g_{jr}}{\partial u^i} - \frac{\partial g_{ij}}{\partial u^r} + \frac{\partial}{\partial u^j}\langle \boldsymbol{x}_r, \boldsymbol{x}_i\rangle - \langle \boldsymbol{x}_r, \boldsymbol{x}_{ij}\rangle,
\end{aligned}$$

whence:

$$2\langle \boldsymbol{x}_{ij}, \boldsymbol{x}_r\rangle = \frac{\partial g_{ik}}{\partial u^j} + \frac{\partial g_{jk}}{\partial u^i} - \frac{\partial g_{ij}}{\partial u^r}$$

and finally:

$$\Gamma_{ij}^k = \sum_{r=1}^{2} \frac{1}{2} g^{kr}\left(\frac{\partial g_{ik}}{\partial u^j} + \frac{\partial g_{jk}}{\partial u^i} - \frac{\partial g_{ij}}{\partial u^r}\right),$$

as wanted. □

Remark. Note that the Christoffel symbols are symmetric in the lower indices (actually, we already knew that since $\boldsymbol{x}_{ij} = \boldsymbol{x}_{ji}$). Furthermore, in practice, many of the parametrizations we will encounter have $F = g_{12} = g_{21} = 0$. In this case, the expressions given for the Γ_{ij}^k boil down to

$$\Gamma_{ij}^k = \frac{1}{2} g^{kk}\left(\frac{\partial g_{ik}}{\partial u^j} + \frac{\partial g_{jk}}{\partial u^i} - \frac{\partial g_{ij}}{\partial u^k}\right).$$

In particular, we see that:

- If $i \neq j$ only, since $\Gamma_{ij}^k = \Gamma_{ji}^k$ and $g_{ij} = 0$, we have

  $$\Gamma_{ij}^i = \frac{1}{2} g^{ii}\frac{\partial g_{ii}}{\partial u^j},$$

  that is,

  $$\Gamma_{uv}^u = \Gamma_{vu}^u = \frac{E_v}{2E} \quad \text{and} \quad \Gamma_{uv}^v = \Gamma_{vu}^v = \frac{G_u}{2G}.$$

- If $i = j \neq k$, we have $g_{ik} = g_{jk} = 0$ and so

  $$\Gamma_{ii}^k = -\frac{1}{2} g^{kk}\frac{\partial g_{ii}}{\partial u^k},$$

  or, equivalently,

  $$\Gamma_{uu}^v = -\frac{E_v}{2G} \quad \text{and} \quad \Gamma_{vv}^u = -\frac{G_u}{2E}.$$

- If $i = j = k$, we have:

  $$\Gamma_{ii}^i = \frac{1}{2} g^{ii}\frac{\partial g_{ii}}{\partial u^i},$$

  or,

  $$\Gamma_{uu}^u = \frac{E_u}{2E} \quad \text{and} \quad \Gamma_{vv}^v = \frac{G_v}{2G}.$$

In particular, Proposition 3.2.17 (p. 170) now yields the following:

Corollary 3.6.17. *Let $M_1, M_2 \subseteq \mathbb{R}^3_\nu$ be two non-degenerate regular surfaces, and $\phi\colon M_1 \to M_2$ be a local isometry. If $(U, \boldsymbol{x})$ is a parametrization for M_1 and $(U, \phi \circ \boldsymbol{x})$ is the corresponding parametrization for M_2, we have that*

$$\Gamma^k_{ij}(u,v) = \widetilde{\Gamma}^k_{ij}(u,v),$$

for all $(u,v) \in U$ and $1 \leq i,j,k \leq 2$, where the $\widetilde{\Gamma}^k_{ij}$ denote the Christoffel Symbols of the parametrization $\phi \circ \boldsymbol{x}$.

In terms of the objects above, we may express geodesics in a surface locally as solutions to a second order system of ODEs:

Proposition 3.6.18 (Geodesic Differential Equations). *If $M \subseteq \mathbb{R}^3_\nu$ is a non-degenerate regular surface, $(U, \boldsymbol{x})$ is a parametrization for M, and $\boldsymbol{\alpha}\colon I \to \boldsymbol{x}(U)$ is given in coordinates by $\boldsymbol{\alpha}(t) = \boldsymbol{x}(u^1(t), u^2(t))$, for smooth functions $u^1, u^2\colon I \to \mathbb{R}$, then $\boldsymbol{\alpha}$ is a geodesic if and only if:*

$$\ddot{u}^k + \sum_{i,j=1}^{2} \Gamma^k_{ij}\dot{u}^i\dot{u}^j = 0, \quad k = 1,2,$$

where the $\dot{u}^i$ are evaluated in t and the Γ^k_{ij} are evaluated in $(u^1(t), u^2(t))$.

Proof: Continuing to omit all the points of application, let's obtain an expression for $\boldsymbol{\alpha}''$ in terms of the Christoffel symbols and of the Second Fundamental Form of $\boldsymbol{x}$. The components of $\boldsymbol{\alpha}''$ in the tangent directions will be precisely the expressions in the statement of the proposition. We have that $\boldsymbol{\alpha}' = \sum_{i=1}^2 \dot{u}^i \boldsymbol{x}_i$, and differentiating again:

$$\begin{aligned}
\boldsymbol{\alpha}'' &= \sum_{i=1}^{2}\left(\ddot{u}^i\boldsymbol{x}_i + \dot{u}^i\sum_{j=1}^{2}\dot{u}^j\boldsymbol{x}_{ij}\right)\\
&= \sum_{i=1}^{2}\left(\ddot{u}^i\boldsymbol{x}_i + \dot{u}^i\sum_{j=1}^{2}\dot{u}^j\left(\sum_{k=1}^{2}\Gamma^k_{ij}\boldsymbol{x}_k + \epsilon_M h_{ij}\boldsymbol{N}\right)\right)\\
&= \sum_{k=1}^{2}\ddot{u}^k\boldsymbol{x}_k + \sum_{i,j,k=1}^{2}\Gamma^k_{ij}\dot{u}^i\dot{u}^j\boldsymbol{x}_k + \sum_{i,j=1}^{2}\epsilon_M\dot{u}^i\dot{u}^j h_{ij}\boldsymbol{N}\\
&= \sum_{k=1}^{2}\left(\ddot{u}^k + \sum_{i,j=1}^{2}\Gamma^k_{ij}\dot{u}^i\dot{u}^j\right)\boldsymbol{x}_k + \sum_{i,j=1}^{2}\epsilon_M\dot{u}^i\dot{u}^j h_{ij}\boldsymbol{N},
\end{aligned}$$

and $\boldsymbol{\alpha}''$ is normal to M if and only if $\ddot{u}^k + \sum_{i,j=1}^2 \Gamma^k_{ij}\dot{u}^i\dot{u}^j = 0$, for $k = 1,2$. □

Remark. In particular, the above proof shows that, in coordinates, the covariant derivative of $\boldsymbol{\alpha}'$ is given by

$$\frac{D\boldsymbol{\alpha}'}{dt} = \sum_{k=1}^{2}\left(\ddot{u}^k + \sum_{i,j=1}^{2}\Gamma^k_{ij}\dot{u}^i\dot{u}^j\right)\boldsymbol{x}_k.$$

See the analogous expression for general vector fields along $\boldsymbol{\alpha}$ in Exercise 3.6.17.

Combining the above with the previous Corollary 3.6.17, we have the:

Corollary 3.6.19. *Let $M_1, M_2 \subseteq \mathbb{R}^3_\nu$ be two non-degenerate regular surfaces, and $\phi\colon M_1 \to M_2$ be a local isometry. If $\boldsymbol{\alpha}\colon I \to M_1$ is a geodesic of M_1, then $\phi \circ \boldsymbol{\alpha}\colon I \to M_2$ is a geodesic of M_2.*

Proof: Suppose without loss of generality that $\boldsymbol{\alpha}(I) \subseteq \boldsymbol{x}(U)$ for some parametrization $(U, \boldsymbol{x})$ for M_1. If $\boldsymbol{\alpha}(t) = \boldsymbol{x}(u^1(t), u^2(t))$, then $\phi(\boldsymbol{\alpha}(t)) = (\phi \circ \boldsymbol{x})(u^1(t), u^2(t))$. Keeping notation from Corollary 3.6.17, we see that the geodesic differential equations are satisfied for $\phi \circ \boldsymbol{\alpha}$:

$$\ddot{u}^k + \sum_{i,j=1}^{2} \widetilde{\Gamma}_{ij}^k \dot{u}^i \dot{u}^j = \ddot{u}^k + \sum_{i,j=1}^{2} \Gamma_{ij}^k \dot{u}^i \dot{u}^j = 0, \quad (1 \leq k \leq 2)$$

as wanted. □

Furthermore, the geodesic differential equations also give us the:

Theorem 3.6.20. *Let $M \subseteq \mathbb{R}^3_\nu$ be a non-degenerate regular surface, $\boldsymbol{p} \in M$ and $\boldsymbol{v} \in T_{\boldsymbol{p}}M$. Then there is an interval I containing 0 and a geodesic $\boldsymbol{\alpha}\colon I \to M$ with $\boldsymbol{\alpha}(0) = \boldsymbol{p}$ and $\boldsymbol{\alpha}'(0) = \boldsymbol{v}$.*

Proof: Take a parametrization $(U, \boldsymbol{x})$ around the point $\boldsymbol{p} = \boldsymbol{x}(u_0, v_0)$ and, for certain $a, b \in \mathbb{R}$, write $\boldsymbol{v} = a\boldsymbol{x}_u(u_0, v_0) + b\boldsymbol{x}_v(u_0, v_0)$. The usual existence and uniqueness result from the theory of ODEs applied to the initial value problem

$$\begin{cases} \ddot{u}^k + \displaystyle\sum_{i,j=1}^{2} \Gamma_{ij}^k \dot{u}^i \dot{u}^j = 0, \quad k = 1, 2 \\ u^1(0) = u_0, \quad u^2(0) = v_0, \quad \dot{u}^1(0) = a, \quad \dot{u}^2(0) = b \end{cases}$$

provides an interval I containing 0 and unique functions $u^1, u^2\colon I \to \mathbb{R}$ satisfying both equations above. We then define $\boldsymbol{\alpha}\colon I \to M$ by $\boldsymbol{\alpha}(t) = \boldsymbol{x}(u^1(t), u^2(t))$. Clearly $\boldsymbol{\alpha}$ satisfies all the requirements. □

Remark. In the above proof, the image of the curve lies in the image of the parametrization initially fixed, but the theorem may be stated so that the interval I is maximal, in the following sense: if $I \subseteq J$ and $\boldsymbol{\beta}\colon J \to M$ is another geodesic with $\boldsymbol{\beta}(0) = \boldsymbol{p}$ and $\boldsymbol{\beta}'(0) = \boldsymbol{v}$, then $J = I$ and $\boldsymbol{\beta} = \boldsymbol{\alpha}$. In particular, the geodesic $\boldsymbol{\alpha}$ is now unique under these conditions.

Corollary 3.6.21. *Let $M \subseteq \mathbb{R}^3_\nu$ be a non-degenerate regular surface, and $\phi\colon M \to M$ an isometry. Let $\boldsymbol{\alpha}\colon I \to M$ be a regular curve whose image is precisely the set of points fixed by ϕ: $\phi(\boldsymbol{\alpha}(t)) = \boldsymbol{\alpha}(t)$ for all $t \in I$. Then $\boldsymbol{\alpha}$ is a pre-geodesic.*

Proof: Suppose without loss of generality that $0 \in I$; let's show that $\boldsymbol{\alpha}$ restricted to a neighborhood of 0 is a pre-geodesic. The previous theorem says that there is $\epsilon > 0$ and a unique geodesic $\boldsymbol{\beta}\colon]-\epsilon, \epsilon[\to M$ such that $\boldsymbol{\beta}(0) = \boldsymbol{\alpha}(0)$ and $\boldsymbol{\beta}'(0) = \boldsymbol{\alpha}'(0)$. By Corollary 3.6.19 above, we know that $\phi \circ \boldsymbol{\beta}$ is also a geodesic. And, moreover, we have that

$$(\phi \circ \boldsymbol{\beta})(0) = \phi(\boldsymbol{\beta}(0)) = \phi(\boldsymbol{\alpha}(0)) = \boldsymbol{\alpha}(0) = \boldsymbol{\beta}(0),$$

as well as

$$(\phi \circ \boldsymbol{\beta})'(0) = \mathrm{d}\phi_{\boldsymbol{\beta}(0)}(\boldsymbol{\beta}'(0)) = \mathrm{d}\phi_{\boldsymbol{\alpha}(0)}(\boldsymbol{\alpha}'(0)) = (\phi \circ \boldsymbol{\alpha})'(0) = \boldsymbol{\alpha}'(0) = \boldsymbol{\beta}'(0).$$

The remark following the previous theorem also ensures that a geodesic is determined by its initial conditions, whence we conclude that $\phi \circ \boldsymbol{\beta} = \boldsymbol{\beta}$. Since the image of $\boldsymbol{\beta}$ is fixed by ϕ, it follows that the image of $\boldsymbol{\beta}$ is contained in the image of $\boldsymbol{\alpha}$, and so $\boldsymbol{\beta}$ is a reparametrization of the restriction $\boldsymbol{\alpha}\big|_{]-\epsilon,\epsilon[}$. □

3.6.3 Critical points of the energy functional

To proceed in a more efficient way with our study of geodesics, it will be convenient to present a brief introduction to *Variational Calculus*: where the independent variables are functions instead of real numbers. We'll study *functionals*, in general given by certain integrals, with the goal to find its extremizers.

Variational Calculus also has several applications in areas other than Mathematics itself, such as Physics, Engineering, Economics, and Control Theory, among others. Our presentation here will barely touch the tip of the iceberg, and so we'll also recommend [24] and [70] for more details.

Previously, we have characterized geodesics in terms of the geometry of the surface on where they lie, but we can also provide a variational characterization. For this end, consider at first the problem of finding a function $y\colon [a,b] \to \mathbb{R}$, of class $\mathscr{C}^1$, with fixed endpoints $y(a) = y_0$ and $y(b) = y_1$, which minimizes the value of the integral:

$$J[y] \doteq \int_a^b L(t, y(t), y'(t))\, dt,$$

for some prescribed class $\mathscr{C}^2$ map $L\colon [a,b] \times \mathbb{R}^2 \to \mathbb{R}$. The above situation is known as a *variational problem*, the map J is called a *functional* (objective, action), and L is called the *Lagrangian* for the variational problem.

We will say that y is a *local minimum* for J if $J[y] \leq J[y + s\eta]$ for all $\eta\colon [a,b] \to \mathbb{R}$ of class $\mathscr{C}^1$ with $\eta(a) = \eta(b) = 0$ and $s \in \mathbb{R}$ sufficiently small. The boundary condition $\eta(a) = \eta(b) = 0$ is necessary since we want $y + s\eta$ to also be a legitimate candidate for solution for the fixed endpoints variational problem. Moreover, we will say that such an η is an *admissible variation*. Similarly, one defined a *local maximum* for J, and we'll say that y is a *critical point* of J if it is either a local minimum or local maximum.

Just like what happens for smooth functions of a single real variable, where the derivative vanishes at critical points in the interior of its domain, we have the following "first derivative test" for functionals:

Theorem 3.6.22 (Euler-Lagrange). *If $y\colon [a,b] \to \mathbb{R}$ is a function of class $\mathscr{C}^1$, which is a critical point for the variational problem*

$$\begin{cases} J[y] = \displaystyle\int_a^b L(t, y(t), y'(t))\, dt \\ y(a) = y_0, \quad y(b) = y_1, \end{cases}$$

where $L\colon [a,b] \times \mathbb{R}^2 \to \mathbb{R}$ is a given class $\mathscr{C}^2$ Lagrangian, then y satisfies the Euler-Lagrange equation:

$$\frac{\partial L}{\partial y} - \frac{d}{dt}\left(\frac{\partial L}{\partial y'}\right) = 0,$$

where the partial derivatives of L are evaluated in $(t, y(t), y'(t))$.

Remark. To emphasize the similarity with functions of a single real variable, the quantity

$$\delta J[y] \doteq \frac{\partial L}{\partial y} - \frac{d}{dt}\left(\frac{\partial L}{\partial y'}\right)$$

is also known as the *variational derivative of J*, or the *first variation of J*.

Proof: Fixed an *arbitrary* admissible variation η and introducing a real parameter s, we look for a critical point of a single variable function (one function for each η), $s \mapsto J[y+s\eta]$, whence:

$$\left.\frac{\mathrm{d}}{\mathrm{d}s}\right|_{s=0} \int_a^b L(t, y(t)+s\eta(t), y'(t)+s\eta'(t))\,\mathrm{d}t = 0.$$

Differentiating under the integral sign, we have that:

$$\int_a^b \eta(t)\frac{\partial L}{\partial y}(t, y(t), y'(t)) + \eta'(t)\frac{\partial L}{\partial y'}(t, y(t), y'(t))\,\mathrm{d}t = 0.$$

To eliminate $\eta'(t)$ from this expression, we'll integrate the last term by parts, obtaining:

$$\int_a^b \eta(t)\left(\frac{\partial L}{\partial y}(t, y(t), y'(t)) - \frac{\mathrm{d}}{\mathrm{d}t}\left(\frac{\partial L}{\partial y'}(t, y(t), y'(t))\right)\right)\mathrm{d}t + \left.\eta(t)\frac{\partial L}{\partial y'}(t, y(t), y'(t))\right|_a^b = 0$$

Using $\eta(a) = \eta(b) = 0$, this boils down to:

$$\int_a^b \eta(t)\left(\frac{\partial L}{\partial y}(t, y(t), y'(t)) - \frac{\mathrm{d}}{\mathrm{d}t}\left(\frac{\partial L}{\partial y'}(t, y(t), y'(t))\right)\right)\mathrm{d}t = 0$$

Since η was arbitrary, it follows that:

$$\frac{\partial L}{\partial y}(t, y(t), y'(t)) - \frac{\mathrm{d}}{\mathrm{d}t}\left(\frac{\partial L}{\partial y'}(t, y(t), y'(t))\right) = 0.$$

□

As a curiosity, there are many generalizations of this result for more complicated variational problems. See a few examples below:

Example 3.6.23 (More variational problems).

(1) We seek $y\colon [a,b] \to \mathbb{R}$ of class $\mathscr{C}^n$ optimizing the integral:

$$J[y] \doteq \int_a^b L(t, y(t), \ldots, y^{(n)}(t))\,\mathrm{d}t,$$

for a prescribed class $\mathscr{C}^{n+1}$ Lagrangian, $L\colon [a,b] \times \mathbb{R}^n \to \mathbb{R}$, imposing endpoint conditions on y and on its first $n-1$ derivatives. Repeating the argument given above and integrating by parts more times, the Euler-Lagrange equation takes the form:

$$\frac{\partial L}{\partial y} - \frac{\mathrm{d}}{\mathrm{d}t}\left(\frac{\partial L}{\partial y'}\right) + \frac{\mathrm{d}^2}{\mathrm{d}t^2}\left(\frac{\partial L}{\partial y''}\right) - \cdots + (-1)^n \frac{\mathrm{d}^n}{\mathrm{d}t^n}\left(\frac{\partial L}{\partial y^{(n)}}\right) = 0,$$

or, more concisely:

$$\sum_{k=0}^{n} (-1)^k \frac{\mathrm{d}^k}{\mathrm{d}t^k}\left(\frac{\partial L}{\partial y^{(k)}}\right) = 0.$$

(2) We seek n functions $y_1, \ldots, y_n\colon [a,b] \to \mathbb{R}$ of class $\mathscr{C}^1$ which optimize the integral

$$J[y_1, \ldots, y_n] \doteq \int_a^b L(t, y_1(t), y_1'(t), \ldots, y_n(t), y_n'(t))\,\mathrm{d}t,$$

for a prescribed class $\mathscr{C}^2$ Lagrangian, $L\colon [a,b] \times \mathbb{R}^{2n} \to \mathbb{R}$, imposing endpoint conditions on all functions. Considering admissible variations in the direction of each y_i, one may show that now we'll have one Euler-Lagrange equation for each argument:

$$\frac{\partial L}{\partial y_i} - \frac{\mathrm{d}}{\mathrm{d}t}\left(\frac{\partial L}{\partial y_i'}\right) = 0, \quad i = 1,2,\ldots,n.$$

Each Euler-Lagrange equation should be seen as one component of a "variational gradient" vanishing.

(3) With the two items above in mind, it is natural to consider a variational problem where we look for for n functions, $y_1,\ldots,y_n\colon [a,b] \to \mathbb{R}$, where y_i is of class $\mathscr{C}^{m_i}$, which optimize the integral

$$J[y_1,\ldots,y_n] \doteq \int_a^b L(t,y_1(t),\ldots,y_1^{(m_1)}(t),\ldots,y_n(t),\ldots,y_n^{(m_n)}(t))\,\mathrm{d}t,$$

for a suitable prescribed Lagrangian, imposing endpoint conditions on each y_i and on its $m_i - 1$ first derivatives. We will have n Euler-Lagrange equations:

$$\sum_{k=0}^{m_i}(-1)^k\frac{\mathrm{d}^k}{\mathrm{d}t^k}\left(\frac{\partial L}{\partial y_i^{(k)}}\right) = 0, \quad i = 1,2,\ldots,n.$$

(4) One may also consider a variational problem for which the function we seek depends on more than one real variable. For instance, if $\Omega \subseteq \mathbb{R}^2$ is a bounded open set with regular boundary, we may seek a function $y\colon \Omega \to \mathbb{R}$ of class $\mathscr{C}^1$ optimizing the functional

$$J[y] \doteq \int_\Omega L(u,v,y(u,v),y_u(u,v),y_v(u,v))\,\mathrm{d}u\,\mathrm{d}v,$$

where $L\colon \Omega \times \mathbb{R}^4 \to \mathbb{R}$ is a prescribed class $\mathscr{C}^2$ Lagrangian, and we impose the boundary condition $y\big|_{\partial\Omega} = 0$. Using the Green-Stokes Theorem (which will play the role of integration by parts in this case), it is possible to show that if y is a critical point of J, then y satisfies the following Euler-Lagrange:

$$\frac{\partial L}{\partial y} - \frac{\partial}{\partial u}\left(\frac{\partial L}{\partial y_u}\right) - \frac{\partial}{\partial v}\left(\frac{\partial L}{\partial y_v}\right) = 0.$$

For interesting applications of this version (and a direct generalization in higher dimensions), see Exercises 3.6.27 and 3.6.28.

Variational techniques such as those may be used to model several phenomena in Physics:

Example 3.6.24 (Hamilton's Principle). We may use what was discussed so far to understand a bit better the situation described in Exercise 2.1.15 (p. 76, in Chapter 2). Suppose that a particle with mass $m > 0$ moves in $\mathbb{R}^n$ under the action of a potential $V\colon \mathbb{R}^n \to \mathbb{R}$, with trajectory described by $\boldsymbol{\alpha}\colon I \to \mathbb{R}^n$. Recall that the kinetic energy is the smooth map $T\colon \mathbb{R}^n \to \mathbb{R}$ defined by $T(\boldsymbol{v}) = m\|\boldsymbol{v}\|_E^2/2$. Consider the Lagrangian $L\colon \mathbb{R}^{2n} \to \mathbb{R}$ defined by

$$L(\boldsymbol{p},\boldsymbol{v}) \doteq T(\boldsymbol{v}) - V(\boldsymbol{p}).$$

Lagrangians such as the one above, which do not explicitly depend on t, are called *autonomous*. The *action* of the trajectory between the instants t_0 and t_1 is defined by the integral

$$S[\boldsymbol{\alpha}] \doteq \int_{t_0}^{t_1} L(\boldsymbol{\alpha}(t), \boldsymbol{\alpha}'(t))\, \mathrm{d}t.$$

Then *Hamilton's Principle* states that particles follow trajectories that minimize this action. If $\boldsymbol{\alpha}(t) = (x_1(t), \ldots, x_n(t))$, the derivatives of this Lagrangian satisfy

$$\frac{\partial L}{\partial x_i} = -\frac{\partial V}{\partial x_i} \quad \text{and} \quad \frac{\partial L}{\partial \dot{x}_i} = m\dot{x}_i,$$

from where we see that the Euler-Lagrange equations (for each i) give us Newton's Second Law

$$m\boldsymbol{\alpha}''(t) = -\nabla V(\boldsymbol{\alpha}(t)).$$

For this reason, we will say that the critical points of the action functional are *physical trajectories*, and we'll keep using this terminology even for Lagrangians which are not of the form $T - V$ (called "natural"), such as the one here. See Exercise 3.6.25 for one more contextualization in Physics.

Now, let's analyze the Euler-Lagrange equations for the energy functional, given by

$$E[\boldsymbol{\alpha}] = \frac{1}{2}\int_I \langle \boldsymbol{\alpha}'(t), \boldsymbol{\alpha}'(t)\rangle\, \mathrm{d}t,$$

to obtain the following characterization of geodesics:

Theorem 3.6.25. *Let $M \subseteq \mathbb{R}^3_\nu$ be a non-degenerate regular surface, $(U, \mathbf{x})$ be a parametrization for M, and $\boldsymbol{\alpha}\colon I \to M$ given in coordinates by $\boldsymbol{\alpha}(t) = \mathbf{x}(u^1(t), u^2(t))$. Then the Euler-Lagrange equations for the energy of $\boldsymbol{\alpha}$ are equivalent to the geodesic differential equations. Thus, all the critical points of the energy functional are geodesics.*

Proof: Again omitting the points of evaluation and bearing in mind that $\dot{u}^i = \dot{u}^i(t)$ and $g_{ij} = g_{ij}(u^1(t), u^2(t))$, we write the (autonomous) Lagrangian for the energy of $\boldsymbol{\alpha}$ as:

$$L = \frac{1}{2}\sum_{i,j=1}^{2} g_{ij}\dot{u}^i\dot{u}^j.$$

We have that

$$\frac{\partial L}{\partial u^k} = \frac{\partial}{\partial u^k}\frac{1}{2}\sum_{i,j=1}^{2} g_{ij}\dot{u}^i\dot{u}^j = \sum_{i,j=1}^{2}\frac{1}{2}\frac{\partial g_{ij}}{\partial u^k}\dot{u}^i\dot{u}^j,$$

and also:

$$\begin{aligned}
\frac{\partial L}{\partial \dot{u}^k} &= \frac{\partial}{\partial \dot{u}^k}\frac{1}{2}\sum_{i,j=1}^{2} g_{ij}\dot{u}^i\dot{u}^j = \sum_{i,j=1}^{2}\frac{1}{2}g_{ij}(\dot{u}^i\delta^j_k + \delta^i_k\dot{u}^j) \\
&= \sum_{i,j=1}^{2}\frac{1}{2}g_{ij}\dot{u}^i\delta^j_k + \sum_{i,j=1}^{2}\frac{1}{2}g_{ij}\dot{u}^j\delta^i_k = \sum_{i=1}^{2}\frac{1}{2}g_{ik}\dot{u}^i + \sum_{j=1}^{2}\frac{1}{2}g_{jk}\dot{u}^j \\
&= \sum_{i=1}^{2} g_{ik}\dot{u}^i.
\end{aligned}$$

With this, it follows that:

$$\frac{\mathrm{d}}{\mathrm{d}t}\left(\frac{\partial L}{\partial \dot{u}^k}\right) = \frac{\mathrm{d}}{\mathrm{d}t}\left(\sum_{i=1}^{2} g_{ik}\dot{u}^i\right) = \sum_{i=1}^{2}\frac{\mathrm{d}}{\mathrm{d}t}(g_{ik}\dot{u}^i) = \sum_{i=1}^{2}\sum_{j=1}^{2}\dot{u}^i\dot{u}^j\frac{\partial g_{ik}}{\partial u^j} + \sum_{i=1}^{2} g_{ik}\ddot{u}^i,$$

whence we obtain the Euler-Lagrange equations:

$$\sum_{i=1}^{2} g_{ik}\ddot{u}^i + \sum_{i,j=1}^{2} \frac{\partial g_{ik}}{\partial u^j}\dot{u}^i\dot{u}^j - \frac{1}{2}\sum_{i,j=1}^{2} \frac{\partial g_{ij}}{\partial u^k}\dot{u}^i\dot{u}^j = 0.$$

Observing that

$$\sum_{i,j=1}^{2} \frac{\partial g_{ik}}{\partial u^j}\dot{u}^i\dot{u}^j = \frac{1}{2}\sum_{i,j=1}^{2} \frac{\partial g_{ik}}{\partial u^j}\dot{u}^i\dot{u}^j + \frac{1}{2}\sum_{i,j=1}^{2} \frac{\partial g_{jk}}{\partial u^i}\dot{u}^i\dot{u}^j,$$

we may reorganize the Euler-Lagrange equations as:

$$\sum_{i=1}^{2} g_{ik}\ddot{u}^i + \sum_{i,j=1}^{2} \frac{1}{2}\left(\frac{\partial g_{ik}}{\partial u^j} + \frac{\partial g_{jk}}{\partial u^i} - \frac{\partial g_{ij}}{\partial u^k}\right)\dot{u}^i\dot{u}^j = 0.$$

Multiplying the whole equation by g^{kr}, summing over k, and renaming $r \to k$, we obtain precisely:

$$\ddot{u}^k + \sum_{i,j=1}^{2} \Gamma_{ij}^k \dot{u}^i\dot{u}^j = 0.$$

□

In general, the converse of Theorem 3.6.22 fails: functions that satisfy the Euler-Lagrange equation are not necessarily extremizers for the functional J under discussion. But, when dealing with geodesics, we have the:

Proposition 3.6.26. *Let $M \subseteq \mathbb{R}^3_\nu$ be a non-degenerate regular surface. Then the geodesics of M are critical points of the energy functional.*

Proof: Suppose that $\boldsymbol{\alpha}\colon [a,b] \to M$ is a geodesic. The initial idea would be to show that for every admissible variation $\boldsymbol{\eta}\colon [a,b] \to M$ with $\boldsymbol{\eta}(a) = \boldsymbol{\eta}(b) = \mathbf{0}$, we'd have that 0 is a critical point of

$$\epsilon \mapsto \frac{1}{2}\int_a^b \langle \boldsymbol{\alpha}'(t) + \epsilon\boldsymbol{\eta}'(t), \boldsymbol{\alpha}'(t) + \epsilon\boldsymbol{\eta}'(t)\rangle \, dt,$$

but this argument has a fatal flaw: $\boldsymbol{\alpha} + \epsilon\boldsymbol{\eta}$ may leave the surface M. To remedy this, we will consider, for $r > 0$, admissible variations $f\colon [a,b] \times [-r,r] \to M$ satisfying $f(t,0) = \boldsymbol{\alpha}(t)$ for all t, and $f(a,s) = \boldsymbol{\alpha}(a)$ and $f(b,s) = \boldsymbol{\alpha}(b)$ for all s.

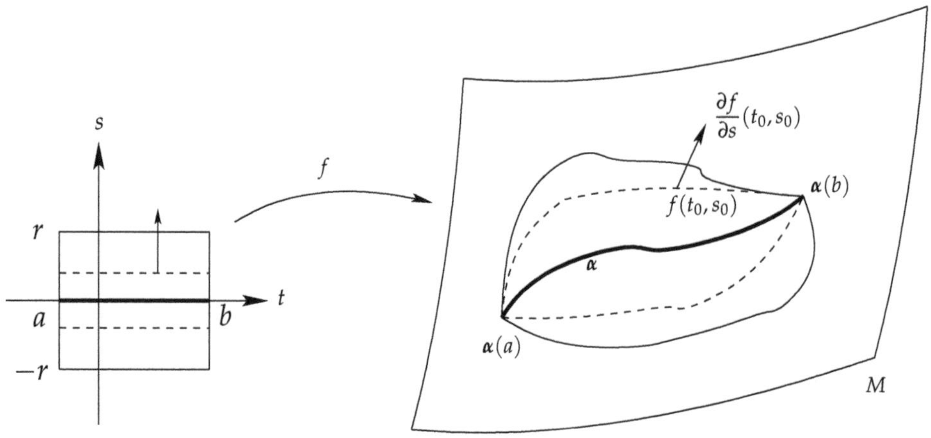

Figure 3.33: The variation of a curve in M.

We then consider

$$E(t) \doteq \frac{1}{2}\int_a^b \left\langle \frac{\partial f}{\partial t}(t,s), \frac{\partial f}{\partial t}(t,s) \right\rangle \, dt.$$

Let's see that $E'(0) = 0$. Observe that the conditions under f ensure that

$$\frac{\partial f}{\partial s}(a,0) = \frac{\partial f}{\partial s}(b,0) = \mathbf{0}.$$

With this, differentiating under the integral sign, we have:

$$\begin{aligned} E'(0) &= \int_a^b \left\langle \frac{\partial^2 f}{\partial s \partial t}(t,0), \boldsymbol{\alpha}'(t) \right\rangle \, dt \\ &\overset{(1)}{=} \int_a^b \frac{\partial}{\partial t}\left(\left\langle \frac{\partial f}{\partial s}(t,0), \boldsymbol{\alpha}'(t) \right\rangle\right) - \left\langle \frac{\partial f}{\partial s}(t,0), \boldsymbol{\alpha}''(t) \right\rangle \, dt \\ &\overset{(2)}{=} \left\langle \frac{\partial f}{\partial s}(t,0), \boldsymbol{\alpha}'(t) \right\rangle \Big|_a^b - \int_a^b \left\langle \frac{\partial f}{\partial s}(t,0), \frac{D\boldsymbol{\alpha}'}{dt}(t) \right\rangle \, dt \\ &= -\int_a^b \left\langle \frac{\partial f}{\partial s}(t,0), \frac{D\boldsymbol{\alpha}'}{dt}(t) \right\rangle \, dt = 0, \end{aligned}$$

where in (1) we integrate by parts and in (2) that $\partial f / \partial s$ is always tangent to M. □

Remark.

- The formula for the *first variation of energy*,

$$E'(0) = -\int_a^b \left\langle \frac{\partial f}{\partial s}(t,0), \frac{D\boldsymbol{\alpha}'}{dt}(t) \right\rangle \, dt,$$

  also provides an alternative proof for Theorem 3.6.25. The advantage of the first proof presented is that it illustrates a simpler way to compute Christoffel symbols, avoiding a direct use of the expression given in terms of the g_{ij}'s and its derivatives, given in Proposition 3.6.16. We will see examples soon.

- Despite the above result, we still cannot guarantee that geodesics are (global) extremizers for the energy functional. For instance, we may consider any point in a surface which admits a closed geodesic[4] passing through that point: the energy is minimized by the degenerate constant curve that never leaves the point, and not by the chosen closed geodesic. If we require regular curves, it suffices to consider two non-antipodal points in a sphere and arcs of great circles with different lengths. But when M is spacelike, we have a local result, see Exercise 3.6.26.

Besides all that, we have seen in Proposition 2.1.11 (p. 68, in Chapter 2) that there is a relation between the energy and the arclength of a curve, given by the Cauchy-Schwarz inequality. Namely, that if $\boldsymbol{\alpha}\colon [a,b] \to M$ is a curve, then the inequality

$$L_a^b[\boldsymbol{\alpha}] \leq \sqrt{2\epsilon_{\boldsymbol{\alpha}}(b-a)E_a^b[\boldsymbol{\alpha}]}$$

holds, with equality if and only if $\boldsymbol{\alpha}$ has constant speed. In particular, this inequality says that a constant speed curve is a critical point of the energy functional if and only if it is a critical point of the arclength functions. It also follows from this that geodesics in spacelike surfaces locally minimize arclength.

[4]Every compact surface in $\mathbb{R}^3$ admits a closed geodesic (this is a particular instance of the so-called *Lyusternik-Fet theorem* – if you know Russian, see [45]).

Example 3.6.27.

(1) Let $\boldsymbol{p} \in \mathbb{R}^3_\nu$, and $\Pi \subseteq \mathbb{R}^3_\nu$ be a non-degenerate plane passing through $\boldsymbol{p}$, with orthonormal basis $\{\boldsymbol{w}_1, \boldsymbol{w}_2\}$. Consider the parametrization $\boldsymbol{x}\colon \mathbb{R}^2 \to \Pi$ given by $\boldsymbol{x}(u,v) = \boldsymbol{p} + u\boldsymbol{w}_1 + v\boldsymbol{w}_2$. For a curve $\boldsymbol{\alpha}\colon I \to \Pi$ given by $\boldsymbol{\alpha}(t) = \boldsymbol{x}(u(t), v(t))$, we have that the energy is

$$E[\boldsymbol{\alpha}] = \frac{1}{2}\int_I \epsilon_{\boldsymbol{w}_1} u'(t)^2 + \epsilon_{\boldsymbol{w}_2} v'(t)^2 \, \mathrm{d}t,$$

and the associated Lagrangian is

$$L(u(t), u'(t), v(t), v'(t)) = \frac{1}{2}\left(\epsilon_{\boldsymbol{w}_1} u'(t)^2 + \epsilon_{\boldsymbol{w}_2} v'(t)^2\right).$$

Directly one sees that the Euler-Lagrange equations are just $u'' = 0$ and $v'' = 0$, so that all the Christoffel symbols vanish. In particular, we conclude yet again that the geodesics in non-degenerate planes are actual straight lines.

(2) Let $\boldsymbol{\sigma}\colon I \to \mathbb{R}^3_\nu$ be an injective and regular smooth plane curve, of the form $\boldsymbol{\sigma}(u) = (f(u), 0, g(u))$, with $f(u) > 0$ for all $u \in I$. Suppose that $\boldsymbol{\sigma}$ is not lightlike. Considering the usual parametrization of revolution, $\boldsymbol{x}\colon I \times \left]0, 2\pi\right[\to \mathbb{R}^3_\nu$ given by $\boldsymbol{x}(u,v) = (f(u)\cos v, f(u)\sin v, g(u))$, we have that its First Fundamental Form is given by

$$\mathrm{d}s^2 = \epsilon_{\boldsymbol{\sigma}} \mathrm{d}u^2 + f(u)^2 \mathrm{d}v^2.$$

If $\boldsymbol{\alpha}\colon I' \to \boldsymbol{x}(I \times \left]0, 2\pi\right[)$ is given in coordinates by $\boldsymbol{\alpha}(t) = \boldsymbol{x}(u(t), v(t))$, the energy of $\boldsymbol{\alpha}$ is:

$$E[\boldsymbol{\alpha}] = \frac{1}{2}\int_{I'} \epsilon_{\boldsymbol{\sigma}} u'(t)^2 + f(u(t))^2 v'(t)^2 \, \mathrm{d}t.$$

Writing the Lagrangian

$$L(u(t), u'(t), v(t), v'(t)) = \frac{1}{2}(\epsilon_{\boldsymbol{\sigma}} u'(t)^2 + f(u(t))^2 v'(t)^2),$$

we have that the Euler-Lagrange equations are:

$$\frac{\partial L}{\partial u} - \frac{\mathrm{d}}{\mathrm{d}t}\left(\frac{\partial L}{\partial u'}\right) = 0 \implies u''(t) - \epsilon_{\boldsymbol{\sigma}} f(u(t)) f'(u(t)) v'(t)^2 = 0$$
$$\frac{\partial L}{\partial v} - \frac{\mathrm{d}}{\mathrm{d}t}\left(\frac{\partial L}{\partial v'}\right) = 0 \implies v''(t) + 2\frac{f'(u(t))}{f(u(t))} u'(t) v'(t) = 0.$$

We then see that the non-vanishing Christoffel symbols are:

$$\Gamma^u_{vv}(u,v) = -\epsilon_{\boldsymbol{\sigma}} f(u) f'(u) \quad \text{and} \quad \Gamma^v_{uv}(u,v) = \Gamma^v_{vu}(u,v) = \frac{f'(u)}{f(u)}.$$

From such equations it is easy to see that the meridians, parametrized by $\boldsymbol{\alpha}(t) = \boldsymbol{x}(at + b, v_0)$, are geodesics in both ambient spaces. Each parallel $u = u_0$, in turn, is a geodesic if and only if $f'(u_0) = 0$. If all the parallels are geodesics, we have that f is a positive constant and, in this case, the surface is part of a straight circular cylinder.

(3) Let's reobtain the geodesics in the cylinder $\mathbb{S}^1(r) \times \mathbb{R} \subseteq \mathbb{R}^3_\nu$ of radius $r > 0$, previously found in Example 3.6.7 (p. 220). Consider again the parametrizations $x\colon]0, 2\pi[\times \mathbb{R} \to \mathbb{S}^1(r) \times \mathbb{R}$, given by $x(u,v) = (r\cos u, r\sin u, v)$. We know that its First Fundamental Form is given by

$$ds^2 = r^2\,du^2 + (-1)^\nu dv^2.$$

The energy of a curve $\alpha\colon I \to \mathbb{S}^1 \times \mathbb{R}$ given in coordinates by $\alpha(t) = x(u(t), v(t))$ is just

$$E[\alpha] = \frac{1}{2}\int_I r^2 u'(t)^2 + (-1)^\nu v'(t)^2\,dt.$$

The Lagrangian is:

$$L(u(t), u'(t), v(t), v'(t)) = \frac{1}{2}(r^2 u'(t)^2 + (-1)^\nu v'(t)^2),$$

and the Euler-Lagrange equations are:

$$\begin{aligned} \frac{\partial L}{\partial u} - \frac{d}{dt}\left(\frac{\partial L}{\partial u'}\right) &= 0 \implies u''(t) = 0 \\ \frac{\partial L}{\partial v} - \frac{d}{dt}\left(\frac{\partial L}{\partial v'}\right) &= 0 \implies v''(t) = 0, \end{aligned}$$

whence all the Christoffel symbols vanish, and the geodesics are given by

$$\alpha(t) = (r\cos(at+b), r\sin(at+b), ct+d),$$

where $a, b, c, d \in \mathbb{R}$. At least when $r = 1$, this was to be expected: x itself is a local isometry, so not only all the Christoffel symbols vanish (see (1) above), but also the geodesics of the cylinder are precisely the images via x of geodesics in the plane, that is, straight lines.

Example 3.6.28 (Geodesics of $\mathbb{S}^2(r)$, $\mathbb{S}^2_1(r)$ and $\mathbb{H}^2_\pm(r)$). Suppose that $M = \mathbb{S}^2(r)$, $\mathbb{S}^2(r)$ or $\mathbb{H}^2_\pm(r)$. Let's see that the geodesics of M are precisely the intersections $M \cap \Pi$, where Π is a plane passing through the origin such that $M \cap \Pi \neq \varnothing$ (such intersection is empty when Π is spacelike or lightlike and $M = \mathbb{H}^2_\pm(r)$). In view of the remark following Theorem 3.6.20 (p. 231), it suffices to show that all the intersections $M \cap \Pi$ may be parametrized by geodesics, and note that given $p \in M$ and $v \in T_pM$, the plane $\Pi = \operatorname{span}\{p, v\}$ has as tangent line at p precisely the line passing through p with direction v.

We already know that in $\mathbb{S}^2_1(r)$, every lightlike curve is a pre-geodesic by Exercise 3.6.10 (p. 227), so assume that Π has a non-lightlike fixed normal vector $n \in \mathbb{R}^3_\nu$. Let $\alpha\colon I \to M \cap \Pi$ be a unit speed curve. For each $s \in I$, write

$$\alpha''(s) = a(s)\alpha(s) + b(s)\alpha'(s) + c(s)n.$$

The condition $\langle \alpha(s), \alpha(s)\rangle = \text{cte.}$ implies that $\langle \alpha'(s), \alpha(s)\rangle = 0$; now α having constant speed gives us that $\langle \alpha''(s), \alpha'(s)\rangle = 0$ and, lastly, the condition $\langle \alpha(s), n\rangle = 0$ yields $\langle \alpha'(s), n\rangle = \langle \alpha''(s), n\rangle = 0$. This way, applying $\langle \cdot, \alpha'(s)\rangle$ and $\langle \cdot, n\rangle$ in the above gives us that $b(s) = c(s) = 0$, whence $\alpha''(s) = a(s)\alpha(s)$. But the position vector of an arbitrary point in M is always normal to M, and so we conclude that $\alpha''(s)$ is always normal M as well. In other words, α is a geodesic, as wanted.

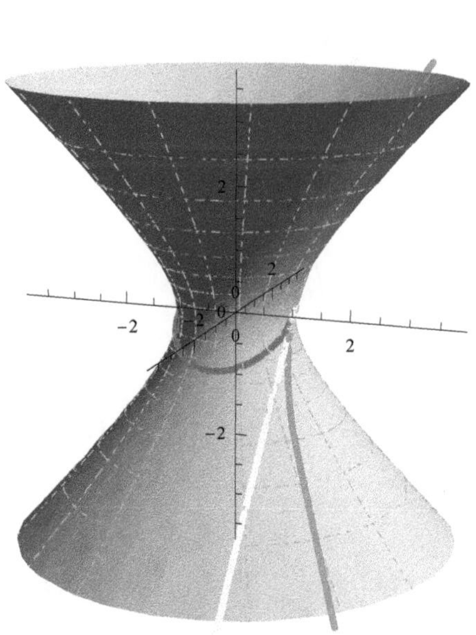

Figure 3.34: The geodesics in $\mathbb{S}^2_1$.

As a last application of the variational techniques presented in this section, let's discuss a specific type of parametrization, convenient for the study of geodesics, and which generalizes the parametrizations of revolution we have encountered so far:

Definition 3.6.29 (Clairaut Parametrizations). Let $M \subseteq \mathbb{R}^3_\nu$ be a non-degenerate regular surface and $(U, \boldsymbol{x})$ be a parametrization for M. If E, F and G denote the coefficients of the First Fundamental Form of M relative to $\boldsymbol{x}$, we'll say that $\boldsymbol{x}$ is a *u-Clairaut parametrization* if $E_v = G_v = F = 0$.

Remark.

- That is, if $\boldsymbol{x}$ is u-Clairaut, its First Fundamental Form written in differential notation is $\mathrm{d}s^2 = E(u)\,\mathrm{d}u^2 + G(u)\,\mathrm{d}v^2$. Moreover, since M is non-degenerate, we always have $EG \neq 0$.
- Clearly the parametrizations of revolution seen so far are u-Clairaut. To emphasize the comparison, we will refer to the coordinate curves $u = u_0$ as *parallels*, and to the coordinate curves $v = v_0$ as *meridians*.
- Similarly, we'll say that $\boldsymbol{x}$ is a *v-Clairaut parametrization* if $E_u = G_u = F = 0$. All the results proven for one case have obvious analogues for the other case.

Lemma 3.6.30. *Let $M \subseteq \mathbb{R}^3_\nu$ be a non-degenerate regular surface, and $(U, \boldsymbol{x})$ be a u-Clairaut parametrization. The geodesic equations for the parametrization $\boldsymbol{x}$ are:*

$$u'' + \frac{E_u}{2E}(u')^2 - \frac{G_u}{2E}(v')^2 = 0 \quad \text{and} \quad v'' + \frac{G_u}{G}u'v' = 0.$$

In particular, the non-vanishing Christoffel symbols are

$$\Gamma^u_{uu}(u,v) = \frac{E_u(u)}{2E(u)}, \quad \Gamma^u_{vv}(u,v) = -\frac{G_u(u)}{2E(u)} \quad \text{and} \quad \Gamma^v_{uv}(u,v) = \Gamma^v_{vu}(u,v) = \frac{G_u(u)}{2G(u)}.$$

Proof: We'll produce the geodesic equations via Euler-Lagrange equations, with the Larangian

$$L(u(t), u'(t), v(t), v'(t)) = \frac{1}{2}(E(u(t))u'(t)^2 + G(u(t))v'(t)^2).$$

Omitting points of evaluation, we have:

$$\begin{aligned}\frac{\partial L}{\partial u} - \frac{\mathrm{d}}{\mathrm{d}t}\left(\frac{\partial L}{\partial u'}\right) &= \frac{1}{2}(E_u(u')^2 + G_u(v')^2) - \frac{\mathrm{d}}{\mathrm{d}t}(Eu') \\ &= \frac{1}{2}E_u(u')^2 + \frac{1}{2}G_u(v')^2 - E_u(u')^2 - Eu'' = 0.\end{aligned}$$

Grouping similar terms and dividing the equation by $-E$ it follows that:

$$u'' + \frac{E_u}{2E}(u')^2 - \frac{G_u}{2E}(v')^2 = 0.$$

For the second equation, we have:

$$\begin{aligned}\frac{\partial L}{\partial v} - \frac{\mathrm{d}}{\mathrm{d}t}\left(\frac{\partial L}{\partial v'}\right) &= -\frac{\mathrm{d}}{\mathrm{d}t}(Gv') \\ &= -G_u u'v' - Gv'' = 0.\end{aligned}$$

Dividing the above by $-G$ we obtain:

$$v'' + \frac{G_u}{G}u'v' = 0.$$

□

Remark. For the geodesic equations in the v-Clairaut case, see Exercise 3.6.33.

The next proposition generalizes what we have discussed before about meridians and parallels in surfaces of revolution:

Proposition 3.6.31. *Let $M \subseteq \mathbb{R}^3_\nu$ be a non-degenerate regular surface and $(U, \boldsymbol{x})$ be a u-Clairaut parametrization. Then:*

(i) The meridians $v = v_0$ are pre-geodesics.

(ii) The parallels $u = u_0$ are geodesics if and only if $G_u(u_0) = 0$.

Proof:

(i) Let's prove that if $\boldsymbol{\alpha}\colon I \to M$ given by $\boldsymbol{\alpha}(u) = \boldsymbol{x}(u, v_0)$ parametrizes a meridian, then $(D\boldsymbol{\alpha}'/\mathrm{d}u)(u)$ is proportional to $\boldsymbol{\alpha}'(u)$, for all $u \in I$. Indeed, we have that $\boldsymbol{\alpha}'(u) = \boldsymbol{x}_u(u, v_0)$ and $\boldsymbol{\alpha}''(u) = \boldsymbol{x}_{uu}(u, v_0)$. By definition of covariant derivative we have:

$$\frac{D\boldsymbol{\alpha}'}{\mathrm{d}u}(u) = \Gamma^u_{uu}(u, v_0)\boldsymbol{x}_u(u, v_0) + \Gamma^v_{uu}(u, v_0)\boldsymbol{x}_v(u, v_0) = \frac{E_u(u)}{2E(u)}\boldsymbol{\alpha}'(u).$$

By Proposition 3.6.9 (p. 221), $\boldsymbol{\alpha}$ is a pre-geodesic, as wanted.

(ii) Let $\boldsymbol{\alpha}\colon I \to M$, $\boldsymbol{\alpha}(v) = \boldsymbol{x}(u_0, v)$, be a parametrization of a parallel. We have $\boldsymbol{\alpha}'(v) = \boldsymbol{x}_v(u_0, v)$ and $\boldsymbol{\alpha}''(v) = \boldsymbol{x}_{vv}(u_0, v)$. As before, we have that:

$$\frac{D\boldsymbol{\alpha}'}{dv}(v) = \Gamma^u_{vv}(u_0, v)\boldsymbol{x}_u(u_0, v) + \Gamma^v_{vv}(u_0, v)\boldsymbol{x}_v(u_0, v) = -\frac{G_u(u_0)}{2E(u_0)}\boldsymbol{x}_u(u_0, v),$$

which vanishes if and only if $G_u(u_0) = 0$.

□

Theorem 3.6.32 (Clairaut Relation). *Let $M \subseteq \mathbb{R}^3_\nu$ be a regular spacelike surface and $(U, \boldsymbol{x})$ be a u-Clairaut parametrization for M. If $\boldsymbol{\alpha}\colon I \to \boldsymbol{x}(U)$ is a unit speed geodesic given in coordinates by $\boldsymbol{\alpha}(s) = \boldsymbol{x}(u(s), v(s))$, there is $c(\boldsymbol{\alpha}) \in \mathbb{R}$ such that*

$$\sqrt{G(u(s))}\sin\theta(s) = c(\boldsymbol{\alpha}), \quad \textit{for all } s \in I,$$

where $\theta(s)$ is the angle formed between $\boldsymbol{x}_u(u(s), v(s))$ and $\boldsymbol{\alpha}'(s)$. In particular, the image of $\boldsymbol{\alpha}$ in M does not leave the region in M where $G \geq c(\boldsymbol{\alpha})^2$.

Proof: Since $F = 0$, we have that the angle between $\boldsymbol{x}_v(u(s), v(s))$ and $\boldsymbol{\alpha}'(s)$ is exactly $\frac{\pi}{2} - \theta(s)$. Using that $G_v = 0$, we may integrate the second geodesic equation:

$$v'' + \frac{G_u}{G}u'v' = 0 \implies Gv'' + G_u u'v' = 0 \implies (Gv')' = 0 \implies Gv' = c(\boldsymbol{\alpha}),$$

for some $c(\boldsymbol{\alpha}) \in \mathbb{R}$. With this, on one hand (omitting points of application in $\boldsymbol{x}_u$ and $\boldsymbol{x}_v$):

$$\langle \boldsymbol{\alpha}'(s), \boldsymbol{x}_v \rangle = \langle u'(s)\boldsymbol{x}_u + v'(s)\boldsymbol{x}_v, \boldsymbol{x}_v \rangle = G(u(s))v'(s) = c(\boldsymbol{\alpha}),$$

and, on the other hand,

$$\langle \boldsymbol{\alpha}'(s), \boldsymbol{x}_v \rangle = \|\boldsymbol{\alpha}'(s)\|\|\boldsymbol{x}_v\| \cos\left(\frac{\pi}{2} - \theta(s)\right) = \sqrt{G(u(s))}\sin\theta(s).$$

Thus, we have $\sqrt{G(u(s))}\sin\theta(s) = c(\boldsymbol{\alpha})$ for all $s \in I$. Finally, we obtain the inequality $c(\boldsymbol{\alpha})^2 = G(u(s))\sin^2\theta(s) \leq G(u(s))$, for all $s \in I$. □

For timelike surfaces the situation is slightly more complicated, once we do not have a general notion of angle between vectors of arbitrary causal type. Imposing a few extra restrictions on $\boldsymbol{x}$, we may obtain something similar to the above result, just with the tools we have so far:

Proposition 3.6.33. *Let $M \subseteq \mathbb{L}^3$ be a regular timelike surface, and $(U, \boldsymbol{x})$ be a u-Clairaut parametrization for M such that $\boldsymbol{x}_u$ is always spacelike. If $\boldsymbol{\alpha}\colon I \to M$ is a proper time parametrized timelike geodesic given in coordinates by $\boldsymbol{\alpha}(t) = \boldsymbol{x}(u(t), v(t))$, there is a constant $c(\boldsymbol{\alpha}) \in \mathbb{R}$, such that*

$$\sqrt{|G(u(t))|}\cosh\varphi_\pm(t) = \mp c(\boldsymbol{\alpha}), \quad \textit{for all } t \in I,$$

where the sign $\pm$ indicates which among $\boldsymbol{\alpha}'(t)$ and $-\boldsymbol{\alpha}'(t)$ lies in the same timecone as $\boldsymbol{x}_v(u(t), v(t))$, and $\varphi_\pm(t)$ denotes the corresponding hyperbolic angle. In particular, the image of $\boldsymbol{\alpha}$ does not leave the region in M where $G \geq -c(\boldsymbol{\alpha})^2$.

Proof: As done before, we integrate the second geodesic equation to obtain $G(u(t))v'(t) = c(\boldsymbol{\alpha})$ for some $c(\boldsymbol{\alpha}) \in \mathbb{R}$ and, omitting the point of evaluation for $\boldsymbol{x}_v$, we have that $\langle \boldsymbol{\alpha}'(t), \boldsymbol{x}_v \rangle_L = c(\boldsymbol{\alpha})$. Let's do the case where $\boldsymbol{\alpha}'(t)$ is in the same timecone as $\boldsymbol{x}_v$. If $\varphi_+(t)$ is the hyperbolic angle between them, we have

$$\langle \boldsymbol{\alpha}'(t), \boldsymbol{x}_v \rangle_L = -\|\boldsymbol{\alpha}'(t)\|_L \|\boldsymbol{x}_v\|_L \cosh \varphi_+(t) = -\sqrt{|G(u(t))|} \cosh \varphi_+(t),$$

and thus $\sqrt{|G(u(t))|} \cosh \varphi_+(t) = -c(\boldsymbol{\alpha})$ for all $t \in I$. When $\boldsymbol{\alpha}'(t)$ and $\boldsymbol{x}_v$ are in opposite timecones, the calculation is similar. So

$$c(\boldsymbol{\alpha})^2 = |G(u(t))| \cosh^2 \varphi_\pm(t) \geq |G(u(t))| = -G(u(t))$$

gives us that $G(u(t)) \geq -c(\boldsymbol{\alpha})^2$ for all $t \in I$, as wanted. □

Remark. The constant $c(\boldsymbol{\alpha})$ in the previous two results is called the *slant* of $\boldsymbol{\alpha}$ in M.

One of the main advantages of working with Clairaut parametrizations is that the geodesic equations together boil down to a single ODE, as the next two results show:

Theorem 3.6.34 (Clairaut). *Let $M \subseteq \mathbb{R}^3_\nu$ be a regular spacelike surface, $(U, \boldsymbol{x})$ be a u-Clairaut parametrization for M, and $\boldsymbol{\alpha}\colon I \to \boldsymbol{x}(U)$ be a unit speed geodesic which never crosses the meridians of $\boldsymbol{x}$ orthogonally. Then $\boldsymbol{\alpha}$ admits a reparametrization of the form $\boldsymbol{\beta}(u) = \boldsymbol{x}(u, v(u))$, where v is a solution for*

$$\frac{dv}{du} = \pm \frac{c(\boldsymbol{\alpha})\sqrt{E(u)}}{\sqrt{G(u)}\sqrt{G(u) - c(\boldsymbol{\alpha})^2}}.$$

Remark. In the above statement, the signs $\pm$ only indicate reverse parametrizations.

Proof: Start writing $\boldsymbol{\alpha}(s) = \boldsymbol{x}(u(s), v(s))$. Omitting points of application, substituting $Gv' = c(\boldsymbol{\alpha})$ into $E(u')^2 + G(v')^2 = 1$ yields

$$(u')^2 = \frac{G - c(\boldsymbol{\alpha})^2}{EG}.$$

The Clairaut relation seen in Theorem 3.6.32 now says that $G - c(\boldsymbol{\alpha})^2 \geq 0$, but the condition on the meridians ensures strict inequality. So not only we may write

$$u' = \pm \frac{\sqrt{G - c(\boldsymbol{\alpha})^2}}{\sqrt{EG}},$$

we may also use that $u' \neq 0$, so that the Inverse Function Theorem allows us to use u as parameter (explicitly, we'll have $\boldsymbol{\beta} \doteq \boldsymbol{\alpha} \circ u^{-1}$) and conclude that

$$\frac{dv}{du} = \frac{v'}{u'} = \pm \frac{c(\boldsymbol{\alpha})/G}{\sqrt{G - c(\boldsymbol{\alpha})^2}/\sqrt{EG}} = \pm \frac{c(\boldsymbol{\alpha})\sqrt{E}}{\sqrt{G}\sqrt{G - c(\boldsymbol{\alpha})^2}}.$$

□

Lastly, Proposition 3.6.33 allows us to use a similar argument to the above and obtain the:

Theorem 3.6.35. *Let $M \subseteq \mathbb{L}^3$ be a regular timelike surface, $(U, \boldsymbol{x})$ be a u-Clairaut parametrization for M such that $\boldsymbol{x}_u$ is always spacelike, and $\boldsymbol{\alpha}: I \to \boldsymbol{x}(U)$ be a proper time parametrized timelike geodesic which never crosses the meridians of $\boldsymbol{x}$ orthogonally. Then $\boldsymbol{\alpha}$ admits a reparametrization of the form $\boldsymbol{\beta}(u) = \boldsymbol{x}(u, v(u))$, where v is a solution for*

$$\frac{dv}{du} = \pm \frac{c(\boldsymbol{\alpha})\sqrt{E(u)}}{\sqrt{|G(u)|}\sqrt{G(u) + c(\boldsymbol{\alpha})^2}}.$$

We ask you to carry out this last proof in Exercise 3.6.35. For more geometric interpretations of Clairaut parametrizations, see [55].

Exercises

Exercise† 3.6.17. Let $M \subseteq \mathbb{R}^3_\nu$ be a non-degenerate regular surface, $(U, \boldsymbol{x})$ be a parametrization for M, $\boldsymbol{\alpha}: I \to \boldsymbol{x}(U)$ be a regular curve, and $\boldsymbol{V}: I \to \mathbb{R}^3_\nu$ a vector field along $\boldsymbol{\alpha}$. Suppose that, in coordinates, we write

$$\boldsymbol{\alpha}(t) = \boldsymbol{x}(u^1(t), u^2(t)) \quad \text{and} \quad \boldsymbol{V}(t) = \sum_{i=1} a^i(t)\boldsymbol{x}_i(u^1(t), u^2(t)).$$

Show that, omitting points of evaluation, we have

$$\frac{D\boldsymbol{V}}{dt} = \sum_{k=1}^{2}\left(\dot{a}^k + \sum_{i,j=1}^{2}\Gamma^k_{ij}a^i\dot{u}^j\right)\boldsymbol{x}_k.$$

Exercise† 3.6.18 (Parallel Translation)**.** Let $M \subseteq \mathbb{R}^3_\nu$ be a non-degenerate regular surface and $\boldsymbol{\alpha}: [0,1] \to M$ be a regular curve joining $\boldsymbol{p} = \boldsymbol{\alpha}(0)$ to $\boldsymbol{q} = \boldsymbol{\alpha}(1)$.

(a) Show that given $\boldsymbol{v} \in T_{\boldsymbol{p}}M$, there is a unique parallel vector field $\boldsymbol{V}: [0,1] \to \mathbb{R}^3_\nu$ along $\boldsymbol{\alpha}$ such that $\boldsymbol{V}(0) = \boldsymbol{v}$. The vector $\boldsymbol{V}(1) \in T_{\boldsymbol{q}}M$ is called the *parallel translation* of $\boldsymbol{v}$ from $\boldsymbol{p}$ to $\boldsymbol{q}$ via $\boldsymbol{\alpha}$.

Hint. To make it easier, you may assume that the image of $\boldsymbol{\alpha}$ lies in the image of a single parametrization for M.

(b) How does parallel translation work when M is a non-degenerate plane?

(c) If $M \subseteq \mathbb{R}^3$ is the surface described by $z = y^2$, consider $\boldsymbol{\alpha}: [0,1] \to M$ given by $\boldsymbol{\alpha}(t) = (0, t, t^2)$, and compute the parallel translation of $(1,1,0) \in T_{(0,0,0)}M$ to $T_{(0,1,1)}M$ via $\boldsymbol{\alpha}$.

Hint. Work with the obvious parametrization.

(d) In the notation from item (a), show that $P_{\boldsymbol{\alpha}}: T_{\boldsymbol{p}}M \to T_{\boldsymbol{q}}M$ given by $P_{\boldsymbol{\alpha}}(\boldsymbol{v}) = \boldsymbol{V}(1)$ is a linear isometry. What is $(P_{\boldsymbol{\alpha}})^{-1}$?

Remark. Given $\boldsymbol{p} \in M$, the collection of all maps $P_{\boldsymbol{\alpha}}$, where $\boldsymbol{\alpha}$ is a smooth curve in M that starts and ends at $\boldsymbol{p}$, equipped with the operation of composition of functions, is a group (called the *holonomy group of M at $\boldsymbol{p}$*).

Exercise† 3.6.19. Let $M_1, M_2 \subseteq \mathbb{R}^3_\nu$ be two non-degenerate regular surfaces, $\boldsymbol{\alpha}: [0,1] \to M_1$ be a regular curve joining $\boldsymbol{p} = \boldsymbol{\alpha}(0)$ to $\boldsymbol{q} = \boldsymbol{\alpha}(1)$, $\boldsymbol{V}$ a vector field along $\boldsymbol{\alpha}$, and $\boldsymbol{\phi}: M_1 \to M_2$ smooth.

(a) Suppose that ϕ is a local isometry. If $\overline{\boldsymbol{V}}\colon I \to \mathbb{R}^3_\nu$ is the vector field along $\phi \circ \boldsymbol{\alpha}$ defined by $\overline{\boldsymbol{V}}(t) \doteq \mathrm{d}\phi_{\boldsymbol{\alpha}(t)}(\boldsymbol{V}(t))$, show that

$$\mathrm{d}\phi_{\boldsymbol{\alpha}(t)}\left(\frac{D\boldsymbol{V}}{\mathrm{d}t}(t)\right) = \frac{D\overline{\boldsymbol{V}}}{\mathrm{d}t}(t),$$

for all $t \in I$. In particular, (local) isometries take parallel fields to parallel fields.

(b) Suppose that ϕ is an isometry. Show that for all $\boldsymbol{v} \in T_{\boldsymbol{p}}M$ we have that

$$P_{\phi\circ\boldsymbol{\alpha}}(\mathrm{d}\phi_{\boldsymbol{p}}(\boldsymbol{v})) = \mathrm{d}\phi_{\boldsymbol{q}}(P_{\boldsymbol{\alpha}}(\boldsymbol{v})).$$

That is, isometries are compatible with parallel translations.

Exercise 3.6.20 (Fermi-Walker Parallelism). Let $M \subseteq \mathbb{R}^3_\nu$ be a non-degenerate regular surface and $\boldsymbol{\alpha}\colon [0,1] \to M$ be a regular curve joining $\boldsymbol{p} = \boldsymbol{\alpha}(0)$ to $\boldsymbol{q} = \boldsymbol{\alpha}(1)$. In Exercise 3.6.18 above we have seen that parallel translation along $\boldsymbol{\alpha}$ is a linear isometry between different tangent planes and, thus, it takes orthonormal bases of $T_{\boldsymbol{p}}M$ onto orthonormal bases of $T_{\boldsymbol{q}}M$. However, if $\boldsymbol{\alpha}'(0)$ is an element of the initial basis, it is not necessarily true that $\boldsymbol{\alpha}'(1)$ is an element of the basis obtained, unless $\boldsymbol{\alpha}$ itself is a geodesic (to wit, if $\boldsymbol{\alpha}$ is a geodesic, the velocity field $\boldsymbol{\alpha}'$ is parallel along $\boldsymbol{\alpha}$ and so $\boldsymbol{\alpha}'(1)$ would be the parallel translation of $\boldsymbol{\alpha}'(0)$ from $\boldsymbol{p}$ to $\boldsymbol{q}$ via $\boldsymbol{\alpha}$). Suppose that M and $\boldsymbol{\alpha}$ are both timelike, with $\boldsymbol{\alpha}$ parametrized with proper time. Aiming to fix the above deficiency, we define the *FW-derivative* of a smooth vector field $\boldsymbol{V}\colon [0,1] \to \mathbb{R}^3_\nu$ along $\boldsymbol{\alpha}$, by introducing a certain correction term:

$$\frac{F\boldsymbol{V}}{\mathrm{d}t}(t) \doteq \frac{D\boldsymbol{V}}{\mathrm{d}t}(t) + \begin{vmatrix} \dfrac{D\boldsymbol{\alpha}'}{\mathrm{d}t}(t) & \boldsymbol{\alpha}'(t) \\ \left\langle \dfrac{D\boldsymbol{\alpha}'}{\mathrm{d}t}(t), \boldsymbol{V}(t)\right\rangle_L & \langle \boldsymbol{\alpha}'(t), \boldsymbol{V}(t)\rangle_L \end{vmatrix}.$$

We'll say that $\boldsymbol{V}$ is *FW-parallel* if $F\boldsymbol{V}/\mathrm{d}t = \boldsymbol{0}$.

(a) Show that the operator $F/\mathrm{d}t$ is $\mathbb{R}$-linear and satisfies the Leibniz rule, as seen for the operator $D/\mathrm{d}t$ in Exercise 3.6.1 (p. 225). This way, the operator $F/\mathrm{d}t$ also deserves the name of "covariant derivative".

(b) Show that the velocity field $\boldsymbol{\alpha}'$ is always FW-parallel and, moreover, if $\boldsymbol{\alpha}$ is a geodesic, then $F/\mathrm{d}t = D/\mathrm{d}t$. Thus, the new operator $F/\mathrm{d}t$ is indeed a type of generalization of $D/\mathrm{d}t$, which makes all timelike curves "FW-geodesics".

(c) Show that if $\boldsymbol{V}$ is a vector field along $\boldsymbol{\alpha}$ which is always orthogonal to $\boldsymbol{\alpha}'$, then $(F\boldsymbol{V}/\mathrm{d}t)(t)$ is the projection of $(D\boldsymbol{V}/\mathrm{d}t)(t)$ onto the orthogonal complement of $\boldsymbol{\alpha}'(t)$.

(d) If $\boldsymbol{V}$ and $\boldsymbol{W}$ are smooth vector fields along $\boldsymbol{\alpha}$, then

$$\frac{\mathrm{d}}{\mathrm{d}t}\langle \boldsymbol{V}(t), \boldsymbol{W}(t)\rangle_L = \left\langle \frac{F\boldsymbol{V}}{\mathrm{d}t}(t), \boldsymbol{W}(t)\right\rangle_L + \left\langle \boldsymbol{V}(t), \frac{F\boldsymbol{W}}{\mathrm{d}t}(t)\right\rangle_L$$

holds. In particular, it follows that FW-parallel fields also have constant speed and causal type (compare with Exercise 3.6.2, p. 226).

(e) It is possible to show (as done in Exercise 3.6.18) that in these conditions, given $\boldsymbol{v} \in T_pM$, there is a unique FW-parallel vector field $\boldsymbol{V}$ along $\boldsymbol{\alpha}$ such that $\boldsymbol{V}(0) = \boldsymbol{v}$, and the vector $\boldsymbol{V}(1)$ is called the *FW-parallel translation* of $\boldsymbol{v}$ from $\boldsymbol{p}$ to $\boldsymbol{q}$ via $\boldsymbol{\alpha}$. Show that the map $P_{\boldsymbol{\alpha},\mathrm{FW}}\colon T_pM \to T_qM$ so defined is also a linear isometry.

(f) Show that local isometries preserve FW-derivatives and conclude that isometries are compatible with FW-parallel translations (in the sense of the previous exercise).

Remark. For more details, see [48].

Exercise 3.6.21 (Transformation law for Γ_{ij}^k)**.** Let $M \subseteq \mathbb{R}^3_\nu$ be a non-degenerate regular surface, and $(U, \boldsymbol{x})$ and $(\widetilde{U}, \widetilde{\boldsymbol{x}})$ be two parametrizations such that $\boldsymbol{x}(U) \cap \widetilde{\boldsymbol{x}}(\widetilde{U}) \neq \varnothing$. In suitable domains, the change of parameters $\widetilde{\boldsymbol{x}}^{-1} \circ \boldsymbol{x}$ defines inverse smooth relations $\widetilde{u}^i = \widetilde{u}^i(u^1, u^2)$ and $u^i = u^i(\widetilde{u}^1, \widetilde{u}^2)$. Denote by $\widetilde{g}_{ij}$ and $\widetilde{\Gamma}_{ij}^k$ the First Fundamental Form and the Christoffel symbols of $\widetilde{\boldsymbol{x}}$.

(a) Show that the First Fundamental Form transforms as

$$\widetilde{g}_{ij} = \sum_{k,\ell=1}^{2} \frac{\partial u^k}{\partial \widetilde{u}^i} \frac{\partial u^\ell}{\partial \widetilde{u}^j} g_{kl} \quad \text{and} \quad \widetilde{g}^{ij} = \sum_{k,\ell=1}^{2} \frac{\partial \widetilde{u}^i}{\partial u^k} \frac{\partial \widetilde{u}^j}{\partial u^\ell} g^{kl},$$

where all the functions above are evaluated in the correct points.

(b) Show that

$$\widetilde{\Gamma}_{ij}^k = \sum_{r,s,\ell=1}^{2} \frac{\partial u^r}{\partial \widetilde{u}^i} \frac{\partial u^s}{\partial \widetilde{u}^j} \frac{\partial \widetilde{u}^k}{\partial u^\ell} \Gamma_{rs}^\ell + \sum_{\ell=1}^{2} \frac{\partial \widetilde{u}^k}{\partial u^\ell} \frac{\partial^2 u^\ell}{\partial \widetilde{u}^i \partial \widetilde{u}^j}.$$

Remark. The formula in item (b) says that the Christoffel symbols do not represent any type of linear transformation in $\mathbb{R}^2$, no matter which is the fixed index i, j or k. The term involving second order derivatives accounts for the action of the Γ_{ij}^k's as a correction in the local expression for covariant derivatives.

Exercise 3.6.22. Let $M \subseteq \mathbb{R}^3_\nu$ be a connected and non-degenerate regular surface such that all the geodesics of M are plane curves. Show that M is contained in a plane, $\mathbb{S}^2(\boldsymbol{c}, r)$, $\mathbb{S}^2_1(\boldsymbol{c}, r)$ or $\mathbb{H}^2_\pm(\boldsymbol{c}, r)$, for certain $\boldsymbol{c} \in \mathbb{R}^3_\nu$ and $r > 0$.

Hint. Show that M is totally umbilic.

Exercise 3.6.23. Find extremizers for the following variational problems:

(a) $$\begin{cases} J_1[y] = \displaystyle\int_1^2 \frac{y'(x)^2}{x^3}\,\mathrm{d}x \\ y(1) = 0, \quad y(2) = 15. \end{cases}$$

(b) $$\begin{cases} J_2[y] = \displaystyle\int_0^1 y(x)^2 + y'(x)^2 + 2y(x)\mathrm{e}^x\,\mathrm{d}x \\ y(0) = 0, \quad y(1) = \mathrm{e}/2. \end{cases}$$

It is more complicated, in general, to decide whether a given extremizer is a local maximum or minimum for a given functional. In some specific cases, the situation is more treatable. If y is the extremizer found in item (a), show that

$$H_y[\eta_1, \eta_2] \doteq \left.\frac{\partial^2}{\partial \epsilon_1 \partial \epsilon_2}\right|_{\epsilon_1 = \epsilon_2 = 0} J_1[y + \epsilon_1\eta_1 + \epsilon_2\eta_2] = 2\int_1^2 \frac{\eta_1'(x)\eta_2'(x)}{x^3}\,\mathrm{d}x$$

for all admissible variations η_1 and η_2. Note that given an admissible variation η, it holds that $H_y[\eta,\eta] \geq 0$ and $H_y[\eta,\eta] = 0$ if and only if $\eta = 0$. Conclude that y is a strict local minimum for J_1.

Remark. The map H_y is a sort of Hessian, useful as a tool to decide the nature of critical points of J_1.

Exercise† 3.6.24. Repeat the idea given in the proof of Theorem 3.6.22 (p. 232) to show that if $y\colon [a,b] \to \mathbb{R}$ is a class $\mathscr{C}^2$ critical point for the variational problem

$$\begin{cases} J[y] = \displaystyle\int_a^b L(t,y(t),y'(t),y''(t))\,\mathrm{d}t \\ y(a) = y_0, \quad y(b) = y_1, \quad y'(a) = y_0^*, \quad y'(b) = y_1^*, \end{cases}$$

where $L\colon [a,b] \times \mathbb{R}^2 \to \mathbb{R}$ is a prescribed class $\mathscr{C}^3$ Lagrangian, then y satisfies the Euler-Lagrange equation:

$$\frac{\partial L}{\partial y} - \frac{\mathrm{d}}{\mathrm{d}t}\left(\frac{\partial L}{\partial y'}\right) + \frac{\mathrm{d}^2}{\mathrm{d}t^2}\left(\frac{\partial L}{\partial y''}\right) = 0,$$

where the partial derivatives of L are evaluated in $(t,y(t),y'(t),y''(t))$.

Exercise† 3.6.25 (Some conservation laws)**.** Consider a class $\mathscr{C}^2$ Lagrangian $L\colon I \times \mathbb{R}^{2n} \to \mathbb{R}$. The *force* of the Lagrangian in the direction x_i is the quantity $\partial L/\partial x_i$, its *momentum* in the direction x_i is the quantity $\partial L/\partial \dot{x}_i$ and, lastly, the quantity $H \doteq \sum_{i=1}^n \dot{x}_i(\partial L/\partial \dot{x}_i) - L$ is called the associated *Hamiltonian.*

(a) Show that if L does not explicitly depend on x_i, that is, if the force of the Lagrangian vanishes in the direction of x_i, the corresponding momentum is constant along physical trajectories.

(b) Show that if L is autonomous (i.e., does not explicitly depend on t, or yet $\partial L/\partial t = 0$), then the Hamiltonian is constant along physical trajectories.

Remark. Many other conservations laws (such as Clairaut's relation) follow from the celebrated *Noether's Theorem* which, informally, says that if a Lagrangian L is *invariant* under a 1-parameter group of diffeomorphisms[5] of $\mathbb{R}^n$, $(\varphi_s)_{s\in\mathbb{R}}$ (i.e., if fixed $s \in \mathbb{R}$, we have $L(t,\boldsymbol{x}(t),\dot{\boldsymbol{x}}(t)) = L(t,\boldsymbol{y}(t),\dot{\boldsymbol{y}}(t))$ for all t, where $\boldsymbol{y} = \varphi_s \circ \boldsymbol{x}$), then the *Noether charge* $\mathcal{J}\colon I \times \mathbb{R}^{2n} \to \mathbb{R}$ defined by

$$\mathcal{J}(t,\boldsymbol{x}(t),\dot{\boldsymbol{x}}(t)) \doteq \sum_{i=1}^n \frac{\partial L}{\partial \dot{x}_i}(t,\boldsymbol{x}(t),\dot{\boldsymbol{x}}(t)) \left.\frac{\partial \varphi_s^i(\boldsymbol{x}(t))}{\partial s}\right|_{s=0},$$

where φ_s^i is the i-th coordinate of φ_s, is conserved along physical trajectories, that is, $\mathcal{J} = \mathsf{cte}$. In other words, in view of Noether's Theorem, to find out conservation laws we must seek symmetries of the Lagrangian. Can you exhibit suitable 1-parameter groups of diffeomorphisms to conclude again the results from items (a) and (b) above using Noether's Theorem?

[5] A collection of diffeomorphisms $\varphi_s\colon \mathbb{R}^n \to \mathbb{R}^n$ satisfying $\varphi_0 = \mathrm{Id}_{\mathbb{R}^n}$ and $\varphi_{s_1+s_2} = \varphi_{s_1} \circ \varphi_{s_2}$, for all $s_1, s_2 \in \mathbb{R}$.

Exercise† **3.6.26.** Consider the energy functional in $\mathbb{R}^n_\nu$. Show that given $\boldsymbol{\alpha}\colon I \to \mathbb{R}^n_\nu$ and admissible variations $\boldsymbol{\eta}_1, \boldsymbol{\eta}_2$, we have that

$$H_{\boldsymbol{\alpha}}[\boldsymbol{\eta}_1, \boldsymbol{\eta}_2] \doteq \left.\frac{\partial^2}{\partial \epsilon_1 \partial \epsilon_2}\right|_{\epsilon_1=\epsilon_2=0} E[\boldsymbol{\alpha} + \epsilon_1 \boldsymbol{\eta}_1 + \epsilon_2 \boldsymbol{\eta}_2] = \int_I \langle \boldsymbol{\eta}_1'(t), \boldsymbol{\eta}_2'(t)\rangle \, dt.$$

Note that for any admissible variation $\boldsymbol{\eta}$ in $\mathbb{R}^n$, we have $H_{\boldsymbol{\alpha}}[\boldsymbol{\eta}, \boldsymbol{\eta}] \geq 0$ and $H_{\boldsymbol{\alpha}}[\boldsymbol{\eta}, \boldsymbol{\eta}] = 0$ if and only if $\boldsymbol{\eta} = \mathbf{0}$. Conclude that in $\mathbb{R}^n$, geodesics locally minimize the energy functional.

Exercise 3.6.27. Let $f\colon U \subseteq \mathbb{R}^2 \to \mathbb{R}$ be a smooth function.

(a) Consider the functional

$$\mathscr{A}[f] = \int_U \sqrt{1 + f_u(u,v)^2 + f_v(u,v)^2}\, du\, dv.$$

Explicitly write the Euler-Lagrange equation in this case. Conclude (again) that the graph of f is a minimal surface in $\mathbb{R}^3$ if and only if

$$f_{uu}(1 + f_v^2) - 2 f_u f_v f_{uv} + f_{vv}(1 + f_u^2) = 0.$$

(b) Considering another suitable functional, repeat the idea from the previous item, and conclude that the graph of f is a critical surface in $\mathbb{L}^3$ if and only if

$$f_{uu}(-1 + f_v^2) - 2 f_u f_v f_{uv} + f_{vv}(-1 + f_u^2) = 0.$$

Exercise 3.6.28. Let $\phi\colon \mathbb{R}^n \to \mathbb{R}$ be a smooth function with *compact support*, i.e., with $\overline{\{\boldsymbol{x} \in \mathbb{R}^n \mid \phi(\boldsymbol{x}) \neq 0\}}$ compact.

(a) Define the Euclidean *Dirichlet energy* of ϕ as

$$D_E[\phi] \doteq \frac{1}{2}\int_{\mathbb{R}^n} \langle \nabla\phi(\boldsymbol{x}), \nabla\phi(\boldsymbol{x})\rangle_E \, d\boldsymbol{x}.$$

Show that the critical points of the Euclidean Dirichlet energy are harmonic functions, i.e., solutions of the equation $\triangle\phi = 0$, where the *Laplacian* of ϕ is defined by

$$\triangle\phi \doteq \sum_{i=1}^{n} \frac{\partial^2 \phi}{\partial x_i^2}.$$

(b) One may also consider a Lorentzian version of the above::

$$D_L[\phi] \doteq \frac{1}{2}\int_{\mathbb{R}^n} \langle \nabla\phi(\boldsymbol{x}), \nabla\phi(\boldsymbol{x})\rangle_L \, d\boldsymbol{x}.$$

Show that the critical points of this Lorentzian Dirichlet energy are solutions to the wave equation $\Box\phi = 0$, where $\Box$ is the *(stationary) wave operator (d'Alembertian)*, defined by

$$\Box\phi \doteq \sum_{i=1}^{n-1} \frac{\partial^2 \phi}{\partial x_i^2} - \frac{\partial^2 \phi}{\partial x_n^2}.$$

Remark. The assumption that ϕ has compact support is meant just to ensure that the integrals in question are not really improper. The operators $\triangle$ and $\Box$ will play an important role when we study critical surfaces in more detail in Chapter 4.

Exercise 3.6.29 (Surfaces of hyperbolic revolutions – III)**.** Consider again an injective, regular, and smooth curve $\sigma\colon I \to \mathbb{L}^3$ of the form $\sigma(u) = (f(u), 0, g(u))$, with $g(u) > 0$ for all $u \in I$, as well as the surface of hyperbolic revolution it generated around the x-axis, with parametrization $\boldsymbol{x}\colon I \times]0, 2\pi[\to \boldsymbol{x}(I \times]0, 2\pi[) \subseteq \mathbb{L}^3$ given by:

$$\boldsymbol{x}(u,v) = (f(u), g(u)\sinh v, g(u)\cosh v).$$

Suppose that σ has unit speed.

(a) Compute the geodesic equations for $\boldsymbol{x}$, and conclude that the non-vanishing Christoffel symbols are

$$\Gamma^u_{vv}(u,v) = -\epsilon_\sigma g(u)g'(u) \quad \text{and} \quad \Gamma^v_{uv}(u,v) = \Gamma^v_{vu}(u,v) = \frac{g'(u)}{g(u)}.$$

(b) Conclude that the meridians $v = v_0$ are geodesics, and that each parallel $u = u_0$ is a geodesic if and only if $g'(u_0) = 0$. What does the surface look like if all the parallels are geodesics?

Exercise 3.6.30.

(a) Recall that in polar coordinates, the metric in $\mathbb{R}^2 \setminus \{\boldsymbol{0}\}$ is given in differential notation by $\mathrm{d}s^2 = \mathrm{d}r^2 + r^2\,\mathrm{d}\theta^2$. Show that the non-vanishing Christoffel symbols are

$$\Gamma^r_{\theta\theta}(r,\theta) = -r \quad \text{and} \quad \Gamma^\theta_{r\theta}(r,\theta) = \Gamma^\theta_{\theta r}(r,\theta) = 1/r.$$

(b) In Rindler coordinates, the metric in the Rindler wedge is given in differential notation by $\mathrm{d}s^2 = \mathrm{d}\rho^2 - \rho^2\,\mathrm{d}\varphi^2$. Show that the non-vanishing Christoffel symbols are

$$\Gamma^\rho_{\varphi\varphi}(\rho,\varphi) = \rho \quad \text{and} \quad \Gamma^\varphi_{\rho\varphi}(\rho,\varphi) = \Gamma^\varphi_{\varphi\rho}(\rho,\varphi) = 1/\rho.$$

(c) Solve the geodesic equations and conclude (again) that the geodesics in each case are straight lines.

Exercise 3.6.31. Let $h\colon \mathbb{R}_{>0} \to \mathbb{R}$ be a smooth function and consider the injective and regular parametrized surface $\boldsymbol{x}\colon \mathbb{R}_{>0} \times]0, 2\pi[\to \mathbb{R}^3_\nu$ given by

$$\boldsymbol{x}(u,v) = (u\cos v, u\sin v, h(u)).$$

Show that the non-vanishing Christoffel symbols for $\boldsymbol{x}$ are

$$\Gamma^u_{uu}(u,v) = (-1)^\nu \frac{h'(u)h''(u)}{1 + (-1)^\nu h'(u)^2}, \quad \Gamma^u_{vv}(u,v) = \frac{-u}{1 + (-1)^\nu h'(u)^2},$$

and

$$\Gamma^v_{uv}(u,v) = \Gamma^v_{vu}(u,v) = \frac{1}{u}.$$

Exercise 3.6.32. Fix $0 < \alpha_0 < \pi/2$ and consider $\boldsymbol{x}\colon \mathbb{R}_{>0} \times]0, 2\pi[\to \mathbb{R}^3_\nu$ given by

$$\boldsymbol{x}(u,v) = (u\cos v\tan\alpha_0, u\sin v\tan\alpha_0, u).$$

We have seen in Exercise 3.2.2 (p. 164) that $M \doteq \boldsymbol{x}(\mathbb{R}_{>0} \times]0, 2\pi[)$ is a regular surface, and that for $\alpha_0 \neq \pi/4$ in $\mathbb{L}^3$, M is non-degenerate. Determine the Christoffel symbols for $\boldsymbol{x}$, and solve the geodesic equations to find the geodesics of M.

Exercise 3.6.33. Let $M \subseteq \mathbb{R}^3_\nu$ be a non-degenerate regular surface and $(U, \boldsymbol{x})$ be a v-Clairaut parametrization for M. Show that as geodesic equations in this case are:

$$u'' + \frac{E_v}{E}u'v' = 0 \quad \text{and} \quad v'' - \frac{E_v}{2G}(u')^2 + \frac{G_v}{2G}(v')^2 = 0.$$

Also read the Christoffel symbols from the above equations.

Exercise 3.6.34. State and prove a result analogous to Proposition 3.6.31 (p. 241) for v-Clairaut parametrizations.

Exercise† 3.6.35. Prove Theorem 3.6.35 (p. 244).

3.7 THE FUNDAMENTAL THEOREM OF SURFACES

In the beginning of this chapter, we promised a surface analogue to the Fundamental Theorem of Curves, seen in Chapter 2 (Theorem 2.3.20, p. 112) and, moreover, at this point we still owe you a proof of Gauss' *Theorema Egregium* (Theorem 3.3.14, p. 186). In the brief discussion about geodesics we had in the previous section, one of the main concepts used to describe geodesics in coordinates were the Christoffel symbols — which now will be used to pay all our debts.

3.7.1 The compatibility equations

In the Fundamental Theorem of Curves, we have used the curvature and torsion of an admissible curve to "reconstruct it", up to a rigid motion of the ambient space. More precisely, we solve the Frenet-Serret system of the curve, to then determine the curve completely from given initial conditions, by using the existence and uniqueness theorem for systems of ODEs.

The natural attack strategy in this case would be to try and reconstruct a non-degenerate regular surface M from its Gaussian and mean curvatures. However, the same geometric information is already captured by the First and Second Fundamental Forms of M. In our study of curves, we dealt with parametrizations themselves, instead of looking at subsets of the space, as we have done for surfaces. So, aiming to repeat the same strategy, we will use a parametrization $(U, \boldsymbol{x})$ to do a local analysis, and then determine the system that has to be satisfied by the coefficients $(g_{ij})_{1\leq i,j\leq 2}$ and $(h_{ij})_{1\leq i,j\leq 2}$ to then, formally solve it.

The first problem we encounter, though, is that this system will now consist of PDEs instead of ODEs. Thus, we need a more powerful existence and uniqueness theorem for solutions. This means that at this point, our progress depends on the following result, whose proof is out of our current reach:

Theorem 3.7.1 (Frobenius). *Let $F_{ij}\colon \mathbb{R}^{m+n} \to \mathbb{R}$ be class $\mathscr{C}^2$ functions, for $1 \leq i \leq n$ and $1 \leq j \leq m$, satisfying the* compatibility equations

$$\frac{\partial F_{ij}}{\partial x_k} + \sum_{\ell=1}^{n} \frac{\partial F_{ij}}{\partial y_\ell} F_{k\ell} = \frac{\partial F_{ik}}{\partial x_j} + \sum_{\ell=1}^{n} \frac{\partial F_{ik}}{\partial y_\ell} F_{j\ell},$$

for all possibilities of i, j and k. If $\boldsymbol{x} = (x_1, \ldots, x_m)$ and $\boldsymbol{y} = (y_1, \ldots, y_n)$, given $\boldsymbol{a} \in \mathbb{R}^m$

and $\boldsymbol{b} \in \mathbb{R}^n$, *the initial value problem*

$$\begin{cases} \dfrac{\partial y_i}{\partial x_j} = F_{ij}(\boldsymbol{x}, \boldsymbol{y}) \\ \boldsymbol{y}(\boldsymbol{a}) = \boldsymbol{b} \end{cases}$$

has a unique solution in some neighborhood of $\boldsymbol{a}$.

Remark. We'll say that a system of PDEs as above is *integrable* if the compatibility equations from the Frobenius Theorem are satisfied.

For a proof see, for instance, [68]. Well, if $\boldsymbol{N}$ is a Gauss map for M, we know that the three identities

$$(\boldsymbol{x}_{uu})_v = (\boldsymbol{x}_{uv})_u, \quad (\boldsymbol{x}_{vv})_u = (\boldsymbol{x}_{uv})_v, \quad \text{and} \quad (\boldsymbol{N} \circ \boldsymbol{x})_{uv} = (\boldsymbol{N} \circ \boldsymbol{x})_{vu}$$

must hold. We may write both sides of all relations as linear combinations of $\boldsymbol{x}_u$, $\boldsymbol{x}_v$ and $\boldsymbol{N} \circ \boldsymbol{x}$, and thus obtain nine relations which must necessarily be satisfied by the coefficients $(g_{ij})_{1 \leq i,j \leq 2}$ and $(h_{ij})_{1 \leq i,j \leq 2}$. Before highlighting some of these relations, note that this tells us explicitly that given smooth maps $g_{ij}, h_{ij} \colon U \to \mathbb{R}$, for $1 \leq i,j \leq 2$, which are symmetric in the indices i and j, such that $\det\left((g_{ij})_{1 \leq i,j \leq 2}\right) \neq 0$, *it is not necessarily true that they represent the Fundamental Forms of a surface.* We will bring this up again soon. In any case, we move on:

Proposition 3.7.2 (Compatibility). *Let* $M \subseteq \mathbb{R}^3_\nu$ *be a non-degenerate regular surface and* $(U, \boldsymbol{x})$ *be a parametrization for* M. *Then the* Gauss equations *hold:*

$$\begin{cases} EK = (\Gamma_{11}^2)_v - (\Gamma_{12}^2)_u + \Gamma_{11}^1\Gamma_{12}^2 + \Gamma_{11}^2\Gamma_{22}^2 - \Gamma_{12}^1\Gamma_{11}^2 - \Gamma_{12}^2\Gamma_{21}^2 \\ FK = (\Gamma_{12}^1)_u - (\Gamma_{11}^1)_v + \Gamma_{12}^2\Gamma_{12}^1 - \Gamma_{11}^2\Gamma_{22}^1 \\ FK = (\Gamma_{12}^2)_v - (\Gamma_{22}^2)_u + \Gamma_{12}^1\Gamma_{12}^2 - \Gamma_{22}^1\Gamma_{11}^2 \\ GK = (\Gamma_{22}^1)_u - (\Gamma_{12}^1)_v + \Gamma_{22}^1\Gamma_{11}^1 + \Gamma_{22}^2\Gamma_{12}^1 - \Gamma_{21}^1\Gamma_{12}^1 - \Gamma_{12}^2\Gamma_{22}^1, \end{cases}$$

as well as the Codazzi-Mainardi equations*:*

$$\begin{cases} e_v - f_u = e\Gamma_{12}^1 + f(\Gamma_{12}^2 - \Gamma_{11}^1) - g\Gamma_{11}^2 \\ f_v - g_u = e\Gamma_{22}^1 + f(\Gamma_{22}^2 - \Gamma_{12}^1) - g\Gamma_{12}^2, \end{cases}$$

where $K \equiv K \circ \boldsymbol{x}$ *denotes the Gaussian curvature of* M.

Remark. These equations are known as the *compatibility equations*, because they are precisely the conditions ensuring that the systems we will encounter in the proof of the Fundamental Theorem of Surfaces are integrable.

Proof: By way of information, let's study the identity $\boldsymbol{x}_{uuv} = \boldsymbol{x}_{uvu}$ to obtain the first Gauss equation and the first Codazzi-Mainardi equation. Identifying $u \leftrightarrow 1$, $v \leftrightarrow 2$, $\boldsymbol{N} \equiv \boldsymbol{N} \circ \boldsymbol{x}$ and omitting points of evaluation as usual, on one hand we have that

$$\begin{aligned} \boldsymbol{x}_{uuv} &= \left(\sum_{k=1}^2 \Gamma_{11}^k \boldsymbol{x}_k + \epsilon_M h_{11} \boldsymbol{N}\right)_v \\ &= \sum_{k=1}^2 (\Gamma_{11}^k)_v \boldsymbol{x}_k + \sum_{k,r=1}^2 \Gamma_{11}^k \Gamma_{k2}^r \boldsymbol{x}_r + \epsilon_M \sum_{k=1}^2 \Gamma_{11}^k h_{k2} \boldsymbol{N} + \epsilon_M (h_{11})_v \boldsymbol{N} - \epsilon_M h_{11} \sum_{r=1}^2 h^r{}_2 \boldsymbol{x}_r \\ &= \sum_{r=1}^2 \left((\Gamma_{11}^r)_v + \sum_{k=1}^2 \Gamma_{11}^k \Gamma_{k2}^r - \epsilon_M h_{11} h^r{}_2\right) \boldsymbol{x}_r + \epsilon_M \left((h_{11})_v + \sum_{k=1}^2 \Gamma_{11}^k h_{k2}\right) \boldsymbol{N}. \end{aligned}$$

On the other hand:

$$\begin{aligned} x_{uvu} &= \left(\sum_{k=1}^{2} \Gamma_{12}^{k} x_k + \epsilon_M h_{12} N \right)_u \\ &= \sum_{k=1}^{2} (\Gamma_{12}^{k})_u x_k + \sum_{k,r=1}^{2} \Gamma_{12}^{k} \Gamma_{k1}^{r} x_r + \epsilon_M \sum_{k=1}^{2} \Gamma_{12}^{k} h_{k1} N + \epsilon_M (h_{12})_u N - \epsilon_M h_{12} \sum_{r=1}^{2} h^{r}{}_{1} x_r \\ &= \sum_{r=1}^{2} \left((\Gamma_{12}^{r})_u + \sum_{k=1}^{2} \Gamma_{12}^{k} \Gamma_{k1}^{r} - \epsilon_M h_{12} h^{r}{}_{1} \right) x_r + \epsilon_M \left((h_{12})_u + \sum_{k=1}^{2} \Gamma_{12}^{k} h_{k1} \right) N. \end{aligned}$$

Now recall that Lemma 3.3.7 (p. 180) gives us that:

$$h^2_2 = \sum_{j=1}^{2} h_{2j} g^{j2} = \frac{-fF + gE}{EG - F^2} \quad \text{and} \quad h^2_1 = \sum_{j=1}^{2} h_{1j} g^{j2} = \frac{-eF + fE}{EG - F^2}.$$

With this, equating the coefficients of x_v on both expressions (i.e., setting $r = 2$), we obtain:

$$(\Gamma_{11}^2)_v + \Gamma_{11}^1 \Gamma_{12}^2 + \Gamma_{11}^2 \Gamma_{22}^2 - \epsilon_M \left(\tfrac{-efF + egE}{EG - F^2} \right) = (\Gamma_{12}^2)_u + \Gamma_{12}^1 \Gamma_{11}^2 + \Gamma_{12}^2 \Gamma_{21}^2 - \epsilon_M \left(\tfrac{-efF + f^2 E}{EG - F^2} \right).$$

Reorganizing the above expression by moving all the Christoffel symbols to the left and all the terms with Fundamental Forms to the right, the Gauss equation

$$(\Gamma_{11}^2)_v - (\Gamma_{12}^2)_u + \Gamma_{11}^1 \Gamma_{12}^2 + \Gamma_{11}^2 \Gamma_{22}^2 - \Gamma_{12}^1 \Gamma_{11}^2 - \Gamma_{12}^2 \Gamma_{21}^2 = EK$$

follows, as wanted. And equating the coefficients of N directly yields

$$e_v - f_u = e\Gamma_{12}^1 + f(\Gamma_{12}^2 - \Gamma_{11}^1) - g\Gamma_{11}^2.$$

□

As an immediate corollary of the above proposition, we have Gauss' *Theorema Egregium*, since the First Fundamental Form, the Christoffel symbols (as well as their derivatives) are preserved by local isometries. Despite this, it is actually possible to prove the *Theorema Egregium* completely bypassing Christoffel symbols (see the steps to do this in Exercise 3.7.4). In particular, we may now formally present the explicit formula for K mentioned in Section 3.3, only in terms of the First Fundamental Form:

Proposition 3.7.3. *Let $M \subseteq \mathbb{R}^3_\nu$ be a non-degenerate regular surface and (U, x) be an* orthogonal parametrization *for M, that is, satisfying $F = 0$. Then*

$$K \circ x = \frac{-1}{\sqrt{|EG|}} \left(\epsilon_u \left(\frac{(\sqrt{|G|})_u}{\sqrt{|E|}} \right)_u + \epsilon_v \left(\frac{(\sqrt{|E|})_v}{\sqrt{|G|}} \right)_v \right).$$

Remark. This formula is also frequently written as

$$K \circ x = \frac{-\epsilon_u \epsilon_v}{2\sqrt{|EG|}} \left(\left(\frac{G_u}{\sqrt{|EG|}} \right)_u + \left(\frac{E_v}{\sqrt{|EG|}} \right)_v \right).$$

For an example, see Exercise 3.7.3.

Proof: Since $F = 0$, we directly have the expressions for the Christoffel symbols in terms of E, G, and their derivatives (see the remark following the proof of Proposition 3.6.16, p. 228). With this, solving for $K \equiv K \circ \boldsymbol{x}$ in the first Gauss equation (given in the statement of Proposition 3.7.2 above) gives us that:

$$K = -\frac{1}{E}\left(\frac{E_v}{2G}\right)_v - \frac{1}{E}\left(\frac{G_u}{2G}\right)_u + \frac{E_u G_u}{4E^2G} - \frac{E_v G_v}{4G^2} + \frac{E_v^2}{4E^2G} - \frac{G_u^2}{4EG^2}.$$

Computing explicitly the expression given in the statement of this result also gives the above after simplifications. □

Back to our main goal: the Fundamental Theorem of Surfaces. Not too long ago we saw that given smooth functions $g_{ij}, h_{ij}\colon U \to \mathbb{R}$, for $1 \leq i,j \leq 2$, symmetric in the indices i and j, such that $\det\left((g_{ij})_{1\leq i,j\leq 2}\right) \neq 0$, they will have to satisfy the compatibility equations where, of course, the Christoffel symbols associated to $(g_{ij})_{1\leq i,j\leq 2}$ are formally defined by the expression given in Proposition 3.6.16:

$$\Gamma_{ij}^k = \sum_{r=1}^{2} \frac{1}{2} g^{kr} \left(\frac{\partial g_{ik}}{\partial u^j} + \frac{\partial g_{jk}}{\partial u^i} - \frac{\partial g_{ij}}{\partial u^r} \right).$$

We know that the compatibility equations are a necessary condition for the $(g_{ij})_{1\leq i,j\leq 2}$ and $(h_{ij})_{1\leq i,j\leq 2}$ to locally determine a regular surface, but are they also *sufficient*? In other words, by differentiating the relations

$$(\boldsymbol{x}_{uu})_v = (\boldsymbol{x}_{uv})_u, \quad (\boldsymbol{x}_{vv})_u = (\boldsymbol{x}_{uv})_v, \quad \text{and} \quad (\boldsymbol{N} \circ \boldsymbol{x})_{uv} = (\boldsymbol{N} \circ \boldsymbol{x})_{vu}$$

even further, can't we obtain more compatibility equations? The negative answer to this question is the:

Theorem 3.7.4 (Bonnet). *Let U be an open subset of $\mathbb{R}^2$ and, for $1 \leq i,j \leq 2$, $g_{ij}, h_{ij}\colon U \to \mathbb{R}$ be smooth functions formally satisfying the Gauss and Codazzi-Mainardi equations, and symmetric: $g_{ij} = g_{ji}$, $h_{ij} = h_{ji}$ for $1 \leq i,j \leq 2$.*

Also, let $(u_0, v_0) \in U$ and $\boldsymbol{p}_0 \in \mathbb{R}^3$ be points, and $(\boldsymbol{x}_{u,0}, \boldsymbol{x}_{v,0}, \boldsymbol{N}_0)$ be a basis for $\mathbb{R}^3_\nu$ such that $g_{ij}(u_0, v_0) = \langle \boldsymbol{x}_{i,0}, \boldsymbol{x}_{j,0} \rangle$ for $1 \leq i,j \leq 2$, with $\boldsymbol{N}_0$ unit and orthogonal to both $\boldsymbol{x}_{u,0}$ and $\boldsymbol{x}_{v,0}$. Suppose in addition that $\det\left((g_{ij}(u,v))_{1\leq i,j\leq 2}\right)$ never vanishes on U and has a constant sign, equal to $(-1)^\nu \epsilon_{\boldsymbol{N}_0}$.

Then there are a neighborhood V of (u_0, v_0) in $\mathbb{R}^2$, a unique injective and regular parametrized surface $\boldsymbol{x}\colon V \to \mathbb{R}^3_\nu$, and a Gauss map $\boldsymbol{N} \equiv \boldsymbol{N} \circ \boldsymbol{x}$ for $M \doteq \boldsymbol{x}(V)$ satisfying:

- $\boldsymbol{x}(u_0, v_0) = \boldsymbol{p}_0$;
- $\dfrac{\partial \boldsymbol{x}}{\partial u^i}(u_0, v_0) = \boldsymbol{x}_{i,0}$, *for* $1 \leq i \leq 2$;
- $\boldsymbol{N}(\boldsymbol{p}_0) = \boldsymbol{N}_0$;
- *the coefficients of the Fundamental Forms of M relative to $\boldsymbol{x}$ are precisely $g_{ij}\big|_V$ and $h_{ij}\big|_V$. In particular, $\epsilon_M = \epsilon_{\boldsymbol{N}_0}$.*

Proof: Introduce new vector variables

$$w_1 = \frac{\partial \boldsymbol{x}}{\partial u^1} \quad \text{and} \quad w_2 = \frac{\partial \boldsymbol{x}}{\partial u^2}.$$

Let's first solve a new system of PDEs for the vector variables $\boldsymbol{w}_1$, $\boldsymbol{w}_2$ and $\boldsymbol{N}$, to then solve the initial system for $\boldsymbol{x}$. Consider:

$$\begin{cases} \dfrac{\partial \boldsymbol{w}_i}{\partial u^j} = \displaystyle\sum_{k=1}^{2} \Gamma_{ij}^k \boldsymbol{w}_k + \epsilon_{\boldsymbol{N}_0} h_{ij} \boldsymbol{N}, \\ \dfrac{\partial \boldsymbol{N}}{\partial u^j} = -\displaystyle\sum_{i,k=1}^{2} g^{ik} h_{kj} \boldsymbol{w}_i, \\ (\boldsymbol{w}_1(u_0,v_0), \boldsymbol{w}_2(u_0,v_0), \boldsymbol{N}(u_0,v_0)) = (\boldsymbol{x}_{u,0}, \boldsymbol{x}_{v,0}, \boldsymbol{N}_0). \end{cases}$$

The compatibility equations for this system are precisely the Gauss and Codazzi-Mainardi equations, which are satisfied by assumption. Thus, the Frobenius Theorem provides a solution $(\boldsymbol{w}_1, \boldsymbol{w}_2, \boldsymbol{N})$ in some neighborhood of (u_0, v_0). The next step is, naturally, to verify that

$$\langle \boldsymbol{N}, \boldsymbol{N} \rangle = \epsilon_{\boldsymbol{N}_0}, \quad \langle \boldsymbol{N}, \boldsymbol{w}_i \rangle = 0, \text{ and } \langle \boldsymbol{w}_i, \boldsymbol{w}_j \rangle = g_{ij}, \quad (1 \leq i, j \leq 2)$$

for all parameters (u, v), and not only at (u_0, v_0). Indeed, using the first system we have that:

$$\begin{cases} \dfrac{\partial}{\partial u^j} \langle \boldsymbol{N}, \boldsymbol{N} \rangle = -2 \displaystyle\sum_{i,k=1}^{2} g^{ik} h_{kj} \langle \boldsymbol{N}, \boldsymbol{w}_i \rangle \\ \dfrac{\partial}{\partial u^j} \langle \boldsymbol{N}, \boldsymbol{w}_i \rangle = -\displaystyle\sum_{\ell,k=1}^{2} g^{\ell k} h_{kj} \langle \boldsymbol{w}_\ell, \boldsymbol{w}_i \rangle + \displaystyle\sum_{k=1}^{2} \Gamma_{ij}^k \langle \boldsymbol{N}, \boldsymbol{w}_k \rangle + \epsilon_{\boldsymbol{N}_0} h_{ij} \langle \boldsymbol{N}, \boldsymbol{N} \rangle \\ \dfrac{\partial}{\partial u^k} \langle \boldsymbol{w}_i, \boldsymbol{w}_j \rangle = \displaystyle\sum_{\ell=1}^{2} \Gamma_{ik}^\ell \langle \boldsymbol{w}_\ell, \boldsymbol{w}_j \rangle + \epsilon_{\boldsymbol{N}_0} h_{ik} \langle \boldsymbol{N}, \boldsymbol{w}_j \rangle + \displaystyle\sum_{\ell=1}^{2} \Gamma_{jk}^\ell \langle \boldsymbol{w}_\ell, \boldsymbol{w}_i \rangle + \epsilon_{\boldsymbol{N}_0} h_{jk} \langle \boldsymbol{N}, \boldsymbol{w}_i \rangle. \end{cases}$$

Note that since this last system was obtained from an integrable system, it is also integrable. Moreover, the functions $\epsilon_{\boldsymbol{N}_0}$, 0 and g_{ij} are also solutions for this system, with the same initial conditions (we ask you to check this in Exercise 3.7.5). Hence the Frobenius Theorem ensures that

$$\langle \boldsymbol{N}, \boldsymbol{N} \rangle = \epsilon_{\boldsymbol{N}_0}, \quad \langle \boldsymbol{N}, \boldsymbol{w}_i \rangle = 0, \text{ and } \langle \boldsymbol{w}_i, \boldsymbol{w}_j \rangle = g_{ij}, \quad (1 \leq i, j \leq 2)$$

on the neighborhood where the solution $(\boldsymbol{w}_1, \boldsymbol{w}_2, \boldsymbol{N})$ is defined. In particular, this ensures that the solution remains a basis for $\mathbb{R}^3_\nu$ in all points.

Let's finally go back to the initial system considered:

$$\boldsymbol{w}_1 = \frac{\partial \boldsymbol{x}}{\partial u^1} \quad \text{and} \quad \boldsymbol{w}_2 = \frac{\partial \boldsymbol{x}}{\partial u^2}.$$

Since $\Gamma_{ij}^k = \Gamma_{ji}^k$ and $h_{ij} = h_{ji}$ for $1 \leq i, j, k \leq 2$, this system is integrable and we obtain a neighborhood V of (u_0, v_0) small enough where all the obtained conditions on $\boldsymbol{w}_1, \boldsymbol{w}_2$ and $\boldsymbol{N}$ so far still hold, and a regular map $\boldsymbol{x}\colon V \to \mathbb{R}^3_\nu$ that solves the system, with initial condition $\boldsymbol{x}(u_0, v_0) = \boldsymbol{p}_0$. Such map may be assumed to be injective, reducing V further if necessary. Now, note that

$$\left\langle \frac{\partial \boldsymbol{x}}{\partial u^i}, \frac{\partial \boldsymbol{x}}{\partial u^j} \right\rangle = \langle \boldsymbol{w}_i, \boldsymbol{w}_j \rangle = g_{ij}, \quad (1 \leq i, j \leq 2)$$

and so the g_{ij} are indeed the coefficients of the First Fundamental Form of $M \doteq \boldsymbol{x}(V)$ relative to $\boldsymbol{x}$. Similarly, we have that

$$\left\langle \boldsymbol{N}, \frac{\partial \boldsymbol{x}}{\partial u^i} \right\rangle = \langle \boldsymbol{N}, \boldsymbol{w}_i \rangle = 0, \quad (1 \leq i \leq 2)$$

and thus $\boldsymbol{N}$ is a Gauss map for M. This allows us to compute the coefficients of the Second Fundamental Form of M relative to $\boldsymbol{x}$ by using the previous systems:

$$\begin{aligned}\left\langle \boldsymbol{N}, \frac{\partial^2 \boldsymbol{x}}{\partial u^i \partial u^j} \right\rangle &= \left\langle \boldsymbol{N}, \frac{\partial \boldsymbol{w}_i}{\partial u^j} \right\rangle \\ &= \left\langle \boldsymbol{N}, \sum_{k=1}^{2} \Gamma_{ij}^k \boldsymbol{w}_k + \epsilon_{\boldsymbol{N}_0} h_{ij} \boldsymbol{N} \right\rangle \\ &= \sum_{k=1}^{2} \Gamma_{ij}^k \langle \boldsymbol{N}, \boldsymbol{w}_k \rangle + \epsilon_{\boldsymbol{N}_0} h_{ij} \langle \boldsymbol{N}, \boldsymbol{N} \rangle = h_{ij}, \quad (1 \leq i,j \leq 2)\end{aligned}$$

as wanted. □

Remark (Flat immersions)**.** One noteworthy situation is when $(g_{ij})_{i,j=1}^n$ is a matrix of constant functions, which forces all the Christoffel symbols to vanish. Then the Gauss equations should impose that $K = 0$ as well. This means that the Gauss equations are satisfied if and only if the matrix

$$\begin{pmatrix} e & f \\ f & g \end{pmatrix}$$

is always singular, while the Codazzi-Mainardi equations are satisfied if and only if we have $e_v = f_u$ and $f_v = g_u$. If this is the case, in particular note that $f = 0$ if and only if $e = 0$ and g depends only on v or $g = 0$ and e depends only on u. Solving the compatibility system in general is very difficult. You can play around in Exercise 3.7.7 with the (reasonable) case where e, f and g are all constant as well to convince yourself of this.

Bonnet's Theorem ensures, under suitable conditions, the existence and uniqueness of the surface, *once initial conditions are given.* In the general case, we have uniqueness up to rigid motions of the ambient space:

Proposition 3.7.5. *Let $\boldsymbol{x}, \widetilde{\boldsymbol{x}}\colon U \to \mathbb{R}^3_\nu$ be two non-degenerate regular parametrized surfaces, with U connected, such that $g_{ij} = \widetilde{g}_{ij}$ and $h_{ij} = \widetilde{h}_{ij}$ on U. Then there is a positive $F \in \mathrm{E}_\nu(3,\mathbb{R})$ such that $\widetilde{\boldsymbol{x}} = F \circ \boldsymbol{x}$.*

Proof: Consider the Gauss maps $\boldsymbol{N}$ and $\widetilde{\boldsymbol{N}}$ compatible with $\boldsymbol{x}$ and $\widetilde{\boldsymbol{x}}$, given by

$$\boldsymbol{N}(u,v) = \frac{\boldsymbol{x}_u(u,v) \times \boldsymbol{x}_v(u,v)}{\|\boldsymbol{x}_u(u,v) \times \boldsymbol{x}_v(u,v)\|},$$

and similarly for $\widetilde{\boldsymbol{N}}$. Since $g_{ij} = \widetilde{g}_{ij}$, in particular we have that $\epsilon_{\boldsymbol{N}(u,v)} = \epsilon_{\widetilde{\boldsymbol{N}}(u,v)} \doteq \epsilon$ and, for each $(u,v) \in U$, the linear map $A(u,v)$ defined by

$$\begin{aligned} A(u,v)(\boldsymbol{x}_u(u,v)) &= \widetilde{\boldsymbol{x}}_u(u,v), \\ A(u,v)(\boldsymbol{x}_v(u,v)) &= \widetilde{\boldsymbol{x}}_v(u,v) \text{ and} \\ A(u,v)(\boldsymbol{N}(u,v)) &= \widetilde{\boldsymbol{N}}(u,v) \end{aligned}$$

is in $\mathrm{SO}_\nu(3,\mathbb{R})$. Let's then check that $A(u,v)$ is actually constant, by verifying that its partial derivatives are the zero map. We identify $u \leftrightarrow u^1$, $v \leftrightarrow u^2$, as always. Omitting points of evaluation everywhere except for $A(u^1,u^2)$, on one hand we have that

$$\begin{aligned}\widetilde{x}_{ij} &= A_j(u^1,u^2)(x_i) + A(u^1,u^2)(x_{ij}) \\ &= A_j(u^1,u^2)(x_i) + A(u^1,u^2)\left(\sum_{k=1}^{2}\Gamma_{ij}^k x_k + \epsilon h_{ij}N\right) \\ &= A_j(u^1,u^2)(x_i) + \sum_{k=1}^{2}\Gamma_{ij}^k \widetilde{x}_k + \epsilon h_{ij}\widetilde{N}.\end{aligned}$$

On the other hand:

$$\widetilde{x}_{ij} = \sum_{k=1}^{2}\widetilde{\Gamma}_{ij}^k \widetilde{x}_k + \epsilon \widetilde{h}_{ij}\widetilde{N}.$$

Note that since $g_{ij} = \widetilde{g}_{ij}$, we also have that $\Gamma_{ij}^k = \widetilde{\Gamma}_{ij}^k$. This, together with the fact that $h_{ij} = \widetilde{h}_{ij}$, allows us to compare both expressions and conclude that

$$A_j(u^1,u^2)(x_i) = \mathbf{0} \quad (1 \leq i,j \leq 2).$$

Moreover, we have that

$$A_j(u^1,u^2)(N) + A(u^1,u^2)(N_j) = \widetilde{N}_j.$$

But, we also have that ${h^i}_j = {\widetilde{h}^i}_j$, and this gives us that $A(u^1,u^2)(N_j) = \widetilde{N}_j$. Hence $A_j(u^1,u^2)(N) = \mathbf{0}$ and, since j was arbitrary and U is connected, we conclude that $A_j(u^1,u^2)$ is the zero map. Thus, $A(u^1,u^2) \equiv A$ is constant. Now, since

$$(\widetilde{x} - A(x))_j = \widetilde{x}_j - A(x_j) = \mathbf{0}$$

for all j, it follows that $\widetilde{x} - A(x) = b \in \mathbb{R}^3_\nu$ is constant, so that $F = T_b \circ A \in \mathrm{E}_\nu(3,\mathbb{R})$ is the positive rigid motion we're looking for. □

Exercises

Exercise† 3.7.1. Let $M \subseteq \mathbb{R}^3_\nu$ be a non-degenerate regular surface, N be a Gauss map for M, and (U,x) be a parametrization for M. Deduce again the Codazzi-Mainardi equations from the relation $(N \circ x)_{uv} = (N \circ x)_{vu}$.

Exercise 3.7.2. Fill the details in the proof of Proposition 3.7.3 (p. 252).

Exercise 3.7.3. Let $M \subseteq \mathbb{R}^3_\nu$ be a non-degenerate regular surface and (U,x) be a parametrization for M. We'll say that x is *isothermal* if $F = 0$ and $|E| = |G| = \lambda^2$, for some non-vanishing smooth function $\lambda\colon U \to \mathbb{R}$. Show that

$$K \circ x = -\frac{\triangle(\log\lambda)}{\lambda^2}$$

if M is spacelike and that

$$K \circ x = -\epsilon_u \frac{\Box(\log\lambda)}{\lambda^2}$$

if M is timelike, where $\triangle$ and $\Box$ denote the Laplacian and d'Alembertian operators, seen previously in Exercise 3.6.28 (p. 248).

Remark. It is possible to define a single operator $\triangle_{\langle\cdot,\cdot\rangle}$ (known as the *Laplace-Beltrami operator*) on functions over M which generalizes the operators $\triangle$ and $\square$. The name "isothermal" is motivated by the fact that such coordinates satisfy $\triangle_{\langle\cdot,\cdot\rangle}\boldsymbol{x} = \boldsymbol{0}$, being stationary solutions to the *heat equation in* M.

Exercise† 3.7.4. Let $M \subseteq \mathbb{R}^3_\nu$ be a non-degenerate regular surface and $\boldsymbol{p} \in M$. In this exercise, we will see an alternative proof for Gauss' *Theorema Egregium* (slightly adapting the argument given by Sternberg in [67]), expressing the Gaussian curvature $K(\boldsymbol{p})$ in terms only of the coefficients of the First Fundamental Form of M relative to an inertial parametrization centered at $\boldsymbol{p}$ (whose existence was shown in Theorem 3.4.9, p. 195). We will treat the situation where M is spacelike, with the timelike case being similar. Recall that rigid motions of the ambient space (that is, elements of $\mathrm{E}_\nu(3,\mathbb{R})$) preserve the Fundamental Forms of M, by Proposition 3.3.13 (p. 185). In particular, the Gaussian curvature is also preserved. And clearly the composition of an inertial parametrization with an isometry is again inertial. Thus, we may assume that we have an inertial parametrization $(U,\boldsymbol{x})$ centered at $\boldsymbol{p} = \boldsymbol{0}$, of the form

$$\boldsymbol{x}(u,v) = (u + a(u,v), v + b(u,v), c(u,v)),$$

such that $\boldsymbol{x}_u(0,0) = (1,0,0)$ and $\boldsymbol{x}_v(0,0) = (0,1,0)$, for certain smooth functions $a,b,c\colon U \to \mathbb{R}$. When M is timelike, one must apply a rigid motion that takes $T_{\boldsymbol{p}}M$ in some coordinate timelike plane in $\mathbb{L}^3$.

(a) Use that $\boldsymbol{x}(0,0) = (0,0,0)$ and $(\boldsymbol{x}_u(0,0), \boldsymbol{x}_v(0,0)) = (\boldsymbol{e}_1, \boldsymbol{e}_2)$ to show that a, b, c and their first order partial derivatives all vanish at the origin.

(b) Omitting points of evaluation, show that

$$\begin{aligned} E &= (1 + a_u)^2 + b_u^2 + (-1)^\nu c_u^2, \\ F &= (1 + a_u)a_v + (1 + b_v)b_u + (-1)^\nu c_u c_v \quad \text{and} \\ G &= a_v^2 + (1 + b_v)^2 + (-1)^\nu c_v^2. \end{aligned}$$

(c) Use that the partial derivatives of E, F and G vanish at the origin (since $\boldsymbol{x}$ is inertial) to show that the second order partial derivatives of a and b also vanish at the origin.

(d) Compute the coefficients $h_{ij}(0,0)$, for $1 \leq i,j \leq 2$, and show that

$$K(\boldsymbol{p}) = \det \operatorname{Hess} c_{(0,0)} = c_{uu}(0,0)c_{vv}(0,0) - c_{uv}(0,0)^2.$$

(e) Show that

$$\begin{aligned} F_{uv}(0,0) &= a_{uvv}(0,0) + b_{uuv}(0,0) + (-1)^\nu(c_{uu}(0,0)c_{vv}(0,0) + c_{uv}(0,0)^2), \\ E_{vv}(0,0) &= 2(a_{uvv}(0,0) + (-1)^\nu c_{uv}(0,0)^2) \quad \text{and} \\ G_{uu}(0,0) &= 2(b_{vuu}(0,0) + (-1)^\nu c_{vu}(0,0)^2). \end{aligned}$$

(f) Show that

$$K(\boldsymbol{p}) = F_{uv}(0,0) - \frac{E_{vv}(0,0)}{2} - \frac{G_{uu}(0,0)}{2}$$

and conclude Gauss' *Theorema Egregium*.

Remark. This proof indeed could have been presented soon after Theorem 3.4.9, but we believe that it would be better appreciated only now. Do you agree?

Exercise† 3.7.5. In the proof of Theorem 3.7.4 (p. 253), verify that the functions ϵ_{N_0}, 0 and g_{ij} are solutions to the system used to show that the relations

$$\langle N,N\rangle = \epsilon_{N_0}, \quad \langle N, w_i\rangle = 0, \text{ and } \langle w_i, w_j\rangle = g_{ij}, \quad (1 \le i,j \le 2)$$

were valid in the domain of definition of the solution (w_1, w_2, N).

Exercise 3.7.6. Show that there is not, in $\mathbb{R}^3_\nu$, any regular parametrized surface with:

(a) $\big(g_{ij}(u,v)\big)_{1\le i,j\le 2} = \begin{pmatrix} 1 & 0 \\ 0 & 1 \end{pmatrix}$ and $\big(h_{ij}(u,v)\big)_{1\le i,j\le 2} = \begin{pmatrix} v & 0 \\ 0 & 1 \end{pmatrix}$.

(b) $\big(g_{ij}(u,v)\big)_{1\le i,j\le 2} = \begin{pmatrix} 1 & 0 \\ 0 & \cos^2 u \end{pmatrix}$ and $\big(h_{ij}(u,v)\big)_{1\le i,j\le 2} = \begin{pmatrix} \cos^2 u & 0 \\ 0 & 1 \end{pmatrix}$.

Hint. Use the Codazzi-Mainardi equations.

Exercise 3.7.7. Determine, when possible, a parametrized regular surface $x\colon U \to \mathbb{R}^3_\nu$ such that:

(a) $\big(g_{ij}(u,v)\big)_{1\le i,j\le 2} = \begin{pmatrix} 1 & 0 \\ 0 & 1 \end{pmatrix}$ and $\big(h_{ij}(u,v)\big)_{1\le i,j\le 2} = \begin{pmatrix} -1 & 0 \\ 0 & 1 \end{pmatrix}$,
with $x(0,0) = (0,0,1)$, $x_u(0,0) = (1,0,0)$ and $x_v(0,1,0)$.

(b) $\big(g_{ij}(u,v)\big)_{1\le i,j\le 2} = \begin{pmatrix} 1 & 0 \\ 0 & 1 \end{pmatrix}$ and $\big(h_{ij}(u,v)\big)_{1\le i,j\le 2} = \begin{pmatrix} 1 & 0 \\ 0 & 0 \end{pmatrix}$,
with $x(0,0) = (0,0,1)$, $x_u(0,0) = (1,0,0)$ and $x_v(0,1,0)$.

(c) $\big(g_{ij}(u,v)\big)_{1\le i,j\le 2} = \begin{pmatrix} -1 & 0 \\ 0 & 1 \end{pmatrix}$ and $\big(h_{ij}(u,v)\big)_{1\le i,j\le 2} = \begin{pmatrix} 0 & 0 \\ 0 & -1 \end{pmatrix}$,
with $x(0,0) = (1,0,0)$, $x_u(0,0) = (0,0,1)$ and $x_v(0,0) = (0,1,0)$.

Exercise 3.7.8. Let $M \subseteq \mathbb{R}^3_\nu$ be a non-degenerate regular surface and (U, x) be a principal parametrization for M. Show that the Codazzi-Mainardi equations boil down to:

(a) $e_v = \dfrac{E_v}{2}\left(\dfrac{e}{E} + \dfrac{g}{G}\right)$ and $g_u = \dfrac{G_u}{2}\left(\dfrac{e}{E} + \dfrac{g}{G}\right)$,

(b) or $\dfrac{1}{\kappa_2 - \kappa_1}\dfrac{\partial \kappa_1}{\partial v} = \dfrac{E_v}{2E}$ and $\dfrac{1}{\kappa_1 - \kappa_2}\dfrac{\partial \kappa_2}{\partial u} = \dfrac{G_u}{2G}$,
under the assumption that M is free of umbilic points.

Remark. Can you see how H is hidden in the formulas in (a)?

CHAPTER 4

Abstract Surfaces and Further Topics

INTRODUCTION

We approach the end of this pleasant excursion through Lorentz Geometry of curves and surfaces with this final chapter, which has the goal of pointing new directions of study for the reader. Here, we will slightly change the dynamics of the text: the Sections 4.2, 4.3, and 4.4 are independent of each other, and may be read in any order. However, here we'll push a few limits, assuming for the first time some familiarity with point-set topology and complex analysis.

In Section 4.1, we seek to free ourselves from the Codazzi-Mainardi equations, seen at the end of the previous chapter, presenting the definition of an abstract surface, which is easily generalized to give us the notion of a differentiable manifold, one of the most important in Differential Geometry. We keep adapting most of the concepts seen for regular surfaces in $\mathbb{R}^3_\nu$ for this new setting, presenting examples. In particular, we have the concept of a *pseudo-Riemannian metric*, that is, a First Fundamental Form which does not come from any ambient space, but is instead prescribed. From this, we again define isometries, geodesics, and Gaussian curvature, which are all illustrated in models for Hyperbolic Geometry.

In Section 4.2, we apply what has been discussed in Section 4.1 to obtain a local classification of the geometric surfaces with constant Gaussian curvature, employing the so-called Fermi parametrizations. We also illustrate some realizations of the obtained metrics in the Lorentzian spaces $\mathbb{L}^3$ and $\mathbb{R}^3_2$.

The long-awaited presentation of the set of split-complex numbers, denoted by $\mathbb{C}'$, is finally given at the beginning of Section 4.3. We briefly discuss the relation between complex and split-complex numbers with critical surfaces in $\mathbb{R}^3_\nu$, via two constructions: the Bonnet rotations, which are isometric deformations between conjugate or Lorentz-conjugate critical surfaces; and the Enneper-Weierstrass representation formulas, which say exactly how critical surfaces are related to triples of holomorphic or split-holomorphic functions satisfying certain conditions. In particular, we'll employ the so-called isothermal parametrizations, which are convenient for the study of critical surfaces, and also the complex Lorentzian space $\mathbb{C}^3_1$.

We'll close the curtains with Section 4.4, where we invite you to proceed with your studies in Riemannian Geometry and General Relativity, presenting a few ideas about completeness in Riemannian manifolds and causality in Lorentzian manifolds. We introduce the notion of intrinsic distance in Riemannian manifolds, and state the famous Hopf-Rinow Theorem, which provides a bridge between the geometry and the topology of the manifold. Motivated to find an analogous result for Lorentzian manifolds, we present the formal definition of a spacetime, taking the first step towards General Relativity. The time separation in a spacetime plays the same role as the intrinsic distance in a Riemannian manifold, and it is closely related to the causality of the spacetime itself. With this, we present the several conditions defining the causal hierarchy of spacetimes, and we conclude the discussion with the statement of the Avez-Seifert Theorem: the analogue of the Hopf-Rinow Theorem for globally hyperbolic spacetimes.

"The views of space and time which I wish to lay before you have sprung from the soil of experimental physics, and therein lies their strength. They are radical. Henceforth space by itself, and time by itself, are doomed to fade away into mere shadows, and only a kind of union of the two will preserve an independent reality." – Hermann Minkowski (address to the 80th *Assembly of German Natural Scientists and Physicians, September* 21st, 1908*).*

4.1 PSEUDO-RIEMANNIAN METRICS

In the previous chapter we have seen that surfaces in $\mathbb{R}^3_\nu$ must satisfy the compatibility equations given in Proposition 3.7.2 (p. 251). The Codazzi–Mainardi equations can be quite restrictive and, since they involve the coefficients of the Second Fundamental Form, are not intrinsic to the surfaces. In this section we will study a more general notion of surface, eliminating the dependence of an ambient space and, consequently, of those equations.

Definition 4.1.1 (Abstract surface). An *abstract surface* is a set M equipped with a collection

$$\mathfrak{A} = \{\boldsymbol{x}_\alpha \colon U_\alpha \subseteq \mathbb{R}^2 \to \boldsymbol{x}_\alpha(U_\alpha) \subseteq M \mid \alpha \in A\}$$

of *abstract parametrizations* (i.e., injective maps defined in open subsets of $\mathbb{R}^2$) satisfying the following conditions:

(i) $\bigcup_{\alpha \in A} \boldsymbol{x}_\alpha(U_\alpha) = M$;

(ii) given $\alpha, \beta \in A$, $\boldsymbol{x}_\beta^{-1} \circ \boldsymbol{x}_\alpha$ is a smooth map between open subsets of $\mathbb{R}^2$ (in the correct domain), whenever $\boldsymbol{x}_\alpha(U_\alpha) \cap \boldsymbol{x}_\beta(U_\beta) \neq \varnothing$;

(iii) $\mathfrak{A}$ is maximal relative to (ii): if $\boldsymbol{y} : V \to \boldsymbol{y}(V) \subseteq M$ is an abstract parametrization satisfying (ii), then $\boldsymbol{y} \in \mathfrak{A}$.

The collection $\mathfrak{A}$ is called a (maximal) *atlas* for M.

Remark.

- If U is an open subset of $\mathbb{R}^n$ instead of $\mathbb{R}^2$, we get what is called a *differentiable manifold of dimension n*.
- We say that $W \subseteq M$ is *open* if $\boldsymbol{x}^{-1}(W)$ is an open subset of $\mathbb{R}^2$, for every $\boldsymbol{x} \in \mathfrak{A}$. This indeed defines a topology on M, which turns all the abstract parametrizations into homeomorphisms. Note that it still does not make sense to ask whether such parametrizations are diffeomorphisms.
- The topology on M induced by $\mathfrak{A}$ is not necessarily Hausdorff or second-countable. One usually assumes from the beginning that all the manifolds in the discussion have those properties. Being Hausdorff is very natural, and second-countability essentially says that the manifold does not have "too many" open sets, avoiding pathological examples and ensuring the existence of the so-called partitions of unity.

Example 4.1.2.

(1) Open subsets of $\mathbb{R}^2$ and regular surfaces $M \subseteq \mathbb{R}^3_\nu$ are the first examples of abstract surfaces.

(2) If $\boldsymbol{x} : U \subseteq \mathbb{R}^2 \to \mathbb{R}^n$ is injective and differentiable, such that $D\boldsymbol{x}(u,v)$ has full rank, then $\boldsymbol{x}(U)$ is an abstract surface, when equipped with the maximal atlas containing $\boldsymbol{x}$ itself.

(3) If on the sphere $\mathbb{S}^2 \subseteq \mathbb{R}^3$ we define the equivalence relation $\sim$ by setting $\boldsymbol{p} \sim \boldsymbol{q}$ if and only if $\boldsymbol{q} \in \{\boldsymbol{p}, -\boldsymbol{p}\}$ (i.e., identifying antipodal points), we may define an atlas in the quotient $\mathbb{R}P^2 \doteq \mathbb{S}^2/_\sim$, which turns it into an abstract surface as well. For more details, see [16].

Remark. A rich class of examples of differentiable manifolds consists of matrix groups. For example, one may show that $\mathrm{O}(n,\mathbb{R})$ and $\mathrm{SO}(n,\mathbb{R})$ are differentiable manifolds of dimension $n(n-1)/2$.

To study geometry in M, we need some notion of Differential Calculus on M. We define what it means for a function $f\colon M \to \mathbb{R}$ to be smooth by copying Definition 3.1.16 (p. 141) word by word, mimicking the situation regarding surfaces in $\mathbb{R}^3_\nu$. Similarly, we define what it means for a function between two abstract surfaces to be smooth and, in particular, we now know what is a diffeomorphism between abstract surfaces.

The next step would be to define what is the tangent plane at a point $p \in M$. Before, however, we defined the tangent plane as a certain subspace of $\mathbb{R}^3_\nu$ (according to Definition 3.1.12, p. 139), so we must be careful and take as a definition something which does not depend on the (now non-existent) ambient space. The characterization given in Proposition 3.1.14 (p. 140) sounds like a good idea, once we define the velocity vector of a curve in M. Differentiating component by component is no longer a valid option, as M is, a priori, an abstract set. Bearing in mind that when we have the ambient space $\mathbb{R}^3_\nu$, tangent vectors act in functions $f \in \mathscr{C}^\infty(M)$ via directional derivatives, we have the:

Definition 4.1.3. Let M be an abstract surface, $p \in M$ and $\boldsymbol{\alpha}\colon I \to M$ be a smooth curve such that $\boldsymbol{\alpha}(t_0) = p$. The *tangent vector* to $\boldsymbol{\alpha}$ at t_0 is the map $\boldsymbol{\alpha}'(t_0)\colon \mathscr{C}^\infty(M) \to \mathbb{R}$ defined by

$$\boldsymbol{\alpha}'(t_0)(f) \doteq \left.\frac{\mathrm{d}}{\mathrm{d}t}\right|_{t=t_0} f(\boldsymbol{\alpha}(t)).$$

Remark. The tangent vector to $\boldsymbol{\alpha}$ at t_0 is also called the *velocity vector* of $\boldsymbol{\alpha}$ at t_0, like before.

With this, we finally define the tangent plane to M at p, T_pM, by Proposition 3.1.14. Similarly, we define the *tangent space* to a manifold M of any dimension n, in any point p, as the space of velocities of curves which start at p. Thus, Definition 3.1.23 (p. 145) of *differential* of a smooth function between regular surfaces now makes sense for abstract surfaces as well. So, if $(U, \boldsymbol{x})$ is a parametrization for M, we write

$$\boldsymbol{x}_u(u,v) = \mathrm{d}\boldsymbol{x}_{(u,v)}(1,0) \quad \text{and} \quad \boldsymbol{x}_v(u,v) = \mathrm{d}\boldsymbol{x}_{(u,v)}(0,1),$$

and continue to identify $u \leftrightarrow u^1$ and $v \leftrightarrow u^2$ whenever convenient.

However, we have no means to directly define the First Fundamental Form of M, as we no longer have the ambient scalar product available. This is the main reason we presented the definition of abstract surface: now the geometry of M will not come from $\mathbb{R}^3_\nu$, but will be prescribed:

Definition 4.1.4 (Pseudo-Riemannian metric). Let M be an abstract surface. A *pseudo-Riemannian metric* on M is a choice of non-degenerate scalar products $\langle \cdot, \cdot \rangle_p$ in each tangent plane T_pM, all with the same index, such that for every parametrization $\boldsymbol{x}$ of M,

$$g_{ij}(u,v) \doteq \langle \boldsymbol{x}_i(u,v), \boldsymbol{x}_j(u,v) \rangle_{\boldsymbol{x}(u,v)}$$

is a smooth function of (u,v). We say that $(M, \langle \cdot, \cdot \rangle)$ is a *geometric surface*.

Remark.

- We say that the metric is *Riemannian* if every scalar product is positive-definite, and that the metric is *Lorentzian* if the index of each scalar product is equal to 1. For geometric surfaces, those two types can be characterized by the sign of the determinant of the Gram matrix $(g_{ij})_{1 \leq i,j \leq 2}$.

- It is usual to omit p in $\langle \cdot, \cdot \rangle_p$, when there's no risk of confusion. Moreover, if $\boldsymbol{v} \in T_pM$, we write $\|\boldsymbol{v}\| \doteq \sqrt{|\langle \boldsymbol{v}, \boldsymbol{v} \rangle|}$. When the metric is Riemannian, we have norms in all the tangent planes.

- When the metric is Lorentzian, we define the causal type and the indicator of $\boldsymbol{v} \in T_pM$ by the sign of $\langle \boldsymbol{v}, \boldsymbol{v} \rangle$, like in Definition 1.1.3 (p. 3, Chapter 1).

- Like before, we'll also express a pseudo-Riemannian metric in terms of coordinates with the differential notation

$$\mathrm{d}s^2 = \sum_{i,j=1}^{2} g_{ij}\, \mathrm{d}u^i\, \mathrm{d}u^j = E(u,v)\, \mathrm{d}u^2 + 2F(u,v)\, \mathrm{d}u\, \mathrm{d}v + G(u,v)\, \mathrm{d}v^2.$$

- Non-degeneracy of the metric ensures the existence of the inverse coefficients $(g^{ij})_{1 \leq i,j \leq 2}$, of fundamental importance for the development of the theory.

- The above definition naturally extends to differentiable manifolds of any dimension. We say that $(M, \langle \cdot, \cdot \rangle)$ is a *Riemannian manifold* or a *Lorentzian manifold*, according to whether the index of all the scalar products is 0 or 1. The names "Riemann surface" and "Lorentz surface", in turn, have precise meanings in other areas of geometry, and should be avoided in this setting.

Example 4.1.5.

(1) The First Fundamental Form of a non-degenerate regular surface $M \subseteq \mathbb{R}^3_\nu$ is a pseudo-Riemannian metric on M (regarded as an abstract surface). Such metric is Riemannian if M is spacelike, and Lorentzian if timelike.

(2) The above idea may be generalized by considering surfaces inside $\mathbb{R}^n_\nu$, and taking the restriction of the ambient scalar product in each tangent plane, whenever non-degenerate. The only difference is that while each tangent plane T_pM has dimension 2, the *normal space* $T_pM^\perp$ is no longer a line for $n > 3$. It is possible to study how the geometry of M relates to the geometry of $\mathbb{R}^n_\nu$ by using a vector-valued Second Fundamental Form $\mathrm{I\!I}$, but the theory becomes more sophisticated, and it is better approached with tools beyond the scope of this text. See, for example, [54].

Bearing this in mind, all the concepts previously studied which are intrinsic to the surface may be studied for abstract surfaces. For example, if $\boldsymbol{\alpha}\colon I \to M$ is given, we define the arclength and the energy of $\boldsymbol{\alpha}$ according to the pseudo-Riemannian metric via the usual formulas

$$L[\boldsymbol{\alpha}] = \int_I \|\boldsymbol{\alpha}'(t)\|_{\boldsymbol{\alpha}(t)}\,dt \quad \text{and} \quad E[\boldsymbol{\alpha}] = \frac{1}{2}\int_I \langle \boldsymbol{\alpha}'(t), \boldsymbol{\alpha}'(t)\rangle_{\boldsymbol{\alpha}(t)}\,dt.$$

We also define the area of a region $R \subseteq M$ according to Definition 3.2.12 (p. 161).

Example 4.1.6 (Warm-up). In $M = \mathbb{R} \times \mathbb{R}_{>0}$, take the abstract parametrization given by the identity.

(1) Consider the Riemannian metric $ds^2 = dx^2 + y\,dy^2$, given in Cartesian coordinates. All the tangent planes to M are (naturally) isomorphic to $\mathbb{R}^2$ (as vector spaces). For instance, we have that

$$\langle (1,3), (-2,3)\rangle_{(0,2)} = 1\cdot(-2) + 2\cdot 3\cdot 3 = 16,$$

but

$$\langle (1,3), (-2,3)\rangle_{(0,1)} = 1\cdot(-2) + 1\cdot 3\cdot 3 = 7.$$

If $\boldsymbol{\alpha}\colon\,]0,1[\, \to M$ is given by $\boldsymbol{\alpha}(t) = (0,t)$, we have that

$$L[\boldsymbol{\alpha}] = \int_0^1 \sqrt{0^2 + t\cdot 1^1}\,dt = \int_0^1 \sqrt{t}\,dt = \frac{2}{3}.$$

In particular, $\widetilde{\boldsymbol{\alpha}}(s) = (0, (3s/2)^{2/3})$ is a unit speed reparametrization of $\boldsymbol{\alpha}$. Lastly, let's see what the area of the square $R \doteq\,]0,1[^2$ is according to this metric. We have that

$$\mathcal{A}(R) = \int_R \sqrt{1\cdot y - 0^2}\,dx\,dy = \int_0^1\int_0^1 \sqrt{y}\,dx\,dy = \frac{2}{3}.$$

(2) Consider now the Lorentzian metric $ds^2 = -dx^2 + y\,dy^2$, given in Cartesian coordinates. We have that

$$\begin{aligned}
\langle (1,1), (1,1)\rangle_{(0,2)} &= -1^2 + 2\cdot 1\cdot 1 = 1\\
\langle (1,1), (1,1)\rangle_{(1,1)} &= -1^2 + 1\cdot 1\cdot 1 = 0\\
\langle (1,1), (1,1)\rangle_{(-2,1/2)} &= -1^2 + \frac{1}{2}\cdot 1\cdot 1 = -\frac{1}{2}.
\end{aligned}$$

Thus, we see that the causal type of a tangent vector may depend on its base point.

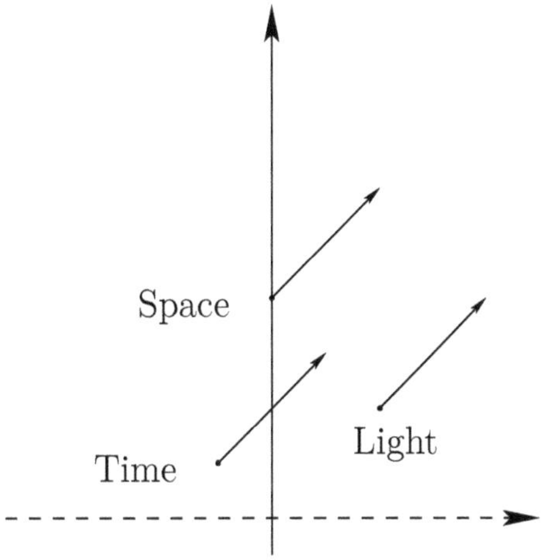

Figure 4.1: Our Euclidean eyes see the vector $(1,1)$ always in the same way, but its length depends on its location.

For a Lorentzian metric, there might or might not exist a vector which is timelike at all points. In this example, we see that $(1,0)$ satisfies $\langle (1,0),(1,0)\rangle_{(x,y)} = -1$, for all $(x,y) \in M$. We say that $(1,0)$ induces a *time orientation* on M. To understand how this happens, let's determine the lightlike curves on M: let $\boldsymbol{\alpha}\colon I \to M$, given by $\boldsymbol{\alpha}(t) = (x(t),y(t))$, be a lightlike curve. Its coordinates then satisfy the differential equation

$$-x'(t)^2 + y(t)y'(t)^2 = 0.$$

If $y'(t) = 0$, then $x'(t) = 0$ and thus $\boldsymbol{\alpha}$ degenerates to a point. Otherwise, the Inverse Function Theorem allows us to use y as a parameter, and assume that the curve has the form $\boldsymbol{\alpha}(y) = (x(y),y)$. We then have that

$$-x'(y)^2 + y = 0,$$

whence $x'(y) = \pm\sqrt{y}$, and so $x(y) = \pm 2y^{3/2}/3 + c$, for some constant c.

The lightlike curves passing through a given point play the same role as the diagonals in the plane $\mathbb{L}^2$, and the vector $(1,0)$ will determine which one of the "timecones" will be the future. For example, for $(1,1) \in M$, if we say that a timelike vector $\boldsymbol{v} \in T_{(1,1)}M$ is future-directed if $\langle \boldsymbol{v},(1,0)\rangle_{(1,1)} < 0$, the future timecone of $(1,1)$ is indicated as follows:

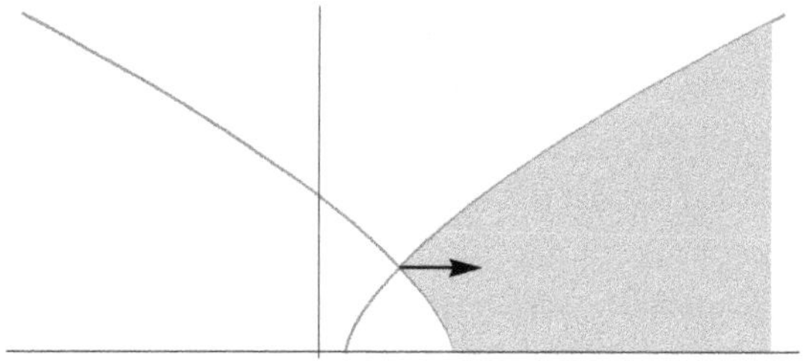

Figure 4.2: The future of $(1,1)$ according to this metric.

In other words, pseudo-Riemannian metrics will give us a new way of measuring lengths and areas. And with a "suitable" Lorentzian metric, one may even define notions of future and past in M. We will discuss this further still in this chapter.

Let's continue to translate the results already established for our new setting:

Definition 4.1.7. Let $(M_1, \langle\cdot,\cdot\rangle_1)$ and $(M_2, \langle\cdot,\cdot\rangle_2)$ be geometric surfaces. We say that that a (local) diffeomorphism $\phi\colon M_1 \to M_2$ is a (local) isometry if given any $p \in M_1$, and $\boldsymbol{v},\boldsymbol{w} \in T_pM_1$, it holds that

$$\langle \boldsymbol{v},\boldsymbol{w}\rangle_1 = \langle d\phi_p(\boldsymbol{v}), d\phi_p(\boldsymbol{w})\rangle_2.$$

Remark. Clearly there can be no local isometry if one of the metrics is Riemannian and the other is Lorentzian. Furthermore, Propositions 3.2.17, 3.2.18, 3.2.19, 3.2.20 and 3.2.21 (p. 170, 171 and 172) all remain valid, and so the verification that certain maps are isometries may be simplified by employing differential notation, just as before.

Example 4.1.8 (Models of Hyperbolic Geometry)**.** In the mid-19th century, Bolyai and Lobachevsky found out that Euclid's fifth postulate (the parallel postulate) is in fact independent of the remaining ones — giving birth to non-Euclidean geometries, with *hyperbolic geometry* being one of the most important ones. With the tools we have developed so far, we can present a few models:

(1) The *Poincaré disk* is the disk $\mathbb{D}^2 \doteq \{(x,y) \in \mathbb{R}^2 \mid x^2+y^2<1\}$, equipped with the Riemannian metric

$$ds_P^2 \doteq \frac{4(dx^2+dy^2)}{(1-x^2-y^2)^2}.$$

Let's see that $\mathbb{D}^2$ and $\mathbb{H}^2$ are isometric, mimicking the idea of the second map seen in Example 3.1.21 (p. 143). The idea is to project $\mathbb{D}^2$ onto $\mathbb{H}^2$, now using the point $(0,0,-1)$ as the base of the projection. Seeing the disk $\mathbb{D}^2$ inside $\mathbb{L}^3$, in the plane $z=0$ and centered at the origin, consider the intersection of the ray starting at $(0,0,-1)$ and passing through $(x,y,0)$ with the hyperbolic plane $\mathbb{H}^2$.

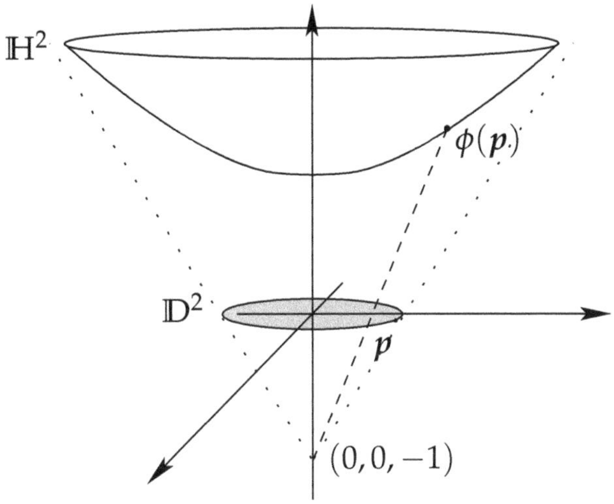

Figure 4.3: The isometry between $\mathbb{D}^2$ and $\mathbb{H}^2$.

This ray can be parametrized by $\boldsymbol{r}(t)=(tx,ty,-1+t)$, and so we seek $t>0$ such that $\langle \boldsymbol{r}(t),\boldsymbol{r}(t)\rangle_L=-1$. Directly, we obtain $t=2/(1-x^2-y^2)$, and so we have $\phi\colon \mathbb{D}^2 \to \mathbb{H}^2$ given by

$$\phi(x,y)=\left(\frac{2x}{1-x^2-y^2},\frac{2y}{1-x^2-y^2},\frac{1+x^2+y^2}{1-x^2-y^2}\right).$$

We claim that ϕ is an isometry. Clearly ϕ is a diffeomorphism, so let's just check that it preserves the metrics. Consider the parametrizations $\boldsymbol{x}\colon\,]0,1[\times]0,2\pi[\to \mathbb{D}^2$ and $\widetilde{\boldsymbol{x}}\colon \mathbb{R}_{>0}\times]0,2\pi[\to\mathbb{H}^2$ given by

$$\begin{aligned}\boldsymbol{x}(u,v)&\doteq(u\cos v,u\sin v)\\ \widetilde{\boldsymbol{x}}(\widetilde{u},\widetilde{v})&\doteq(\sinh\widetilde{u}\cos\widetilde{v},\sinh\widetilde{u}\sin\widetilde{v},\cosh\widetilde{u}).\end{aligned}$$

Thus, we have that the metrics on $\mathbb{D}^2$ and $\mathbb{H}^2$ are expressed, respectively, as

$$\frac{4(du^2+u^2\,dv^2)}{(1-u^2)^2}\quad\text{and}\quad d\widetilde{u}^2+\sinh^2\widetilde{u}\,d\widetilde{v}^2,$$

while ϕ is expressed as

$$\phi(\boldsymbol{x}(u,v))=\widetilde{\boldsymbol{x}}\left(\operatorname{arcsinh}\left(\frac{2u}{1-u^2}\right),v\right).$$

This way, if $\widetilde{u} = \operatorname{arcsinh}(2u/(1-u^2))$ and $\widetilde{v} = v$, we have that

$$\mathrm{d}\widetilde{u} = \frac{2}{1-u^2}\,\mathrm{d}u \quad \text{and} \quad \mathrm{d}\widetilde{v} = \mathrm{d}v.$$

So:

$$\begin{aligned}
\mathrm{d}\widetilde{u}^2 + \sinh^2\widetilde{u}\,\mathrm{d}\widetilde{v}^2 &= \left(\frac{2}{1-u^2}\,\mathrm{d}u\right)^2 + \left(\frac{2u}{1-u^2}\right)^2 \mathrm{d}v^2 \\
&= \frac{4\,\mathrm{d}u^2}{(1-u^2)^2} + \frac{4u^2\,\mathrm{d}v^2}{(1-u^2)^2} \\
&= \frac{4(\mathrm{d}u^2 + u^2\,\mathrm{d}v^2)}{(1-u^2)^2},
\end{aligned}$$

as wanted.

(2) The *Poincaré half-plane* is the half-plane $\{(x,y) \in \mathbb{R}^2 \mid y > 0\}$ equipped with the Riemannian metric

$$\mathrm{d}s_H^2 = \frac{\mathrm{d}x^2 + \mathrm{d}y^2}{y^2},$$

and it is also denoted by $\mathbb{H}^2$. The Poincaré half-plane is also isometric to the Poincaré disk and the hyperbolic plane in $\mathbb{L}^3$. The most efficient way to exhibit the isometry is to explore the relation of $\mathbb{R}^2$ with the complex numbers $\mathbb{C}$, writing

$$\mathbb{H}^2 = \{z \in \mathbb{C} \mid \operatorname{Im}(z) > 0\} \quad \text{and} \quad \mathbb{D}^2 = \{w \in \mathbb{C} \mid |w| < 1\},$$

with metrics given, respectively, by

$$\mathrm{d}s_H^2 = \frac{\mathrm{d}z\,\mathrm{d}\overline{z}}{\operatorname{Im}(z)^2} \quad \text{and} \quad \mathrm{d}s_P^2 = \frac{4\,\mathrm{d}w\,\mathrm{d}\overline{w}}{(1-|w|^2)^2},$$

where we use $\mathrm{d}z = \mathrm{d}x + \mathrm{i}\,\mathrm{d}y$, $\mathrm{d}\overline{z} = \mathrm{d}x - \mathrm{i}\,\mathrm{d}y$, etc. Define the so-called *Cayley transform* $\psi\colon \mathbb{H}^2 \to \mathbb{D}^2$ by

$$\psi(z) = \frac{z - \mathrm{i}}{z + \mathrm{i}}.$$

Firstly, we have to check that if $\operatorname{Im}(z) > 0$, then $|\psi(z)| < 1$, so that the mapping does in fact take values in $\mathbb{D}^2$. Indeed, using the known expression $z - \overline{z} = 2\mathrm{i}\operatorname{Im}(z)$, we have that

$$|\psi(z)|^2 = \left|\frac{z-\mathrm{i}}{z+\mathrm{i}}\right|^2 = \frac{(z-\mathrm{i})(\overline{z}+\mathrm{i})}{(z+\mathrm{i})(\overline{z}-\mathrm{i})} = \frac{1+|z|^2 - 2\operatorname{Im}(z)}{1+|z|^2+2\operatorname{Im}(z)} < 1.$$

In fact, since ψ is a Möbius transformation, it maps the asymptotic boundary of $\mathbb{H}^2$ (the real axis) into the asymptotic boundary of $\mathbb{D}^2$ (the unit circle): the above calculation also says that $\operatorname{Im}(z) = 0$ if and only if $|\psi(z)| = 1$. One can show that ψ is a diffeomorphism by directly exhibiting its inverse, for instance. Let's see now that ψ preserves the metrics. Set $w = \psi(z)$. With this, we have

$$w = \frac{z-\mathrm{i}}{z+\mathrm{i}} \quad \text{and} \quad \overline{w} = \frac{\overline{z}+\mathrm{i}}{\overline{z}-\mathrm{i}},$$

whence

$$\mathrm{d}w = \frac{2\mathrm{i}}{(z+\mathrm{i})^2}\,\mathrm{d}z \quad \text{and} \quad \mathrm{d}\overline{w} = -\frac{2\mathrm{i}}{(\overline{z}-\mathrm{i})^2}\,\mathrm{d}\overline{z}.$$

Thus:

$$\begin{aligned}\frac{4\,\mathrm{d}w\,\mathrm{d}\overline{w}}{(1-|w|^2)^2} &= \frac{4}{\left(1-\dfrac{1+|z|^2-2\,\mathrm{Im}(z)}{1+|z|^2+2\,\mathrm{Im}(z)}\right)^2}\left(\frac{2\mathrm{i}}{(z+\mathrm{i})^2}\,\mathrm{d}z\right)\left(-\frac{2\mathrm{i}}{(\overline{z}-\mathrm{i})^2}\,\mathrm{d}\overline{z}\right)\\ &= \frac{16}{\left(\dfrac{4\,\mathrm{Im}(z)}{|z+\mathrm{i}|^2}\right)^2}\frac{\mathrm{d}z\,\mathrm{d}\overline{z}}{|z+\mathrm{i}|^4}\\ &= \frac{\mathrm{d}z\,\mathrm{d}\overline{z}}{\mathrm{Im}(z)^2},\end{aligned}$$

as wanted.

For further models, see Exercises 4.1.9 and 4.1.10.

The adaptations to be made go on:

Definition 4.1.9. Let $(M, \langle\cdot,\cdot\rangle)$ be a geometric surface, and consider a parametrization $(U, \boldsymbol{x})$ for M. For $1 \le i, j, k \le 2$, the *Christoffel symbols* of $\boldsymbol{x}$ are defined by

$$\Gamma_{ij}^k = \sum_{r=1}^{2} \frac{1}{2} g^{kr} \left(\frac{\partial g_{ik}}{\partial u^j} + \frac{\partial g_{jk}}{\partial u^i} - \frac{\partial g_{ij}}{\partial u^r} \right).$$

Definition 4.1.10. Let $(M, \langle\cdot,\cdot\rangle)$ be a geometric surface and $\boldsymbol{\alpha}\colon I \to M$ be a curve on M. We say that $\boldsymbol{\alpha}$ is a *geodesic* of M if $\boldsymbol{\alpha}$ is a critical point of the energy functional associated to the metric $\langle\cdot,\cdot\rangle$.

Those two definitions validate all the discussion done in Section 3.6, so that we may continue to use variational methods to compute Christoffel symbols and geodesics in a geometric surface. All the relevant results remain true, in particular, the characterization given in coordinates by the geodesic differential equations

$$\ddot{u}^k + \sum_{i,j=1}^{2} \Gamma_{ij}^k \dot{u}^i \dot{u}^j = 0,$$

and the results regarding Clairaut parametrizations.

Example 4.1.11.

(1) Let's find the Christoffel symbols for the Riemannian metric given in coordinates by $\mathrm{d}s^2 = y^2\,\mathrm{d}x^2 + x^2\,\mathrm{d}y^2$ in $\mathbb{R}^2 \setminus \{(x,y) \in \mathbb{R}^2 \mid xy \neq 0\}$. Define the Lagrangian

$$L(t, x(t), x'(t), y(t), y'(t)) = \frac{1}{2}(y(t)^2 x'(t)^2 + x(t)^2 y'(t)^2).$$

Omitting points of evaluation, we have:

$$\begin{aligned} 0 &= \frac{\partial L}{\partial x} - \frac{\mathrm{d}}{\mathrm{d}t}\left(\frac{\partial L}{\partial x'}\right) = xy'^2 - (y^2x')' = xy'^2 - x''y^2 - 2yx'y' = 0\\ 0 &= \frac{\partial L}{\partial y} - \frac{\mathrm{d}}{\mathrm{d}t}\left(\frac{\partial L}{\partial y'}\right) = yx'^2 - (x^2y')' = yx'^2 - y''x^2 - 2xx'y' = 0,\end{aligned}$$

so that the geodesic equations are

$$x'' + \frac{2}{y}x'y' - \frac{x}{y^2}y'^2 = 0 \quad \text{and} \quad y'' - \frac{y}{x^2}x'^2 + \frac{2}{x}x'y' = 0,$$

and the non-zero Christoffel symbols are

$$\Gamma^x_{xy}(x,y) = \Gamma^x_{yx}(x,y) = \frac{1}{y}, \quad \Gamma^x_{yy}(x,y) = -\frac{x}{y^2}$$
$$\Gamma^y_{xy}(x,y) = \Gamma^y_{yx}(x,y) = \frac{1}{x} \quad \text{and} \quad \Gamma^y_{xx}(x,y) = -\frac{y}{x^2}.$$

We will not bother to solve such differential equations. Surprisingly, this metric is the usual Euclidean one after a convenient change of coordinates $\widetilde{x} = \widetilde{x}(x,y), \widetilde{y} = \widetilde{y}(x,y)$. To deduce such change is non-trivial (convince yourself of this), but we'll register it here: if

$$\widetilde{x} = \frac{\sqrt{2}}{2}xy \quad \text{and} \quad \widetilde{y} = \frac{\sqrt{2}}{4}y^2 - \frac{\sqrt{2}}{4}x^2,$$

then $\mathrm{d}\widetilde{x}^2 + \mathrm{d}\widetilde{y}^2 = y^2\,\mathrm{d}x^2 + x^2\,\mathrm{d}y^2$ holds (see Exercise 4.1.11).

(2) Let's determine the geodesics in the Poincaré disk $\mathbb{D}^2$, by exploring some of its isometries. Recall that the metric is given by

$$\mathrm{d}s^2_P = \frac{4(\mathrm{d}x^2 + \mathrm{d}y^2)}{(1 - x^2 - y^2)^2}.$$

From such expression, it is easy to see that the reflection $(x,y) \mapsto (-x,y)$ is an isometry of $\mathbb{D}^2$. With this, Corollary 3.6.21 (p. 231) says that the vertical diameter is a geodesic. Moreover, rotations about the origin are also isometries of $\mathbb{D}^2$ (to wit, they are Euclidean isometries which preserve the quantity $x^2 + y^2$), from where we conclude that actually any diameter is a geodesic. In particular, we now know that every geodesic that passes through the origin is (part of) a diameter.

In Exercise 4.1.7, we ask you to verify that, for each $z_0 \in \mathbb{D}^2$ and $\theta \in \mathbb{R}$, the map $F\colon \mathbb{D}^2 \to \mathbb{D}^2$ given by

$$F(z) = \mathrm{e}^{\mathrm{i}\theta}\frac{z - z_0}{1 - z\overline{z}_0}$$

is an isometry of $\mathbb{D}^2$. In particular, F and its inverse are Möbius transformations, that send lines and circles into lines or circles. Well, if $\boldsymbol{\alpha}$ is a geodesic that passes through z_0, $F \circ \boldsymbol{\alpha}$ is a geodesic that passes through the origin, and thus is a diameter. Hence, $\boldsymbol{\alpha} = F^{-1} \circ (F \circ \boldsymbol{\alpha})$ is necessarily a line segment (if $z_0 = 0$) or an arc of circumference contained in $\mathbb{D}^2$. Furthermore, such an arc crosses the boundary of $\mathbb{D}^2$ orthogonally: indeed, since it is an isometry, F preserves Euclidean angles (to wit, it preserves angles measured according to the Poincaré disk metric, which is conformal to the Euclidean one) — and diameters do cross the boundary orthogonally.

Conclusion: the geodesics in $\mathbb{D}^2$ are its diameters, and arcs of circumferences contained in $\mathbb{D}^2$ which cross its boundary orthogonally.

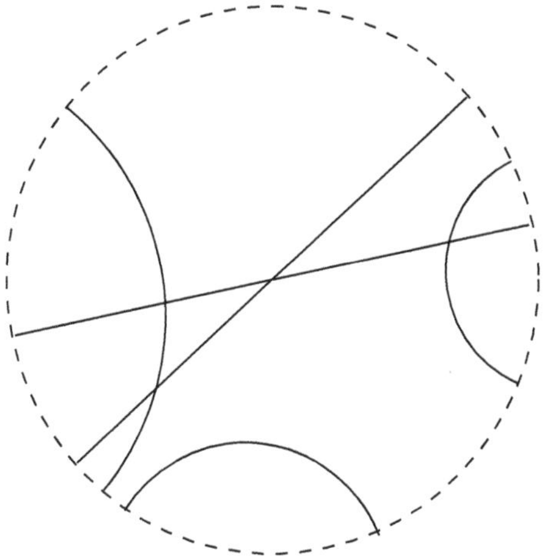

Figure 4.4: Geodesics in the Poincaré disk.

Of course, to conclude that those are the geodesics of $\mathbb{D}^2$, we could also geometrically see the images of geodesics in $\mathbb{H}^2 \subseteq \mathbb{L}^3$ under the inverse of the isometry ϕ seen in Example 4.1.8 above. Also, Clairaut parametrizations are efficient to find geodesics in other models, for example, in the Poincaré half-plane (see Exercise 4.1.12).

Among the intrinsic concepts studied for surfaces in $\mathbb{R}^3_\nu$, we still have to see what happens with the Gaussian curvature in our new setting. The Gauss equations, seen in Proposition 3.7.2 (p. 251), which do not depend on the Second Fundamental Form, give us expressions for the Gaussian curvature $K \circ \boldsymbol{x}$, where $(U, \boldsymbol{x})$ is any parametrization for M. It is true, and we should verify it, that the local expressions obtained for K don't actually depend on the chosen parametrization and so can be used to define the Gaussian curvature of a geometric surface $(M, \langle \cdot, \cdot \rangle)$.

We have seen that these expressions are much simpler when the parametrization is orthogonal. In practice, we will keep using the expression given in Proposition 3.7.3 (p. 252):

$$K(p) = \frac{-1}{\sqrt{|EG|}} \left(\epsilon_u \left(\frac{(\sqrt{|G|})_u}{\sqrt{|E|}} \right)_u + \epsilon_v \left(\frac{(\sqrt{|E|})_v}{\sqrt{|G|}} \right)_v \right),$$

where $p = \boldsymbol{x}(u_0, v_0)$, and all of the coefficients in the right side are evaluated in (u_0, v_0), since given any point in a geometric surface, there is a parametrization around it with $F = 0$. For a proof of this result, which does not depend on the index of $\langle \cdot, \cdot \rangle$, see [17].

With this definition, K is automatically invariant under isometries.

Example 4.1.12.

(1) Since the Gaussian curvature of the hyperbolic plane $\mathbb{H}^2 \subseteq \mathbb{L}^3$ is constant and equal to -1, we immediately know that the curvatures of the Poincaré disk $\mathbb{D}^2$ and of the Poincaré half-plane $\mathbb{H}^2$ are also equal to -1. This can be verified directly. For

instance, for the half-plane we have:

$$\begin{aligned}
K &= \frac{-1}{\sqrt{1/y^4}}\left(\left(\frac{(\sqrt{1/y^2})_x}{\sqrt{1/y^2}}\right)_x + \left(\frac{(\sqrt{1/y^2})_y}{\sqrt{1/y^2}}\right)_y\right) \\
&= -y^2\left(y\left(\frac{1}{y}\right)_y\right)_y \\
&= -y^2\left(-\frac{1}{y}\right)_y = -y^2\frac{1}{y^2} \\
&= -1.
\end{aligned}$$

See in Exercise 4.1.15 how to compute the Gaussian curvature of any metric proportional (conformal) to the usual metrics in $\mathbb{R}^2$ and $\mathbb{L}^2$.

(2) Consider $r_1, r_2 > 0$, and the product of circles $M \doteq \mathbb{S}^1(r_1) \times \mathbb{S}^1(r_2) \subseteq \mathbb{R}^4$. When $r_1^2 + r_2^2 = 1$, M is called a *Clifford torus*. The usual inner product in $\mathbb{R}^4$ restricted to the tangent planes to M defines a Riemannian metric on M. If $\boldsymbol{\alpha}_1$ and $\boldsymbol{\alpha}_2$ are the usual parametrizations for the given circles, we define a product parametrization $\boldsymbol{x}\colon\,]0,2\pi[^2 \to \boldsymbol{x}(]0,2\pi[^2) \subseteq \mathbb{S}^1(r_1) \times \mathbb{S}^1(r_2)$ by

$$\boldsymbol{x}(u,v) \equiv (\boldsymbol{\alpha}_1 \oplus \boldsymbol{\alpha}_2)(u,v) \doteq (r_1 \cos u, r_1 \sin u, r_2 \cos v, r_2 \sin v).$$

Let's see how to express the metric induced by $\mathbb{R}^4$ on these coordinates. We have

$$\begin{aligned}
\boldsymbol{x}_u(u,v) &= (-r_1 \sin u, r_1 \cos u, 0, 0) \quad \text{and} \\
\boldsymbol{x}_v(u,v) &= (0,0,-r_2 \sin v, r_2 \cos v),
\end{aligned}$$

whence $\mathrm{d}\boldsymbol{x}_{(u,v)}$ has full rank, for all $(u,v) \in\,]0,2\pi[^2$. Evaluating the coefficients $E(u,v) = \langle \boldsymbol{x}_u(u,v), \boldsymbol{x}_u(u,v)\rangle_E$, etc., we have that

$$\mathrm{d}s^2 = r_1^2\,\mathrm{d}u^2 + r_2^2\,\mathrm{d}v^2$$

along the image of $\boldsymbol{x}$. As the image of $\boldsymbol{x}$ is dense in M, K is a continuous function, and all the coefficients above are all constants, we have that $K(\boldsymbol{p}) = 0$ for all $\boldsymbol{p} \in M$.

(3) In the previous example, we have considered a "direct product" of the circles. If $\boldsymbol{\alpha}_1$ and $\boldsymbol{\alpha}_2$ are given as above, we now consider their "tensor product", i.e., the map $\boldsymbol{x}\colon\,]0,2\pi[^2 \to \mathbb{R}^4$ given by

$$\begin{aligned}
\boldsymbol{x}(u,v) &\equiv (\boldsymbol{\alpha}_1 \otimes \boldsymbol{\alpha}_2)(u,v) \\
&\equiv \boldsymbol{\alpha}_1(u) \otimes \boldsymbol{\alpha}_2(v) \\
&\doteq \begin{pmatrix} r_1 \cos u \\ r_1 \sin u \end{pmatrix} \begin{pmatrix} r_2 \cos v & r_2 \sin v \end{pmatrix} \\
&= \begin{pmatrix} r_1 r_2 \cos u \cos v & r_1 r_2 \cos u \sin v \\ r_1 r_2 \sin u \cos v & r_1 r_2 \sin u \sin v \end{pmatrix} \\
&\equiv (r_1 r_2 \cos u \cos v, r_1 r_2 \cos u \sin v, r_1 r_2 \sin u \cos v, r_1 r_2 \sin u \sin v),
\end{aligned}$$

using the identification $\mathrm{Mat}(2,\mathbb{R}) \cong \mathbb{R}^4$. Let's see what the Riemannian metric induced by the usual inner product of $\mathbb{R}^4$ looks like. We have:

$$\begin{aligned}
\boldsymbol{x}_u(u,v) &= r_1 r_2(-\sin u \cos v, -\sin u \sin v, \cos u \cos v, \cos u \sin v) \text{ and} \\
\boldsymbol{x}_v(u,v) &= r_1 r_2(-\cos u \sin v, \cos u \cos v, -\sin u \sin v, \sin u \cos v),
\end{aligned}$$

and one can verify that $\mathbf{d}\boldsymbol{x}_{(u,v)}$ always has full rank, just as above. So, it follows that

$$ds^2 = r_1^2 r_2^2(du^2 + dv^2),$$

and the Gaussian curvature again vanishes. For more about tensor products in this setting, see [71].

A natural continuation of the ideas presented in this text so far will lead us to an area of Geometry called *Pseudo-Riemannian Geometry*, which consists of the study of the geometry of differentiable manifolds of arbitrary dimension, equipped with metrics of arbitrary index. A standard reference for this subject is [54], for example.

Exercises

Exercise 4.1.1. Let $0 < q < 1$ and consider in $\mathbb{R}_{>0} \times \mathbb{R}$ the Lorentzian metric given by

$$ds^2 = -dt^2 + t^{2q}\, dx^2.$$

Determine the lightlike curves for this metric.

Remark. In higher dimensions, metrics of the form

$$-dt^2 + a(t)^2(dx^2 + dy^2 + dz^2)$$

model universes for which space, in each fixed instant of time, is a Euclidean space with curvature, but that expands as a function of t. Of particular interest in Physics are the *scales* of the form $a(t) = t^q$, with $0 < q < 1$. For more details, see [10].

Exercise 4.1.2. Let's rotate the lightcones in $\mathbb{L}^2$. Consider in $\mathbb{R}^2$ the metric

$$ds^2 = \cos(2\pi x)(dx^2 - dy^2) - 2\sin(2\pi x)\, dx\, dy.$$

(a) Verify that this metric is Lorentzian.

(b) For every non-zero $\boldsymbol{v} \in \mathbb{R}^2$ there is $(x, y) \in \mathbb{R}^2$ (actually, infinitely many of them) such that $\langle \boldsymbol{v}, \boldsymbol{v} \rangle_{(x,y)}$ is negative, zero, or positive. In other words, according to this metric, every vector assumes all causal types for convenient base points, infinitely many times.

Hint. Since $\boldsymbol{v}$ is non-zero, assume without loss of generality that $\boldsymbol{v} = (\cos\theta, \sin\theta)$ for some $\theta \in [0, 2\pi[$, and choose x in terms of θ.

(c) Exhibit non-vanishing smooth vector fields $\boldsymbol{V}\colon \mathbb{R}^2 \to \mathbb{R}^2$ such that the product $\langle \boldsymbol{V}(x,y), \boldsymbol{V}(x,y) \rangle_{(x,y)}$ is always negative, zero, or positive, for all $(x, y) \in \mathbb{R}^2$. That is, despite item (b), there are fields with constant causal type.

Hint. Try to rotate the vectors $(0, 1)$, $(1, 1)$ and $(1, 0)$ "together with the metric of $\mathbb{L}^2$".

Remark. This metric passes to the quotient, defining a time-orientable Lorentzian metric in the Möbius strip.

Exercise 4.1.3. Consider in $\mathbb{R}^2$ the Riemannian metric given by

$$ds^2 = \frac{dx^2 + dy^2}{1 + (x^2 + y^2)^2}.$$

Show that, according to this metric, $\mathscr{A}(\mathbb{R}^2) < +\infty$.

Exercise 4.1.4. Let $0 < r < 1$ and consider the region

$$R \doteq \{(x,y) \in \mathbb{R}^2 \mid x^2 + y^2 < r^2\}$$

in the Poincaré disk $\mathbb{D}^2$. Compute $\mathscr{A}(R)$. What happens when $r \to 1^-$?

Remark. Compare this with Exercise 3.2.14 (p. 168).

Exercise 4.1.5 (Pseudo-sphere). The image of the regular parametrized surface $\boldsymbol{x}\colon \mathbb{R}_{>0} \times]0,2\pi[\to \mathbb{R}^3$ given by

$$\boldsymbol{x}(u,v) = (\operatorname{sech} u \cos v, \operatorname{sech} u \sin v, u - \tanh u)$$

is known as the *pseudo-sphere.*

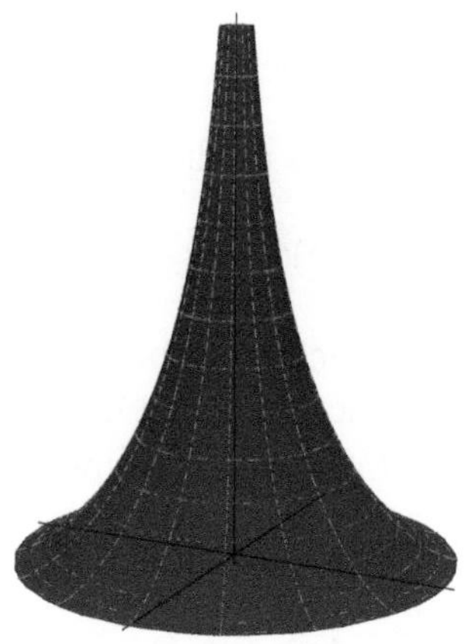

Figure 4.5: The pseudo-sphere.

Consider the Poincaré half-plane $\mathbb{H}^2$, with Riemannian metric

$$ds_H^2 = \frac{dx^2 + dy^2}{y^2}.$$

Show that $F\colon \boldsymbol{x}\left(]0,\pi/2[\times]0,2\pi[\right) \to \mathbb{H}^2$ given by $F(\boldsymbol{x}(u,v)) = (v, \cosh u)$ is an isometry onto its image.

Exercise 4.1.6. Consider the Poincaré half-plane $\mathbb{H}^2$ with the Riemannian metric given in complex coordinates by

$$ds_H^2 = \frac{dz\,d\bar{z}}{\operatorname{Im}(z)^2}.$$

If we have numbers $a,b,c,d \in \mathbb{R}$ with $ad - bc = 1$, show that the *Möbius transformation* $T\colon \mathbb{H}^2 \to \mathbb{H}^2$ given by

$$w = T(z) = \frac{az + b}{cz + d}$$

is an isometry of $\mathbb{H}^2$ onto itself.

Hint. Don't be afraid of $\mathrm{d}z$ and $\mathrm{d}\bar{z}$, and do the computation just like in Example 4.1.8 (p. 264).

Exercise† 4.1.7. Let $\theta \in \mathbb{R}$ and $z_0 \in \mathbb{D}^2$. Consider $F\colon \mathbb{C} \to \mathbb{C}$ given by

$$F(z) \doteq \mathrm{e}^{\mathrm{i}\theta}\frac{z - z_0}{1 - z\overline{z_0}}.$$

(a) Compute $|F(z)|^2$ in terms of $|z|$ and conclude not only that $F(\mathbb{S}^1) \subseteq \mathbb{S}^1$, but also that $F(\mathbb{D}^2) \subseteq \mathbb{D}^2$.

(b) Show that F is bijective, with inverse given by

$$F^{-1}(w) = \mathrm{e}^{-\mathrm{i}\theta}\frac{w + z_0\mathrm{e}^{\mathrm{i}\theta}}{1 + \mathrm{e}^{-\mathrm{i}\theta}\overline{z_0}w}.$$

In particular, we have the equalities $F(\mathbb{S}^1) = \mathbb{S}^1$ and $F(\mathbb{D}^2) = \mathbb{D}^2$.

(c) Show that $F|_{\mathbb{D}^2}$ is an isometry of the Poincaré disk, recalling that its metric in complex coordinates is given by

$$\mathrm{d}s^2 = \frac{4\,\mathrm{d}z\,\mathrm{d}\bar{z}}{(1 - |z|^2)^2}.$$

Exercise† 4.1.8. Consider the hyperbolic plane $\mathbb{H}^2 \subseteq \mathbb{L}^3$, and also the Poincaré half-plane, denoted only in this exercise by $\mathbb{H}^2_{\mathrm{hp}}$. Show that the map $F\colon \mathbb{H}^2 \to \mathbb{H}^2_{\mathrm{hp}}$ given by

$$F(x,y,z) = \left(\frac{x}{z-y}, \frac{1}{z-y}\right)$$

is an isometry.

Hint. Consider, as usual, the revolution parametrization $\boldsymbol{x}\colon \mathbb{R}_{\geq 0} \times [0, 2\pi[\to \mathbb{H}^2$ given by $\boldsymbol{x}(u,v) = (\sinh u \cos v, \sinh u \sin v, \cosh u)$ and compute the products between the derivatives of $F(\boldsymbol{x}(u,v))$ using the half-plane metric.

Exercise† 4.1.9 (Klein disk)**.** We have one more model for hyperbolic geometry: the *Klein disk* is the set $\mathbb{D}^2$ equipped with the Riemannian metric

$$\mathrm{d}s_K^2 = \frac{\mathrm{d}x^2 + \mathrm{d}y^2}{1 - x^2 - y^2} + \frac{(x\,\mathrm{d}x + y\,\mathrm{d}y)^2}{(1 - x^2 - y^2)^2}.$$

(a) Show that this metric is indeed Riemannian (positive-definite).

Hint. Sylvester's Criterion.

(b) Show that if $\boldsymbol{x}\colon]0,1[\times]0,2\pi[\to \mathbb{D}^2 \setminus \{(0,0)\}$ is the usual polar parametrization $\boldsymbol{x}(u,v) = (u\cos v, u\sin v)$, then

$$\mathrm{d}s_K^2 = \frac{1}{(1-u^2)^2}\,\mathrm{d}u^2 + \frac{u^2}{1-u^2}\,\mathrm{d}v^2.$$

(c) Determine an isometry between the Klein disk and the hyperbolic plane $\mathbb{H}^2 \subseteq \mathbb{L}^3$.

Hint. One possible isometry already appeared in the text. Can you find which one it is?

Remark. The Klein disk is less used than the Poincaré disk, since its metric is not conformal to the usual metric in $\mathbb{R}^2$, and so we see distorted angles, while this does not occur in the Poincaré disk.

Exercise† 4.1.10 (Hemisphere)**.** Consider the *Hemisphere*

$$\mathbb{J}^2 \doteq \{(x,y,z) \in \mathbb{R}^3 \mid x^2+y^2+z^2=1 \text{ and } z>0\},$$

equipped with the Riemannian metric

$$ds_J^2 \doteq \frac{dx^2+dy^2+dz^2}{z^2}.$$

(a) Show that if $\boldsymbol{x}\colon\,]0,\pi/2[\, \times\,]0,2\pi[\, \to \mathbb{J}^2$ is the usual parametrization of revolution given by $\boldsymbol{x}(u,v) = (\cos u \cos v, \cos u \sin v, \sin u)$, then

$$ds_J^2 = \frac{du^2+\cos^2 u\, dv^2}{\sin^2 u}.$$

(b) The Hemisphere equipped with this metric may also be used as a model for hyperbolic geometry: show that $\mathbb{J}^2$ is isometric to the hyperbolic plane $\mathbb{H}^2 \subseteq \mathbb{L}^3$ via the central projection based on $(0,0,-1)$, $\Pi\colon \mathbb{H}^2 \to \mathbb{J}^2$ given by

$$\Pi(x,y,z) = \left(\frac{x}{z}, \frac{y}{z}, \frac{1}{z}\right).$$

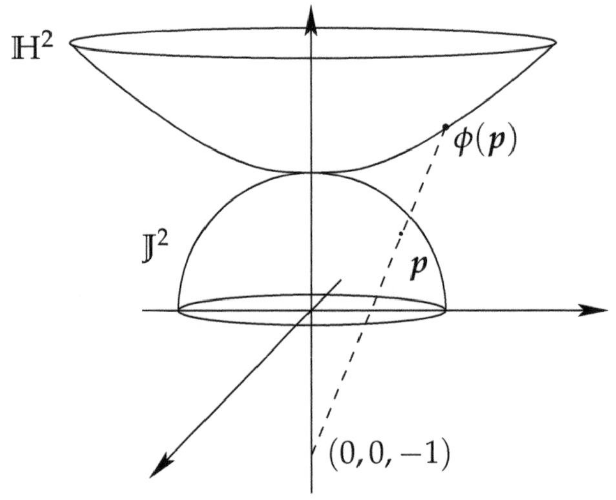

Figure 4.6: The isometry $\phi = \Pi^{-1}$ between $\mathbb{J}^2$ and $\mathbb{H}^2$.

Hint. If $\widetilde{\boldsymbol{x}}$ is the usual parametrization for $\mathbb{H}^2$, used in Example 4.1.8 (p. 264), show that $\Pi(\widetilde{\boldsymbol{x}}(\widetilde{u},\widetilde{v})) = \boldsymbol{x}(\arccos(\tanh\widetilde{u}),\widetilde{v})$.

Exercise 4.1.11 (Parabolic coordinates in $\mathbb{R}^2$)**.** Following Example 4.1.11 (p. 267), show that in $\mathbb{R}^2 \setminus \{(x,y) \in \mathbb{R}^2 \mid xy=0\}$, if

$$\widetilde{x} = \frac{\sqrt{2}}{2}xy \quad \text{and} \quad \widetilde{y} = \frac{\sqrt{2}}{4}y^2 - \frac{\sqrt{2}}{4}x^2,$$

then $d\widetilde{x}^2 + d\widetilde{y}^2 = y^2\, dx^2 + x^2\, dy^2$.

Exercise 4.1.12. Consider the Poincaré half-plane $\mathbb{H}^2$ with the Riemannian metric

$$\mathrm{d}s_H^2 = \frac{\mathrm{d}x^2 + \mathrm{d}y^2}{y^2}.$$

Observing that the identity is a y-Clairaut parametrization, use Theorem 3.6.34 (p. 243) and show that all the geodesics in $\mathbb{H}^2$ are vertical lines and arcs of circumferences which orthogonally cross the axis $y = 0$ (the so-called *asymptotic boundary* of $\mathbb{H}^2$, denoted on literature by $\partial_\infty \mathbb{H}^2$).

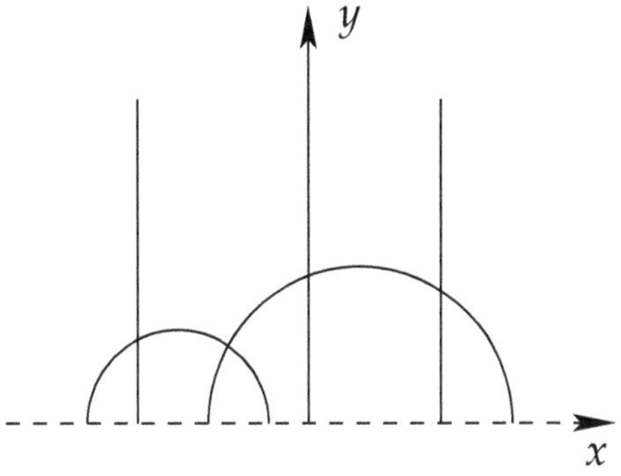

Figure 4.7: Geodesics in the Poincaré half-plane.

Hint. In the statement of Theorem 3.6.34, make $u = y$ and $v = x$. Since in this case we have $E = G$, there's no risk of switching them.

Exercise 4.1.13. Let $(M, \langle \cdot, \cdot \rangle)$ be a geometric surface equipped with a Lorentzian metric and $(U, \boldsymbol{x})$ be a parametrization of M for which both $\boldsymbol{x}_u$ and $\boldsymbol{x}_v$ are always lightlike.

(a) Show that

$$K \circ \boldsymbol{x} = -\frac{1}{F}\left(\frac{F_u}{F}\right)_v = -\frac{1}{F}\left(\frac{F_v}{F}\right)_u.$$

(b) Show that the geodesic differential equations simply boil down to

$$u'' + \frac{F_u}{F}u'^2 = 0 \quad \text{and} \quad v'' + \frac{F_v}{F}v'^2 = 0.$$

Exercise† 4.1.14. Show that switching the sign of a Lorentzian metric also switches the sign of its Gaussian curvature. Thus, Lorentzian metrics with $K > 0$ differ from those with $K < 0$ only in causal type.

Exercise† 4.1.15. Let $U \subseteq \mathbb{R}^2$ be an open subset and $h \colon U \to \mathbb{R}$ be a nowhere vanishing smooth function.

(a) Show that the curvature of the Riemannian metric $(\mathrm{d}x^2 + \mathrm{d}y^2)/h(x,y)^2$ is given by

$$K = h(h_{xx} + h_{yy}) - (h_x^2 + h_y^2).$$

(b) Show that the curvature of the Lorentzian metric $(\mathrm{d}x^2 - \mathrm{d}y^2)/h(x,y)^2$ is given by

$$K = h(h_{xx} - h_{yy}) - (h_x^2 - h_y^2).$$

Exercise 4.1.16 (The Schwarzschild half-plane). Let $\mathrm{M} > 0$ be a constant. Consider in $P_I \doteq \{(t,r) \in \mathbb{R}^2 \mid r > 2\mathrm{M}\}$ the Lorentzian metric given by

$$\mathrm{d}s^2 = -\hbar(r)\,\mathrm{d}t^2 + \hbar(r)^{-1}\,\mathrm{d}r^2,$$

where

$$\hbar(r) \doteq 1 - \frac{2\mathrm{M}}{r}$$

is the *Schwarzschild horizon function*. Note that in P_I, this horizon function is positive.

(a) Show that the Gaussian curvature of P_I is always positive, given by the expression $K(t,r) = 2\mathrm{M}/r^3 > 0$.

(b) Show that the only non-zero Christoffel symbols are:

$$\Gamma^t_{tr}(t,r) = \Gamma^t_{rt}(t,r) = -\Gamma^r_{rr}(t,r) = \frac{\mathrm{M}}{r^2 - 2\mathrm{M}r}, \quad \text{and} \quad \Gamma^r_{tt}(t,r) = \frac{\mathrm{M}r - 2\mathrm{M}^2}{r^3}.$$

(c) Show that $\boldsymbol{\alpha}\colon \mathbb{R}_{>0} \to P_I$ given by $\boldsymbol{\alpha}(s) = (s + 2\mathrm{M}\log s, s + 2\mathrm{M})$ is a lightlike geodesic in P_I.

(d) Verify that horizontal translations and time reflection (i.e., the maps of the form $P_I \ni (t,r) \mapsto (\pm t + b, r) \in P_I$) are isometries of P_I. Conclude that *all* lightlike geodesics in P_I are images of the curve given in item (b) under transformations of this type.

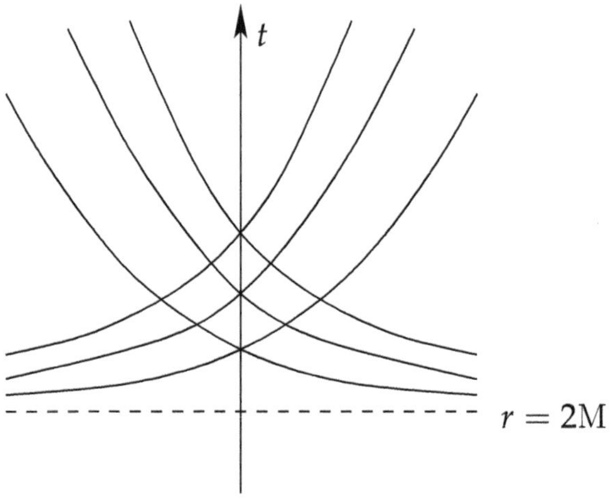

Figure 4.8: Light rays in P_I.

Remark. The Schwarzschild half-plane has several applications in General Relativity. It is a bidimensional slice of a model of a 4-dimensional spacetime, which takes into account gravitational forces and, thus, has curvature. The constant M is interpreted as the mass of a massive particle. We have two important applications. The Schwarzschild space may model:

- the solar system in a more precise way than the models adopted in Newtonian mechanics, when such particle is regarded as the Sun;
- the proximities of a black hole, situation on which M is its mass, the region $r < 2\mathrm{M}$ (where the sign of the metric is reversed) is the interior of the black hole, the region $r > 2\mathrm{M}$ is its exterior, and $r = 2\mathrm{M}$ is the event horizon (where $\hbar$ is singular).

Conveniently altering the function $\hbar$, we obtain other models of spacetimes. For example, the metric in the *Reissner-Nordström space* uses the slightly more sophisticated horizon function

$$\hbar(r) = 1 - \frac{2M}{r} + \frac{q^2}{r^2},$$

taking into account some electric charge q that the black hole may have. For more details, see [10].

Exercise 4.1.17 (Lorentz-Poincaré half-plane)**.** We may consider a Lorentzian version of the Poincaré half-plane. Take, in $M = \mathbb{R} \times \mathbb{R}_{>0}$, the Lorentzian metric given by

$$ds^2 = \frac{dx^2 - dy^2}{y^2}.$$

(a) Verify that its Gaussian curvature is constant, $K = 1$;

(b) Show that all non-zero Christoffel symbols are

$$\Gamma^x_{xy}(x,y) = \Gamma^x_{yx}(x,y) = \Gamma^y_{xx}(x,y) = \Gamma^y_{yy}(x,y) = -\frac{1}{y}.$$

(c) Show that:

- $\alpha\colon \mathbb{R} \to M$ given by $\alpha(t) = (0, e^t)$ is a timelike geodesic, parametrized with proper time;
- the curve $\beta\colon\,]-\pi/2, \pi/2[\, \to M$ given by $\beta(t) = (r \sec t, r \tan t)$, for each $r > 0$, is a timelike geodesic, parametrized with proper time;
- the curve $\gamma\colon\,]-\pi/2, \pi/2[\, \to M$ given by $\gamma(s) = (r \tan s, r \sec s)$, for each $r > 0$, is a spacelike geodesic, parametrized with unit speed;
- the lightlike curves in M are precisely lines with slope $\pi/4$.

(d) Verify that horizontal translations and horizontal reflections are isometries and conclude, in analogy with the Poincaré half-plane, that the geodesics in M are branches of hyperbolas and light rays, while the hyperbola branches are spacelike and crossing the axis $y = 0$ orthogonally.

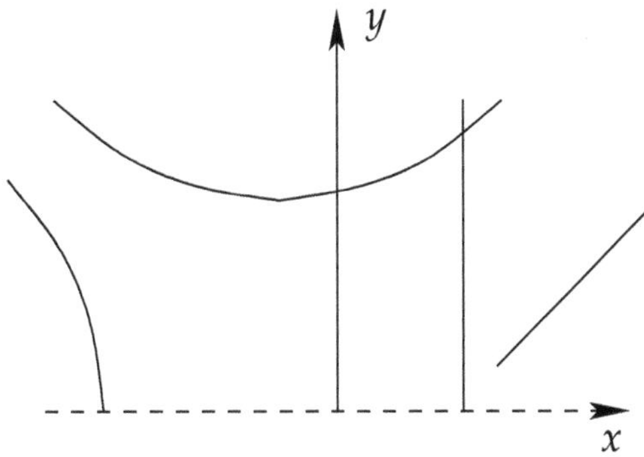

Figure 4.9: Geodesics in the Lorentz-Poincaré half-plane.

(e) In the same fashion that the Poincaré half-plane is related to the hyperbolic plane $\mathbb{H}^2 \subseteq \mathbb{L}^3$, M is related to de Sitter space $\mathbb{S}^2_1$. Consider $\zeta\colon M \to \mathbb{S}^2_1$ given by

$$\zeta(x,y) = \left(-\frac{x}{y}, \frac{1 - x^2 + y^2}{2y}, \frac{1 + x^2 - y^2}{2y}\right).$$

Verify that ζ indeed takes values in $\mathbb{S}^2_1$ and that it is an isometry onto its image (determine it).

Hint. In this case, it is much simpler to just compute the both derivatives $d\zeta_{(x,y)}(1,0) = (\partial\zeta/\partial x)(x,y)$ and $d\zeta_{(x,y)}(0,1) = (\partial\zeta/\partial y)(x,y)$, and verify that

$$\begin{aligned}\langle d\zeta_{(x,y)}(1,0), d\zeta_{(x,y)}(1,0)\rangle_L &= \frac{1}{y^2},\\ \langle d\zeta_{(x,y)}(0,1), d\zeta_{(x,y)}(0,1)\rangle_L &= -\frac{1}{y^2} \text{ and}\\ \langle d\zeta_{(x,y)}(1,0), d\zeta_{(x,y)}(0,1)\rangle_L &= 0.\end{aligned}$$

Remark. Another strategy to find the timelike geodesics is to mimic what was done in Exercise 4.1.12 above, using Theorem 3.6.35 (p. 244). For more details about this metric, see [53].

Exercise† 4.1.18 (Anti-de Sitter)**.** Consider the space $\mathbb{R}^3_2 = (\mathbb{R}^3, \langle\cdot,\cdot\rangle_2)$, with the product $\langle\cdot,\cdot\rangle_2$ is defined by

$$\langle \boldsymbol{v}, \boldsymbol{w}\rangle_2 \doteq v_1w_1 - v_2w_2 - v_3w_3,$$

where $\boldsymbol{v} = (v_1, v_2, v_3)$ and $\boldsymbol{w} = (w_1, w_2, w_3)$. The *anti-de Sitter space* is defined by

$$\mathbb{H}^2_1 \doteq \{\boldsymbol{v} \in \mathbb{R}^3_2 \mid \langle \boldsymbol{v}, \boldsymbol{v}\rangle_2 = -1\}.$$

Show that the restriction of $\langle\cdot,\cdot\rangle_2$ to the tangent planes of $\mathbb{H}^2_1$ defines a Lorentzian metric in $\mathbb{H}^2_1$, whose Gaussian curvature is constant and equal to -1.

Hint. Try a parametrization of revolution around the x-axis.

Remark. In the same way that $\mathbb{S}^2_1$ is a Lorentzian version of $\mathbb{S}^2$, $\mathbb{H}^2_1$ is a Lorentzian version of $\mathbb{H}^2$. As subsets of $\mathbb{R}^3$, the spaces $\mathbb{S}^2_1$ and $\mathbb{H}^2_1$ differ only by a permutation of axes. Such phenomenon is particular to dimension 2, and the higher dimensional versions of the anti-de Sitter have applications in relativity, serving as models for spacetimes which take into account gravitational forces.

Exercise 4.1.19 (What if?)**.** Consider $M = \mathbb{R} \times \mathbb{R}_{>0}$. Instead of searching a Lorentzian version of the Poincaré half-plane metric by dividing the metric of $\mathbb{L}^2$ by a timelike coordinate, like in Exercise 4.1.17, let's see what happens if we divide by a spacelike coordinate instead. Take, in M, the Lorentzian metric given by

$$ds^2 = \frac{-dx^2 + dy^2}{y^2}.$$

By Exercise 4.1.14 above, we know that this metric has constant Gaussian curvature, equal to -1. Moreover, the geodesics are the same as the geodesics for the metric studied in Exercise 4.1.17, but with flipped causal types.

The next question that occurs is if M, equipped with such a metric, is isometric to an open subset of any of our known surfaces. With a Lorentzian metric of constant and negative curvature, we have a natural candidate: the anti-de Sitter space $\mathbb{H}^2_1$. Seems reasonable to expect, then, that the map we look for is similar to the isometry ζ given in Exercise 4.1.17, switching the axes.

Show that $\xi\colon M \to \mathbb{H}^2_1$ given by

$$\xi(x,y) = \left(\frac{1+x^2-y^2}{2y}, \frac{1-x^2+y^2}{2y}, -\frac{x}{y}\right)$$

indeed takes values in $\mathbb{H}^2_1$, and it is an isometry onto its image.

Remark. For more details about this metric, see [47].

Exercise 4.1.20. Let $f\colon U \subseteq \mathbb{C} \to \mathbb{C}$ be a holomorphic function. Writing $z = x + \mathrm{i}y$ and $f(z) = \phi(x,y) + \mathrm{i}\psi(x,y)$, consider the graph of f:

$$M \doteq \{(x, y, \phi(x,y), \psi(x,y)) \in \mathbb{R}^4 \mid (x,y) \in U\}.$$

Let's compute the Gaussian curvature of the metric in M induced by $\langle \cdot, \cdot \rangle_E$. Omit points of evaluation in what follows.

(a) Show that the induced metric is given by

$$\mathrm{d}s^2 = (1 + \phi_x^2 + \phi_y^2)(\mathrm{d}x^2 + \mathrm{d}y^2).$$

(b) If $h \doteq (1 + \phi_x^2 + \phi_y^2)^{-1/2}$, show that

$$h_x = -h^3(\phi_x\phi_{xx} + \phi_y\phi_{yx}), \quad h_y = -h^3(\phi_x\phi_{xy} + \phi_y\phi_{yy}),$$

and that

$$h_x^2 + h_y^2 = h^6(\phi_x^2 + \phi_y^2)(\phi_{xx}^2 + \phi_{xy}^2).$$

(c) Verify that

$$h_{xx} = \frac{3h_x^2}{h} - h^3(\phi_{xx}^2 + \phi_x\phi_{xxx} + \phi_{yx}^2 + \phi_y\phi_{yxx}),$$
$$h_{yy} = \frac{3h_y^2}{h} - h^3(\phi_{xy}^2 + \phi_x\phi_{xyy} + \phi_{yy}^2 + \phi_y\phi_{yyy}),$$

and that

$$h_{xx} + h_{yy} = \frac{3(h_x^2 + h_y^2)}{h} - 2h^3(\phi_{xx}^2 + \phi_{xy}^2).$$

(d) Put all the previous items together and finally conclude that

$$K((x, y, \phi(x,y), \psi(x,y))) = \frac{-2(\phi_{xx}^2 + \phi_{xy}^2)}{(1 + \phi_x^2 + \phi_y^2)^3}.$$

Hint. Use the Cauchy-Riemann equations and recall that if f is holomorphic, ϕ is harmonic.

Remark.

- In the theory of submanifolds of $\mathbb{R}^n$, we again have the notion of mean curvature vector. One may show that every graph of a holomorphic function, as above, has zero mean curvature vector (and thus is a minimal surface in $\mathbb{R}^4$). Compare this with the situation on $\mathbb{R}^3$, where minimal surfaces have non-positive Gaussian curvature in all points.

- There's a Lorentzian analogue of the calculations done in this exercise, considering instead of graphs of holomorphic functions in $\mathbb{R}^4$, graphs of "split-holomorphic" functions (defined in $\mathbb{C}'$) in $\mathbb{L}^4$. We will formally introduce $\mathbb{C}'$ soon.

Exercise 4.1.21. Consider the *Artin space* $\mathbb{R}^4_2 = (\mathbb{R}^4, \langle\cdot,\cdot\rangle_2)$, with

$$\langle \boldsymbol{v}, \boldsymbol{w}\rangle_2 \doteq v_1w_1 + v_2w_2 - v_3w_3 - v_4w_4,$$

where $\boldsymbol{v} = (v_1, v_2, v_3, v_4)$ and $\boldsymbol{w} = (w_1, w_2, w_3, w_4)$. The product $\langle\cdot,\cdot\rangle_2$ is called the *neutral product* in $\mathbb{R}^4$. Consider the tensor product of the usual parametrizations for the right branch of $\mathbb{S}^1_1$, and $\mathbb{S}^1$: $\boldsymbol{x}\colon \mathbb{R}\times]0,2\pi[\to \mathbb{R}^4_2$ given by

$$\boldsymbol{x}(u,v) = (\cosh u \cos v, \cosh u \sin v, \sinh u \cos v, \sinh u \sin v).$$

If M is the image of $\boldsymbol{x}$, show that the restriction of the neutral product to the tangent planes to M defines a Lorentzian metric on M, which makes it isometric to $\mathbb{R}\times]0,2\pi[$, equipped with the usual metric from $\mathbb{L}^2$.

Remark.

- Even though the codomain of $\boldsymbol{x}$ is $\mathbb{R}^4_2$ instead of $\mathbb{R}^3_\nu$, the derivative $D\boldsymbol{x}(u,v)$ still has full rank for all $(u,v) \in \mathbb{R}\times]0,2\pi[$. This is the reason why we still use the terminology of *regular parametrized surfaces* even in this case.

- The Artin space also has applications in Physics, serving as the ambient space for spacetime models other than Lorentz-Minkowski space, for example, the tridimensional version of the *anti-de Sitter* space,

$$\mathbb{H}^3_1 \doteq \{\boldsymbol{v} \in \mathbb{R}^4_2 \mid \langle \boldsymbol{v},\boldsymbol{v}\rangle_2 = -1\}.$$

Exercise 4.1.22. Determine the Gaussian curvature of the metric induced in the surface M of the previous exercise, by the usual scalar product in $\mathbb{R}^4$.

4.2 RIEMANN'S CLASSIFICATION THEOREM

Up until this moment, we have seen some surfaces with constant Gaussian curvature. Namely, we have met:

- the planes $\mathbb{R}^2$ and $\mathbb{L}^2$, with $K = 0$;
- the sphere $\mathbb{S}^2$ and the de Sitter space $\mathbb{S}^2_1$, with $K = 1$;
- the hyperbolic plane $\mathbb{H}^2$ and the anti-de Sitter $\mathbb{H}^2_1$, with $K = -1$.

Our goal here is to show that, locally, every surface with constant K "is" one of those surfaces described above. More precisely, we want to prove the:

Theorem 4.2.1 (Riemann). *Let $(M, \langle\cdot,\cdot\rangle)$ be a geometric surface with constant Gaussian curvature $K \in \{-1,0,1\}$. Then:*

(A) if the metric is Riemannian, every point in M has a neighborhood isometric to an open subset of

(i) $\mathbb{R}^2$, *if* $K = 0$*;*

(ii) $\mathbb{S}^2$, *if* $K = 1$*;*

(iii) $\mathbb{H}^2$, *if* $K = -1$*,*

(B) if the metric is Lorentzian, to an open subset of

(i) $\mathbb{L}^2$, *if* $K = 0$*;*

(ii) $\mathbb{S}^2_1$, *if* $K = 1$*;*

(iii) $\mathbb{H}^2_1$, *if* $K = -1$.

The proof strategy consists of constructing parametrizations for which the metric assumes a simple form. To actually do this, we will use *geodesics*, which are known to be plentiful in any geometric surface. To avoid singularities, we won't consider lightlike geodesics.

Thus, we fix throughout this section a geometric surface $(M, \langle\cdot,\cdot\rangle)$*, with metric tensor of index* $\nu \in \{0,1\}$*, and a unit speed geodesic* $\gamma\colon I \to M$. For each $v \in I$, consider a unit speed geodesic $\gamma_v\colon J_v \to M$, which crosses γ orthogonally at the point $\gamma_v(0) \doteq \gamma(v)$. Setting

$$U \doteq \{(u,v) \in \mathbb{R}^2 \mid v \in I \text{ and } u \in J_v\},$$

define $\boldsymbol{x}\colon U \to \boldsymbol{x}(U) \subseteq M$ by $\boldsymbol{x}(u,v) = \gamma_v(u)$.

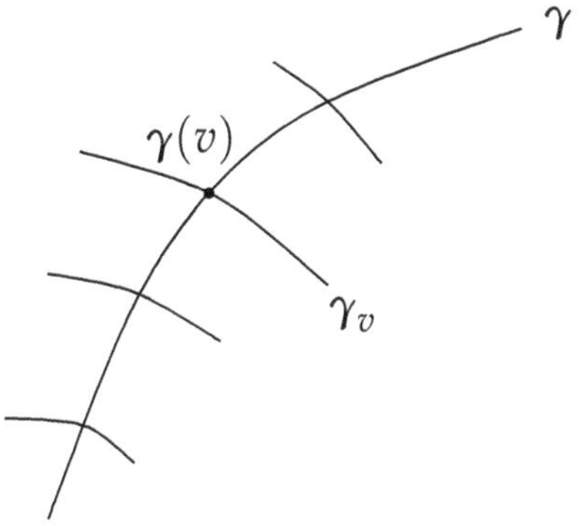

Figure 4.10: Construction of a Fermi chart $\boldsymbol{x}$.

Definition 4.2.2. The above chart $\boldsymbol{x}$ is called a *Fermi chart* for M, centered at γ.

We'll also fix, until the end of the section, this Fermi chart $(U, \boldsymbol{x})$ so constructed.

Remark.

- When $\langle\cdot,\cdot\rangle$ is Lorentzian, we'll have two types of Fermi charts, according to the causal character of γ. Moreover, recalling that geodesics have automatically constant causal character (hence determined by a single velocity vector), it follows that if γ is spacelike (resp. timelike), then all the γ_v are timelike (resp. spacelike), since $\{\gamma'(v), \gamma_v'(0)\}$ is an orthonormal basis of $T_{\gamma(v)}M$, for all $v \in I$.

- When necessary, if γ is timelike, we might denote the coordinates by (τ, ϑ) instead of (u,v).

Proposition 4.2.3. *The Fermi chart* $\boldsymbol{x}$ *is indeed regular in a neighborhood of* $\{0\} \times I$ *(so that reducing* U *if necessary, we may assume that* $(U, \boldsymbol{x})$ *itself is regular).*

Proof: We'll show that for all $v \in I$, the vectors $\boldsymbol{x}_u(0,v)$ and $\boldsymbol{x}_v(0,v)$ are orthogonal. To wit, we have by construction that

$$\langle \boldsymbol{x}_u(0,v), \boldsymbol{x}_v(0,v)\rangle = \langle \gamma_v'(0), \gamma'(v)\rangle = 0.$$

Since none of those vectors is lightlike, orthogonality implies linear independence. By continuity of $\boldsymbol{x}$, the vectors $\boldsymbol{x}_u(u,v)$ and $\boldsymbol{x}_v(u,v)$ remain linearly independent for small enough values of u. □

Proposition 4.2.4. *The coordinate expression of $\langle\cdot,\cdot\rangle$ with respect to the Fermi chart $(U, \boldsymbol{x})$ is*

$$ds^2 = (-1)^\nu \epsilon_\gamma \, du^2 + G(u,v)\, dv^2.$$

Proof: All the γ_v are unit speed curves with the same indicator ϵ_{γ_v}. We have that

$$E(u,v) = \langle \boldsymbol{x}_u(u,v), \boldsymbol{x}_u(u,v)\rangle = \langle \gamma_v'(u), \gamma_v'(u)\rangle = \epsilon_{\gamma_v}.$$

Now, $\epsilon_\gamma \epsilon_{\gamma_v} = (-1)^\nu$ for all $v \in I$, whence $E(u,v) = (-1)^\nu \epsilon_\gamma$.

Proceeding, we see that by construction, $F(0,v) = 0$ for all $v \in I$, so that it suffices to check that F does not depend on the variable u. Fixed $v_0 \in I$, we have the expression $\boldsymbol{x}(u, v_0) = \gamma_{v_0}(u)$, and so the second geodesic equation for γ_{v_0} yields $\Gamma_{11}^2(u,v_0) = 0$. Since v_0 was arbitrary, it follows that $\Gamma_{11}^2 = 0$. On the other hand, by definition of Γ_{11}^2, we have

$$\Gamma_{11}^2(u,v) = \frac{(-1)^\nu \epsilon_\gamma}{(-1)^\nu \epsilon_\gamma G(u,v) - F(u,v)^2} F_u(u,v),$$

so that $F_u(u,v) = 0$, and we conclude that $F(u,v) = 0$ for all $(u,v) \in U$, as desired. □

Remark. Since $G(0,v) = \epsilon_\gamma \neq 0$, the continuity of G allows us to assume, by reducing U again if necessary, that $G(u,v)$ has the same sign as ϵ_γ for all $(u,v) \in U$.

Corollary 4.2.5. *The Gaussian curvature of $(M, \langle\cdot,\cdot\rangle)$ is expressed in terms of the Fermi chart $(U, \boldsymbol{x})$ by*

$$K \circ \boldsymbol{x} = (-1)^{\nu+1} \epsilon_\gamma \frac{(\sqrt{|G|})_{uu}}{\sqrt{|G|}}.$$

Before starting the proof of Theorem 4.2.1 (p. 280), we only need to get one more technical lemma out of the way:

Lemma 4.2.6 (Boundary conditions). *The Fermi chart $(U, \boldsymbol{x})$ satisfies $G_u(0,v) = 0$, for all $v \in I$.*

Proof: As $\gamma(v) = \boldsymbol{x}(0,v)$, the first geodesic equation for γ boils down to $\Gamma_{22}^1(0,v) = 0$, for all $v \in I$. Since $F(0,v) = 0$, it directly follows that

$$\Gamma_{22}^1(0,v) = -\frac{G_u(0,v)}{2\epsilon_\gamma},$$

whence $G_u(0,v) = 0$, as desired. □

Finally:

Proof (of Theorem 4.2.1)**:** In all possible cases, the coefficient G must satisfy the following differential equation:

$$(\sqrt{|G|})_{uu} + (-1)^\nu \epsilon_\gamma K \sqrt{|G|} = 0.$$

Now, we solve this equation (in each case) for $\sqrt{|G|}$, and use the boundary conditions $G(0,v) = \epsilon_\gamma$ and $G_u(0,v) = 0$ to determine G explicitly.

(A) Assume that $\langle\cdot,\cdot\rangle$ is Riemannian.

(i) For $K = 0$, we have $(\sqrt{G})_{uu} = 0$, and so $\sqrt{G(u,v)} = A(v)u + B(v)$. The boundary conditions then give $A(v) = 0$ and $B(v) = 1$, so that $G(u,v) = 1$ for all $(u,v) \in U$, and $\mathrm{d}s^2 = \mathrm{d}u^2 + \mathrm{d}v^2$.

(ii) When $K = 1$, we have $(\sqrt{G})_{uu} + \sqrt{|G|} = 0$, whose solutions are of the form $\sqrt{G(u,v)} = A(v)\cos u + B(v)\sin u$. Now, the boundary conditions give $A(v) = 1$ and $B(v) = 0$, and so $G(u,v) = \cos^2 u$, and it follows that $\mathrm{d}s^2 = \mathrm{d}u^2 + \cos^2 u\, \mathrm{d}v^2$: the metric in $\mathbb{S}^2$.

(iii) If $K = -1$, the equation to be solved is $(\sqrt{G})_{uu} - \sqrt{|G|} = 0$. We have that $\sqrt{G(u,v)} = A(v)\mathrm{e}^u + B(v)\mathrm{e}^{-u}$, and now the boundary conditions give $A(v) = B(v) = 1/2$, whence $G(u,v) = \cosh^2 u$ and we obtain the local expression $\mathrm{d}s^2 = \mathrm{d}u^2 + \cosh^2 u\, \mathrm{d}v^2$. To recognize this in an easier way as the metric in $\mathbb{H}^2$, we may let $x = \mathrm{e}^v \tanh u$ and $y = \mathrm{e}^v \operatorname{sech} u$, so that

$$\mathrm{d}s^2 = \frac{\mathrm{d}x^2 + \mathrm{d}y^2}{y^2},$$

as desired.

(B) Assume now that $\langle\cdot,\cdot\rangle$ is Lorentzian.

(i) For $K = 0$, just like above, we have $\mathrm{d}s^2 = -\mathrm{d}u^2 + \mathrm{d}v^2 = \mathrm{d}\tau^2 - \mathrm{d}\vartheta^2$.

(ii) If $K = 1$, we now have two cases to discuss. If γ is spacelike, we again obtain that $(\sqrt{G})_{uu} - \sqrt{G} = 0$, from where it follows that $G(u,v) = \cosh^2 u$ and we get the $\mathbb{S}^2_1$ metric: $\mathrm{d}s^2 = -\mathrm{d}u^2 + \cosh^2 u\, \mathrm{d}v^2$ (expressed in the usual revolution parametrization). If γ is timelike instead, we have $(\sqrt{-G})_{\tau\tau} + \sqrt{-G} = 0$, whose solution is $G(\tau,\vartheta) = -\cos^2 \tau$, and so $\mathrm{d}s^2 = \mathrm{d}\tau^2 - \cos^2 \tau\, \mathrm{d}\vartheta^2$.

(iii) If $K = -1$, the situation is dual to the previous one, switching "spacelike" and "timelike", and also the signs of the metric expressions. Omitting repeated calculations, we obtain

$$\mathrm{d}s^2 = -\mathrm{d}u^2 + \cos^2 u\, \mathrm{d}v^2 = \mathrm{d}\tau^2 - \cosh^2 \tau\, \mathrm{d}\vartheta^2,$$

which is the metric of $\mathbb{H}^2_1$ in suitable coordinates. □

We will conclude the section by presenting surfaces in the ambient spaces $\mathbb{L}^3$ and $\mathbb{R}^3_2$ whose metric's coordinate expressions are the ones discovered in the proof above. For $K = 0$ the situation is completely uninteresting. But for $K \neq 0$ we have the following:

Example 4.2.7.

(1) $K = 1$:

- The metric $\mathrm{d}s^2 = -\mathrm{d}u^2 + \cosh^2 u\, \mathrm{d}v^2$ may be realized by the usual revolution parametrization $\boldsymbol{x}\colon \mathbb{R}^2 \to \mathbb{S}^2_1 \subseteq \mathbb{L}^3$ given by

$$\boldsymbol{x}(u,v) = (\cosh u \cos v, \cosh u \sin v, \sinh u),$$

and also by $\boldsymbol{y}\colon \cosh^{-1}\left(]1,\sqrt{2}[\right) \times \mathbb{R} \to \mathbb{R}^3_2$ given by

$$\boldsymbol{y}(u,v) = \left(\cosh u \cosh v, \cosh u \sinh v, \int_0^u \sqrt{2 - \cosh^2 t}\, \mathrm{d}t\right).$$

- For $\mathrm{d}s^2 = \mathrm{d}\tau^2 - \cos^2\tau\,\mathrm{d}\vartheta^2$, consider $\boldsymbol{x}\colon\ ]0,2\pi[\times\mathbb{R}\to\mathbb{S}^2_1\subseteq\mathbb{L}^3$ given by

$$\boldsymbol{x}(\tau,\vartheta) = (\sin\tau, \cos\tau\cosh\vartheta, \cos\tau\sinh\vartheta),$$

and also by $\boldsymbol{y}\colon\ ]-\pi/2,\pi/2[\times\mathbb{R}\to\mathbb{R}^3_2$, given by

$$\boldsymbol{y}(\tau,\vartheta) = \left(\int_0^\tau \sqrt{1+\sin^2 t}\,\mathrm{d}t, \cos\tau\cos\vartheta, \cos\tau\sin\vartheta\right).$$

Remark. The periodicity condition $\boldsymbol{y}(\tau,\vartheta) = \boldsymbol{y}(\tau+\pi,\vartheta)$ in the last given parametrization along with the fact that translations are isometries in $\mathbb{R}^3_2$ allow us to restrict everything to the given domains, which is maximal for non-degenerability.

To summarize, when $K = 1$ we have the following visualizations:

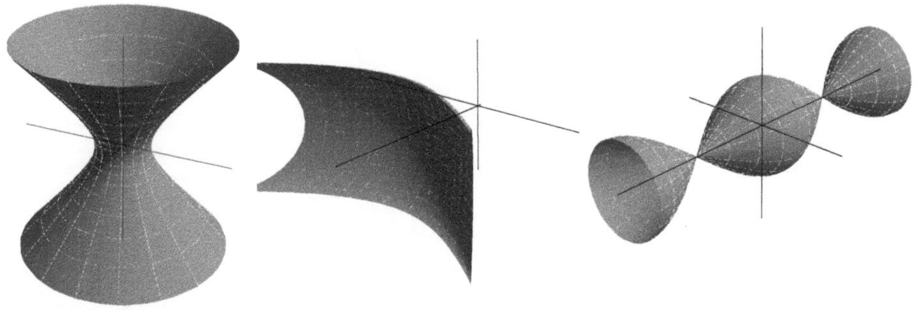

(a) In $\mathbb{L}^3$ (b) In $\mathbb{R}^3_2$

Figure 4.11: Constant Gaussian curvature $K = 1$.

(2) $K = -1$:

- The metric $\mathrm{d}s^2 = -\mathrm{d}u^2 + \cos^2 u\,\mathrm{d}v^2$ may be realized by the parametrization $\boldsymbol{x}\colon\ ]-\pi/2,\pi/2[\times\mathbb{R}\to\mathbb{L}^3$, given by

$$\boldsymbol{x}(u,v) = \left(\cos u\cos v, \cos u\sin v, \int_0^u \sqrt{1+\sin^2 t}\,\mathrm{d}t\right),$$

and also by $\boldsymbol{y}\colon\ ]0,2\pi[\times\mathbb{R}\to\mathbb{H}^2_1\subseteq\mathbb{R}^3_2$:

$$\boldsymbol{y}(u,v) = (\cos u\sinh v, \cos u\cosh v, \sin u).$$

In this case, the same remark made for $\boldsymbol{y}$ in the case $K = 1$ holds for $\boldsymbol{x}$ here.

- The metric $\mathrm{d}s^2 = \mathrm{d}\tau^2 - \cosh^2\tau\,\mathrm{d}\vartheta^2$ may be realized by the parametrization $\boldsymbol{x}\colon \cosh^{-1}(]1,\sqrt{2}[)\times\mathbb{R}\to\mathbb{L}^3$ given by

$$\boldsymbol{x}(\tau,\vartheta) = \left(\int_0^\tau \sqrt{2-\cosh^2 t}\,\mathrm{d}t, \cosh\tau\cosh\vartheta, \cosh\tau\sinh\vartheta\right)$$

and by $\boldsymbol{y}\colon \mathbb{R}\times\,]0,2\pi[\,\to\mathbb{H}^2_1\subseteq\mathbb{R}^3_2$,

$$\boldsymbol{y}(\tau,\vartheta) = (\sinh\tau, \cosh\tau\cos\vartheta, \cosh\tau\sin\vartheta).$$

So in this case, we have:

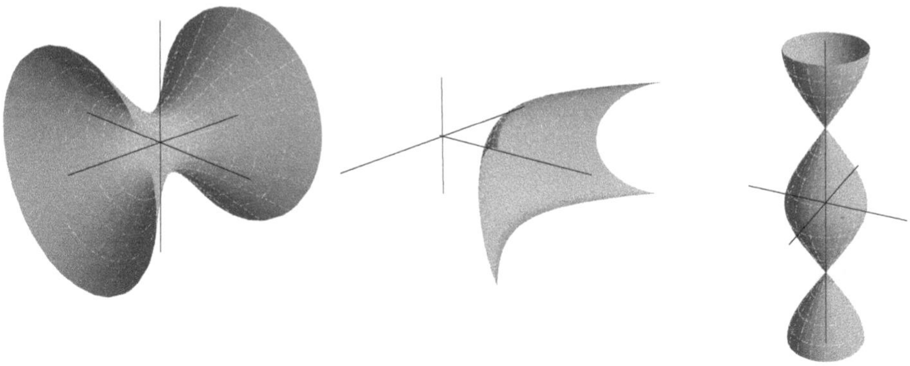

(a) In $\mathbb{R}^3_2$ (b) In $\mathbb{L}^3$

Figure 4.12: Constant Gaussian curvature $K = -1$.

Lastly, we observe that the surfaces in Figures 4.11(a) and 4.12(a) are isometric when equipped with the metrics induced by $\mathbb{R}^3$ but, on the pseudo-Riemannian ambient spaces considered, they have rotational symmetry along axes of distinct causal characters. The same holds for the surfaces given in Figures 4.11(b) and 4.12(b). Furthermore, note that $\mathbb{S}^2_1$ and $\mathbb{H}^2_1$ "fit better" in $\mathbb{L}^3$ and $\mathbb{R}^3_2$, respectively — switching the ambient spaces requires the use of parametrizations depending on certain elliptic integrals.

Exercises

Exercise 4.2.1 (Riemann's Formula). Let $(M, \langle\cdot,\cdot\rangle)$ be a geometric surface equipped with a Riemannian metric, and $(U, \boldsymbol{x})$ be a Fermi chart for M (on which the metric is expressed by $\mathrm{d}s^2 = \mathrm{d}u^2 + G(u,v)\,\mathrm{d}v^2$). In some adequate domain, consider the reparametrization $x = u\cos v$ and $y = u\sin v$. Show that

$$\mathrm{d}s^2 = \mathrm{d}x^2 + \mathrm{d}y^2 + H(x,y)(x\,\mathrm{d}y - y\,\mathrm{d}x)^2,$$

where $H(x,y) = (G(u,v) - u^2)/u^4$.

Hint. Differentiating the three relations $u^2 = x^2 + y^2$, $x = u\cos v$ and $y = u\sin v$, respectively, we obtain that $u\,\mathrm{d}u = x\,\mathrm{d}x + y\,\mathrm{d}y$, $\mathrm{d}x = \cos v\,\mathrm{d}u - u\sin v\,\mathrm{d}v$ and finally $\mathrm{d}y = \sin v\,\mathrm{d}u + u\cos v\,\mathrm{d}v$. Proceed from there.

Remark. For metrics of the above form, with H defined in some open set around the origin (it is not the case here), one can show by using the so-called *Brioschi formula* (see for example [56]), that the Gaussian curvature of the metric in the point of coordinates $(x,y) = (0,0)$ is $-3H(0,0)$. That is, this metric is "Euclidean up to second order, near the origin", and the function H measures this deviation.

Exercise 4.2.2 (Revolution surfaces with constant K). Let $\boldsymbol{\alpha}\colon I \to \mathbb{R}^3_\nu$ be smooth, regular, non-degenerate, injective and of the form $\boldsymbol{\alpha}(u) = (f(u), 0, g(u))$, for certain functions f and g with $f(u) > 0$ for all $u \in I$, and let M be the revolution surface spanned by $\boldsymbol{\alpha}$, around the z-axis. Assume that $\boldsymbol{\alpha}$ has unit speed, M has constant Gaussian curvature K, and consider the parametrization $\boldsymbol{x}\colon I \times]0, 2\pi[\to I \to \boldsymbol{x}(U) \subseteq M$ given by

$$\boldsymbol{x}(u,v) = (f(u)\cos v, f(u)\sin v, g(u)).$$

(a) Show that, in general, f and g satisfy

$$f''(u) + \epsilon_\alpha K f(u) = 0 \quad \text{and} \quad g(u) = \int \sqrt{(-1)^\nu(\epsilon_\alpha - f'(u)^2)}\, du.$$

(b) Verify that

$$f(u) = \begin{cases} A\cos(\sqrt{\epsilon_\alpha K}u) + B\sin(\sqrt{\epsilon_\alpha K}u), & \text{if } \epsilon_\alpha K > 0 \\ Au + B, & \text{if } K = 0, \\ A\cosh(\sqrt{-\epsilon_\alpha K}u) + B\sinh(\sqrt{-\epsilon_\alpha K}u) & \text{if } \epsilon_\alpha K < 0, \end{cases}$$

where, in the case $K = 0$, we necessarily have $|A| \leq 1$ if the ambient space is $\mathbb{R}^3$, while $|A| \geq 1$ if the curve is spacelike in $\mathbb{L}^3$ (for timelike curves there are no restrictions).

(c) Identify all the revolution surfaces with constant Gaussian curvature $K \in \{-1, 0, 1\}$.

Hint. Draw sketches of the generating curve on the plane $y = 0$. For the discussion in $\mathbb{R}^3$, see [39].

Exercise 4.2.3 (Hyperbolic revolution surfaces with constant K)**.** Investigate the hyperbolic revolution surfaces in $\mathbb{L}^3$ with constant Gaussian curvature K, following the idea of the previous exercise.

4.3 SPLIT-COMPLEX NUMBERS AND CRITICAL SURFACES

In this section we will briefly present the strong relation between complex variables and spacelike surfaces in $\mathbb{R}^3_\nu$. For timelike surfaces in $\mathbb{L}^3$, we will need the set of *split-complex numbers.* Thus, we will start the discussion formalizing this concept and presenting basic facts about Calculus in a single split-complex variable. Next, we will apply this to study Bonnet rotations and the Enneper-Weierstrass representation formulas.

4.3.1 A brief introduction to split-complex numbers

Let's recall one possible construction of the complex numbers: define in $\mathbb{R}^2$ the operations

$$(a,b) + (c,d) \doteq (a+c, b+d) \text{ and}$$
$$(a,b)(c,d) \doteq (ac - bd, ad + bc).$$

Such operations turn $\mathbb{R}^2$ into a *field*, which is then denoted by $\mathbb{C}$. Since we have that $(a,b) = (a,0) + (b,0)(0,1)$ and $(0,1)^2 = (-1,0)$, we may identify $\mathbb{R}$ with the set $\{(a,0) \in \mathbb{R}^2 \mid a \in \mathbb{R}\}$, write $\mathrm{i} \doteq (0,1)$, to finally recover the usual description

$$\mathbb{C} = \{a + b\mathrm{i} \mid a, b \in \mathbb{R} \text{ and } \mathrm{i}^2 = -1\}.$$

Given $z = a + b\mathrm{i} \in \mathbb{C}$, the projections $\mathrm{Re}(z) \doteq a$ and $\mathrm{Im}(z) \doteq b$ are called the *real and imaginary parts of* z. The *conjugate* of z is defined as $\bar{z} \doteq a - b\mathrm{i}$, and the *absolute value* of z as $|z| \doteq \sqrt{a^2 + b^2} = \|(a,b)\|_E$. Moreover, if $z_1 = a_1 + b_1\mathrm{i}$ and $z_2 = a_2 + b_2\mathrm{i}$ are two complex numbers, we have that

$$\mathrm{Re}(z_1\overline{z_2}) = \langle (a_1, b_1), (a_2, b_2) \rangle_E,$$

which says that $\mathbb{C}$ captures the geometry of the usual inner product in $\mathbb{R}^2$. With this in place, one proceeds to develop the theory of Calculus in a complex variable, with which we will assume some familiarity (we recommend [40] and [61], for good measure).

Our goals, then, are to construct a Lorentzian version of $\mathbb{C}$, based in the above brief review; and to understand the basics about how Calculus works in this new setting.

Definition 4.3.1 (Split-complex numbers). The set of *split-complex numbers*, denoted by $\mathbb{C}'$, is the space $\mathbb{L}^2$ equipped with the operations

$$\begin{aligned}(a,b)+(c,d) &\doteq (a+c,b+d) \text{ and}\\ (a,b)(c,d) &\doteq (ac+bd,ad+bc).\end{aligned}$$

Remark. The split-complex numbers are also known as *hyperbolic numbers*. For a reason for this terminology, see Exercise 4.3.6.

The verification that such operations turn $\mathbb{C}'$ into a commutative ring with unit is straightforward. Since we have that $(a,b)=(a,0)+(b,0)(0,1)$ and also $(0,1)^2=(1,0)$, we may again identify $\mathbb{R}$ with $\{(a,0)\in\mathbb{L}^2 \mid a\in\mathbb{R}\}$ and write $\mathrm{h}\doteq(0,1)$ to obtain a description similar to the one previously given for $\mathbb{C}$:

$$\mathbb{C}'=\{a+b\mathrm{h} \mid a,b\in\mathbb{R} \text{ and } \mathrm{h}^2=1\}.$$

Definition 4.3.2. Let $w=a+b\mathrm{h}\in\mathbb{C}'$.

(i) The *split-conjugate* of w is given by $\overline{w}\doteq a-b\mathrm{h}$.

(ii) The *split-complex absolute value* of w is given by $|w|\doteq\sqrt{|a^2-b^2|}=\|(a,b)\|_L$.

(iii) The *real part* of w is given by $\mathrm{Re}(w)\doteq a$, and its *imaginary part* by $\mathrm{Im}(w)\doteq b$.

Remark. Rigorously, the symbol $|\cdot|$ appears in (ii) with two different meanings, but the context should prevent any risk of confusion.

Let's register a few basic properties of split-complex conjugation and absolute value:

Proposition 4.3.3. *Let $w,w_1,w_2\in\mathbb{C}'$.*

(i) $\overline{w_1+w_2}=\overline{w_1}+\overline{w_2}$, $\overline{w_1w_2}=\overline{w_1}\,\overline{w_2}$, $\overline{\overline{w}}=w$, and $\overline{w}=w$ if and only if $w\in\mathbb{R}$. In other words, conjugation in $\mathbb{C}'$ is still an involutive automorphism which preserves $\mathbb{R}$;

(ii) if the inverse $1/w$ exists, then $\overline{1/w}=1/\overline{w}$;

(iii) $|w|=|\overline{w}|$, $|w\overline{w}|=|w|^2$;

(iv) $|w_1w_2|=|w_1||w_2|$ and, if the inverse $1/w$ exists, $|1/w|=1/|w|$. In particular, if the inverse $1/w$ exists, we necessarily have $|w|\neq 0$.

Remark. The proof is an instructive warm-up, and we ask you to do it in Exercise 4.3.1.

The strength of $\mathbb{C}'$ resides in the fact that if $w_1=a_1+b_1\mathrm{h}$ and $w_2=a_2+b_2\mathrm{h}$ are two split-complex numbers, it holds that

$$\mathrm{Re}(w_1\overline{w_2})=\langle(a_1,b_1),(a_2,b_2)\rangle_L,$$

so that $\mathbb{C}'$ captures the geometry of $\mathbb{L}^2$ in the same fashion that $\mathbb{C}$ captures the geometry of $\mathbb{R}^2$. This observation gives us a geometric intuition for the fact that $\mathbb{C}'$ is not a field, while $\mathbb{C}$ is: zero divisors in $\mathbb{C}'$ correspond precisely to lightlike directions in $\mathbb{L}^2$ (we ask that you verify this in Exercise 4.3.3).

To proceed, we take in $\mathbb{C}'$ the usual topology of $\mathbb{C}$. That is, open subsets of $\mathbb{C}'$ are the same ones as in $\mathbb{C}$, and continuous functions remain the same. In particular, if $U \subseteq \mathbb{C}'$ is open and $f\colon U \to \mathbb{C}'$ is written in the form

$$f(x+\mathrm{h}y) = \phi(x,y) + \mathrm{h}\psi(x,y)$$

for certain $\phi, \psi\colon U \to \mathbb{R}$, then f is continuous if and only if both ϕ and ψ are. In this aspect, there is nothing new. But to define holomorphicity in $\mathbb{C}'$, we'll again mimic the definition adopted in $\mathbb{C}$, being careful to avoid taking the limit along light rays in the definition of derivative:

Definition 4.3.4. Let $U \subseteq \mathbb{C}'$ be an open subset, $w_0 \in U$, and $f\colon U \to \mathbb{C}'$ be a function. We'll say that f is $\mathbb{C}'$-differentiable at w_0 if the limit

$$f'(w_0) \doteq \lim_{\substack{\Delta w \to 0 \\ \Delta w \notin C_L(\mathbf{0})}} \frac{f(w_0+\Delta w) - f(w_0)}{\Delta w}$$

exists. This being the case, $f'(w_0)$ is called the *derivative* of f at w_0. Moreover, f is called *split-holomorphic* at w_0 if it is $\mathbb{C}'$-differentiable at all points in some neighborhood of w_0.

The usual Calculus rules hold, with the same proofs (which are then omitted):

Proposition 4.3.5. *Let $U \subseteq \mathbb{C}'$ be an open subset, $w_0 \in U$, and $f, g\colon U \to \mathbb{C}'$ be two $\mathbb{C}'$-differentiable functions at w_0. Then:*

(i) $f+g$ is $\mathbb{C}'$-differentiable at w_0 and $(f+g)'(w_0) = f'(w_0) + g'(w_0)$;

(ii) fg is $\mathbb{C}'$-differentiable at w_0 and $(fg)'(w_0) = g(w_0)f'(w_0) + f(w_0)g'(w_0)$;

(iii) if g never takes values in the light rays, then f/g is $\mathbb{C}'$-differentiable at w_0 and $(f/g)'(w_0) = (f'(w_0)g(w_0) - f(w_0)g'(w_0))/g(w_0)^2$.

Example 4.3.6. It is easy to see that the identity mapping $\mathrm{id}_{\mathbb{C}'}\colon \mathbb{C}' \to \mathbb{C}'$ and constant maps are split-holomorphic, with constant derivatives equal to 1 and 0, respectively. Hence, the above result says that all polynomials in the variable w are split-holomorphic, as well as all rational functions (quotients of polynomials, when the denominator does not take values in the light rays), with derivatives given by the usual formulas.

Proposition 4.3.7 (Chain rule). *Let $U_1, U_2 \subseteq \mathbb{C}'$ be open, and also $f\colon U_1 \to \mathbb{C}'$, $g\colon U_2 \to \mathbb{C}'$ be functions with $f(U_1) \subseteq U_2$. If f is $\mathbb{C}'$-differentiable at w_0 and g is $\mathbb{C}'$-differentiable at $f(w_0)$, then $g \circ f$ is $\mathbb{C}'$-differentiable at w_0 and we have the relation $(g \circ f)'(w_0) = g'(f(w_0))f'(w_0)$.*

In the usual theory of Calculus in a single complex variable, we know that the real and imaginary parts of a holomorphic function must satisfy the Cauchy-Riemann equations. Let's see what these equations become for split-holomorphic functions:

Proposition 4.3.8 (Revised Cauchy-Riemann). *Let $U \subseteq \mathbb{C}'$ be an open subset and $w_0 \in U$ be a point. If $f\colon U \to \mathbb{C}'$ is $\mathbb{C}'$-differentiable at w_0, and we write the function as $f(x+\mathrm{h}y) = \phi(x,y) + \mathrm{h}\psi(x,y)$, then we have that*

$$\frac{\partial \phi}{\partial x}(w_0) = \frac{\partial \psi}{\partial y}(w_0) \quad \textit{and} \quad \frac{\partial \phi}{\partial y}(w_0) = \frac{\partial \psi}{\partial x}(w_0).$$

Proof: Writing $\Delta w = \Delta w_1 + \mathrm{h}\Delta w_2$, we have that $\Delta w \to 0$ if and only if $\Delta w_1 \to 0$ and $\Delta w_2 \to 0$. We know that the limit in the definition of the derivative of f at w_0 exists, and so we may compute it by using any path away from the light rays in $\mathbb{L}^2$. In particular, we may consider the paths $\Delta w_1 = 0$ and $\Delta w_2 = 0$. On one hand, we have that

$$\begin{aligned} f'(w_0) &= \lim_{\Delta w_1 \to 0} \frac{\phi(w_0 + \Delta w_1) - \phi(w_0) + \mathrm{h}(\psi(w_0 + \Delta w_1) - \psi(w_0))}{\Delta w_1} \\ &= \frac{\partial \phi}{\partial x}(w_0) + \mathrm{h}\frac{\partial \psi}{\partial x}(w_0), \end{aligned}$$

and on the other hand that

$$\begin{aligned} f'(w_0) &= \lim_{\Delta w_2 \to 0} \frac{\phi(w_0 + \Delta w_2) - \phi(w_0) + \mathrm{h}(\psi(w_0 + \Delta w_2) - \psi(w_0))}{\mathrm{h}\Delta w_2} \\ &= \mathrm{h}\frac{\partial \phi}{\partial y}(w_0) + \frac{\partial \psi}{\partial y}(w_0). \end{aligned}$$

Equating real and imaginary parts, the result follows. □

Remark. The revised Cauchy-Riemann equations may be expressed in a shorter way by introducing split-complex versions of the *Wirtinger operators*:

$$\frac{\partial}{\partial w} \doteq \frac{1}{2}\left(\frac{\partial}{\partial x} + \mathrm{h}\frac{\partial}{\partial y}\right) \quad \text{and} \quad \frac{\partial}{\partial \overline{w}} \doteq \frac{1}{2}\left(\frac{\partial}{\partial x} - \mathrm{h}\frac{\partial}{\partial y}\right).$$

This way, in analogy with the complex case, the revised Cauchy-Riemann equations become just

$$\frac{\partial f}{\partial \overline{w}} = 0$$

and, being satisfied, imply that

$$f'(w) = \frac{\partial f}{\partial w}(w).$$

Proposition 4.3.9. *Let $U \subseteq \mathbb{C}'$ be an open subset and $\phi, \psi \colon U \to \mathbb{R}$ be two real-differentiable functions, whose partial derivatives satisfy the revised Cauchy-Riemann equations on U. Then the function $f \colon \mathbb{C}' \to \mathbb{C}'$ defined by $f \doteq \phi + \mathrm{h}\psi$ is split-holomorphic.*

These ideas may be interpreted in a convenient way in terms of the total derivative of f, seen as a map on $\mathbb{R}^2$, see Exercise 4.3.7.

Example 4.3.10. Motivated by Euler's formula

$$\mathrm{e}^{x+\mathrm{i}y} = \mathrm{e}^x(\cos y + \mathrm{i}\sin y)$$

in $\mathbb{C}$, we define $\exp_{\mathbb{C}'} \colon \mathbb{C}' \to \mathbb{C}'$ by

$$\exp_{\mathbb{C}'}(w) = \mathrm{e}^x(\cosh y + \mathrm{h}\sinh y),$$

where $w = x + \mathrm{h}y$. It follows from the previous results that $\exp_{\mathbb{C}'}$ is split-holomorphic, with derivative given by $(\exp_{\mathbb{C}'})' = \exp_{\mathbb{C}'}$. When there's no risk of confusion, we denote the split-complex exponential of w simply by e^w.

An important consequence of the revised Cauchy-Riemann equations is the:

Corollary 4.3.11. *Let $U \subseteq \mathbb{C}'$ be an open subset and $f\colon U \to \mathbb{C}'$ be split-holomorphic. If $f = \phi + \mathrm{h}\psi$, then ϕ and ψ are solutions of the wave equation: $\Box\phi = \Box\psi = 0$. In these conditions, we say that ϕ and ψ are* Lorentz-harmonic.

Remark. In terms of the differential operators defined above, we may write the d'Alembertian as

$$\Box = 4\frac{\partial}{\partial \overline{w}}\frac{\partial}{\partial w}.$$

Here we have the first crucial difference between holomorphicity and split-holomorphicity: in contrast to what happens with the heat equation, we can explicitly solve the wave equation in suitable domains, by employing a convenient change of variables. However, we will adopt a slightly simpler strategy to classify split-holomorphic functions on convex domains:

Definition 4.3.12. The *Hadamard product* in $\mathbb{R}^2$ is the coordinatewise multiplication $\star\colon \mathbb{R}^2 \times \mathbb{R}^2 \to \mathbb{R}^2$ defined by

$$(u_1, v_1) \star (u_2, v_2) \doteq (u_1u_2, v_1v_2).$$

Proposition 4.3.13. *The map $\sigma\colon \mathbb{R}^2 \to \mathbb{C}'$ defined by*

$$\sigma(u, v) \doteq u\left(\frac{1+\mathrm{h}}{2}\right) + v\left(\frac{1-\mathrm{h}}{2}\right) = \frac{u+v}{2} + h\frac{u-v}{2}$$

is an isomorphism of $\mathbb{R}$-algebras *when $\mathbb{R}^2$ is equipped with the Hadamard product, that is, it satisfies:*

(i) σ is bijective;

(ii) σ is $\mathbb{R}$-linear;

(iii) $\sigma((u_1, v_1) \star (u_2, v_2)) = \sigma(u_1, v_1)\sigma(u_2, v_2)$.

Moreover, its inverse is given by

$$\sigma^{-1}(x + \mathrm{h}y) = (x + y, x - y).$$

Proof: Let's verify only item (iii). Put $\ell \doteq (1 + \mathrm{h})/2$, so that σ may be written as $\sigma(u, v) = u\ell + v\overline{\ell}$. Noting that $\ell\overline{\ell} = 0$ and $\ell^2 = \ell$, we have:

$$\begin{aligned}
\sigma(u_1, v_1)\sigma(u_2, v_2) &= (u_1\ell + v_1\overline{\ell})(u_2\ell + v_2\overline{\ell}) \\
&= u_1u_2\ell^2 + u_1v_2\ell\overline{\ell} + v_1u_2\overline{\ell}\ell + v_1v_2\overline{\ell}^2 \\
&= u_1u_2\ell + v_1v_2\overline{\ell} \\
&= \sigma(u_1u_2, v_1v_2) \\
&= \sigma((u_1, v_1) \star (u_2, v_2)).
\end{aligned}$$

□

The idea is to "transfer" split-holomorphic functions from $\mathbb{C}'$ to $\mathbb{R}^2$ via σ^{-1}, where their treatment becomes simpler. Let's start with a "static" version of what we intend to do later:

Lemma 4.3.14. *Let $A \in \mathrm{Mat}(2, \mathbb{R})$ be any matrix. So, denoting by* can *both the standard basis of $\mathbb{R}^2$ and also the basis $(1, h)$ of $\mathbb{C}'$, we have that A is of the form*

$$\begin{pmatrix} a & b \\ b & a \end{pmatrix},$$

for certain $a, b \in \mathbb{R}$, if and only if $[\sigma]_{\mathrm{can}}^{-1} A [\sigma]_{\mathrm{can}}$ is diagonal.

Proof: Suppose that A has the form given in the statement above. We have that:

$$\begin{aligned}
[\sigma]_{\mathrm{can}}^{-1} A [\sigma]_{\mathrm{can}} &= \begin{pmatrix} 1 & 1 \\ 1 & -1 \end{pmatrix} \begin{pmatrix} a & b \\ b & a \end{pmatrix} \begin{pmatrix} 1/2 & 1/2 \\ 1/2 & -1/2 \end{pmatrix} \\
&= \begin{pmatrix} a+b & b+a \\ a-b & b-a \end{pmatrix} \begin{pmatrix} 1/2 & 1/2 \\ 1/2 & -1/2 \end{pmatrix} \\
&= \begin{pmatrix} a+b & 0 \\ 0 & a-b \end{pmatrix}.
\end{aligned}$$

Since the map $(a, b) \mapsto (a+b, a-b)$ is bijective, this directly implies that if $[\sigma]_{\mathrm{can}}^{-1} A [\sigma]_{\mathrm{can}} = \mathrm{diag}(c, d)$, then

$$A = \frac{1}{2} \begin{pmatrix} c+d & c-d \\ c-d & c+d \end{pmatrix},$$

as wanted. □

Proposition 4.3.9 above and the characterization given in Exercise 4.3.7 then give us the:

Proposition 4.3.15. *Let $U \subseteq \mathbb{C}'$ be an open subset and $f \colon U \to \mathbb{C}'$ be a function of real-differentiable real and imaginary parts. If*

$$\widetilde{f} \doteq \sigma^{-1} \circ f \circ \sigma \colon \sigma^{-1}(U) \to \mathbb{R}^2,$$

we have that f is split-holomorphic if and only if the Jacobian matrix of $\widetilde{f}$ is diagonal in all points.

As a consequence of this result, we have the:

Theorem 4.3.16. *Let $U \subseteq \mathbb{C}'$ be a connected open set. Then, if $f \colon U \to \mathbb{C}'$ is split-holomorphic, there are differentiable functions of a single real variable, $\widetilde{\phi}$ and $\widetilde{\psi}$, such that*

$$f(x + \mathrm{h}y) = \frac{\widetilde{\phi}(x+y) + \widetilde{\psi}(x-y)}{2} + \mathrm{h}\left(\frac{\widetilde{\phi}(x+y) - \widetilde{\psi}(x-y)}{2}\right),$$

for all $x + \mathrm{h}y \in U$.

Proof: We'll maintain the notation adopted in the statement of the previous proposition. As σ is linear and U is convex, we have that $\sigma^{-1}(U)$ is also convex. So, f being split-holomorphic implies that the Jacobian matrix of $\widetilde{f}$ is diagonal, and $\sigma^{-1}(U)$ being convex now implies that $\widetilde{f}$ has the form $\widetilde{f}(u, v) = (\widetilde{\phi}(u), \widetilde{\psi}(v))$ for certain differentiable functions of a single real variable, $\widetilde{\phi}$ and $\widetilde{\psi}$. With this:

$$\begin{aligned}
f(x + \mathrm{h}y) &= \sigma(\widetilde{f}(\sigma^{-1}(x + \mathrm{h}y))) \\
&= \sigma(\widetilde{f}(x+y, x-y)) \\
&= \sigma(\widetilde{\phi}(x+y), \widetilde{\psi}(x-y)) \\
&= \frac{\widetilde{\phi}(x+y) + \widetilde{\psi}(x-y)}{2} + \mathrm{h}\left(\frac{\widetilde{\phi}(x+y) - \widetilde{\psi}(x-y)}{2}\right),
\end{aligned}$$

as wanted. □

Remark. In particular, note that we may choose $\widetilde{\phi}$ and $\widetilde{\psi}$ to be differentiable, but not twice differentiable. This means that, in contrast to what happens with holomorphic functions in $\mathbb{C}$, split-holomorphic functions are not necessarily of class $\mathscr{C}^\infty$ (not even $\mathscr{C}^2$).

To conclude our discussion about differentiation in $\mathbb{C}'$, we'll introduce the notions of pole and split-meromorphic function, needed for what will be done in Subsection 4.3.3.

Definition 4.3.17. Let $U \subseteq \mathbb{C}'$ be open, $w_0 \in U$ and $f\colon U \setminus \{w_0\} \to \mathbb{C}'$. We say that w_0 is an *order* $k \geq 1$ *pole* of f if k is the *least* integer for which $(w - w_0)^k f(w)$ is split-holomorphic.

Definition 4.3.18. Let $U \subseteq \mathbb{C}'$ be open and $P \subseteq U$ discrete. We say that a split-holomorphic function $f\colon U \setminus P \to \mathbb{C}'$ is *split-meromorphic on* U if P is precisely the set of poles of f.

Let's proceed with integration:

Definition 4.3.19. Let $U \subseteq \mathbb{C}'$ be an open subset, $f\colon U \to \mathbb{C}'$ be a continuous function, and $\gamma\colon I \to U$ be a smooth curve. The *integral of* f *along* γ is defined as

$$\int_\gamma f(w)\,\mathrm{d}w \doteq \int_I f(\gamma(t))\gamma'(t)\,\mathrm{d}t.$$

Remark.

- The expression $f(\gamma(t))\gamma'(t)$ is a product of two split-complex numbers.
- Suppose that $f = \phi + \mathrm{h}\psi$. Noting that $w = x + \mathrm{h}y$ implies that $\mathrm{d}w = \mathrm{d}x + \mathrm{h}\,\mathrm{d}y$, and that if $\gamma(t) = x(t) + \mathrm{h}y(t)$ then $\mathrm{d}x = x'(t)\,\mathrm{d}t$ and $\mathrm{d}y = y'(t)\,\mathrm{d}t$ along γ, we have that

$$\int_\gamma f(w)\,\mathrm{d}w \doteq \int_\gamma \phi(x,y)\,\mathrm{d}x + \psi(x,y)\,\mathrm{d}y + \mathrm{h}\int_\gamma \psi(x,y)\,\mathrm{d}x + \phi(x,y)\,\mathrm{d}y,$$

  where the integrals on the right side are usual real line integrals.
- When γ is closed, we denote the integral by $\oint_\gamma f(w)\,\mathrm{d}w$, as usual.
- This definition is naturally extended to the case where γ is piecewise smooth.

It follows directly from the definition that this integral operator is $\mathbb{R}$-linear over functions. And more importantly, we have the:

Theorem 4.3.20 (Fundamental Theorem of Calculus). *Let $U \subseteq \mathbb{C}'$ be open, $f\colon U \to \mathbb{C}'$ be continuous, and $\gamma\colon [a,b] \to U$ be a piecewise $\mathscr{C}^1$ curve. If $F\colon U \to \mathbb{C}'$ is an* anti-derivative *for f (that is, F is split-holomorphic and satisfies $F' = f$), then*

$$\int_\gamma f(w)\,\mathrm{d}w = F(\gamma(b)) - F(\gamma(a)).$$

Proof: Suppose without loss of generality that γ is smooth. By the chain rule, we have that $F \circ \gamma$ is real-differentiable, and an anti-derivative for $(f \circ \gamma)\gamma'$, so that the Fundamental Theorem of Calculus for real functions gives us that

$$\int_\gamma f(w)\,\mathrm{d}w = \int_a^b f(\gamma(t))\gamma'(t)\,\mathrm{d}t = F(\gamma(b)) - F(\gamma(a)),$$

as wanted. □

For what's next, we'll need integrals of split-holomorphic functions along curves, under some conditions, to depend only on the endpoints of the curve, just like what we had in $\mathbb{C}$. This follows from the:

Theorem 4.3.21 (Revised Cauchy-Goursat). *Let $U \subseteq \mathbb{C}'$ be a simply connected open set, $f\colon U \to \mathbb{C}'$ be a split-holomorphic function with continuous derivative, and $\gamma\colon [a,b] \to U$ be a piecewise $\mathscr{C}^1$ curve, closed and injective on $]a,b[$. Then*

$$\oint_\gamma f(w)\,\mathrm{d}w = 0.$$

Proof: The strategy is to apply the Green-Stokes Theorem twice, together with the revised Cauchy-Riemann equations. Let R be the interior of the region in the plane bounded by γ. Writing $f = \phi + \mathrm{h}\psi$, we have:

$$\begin{aligned}\oint_\gamma f(w)\,\mathrm{d}w &= \int_\gamma \phi(x,y)\,\mathrm{d}x + \psi(x,y)\,\mathrm{d}y + \mathrm{h}\int_\gamma \psi(x,y)\,\mathrm{d}x + \phi(x,y)\,\mathrm{d}y \\ &= \int_R \left(\frac{\partial\psi}{\partial x}(x,y) - \frac{\partial\phi}{\partial y}(x,y)\right)\mathrm{d}x\,\mathrm{d}y + \mathrm{h}\int_R \left(\frac{\partial\phi}{\partial x}(x,y) - \frac{\partial\psi}{\partial y}(x,y)\right)\mathrm{d}x\,\mathrm{d}y \\ &= 0 + \mathrm{h}0 = 0.\end{aligned}$$

□

Remark. The assumptions over γ and U allow us to apply Green-Stokes. If U is not simply connected, the region R bounded by γ may contain points outside U (think of a punctured disk), and we need f split-holomorphic on all of R to conclude the argument (this is not possible if f is not defined on all of R, to begin with).

Corollary 4.3.22. *The line integral of a split-holomorphic function, in the conditions of the previous theorem, depends only on the endpoints of the curve, and not on the curve itself. In this case, we write*

$$\int_\gamma f(\omega)\,\mathrm{d}\omega = \int_{w_0}^{w} f(\omega)\,\mathrm{d}\omega,$$

where γ joins w_0 to w.

Corollary 4.3.23. *Let $U \subseteq \mathbb{C}'$ be a simply connected open set, $w_0 \in U$ and $f\colon U \to \mathbb{C}'$ be a continuous function. Then $F\colon U \to \mathbb{C}'$ given by*

$$F(w) = \int_{w_0}^{w} f(\omega)\,\mathrm{d}\omega$$

is split-holomorphic and satisfies $F' = f$.

Similar to the complex case, we have the:

Definition 4.3.24. Two functions $\phi, \psi\colon U \subseteq \mathbb{R}^2 \to \mathbb{R}$ (of class $\mathscr{C}^2$) are called *Lorentz-conjugates* if $\phi_u = \psi_v$ and $\phi_v = \psi_u$. Such condition implies that both ϕ and ψ are Lorentz-harmonic.

Theorem 4.3.25. *Let $U \subseteq \mathbb{R}^2 \equiv \mathbb{C}'$ be a simply connected open set, and $\phi\colon U \to \mathbb{R}$ be a $\mathscr{C}^2$ Lorentz-harmonic function. Then there exists a split-holomorphic function $f\colon U \to \mathbb{C}'$ whose real part is ϕ. In particular, there is a function Lorentz-conjugate to ϕ.*

Proof: Define $g \doteq \phi_u + \mathrm{h}\phi_v$. The condition $\Box\phi = 0$ ensures that g is split-holomorphic and, in particular, continuous, so that U being simply connected gives us the existence of an anti-derivative $G = \psi + \mathrm{h}\zeta$ for g. With this in place, using that $G' = g$ with the revised Cauchy-Riemann equations for G yield that

$$\phi_u + \mathrm{h}\phi_v = \psi_u + \mathrm{h}\zeta_u = \psi_u + \mathrm{h}\psi_v.$$

Equating real and imaginary parts, we obtain that $\psi = \phi + c$ for some $c \in \mathbb{R}$. Thus, $f \doteq G - c$ is the split-holomorphic function we seek. □

The next natural step is to look for, in $\mathbb{C}'$, an analogue for the Cauchy integral formula, which has as a consequence the fact that to know the value of a holomorphic function in some point, it suffices to know the value of this function in some circle centered at this point. Thinking in Euclidean terms, we have the:

Example 4.3.26. Let $a \in \mathbb{C}'$, $a \neq 0$, and consider $f\colon \mathbb{C}' \to \mathbb{C}'$ given by $f(w) = w^2 + a$. For $\gamma\colon [0, 2\pi[\to \mathbb{C}'$ given by $\gamma(t) = \cos t + \mathrm{h}\sin t$, one may check (by doing a long calculation with improper integrals, since w takes values in the intersection of γ with the two principal light rays four times) that

$$a = f(0) \neq \frac{1}{2\pi\mathrm{h}} \oint_\gamma \frac{f(w)}{w - 0}\, \mathrm{d}w = 0.$$

For more details, see [35].

Thinking in Lorentzian terms in $\mathbb{C}'$, the "circles" are no longer compact and connected, so that the relevant integrals again become improper. In this case, we do not have a direct analogue for the Cauchy integral formula. The best we get, aiming towards the above conclusion, is the:

Proposition 4.3.27. *Let $U \subseteq \mathbb{C}'$ be a connected open set and $f\colon U \to \mathbb{C}'$ be a split-holomorphic function. Put $\ell = (1 + \mathrm{h})/2$. So, given $s, t \in \mathbb{R}$, for every $w \in U$ such that $w + s\bar{\ell}, w + t\ell \in U$ we have that*

$$f(w) = \bar{\ell} f(w + t\ell) + \ell f(w + s\bar{\ell}).$$

Proof: Fix an arbitrary $w \in U$ and consider $F(s,t) = \bar{\ell} f(w + t\ell) + \ell f(w + s\bar{\ell})$. We have that $F(0,0) = f(w)$, and since $\ell\bar{\ell} = 0$, it follows that

$$\frac{\partial F}{\partial s} = \frac{\partial F}{\partial t} = 0.$$

Thus F is constant and equals $f(w)$. □

For more general facts about split-complex numbers we recommend, for example, [4] and [11].

Exercises

Exercise† 4.3.1.

(a) Show Proposition 4.3.3 (p. 287);

(b) Show that the split-complex absolute value $|\cdot|\colon \mathbb{C}' \to \mathbb{R}$ does *not* satisfy the properties of a norm.

Hint. Think about how $|\cdot|$ is related to the "norm" $\|\cdot\|_L$.

Exercise 4.3.2. If $w = x + \mathrm{h}y \in \mathbb{C}'$ and we assume that all the necessary inverses below exist, express in terms of x and y the real and imaginary parts of:

(a) w^2;

(b) $\dfrac{w-1}{w+1}$;

(c) $1/w^2$.

Exercise[†] 4.3.3. Show that the zero divisors in $\mathbb{C}'$ are precisely the real multiples of $1+\mathrm{h}$ and $1-\mathrm{h}$. That is, zero divisors in $\mathbb{C}'$ correspond to lightlike directions in $\mathbb{L}^2$.

Exercise 4.3.4 (More avatars of $\mathbb{C}'$).

(a) Show that the map $\Phi\colon \mathbb{C}' \to \mathrm{Mat}(2,\mathbb{R})$ given by

$$\Phi(x+\mathrm{h}y) = \begin{pmatrix} x & y \\ y & x \end{pmatrix}$$

is a ring monomorphism. Moreover, note that given any $w \in \mathbb{C}'$, we have that $w\overline{w} = \det \Phi(w)$. This way, $\mathbb{C}'$ may be seen as a collection of matrices.

(b) Consider the *evaluation at* h, $\mathrm{eval}_{\mathrm{h}}\colon \mathbb{R}[x] \to \mathbb{C}'$, given by

$$\mathrm{eval}_{\mathrm{h}}(p(x)) \doteq p(\mathrm{h}).$$

Show that $\mathrm{eval}_{\mathrm{h}}$ is a ring epimorphism, with $\ker \mathrm{eval}_{\mathrm{h}} = (x^2-1)$. Conclude that $\mathbb{C}' \cong \mathbb{R}[x]/(x^2-1)$.

Hint. Use the division algorithm in $\mathbb{R}[x]$.

Exercise 4.3.5. Let $T\colon \mathbb{R}^2 \to \mathbb{C}'$ be a $\mathbb{R}$-linear map, written as

$$T(x,y) = (ax+by) + \mathrm{h}(cx+dy),$$

for certain $a,b,c,d \in \mathbb{R}$. Identifying $(x,y) \equiv w = x + \mathrm{h}y$, show that there are $\alpha,\beta \in \mathbb{C}'$ such that $T(w) = \alpha w + \beta\overline{w}$ and $\alpha\overline{\alpha} - \beta\overline{\beta} = ad - bc$, so that T is an isomorphism if and only if $\alpha\overline{\alpha} \neq \beta\overline{\beta}$.

Exercise 4.3.6 (Generalized complex numbers). Let u be any symbol, and consider the real commutative algebra generated by $\{1,\mathrm{u}\}$, subject to the relation $\mathrm{u}^2 = \alpha + \beta\mathrm{u}$, for certain *structure constants* $\alpha,\beta \in \mathbb{R}$. Denote such algebra by $\mathbb{C}_{\alpha,\beta}$, that is:

$$\mathbb{C}_{\alpha,\beta} = \{a + \mathrm{u}b \mid a,b \in \mathbb{R} \text{ and } \mathrm{u}^2 = \alpha + \beta\mathrm{u}\}.$$

In particular, note that $\mathbb{C}_{-1,0} = \mathbb{C}$ and $\mathbb{C}_{1,0} = \mathbb{C}'$.

(a) Show that an element $a + \mathrm{u}b \in \mathbb{C}_{\alpha,\beta}$ has a multiplicative inverse if and only if $D \doteq a^2 + \beta ab - \alpha b^2 \neq 0$.

(b) Let $a + ub \in \mathbb{C}_{\alpha,\beta}$ be nonzero. If $b \neq 0$, then $a + ub$ has an inverse. In this case, note that $D/b^2 = 0$ may be seen as a quadratic equation in the variable a/b. Verify that the discriminant of such equation is $\Delta \doteq \beta^2 + 4\alpha$.

(c) The position of a point (α, β) in the parameter plane, relative to the parabola $\Delta = 0$, determines the possibility or not of performing divisions in $\mathbb{C}_{\alpha,\beta}$. More precisely, show that:

- if $\Delta < 0$, every nonzero element of $\mathbb{C}_{\alpha,\beta}$ has an inverse;
- if $\Delta = 0$, the zero divisors are the elements $a + ub$ with $a + \beta b/2 = 0$, and all the remaining ones have inverses;
- if $\Delta > 0$, the zero divisors are the elements $a + ub$ with $a + (\beta + \sqrt{\Delta})b/2 = 0$ or $a + (\beta - \sqrt{\Delta})b/2 = 0$, and all the remaining ones have inverses.

(d) According to the three cases listed in the item above, we'll say that $\mathbb{C}_{\alpha,\beta}$ is, respectively, a system of elliptic, parabolic, or hyperbolic numbers. Justify this terminology by describing, in terms of Δ, the conic $x^2 + \beta xy - \alpha y^2 = 0$ in the plane.

Remark.

- Item (a) motivates us to define $\overline{a + ub} \doteq a + \beta b - ub$, thus making D a "squared norm" of $a + ub$. The behavior of this map $D\colon \mathbb{C}_{\alpha,\beta} \to \mathbb{R}$ ends up being controlled by the discriminant Δ given in item (b). It follows from Sylvester's Criterion that D is positive-definite when $\Delta < 0$, degenerate for $\Delta = 0$, and indefinite for $\Delta > 0$. Polarizing D, we have that $\mathbb{C}_{\alpha,\beta}$ is an algebraic model for the geometry of the symmetric bilinear form

$$\langle (a,b), (c,d) \rangle_{\alpha,\beta} \doteq ac + \frac{\beta}{2}ad + \frac{\beta}{2}bc - \alpha bd$$

in $\mathbb{R}^2$.

- Note that item (c) generalizes Exercise 4.3.3 above.
- For $\alpha = \beta = 0$, the set $\mathbb{C}_{0,0} = \{a + b\varepsilon \mid a, b \in \mathbb{R} \text{ and } \varepsilon^2 = 0\}$ is called the set of *dual numbers*, and it also has applications in several areas of Mathematics and Physics. One may show that according to the discriminant Δ, $\mathbb{C}_{\alpha,\beta}$ is isomorphic to either $\mathbb{C}$, $\mathbb{C}'$ or $\mathbb{C}_{0,0}$.

Exercise† 4.3.7.

(a) Let $U \subseteq \mathbb{C}'$ be an open subset, $w_0 = x_0 + \mathrm{h}y_0 \in U$ and $f\colon U \to \mathbb{C}'$ be any function. Show that f is $\mathbb{C}'$-differentiable at w_0 if and only if it is differentiable at (x_0, y_0) when seen as a map from $\mathbb{L}^2$ to $\mathbb{L}^2$ and the total derivative $Df(x_0, y_0)$ is $\mathbb{C}'$-*linear*.

Hint. Saying that $Df(x_0, y_0)$ is $\mathbb{C}'$-linear is equivalent (why?) to saying that

$$Df(x_0, y_0)\begin{pmatrix} 0 & 1 \\ 1 & 0 \end{pmatrix} = \begin{pmatrix} 0 & 1 \\ 1 & 0 \end{pmatrix} Df(x_0, y_0).$$

(b) Show Proposition 4.3.7 (p. 288).

Exercise† 4.3.8. Let $U \subseteq \mathbb{C}'$ be a connected open set and $f\colon U \to \mathbb{C}'$ be split-holomorphic. Show that if $f'(w) = 0$ for all $w \in U$, then f is constant.

Exercise 4.3.9. Let $U \subseteq \mathbb{C}'$ be an open and connected set and $f\colon U \to \mathbb{C}'$ be split-holomorphic.

(a) Show that if f only assumes real values or only purely imaginary values, then f is constant.

(b) Show that if the function $g\colon U \to \mathbb{C}'$ given by $g(w) = f(w)\overline{f(w)}$ is a non-zero constant, then f is constant. Give a counter-example where $g = 0$ and f is non-constant.

Exercise 4.3.10. Let $U \subseteq \mathbb{C}'$ be an open and connected set, and $f\colon U \to \mathbb{C}'$ be split-holomorphic, written as $f = \phi + \mathrm{h}\psi$. If $\xi\colon U \to \mathbb{R}$, show that $\phi + \mathrm{h}\xi$ is split-holomorphic if and only if $\psi - \xi$ is constant.

Exercise 4.3.11. Let $f\colon \mathbb{C}' \to \mathbb{C}'$ be split-holomorphic. Define a function $g\colon \mathbb{C}' \to \mathbb{C}'$ by $g(w) = \overline{f(\overline{w})}$. Show that g is also split-holomorphic, and that its derivative is given by $g'(w) = \overline{f'(\overline{w})}$.

Exercise[†] 4.3.12. Show that:

(a) given an open set $U \subseteq \mathbb{C}'$ and a function $f\colon U \to \mathbb{C}'$, f satisfies the revised Cauchy-Riemann equations if and only if $\dfrac{\partial f}{\partial \overline{w}} = 0$;

(b) if f is split-holomorphic, then $f'(w) = \dfrac{\partial f}{\partial w}(w)$, for each $w \in U$;

(c) $\Box = 4\dfrac{\partial}{\partial \overline{w}}\dfrac{\partial}{\partial w}$.

Exercise 4.3.13. *Liouville's Theorem* claims that every bounded holomorphic function defined on the whole complex plane $\mathbb{C}$ is necessarily constant. In $\mathbb{C}'$, this theorem is lost. Show that $f\colon \mathbb{C}' \to \mathbb{C}'$ given by

$$f(x + \mathrm{h}y) = \frac{1 + \mathrm{h}}{1 + \mathrm{e}^{-x}\mathrm{e}^{-y}}$$

is a counter-example.

Exercise 4.3.14. Let $U \subseteq \mathbb{C}'$ be an open set, $f\colon U \to \mathbb{C}'$ be a continuous map, and $\gamma\colon [a, b] \to U$ be a $\mathscr{C}^1$ curve.

(a) Suppose that the image of f is contained in a single light ray passing through the origin. Show that

$$\left| \int_\gamma f(w)\,\mathrm{d}w \right| = 0.$$

(b) Assume that γ parametrizes a portion of a light ray (possibly affine). Show that

$$\left| \int_\gamma f(w)\,\mathrm{d}w \right| = 0.$$

4.3.2 Bonnet rotations

We have previously seen that some parametrizations of a non-degenerate regular surface may emphasize some of its geometric aspects better than others. For critical surfaces, it is convenient to work with isothermal parametrizations:

Definition 4.3.28. Let $M \subseteq \mathbb{R}^3_\nu$ be a non-degenerate regular surface and $(U, \boldsymbol{x})$ be a parametrization for M. We say that $\boldsymbol{x}$ is *isothermal* if $F = 0$ and $|E| = |G| = \lambda^2$, for some smooth function $\lambda\colon U \to \mathbb{R}$.

Remark.

- We are free to assume, when necessary, that $\boldsymbol{x}_u$ is always spacelike. No generality is lost, since if $M \subseteq \mathbb{L}^3$ and $\boldsymbol{x}_u$ are timelike, we consider the reflected open set $U' = \{(u,v) \in U \mid (v,u) \in U\}$ and the reparametrization $(U', \boldsymbol{y})$ given by $\boldsymbol{y}(u,v) = \boldsymbol{x}(v,u)$, whose image is the same as the image of $\boldsymbol{x}$, now with $\boldsymbol{y}_u$ spacelike.
- Under these conditions, the First Fundamental Form of M is expressed in terms of $\boldsymbol{x}$ by $\mathrm{d}s^2 = \lambda(u,v)^2(\mathrm{d}u^2 + (-1)^\nu\,\mathrm{d}v^2)$.

The existence of such parametrizations is guaranteed by the:

Theorem 4.3.29 (Korn-Lichtenstein). *Let $M \subseteq \mathbb{R}^3_\nu$ be a non-degenerate regular surface. Then, for each point $\boldsymbol{p} \in M$ there is an open set $U \subseteq \mathbb{R}^2$ and an isothermal parametrization $(U, \boldsymbol{x})$ for M around $\boldsymbol{p}$.*

The proof of this result is outside the scope of this text, but it may be found in [3]. Naturally, the first step in what follows is to find expressions for the mean curvature in terms of such parametrizations:

Proposition 4.3.30. *If $M \subseteq \mathbb{R}^3_\nu$ is a non-degenerate regular surface and $(U, \boldsymbol{x})$ is an isothermal parametrization for M, then*

$$H \circ \boldsymbol{x} = (-1)^\nu \frac{\epsilon_v e + \epsilon_u g}{2\lambda^2},$$

where e and g denote the coefficients of the Second Fundamental Form of $\boldsymbol{x}$ (computed relative to a Gauss map compatible with $\boldsymbol{x}$), and ϵ_u and ϵ_v are its partial indicators (recall Definition 3.2.2, p. 156).

Proof: Using the expression for $H \circ \boldsymbol{x}$ given in Proposition 3.3.10 (p. 182), we have that

$$H \circ \boldsymbol{x} = \frac{\epsilon_M}{2} \frac{eG - 2fF + Eg}{EG - F^2} = \frac{\epsilon_M}{2} \frac{e\epsilon_v \lambda^2 + g\epsilon_u \lambda^2}{\epsilon_u \epsilon_v \lambda^4} = (-1)^\nu \frac{\epsilon_v e + \epsilon_u g}{2\lambda^2}.$$

□

Remark. The expressions for the Gaussian curvature have already been registered in Exercise 3.7.3 (p. 256).

The operators $\triangle$ and $\square$ also act on vector-valued functions in a natural way: if $\Phi\colon U \subseteq \mathbb{R}^2 \to \mathbb{R}^3$ is differentiable, the *Laplacian* of Φ is defined by $\triangle\Phi \doteq \Phi_{uu} + \Phi_{vv}$, and the *d'Alembertian* of Φ is defined by $\square\Phi \doteq \Phi_{uu} - \Phi_{vv}$. We say that Φ is *harmonic* if $\triangle\Phi = \mathbf{0}$ and *Lorentz-harmonic* if $\square\Phi = \mathbf{0}$. Moreover, if $\Psi\colon U \subseteq \mathbb{R}^2 \to \mathbb{R}^3$ is another differentiable function, we say that Ψ is conjugate or Lorentz-conjugate to Φ if its components are (see Definition 4.3.24, p. 293). These operators have a close relation with the mean curvature, given in the:

Proposition 4.3.31. *Let $M \subseteq \mathbb{R}^3_\nu$ be a non-degenerate regular surface and $(U, \boldsymbol{x})$ be an isothermal parametrization for M. We have that:*

(i) if M is spacelike, then $\triangle \boldsymbol{x} = 2\lambda^2 \boldsymbol{H}$;

(ii) if M is timelike, then $\square \boldsymbol{x} = 2\lambda^2 \epsilon_u \boldsymbol{H}$,

where we abbreviate the mean curvature vector by $\boldsymbol{H} \equiv \boldsymbol{H} \circ \boldsymbol{x}$.

Proof: Suppose initially that M is spacelike and fix a Gauss map $\boldsymbol{N}$ compatible with $\boldsymbol{x}$. We have that

$$\begin{aligned}\langle \triangle \boldsymbol{x}, \boldsymbol{x}_u \rangle &= \langle \boldsymbol{x}_{uu}, \boldsymbol{x}_u \rangle + \langle \boldsymbol{x}_{vv}, \boldsymbol{x}_u \rangle \\ &= \langle \boldsymbol{x}_{uu}, \boldsymbol{x}_u \rangle - \langle \boldsymbol{x}_v, \boldsymbol{x}_{uv} \rangle \\ &= \frac{1}{2}\frac{\partial}{\partial u}(\langle \boldsymbol{x}_u, \boldsymbol{x}_u \rangle - \langle \boldsymbol{x}_v, \boldsymbol{x}_v \rangle) = 0\end{aligned}$$

and, similarly, $\langle \triangle \boldsymbol{x}, \boldsymbol{x}_v \rangle = 0$. This way, we have that $\triangle \boldsymbol{x}$ is normal to M and, also identifying $\boldsymbol{N} \equiv \boldsymbol{N} \circ \boldsymbol{x}$, we have that $\triangle \boldsymbol{x} = (-1)^\nu \langle \triangle \boldsymbol{x}, \boldsymbol{N} \rangle \boldsymbol{N}$. But:

$$\langle \triangle \boldsymbol{x}, \boldsymbol{N} \rangle = \langle \boldsymbol{x}_{uu}, \boldsymbol{N} \rangle + \langle \boldsymbol{x}_{vv}, \boldsymbol{N} \rangle = e + g = 2(-1)^\nu \lambda^2 H,$$

and the result follows.

Suppose now that M is timelike. Similarly as done above, one may show that $\langle \square \boldsymbol{x}, \boldsymbol{x}_u \rangle_L = \langle \square \boldsymbol{x}, \boldsymbol{x}_v \rangle_L = 0$. Since $\boldsymbol{N}$ is spacelike, we have that $\square \boldsymbol{x} = \langle \square \boldsymbol{x}, \boldsymbol{N} \rangle_L \boldsymbol{N}$ and, moreover, $\epsilon_v = -\epsilon_u$. Thus

$$\langle \square \boldsymbol{x}, \boldsymbol{N} \rangle_L = \langle \boldsymbol{x}_{uu}, \boldsymbol{N} \rangle_L - \langle \boldsymbol{x}_{vv}, \boldsymbol{N} \rangle_L = e - g = 2\lambda^2 \epsilon_u H.$$

□

Corollary 4.3.32. *Let $M \subseteq \mathbb{R}^3_\nu$ be a non-degenerate regular surface and $(U, \boldsymbol{x})$ be an isothermal parametrization for M. Then $\boldsymbol{x}(U)$ has zero mean curvature if and only if $\boldsymbol{x}$ is harmonic or Lorentz-harmonic, depending on whether M is spacelike or timelike.*

Definition 4.3.33. Given parametrized regular surfaces $\boldsymbol{x}, \boldsymbol{y} \colon U \to \mathbb{R}^3_\nu$, we define the 1-parameter families of maps $\boldsymbol{z}^t, \boldsymbol{w}^t : U \to \mathbb{R}^3_\nu$, with $t \in \mathbb{R}$, by:

(i) $\boldsymbol{z}^t \doteq (\cos t)\boldsymbol{x} + (\sin t)\boldsymbol{y}$, if $\boldsymbol{x}$ and $\boldsymbol{y}$ are spacelike and conjugate.

(ii) $\boldsymbol{w}^t \doteq (\cosh t)\boldsymbol{x} + (\sinh t)\boldsymbol{y}$, if $\boldsymbol{x}$ and $\boldsymbol{y}$ are timelike and Lorentz-conjugate.

We say that $\boldsymbol{z}^t$ and $\boldsymbol{w}^t$ are the Euclidean and Lorentzian *Bonnet rotations*, respectively, associated to $\boldsymbol{x}$ and $\boldsymbol{y}$.

Remark. Since the functions cosh and sinh have no real period and cosh never vanishes, the maps $\boldsymbol{x}$ and $\boldsymbol{y}$ play different roles in the definition of the Lorentzian Bonnet rotation $\boldsymbol{w}^t$, while for $\boldsymbol{z}^t$, their order is irrelevant. Let's start looking for properties shared by both situations, despite this loss of periodicity in the timelike case. *In the following proofs, until the end of this subsection, we will work under the assumption that $\boldsymbol{x}_u$ is always spacelike.*

Theorem 4.3.34. *Let $\boldsymbol{x}, \boldsymbol{y} \colon U \to \mathbb{R}^3_\nu$ be two parametrized regular critical surfaces, conjugate and isothermal. The following objects are invariant under a Bonnet rotation associated to $\boldsymbol{x}$ and $\boldsymbol{y}$:*

(i) the coefficients of the First Fundamental Form (and, thus, all the obtained surfaces are also isothermal and pairwise isometric);

(ii) the direction of the Gauss maps $\mathbf{N}^t$ *(and, thus, all the tangent planes for fixed* $(u, v) \in U$ *are parallel).*

Furthermore, all the obtained surfaces are critical.

Proof:

(i) Let's verify the result for Euclidean Bonnet rotations: both $\boldsymbol{x}$ and $\boldsymbol{y}$ have their First Fundamental Form coefficients given by $E = G = \lambda^2$ and $F = 0$ (i.e., the same parameter λ for both). With this:

$$\boldsymbol{z}_u^t = (\cos t)\boldsymbol{x}_u + (\sin t)\boldsymbol{y}_u \quad \text{and} \quad \boldsymbol{z}_v^t = (\cos t)\boldsymbol{x}_v + (\sin t)\boldsymbol{y}_v.$$

Rewriting this only in terms of derivatives of $\boldsymbol{x}$, we get that:

$$\boldsymbol{z}_u^t = (\cos t)\boldsymbol{x}_u - (\sin t)\boldsymbol{x}_v \quad \text{and} \quad \boldsymbol{z}_v^t = (\sin t)\boldsymbol{x}_u + (\cos t)\boldsymbol{x}_v.$$

It follows from this that if E^t, F^t and G^t denote the coefficients of the First Fundamental Form of $\boldsymbol{z}^t$, then

$$E^t = G^t = \lambda^2 \quad \text{and} \quad F^t = 0,$$

as wanted, by using that $F = 0$ and $\cos^2 t + \sin^2 t = 1$. The Lorentzian case is analogous, using $\cosh^2 t - \sinh^2 t = 1$ instead.

(ii) It suffices to show that $\boldsymbol{z}_u^t \times \boldsymbol{z}_v^t$ is always proportional to $\boldsymbol{x}_u \times \boldsymbol{x}_v$, for every t. To wit, they're actually equal. In the Euclidean case, we have:

$$\begin{aligned}\boldsymbol{z}_u^t \times \boldsymbol{z}_v^t &= ((\cos t)\boldsymbol{x}_u - (\sin t)\boldsymbol{x}_v) \times ((\sin t)\boldsymbol{x}_u + (\cos t)\boldsymbol{x}_v) \\ &= (\cos^2 t)\boldsymbol{x}_u \times \boldsymbol{x}_v - (\sin^2 t)\boldsymbol{x}_v \times \boldsymbol{x}_u \\ &= \boldsymbol{x}_u \times \boldsymbol{x}_v.\end{aligned}$$

The other case, again, is similar.

Lastly, we have that:

$$\boldsymbol{z}_{uu}^t = (\cos t)\boldsymbol{x}_{uu} - (\sin t)\boldsymbol{x}_{vu}, \quad \boldsymbol{z}_{vv}^t = (\sin t)\boldsymbol{x}_{uv} + (\cos t)\boldsymbol{x}_{vv},$$

and adding we obtain that $\triangle \boldsymbol{z}^t = (\cos t)\triangle \boldsymbol{x} = \boldsymbol{0}$, since $\boldsymbol{x}$ is critical. In the Lorentzian case we'll have $\Box \boldsymbol{w}^t = (\cosh t)\Box \boldsymbol{x} = \boldsymbol{0}$ instead. $\Box$

Remark. In the Euclidean case, in the same way that $\boldsymbol{x}$ and $\boldsymbol{y}$ are conjugate, so are $\boldsymbol{z}^t$ and $\boldsymbol{z}^{t+\pi/2}$, for all $t \in \mathbb{R}$.

Corollary 4.3.35. *In the above conditions, the Weingarten maps of all the* $\boldsymbol{w}^t$ *will be diagonalizable if the Weingarten map of* $\boldsymbol{x}$ *is.*

Example 4.3.36.

(1) Isometric deformation from the catenoid to the helicoid: consider the catenoid parametrized by $\boldsymbol{x}\colon \mathbb{R}\times\,]0,2\pi[\,\to \mathbb{R}^3$, given by

$$\boldsymbol{x}(u,v) = (\cosh u \cos v, \cosh u \sin v, u).$$

We have that

$$\boldsymbol{x}_u(u,v) = (\sinh u \cos v, \sinh u \sin v, 1) \quad \text{and}$$
$$\boldsymbol{x}_v(u,v) = (-\cosh u \sin v, \cosh u \cos v, 0),$$

whence $\mathrm{d}s^2 = (\cosh^2 u)(\mathrm{d}u^2 + \mathrm{d}v^2)$. Moreover, $\triangle \boldsymbol{x} = \boldsymbol{0}$. Let's find a surface $\boldsymbol{y}$ conjugate to $\boldsymbol{x}$. Integrating $\boldsymbol{x}_u$ with respect to v, we have

$$\boldsymbol{y}(u,v) = (\sinh u \sin v, -\sinh u \cos v, v) + \boldsymbol{c}(u),$$

and differentiating with respect to u, we must have $\boldsymbol{c}'(u) = 0$. This shows that $\boldsymbol{c}$ is a constant, which may take to be $\boldsymbol{0}$. Therefore a conjugate surface to $\boldsymbol{x}$ is given by

$$\boldsymbol{y}(u,v) = (\sinh u \sin v, -\sinh u \cos v, v),$$

which also satisfies $\triangle \boldsymbol{y} = \boldsymbol{0}$ and has the First Fundamental Form expressed by $\mathrm{d}s^2 = (\cosh^2 u)(\mathrm{d}u^2 + \mathrm{d}v^2)$. We have a family of minimal and pairwise isometric surfaces $\boldsymbol{z}^t$, $t \in \mathbb{R}$. Let's illustrate this Bonnet rotation:

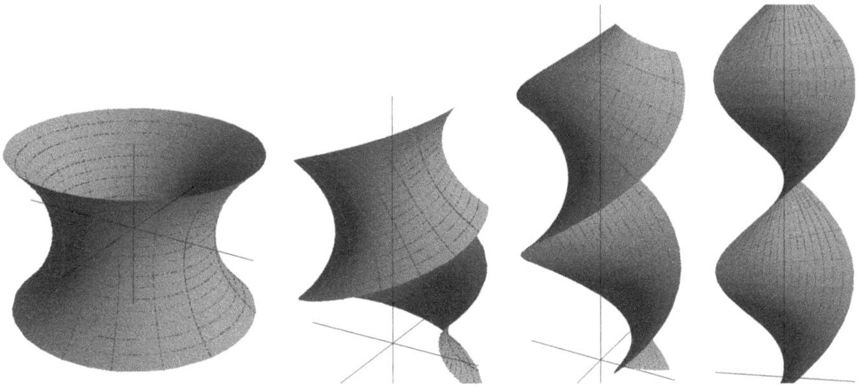

(a) $t = 0$ (b) $t = \pi/6$ (c) $t = \pi/3$ (d) $t = \pi/2$

Figure 4.13: Minimal isometric deformation from the catenoid to the helicoid.

(2) Bonnet rotation for the Lorentzian (hyperbolic catenoid): consider the parametrization $\boldsymbol{x}\colon \mathbb{R}^2 \to \mathbb{L}^3$, given by

$$\boldsymbol{x}(u,v) = (u, \cosh u \cosh v, \cosh u \sinh v).$$

We have that

$$\boldsymbol{x}_u(u,v) = (1, \sinh u \cosh v, \sinh u \sinh v) \quad \text{and}$$
$$\boldsymbol{x}_v(u,v) = (0, \cosh u \sinh v, \cosh u \cosh v),$$

whence $\mathrm{d}s^2 = (\cosh^2 u)(\mathrm{d}u^2 - \mathrm{d}v^2)$. Moreover, we have $\Box \boldsymbol{x} = \boldsymbol{0}$. Let's find a surface $\boldsymbol{y}$ Lorentz-conjugate to $\boldsymbol{x}$. Integrating $\boldsymbol{x}_u$ with respect to v, we have

$$\boldsymbol{y}(u,v) = (v, \sinh u \sinh v, \sinh u \cosh v) + \boldsymbol{c}(u),$$

and differentiating this with respect to u we get that $\boldsymbol{c}'(u) = 0$, whence $\boldsymbol{c}$ is a constant, which we'll take to be zero. Thus, we obtain

$$\boldsymbol{y}(u,v) = (v, \sinh u \sinh v, \sinh u \cosh v).$$

We have that $\boldsymbol{y}$ also satisfies $\Box \boldsymbol{y} = \boldsymbol{0}$ and has its Minkowski First Fundamental Form expressed by $\mathrm{d}s^2 = (\cosh^2 u)(\mathrm{d}u^2 - \mathrm{d}v^2)$. So we have a family $\boldsymbol{w}^t$ of critical and pairwise isometric surfaces, illustrated as follows:

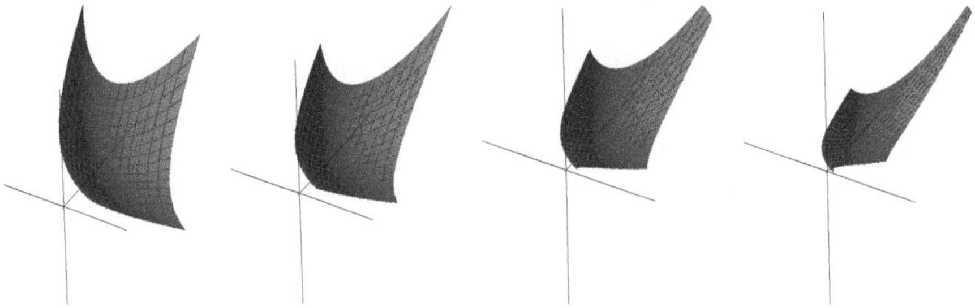

(a) $t = 0$ (b) $t = 1/4$ (c) $t = 1/2$ (d) $t = 3/4$

Figure 4.14: Minimal isometric deformation of a Lorentzian catenoid.

Now, let's see the relation between the Second Fundamental Forms of the surfaces obtained via a Bonnet rotation and the Second Fundamental Form of the initial surfaces:

Lemma 4.3.37. *Let $\boldsymbol{x}, \boldsymbol{y} \colon U \to \mathbb{R}^3_\nu$ be two parametrized regular surfaces, conjugate and isothermal. Denote by e and f the coefficients of the Second Fundamental Forms of $\boldsymbol{x}$ and $\boldsymbol{y}$, and by e^t and f^t the coefficients of the Second Fundamental Form of the surfaces obtained via the Bonnet rotation associated to $\boldsymbol{x}$ and $\boldsymbol{y}$.*

(i) In the Euclidean case, we have that

$$e^t = e \cos t - f \sin t \quad \text{and} \quad f^t = e \sin t + f \cos t.$$

In particular, $e^{t+\pi/2} = -f^t$ and $f^{t+\pi/2} = e^t$.

(ii) In the Lorentzian case, we have that

$$e^t = e \cosh t + f \sinh t \quad \text{and} \quad f^t = e \sinh t + f \cosh t.$$

Remark. Do not confuse e^t with e^t (an exponential).

Proof: Let's verify the Lorentzian case this time. Denoting the common normal direction to all the $\boldsymbol{w}^t$ by $\boldsymbol{N} \equiv \boldsymbol{N} \circ \boldsymbol{w}^t$, we have

$$\begin{aligned} e^t &= \langle \boldsymbol{w}^t_{uu}, \boldsymbol{N} \rangle_L \\ &= \langle (\cosh t)\boldsymbol{x}_{uu} + (\sinh t)\boldsymbol{x}_{uv}, \boldsymbol{N} \rangle_L \\ &= \cosh t \langle \boldsymbol{x}_{uu}, \boldsymbol{N} \rangle_L + \sinh t \langle \boldsymbol{x}_{uv}, \boldsymbol{N} \rangle_L \\ &= e \cosh t + f \sinh t, \end{aligned}$$

and also

$$\begin{aligned} f^t &= \langle \boldsymbol{w}^t_{uv}, \boldsymbol{N}\rangle_L \\ &= \langle (\cosh t)\boldsymbol{x}_{uv} + (\sinh t)\boldsymbol{x}_{vv}, \boldsymbol{N}\rangle_L \\ &= \cosh t \langle \boldsymbol{x}_{uv}, \boldsymbol{N}\rangle_L + \sinh t \langle \boldsymbol{x}_{vv}, \boldsymbol{N}\rangle_L \\ &= f\cosh t + g \sinh t \\ &= e \sinh t + f \cosh t, \end{aligned}$$

by using that $\boldsymbol{x}$ is critical and thus $g = e$. □

The periodicity stated above for the Euclidean case has the following result as a consequence:

Theorem 4.3.38 (Special curves). *Let $\boldsymbol{x}, \boldsymbol{y}\colon U \to \mathbb{R}^3_\nu$ be two parametrized regular spacelike surfaces, conjugate and isothermal, and $\boldsymbol{z}^t$ be the Euclidean Bonnet rotation associated to $\boldsymbol{x}$ and $\boldsymbol{y}$. For each $t \in \mathbb{R}$, consider the curve $\boldsymbol{\alpha}^t\colon I \to \boldsymbol{z}^t(U) \subseteq \mathbb{R}^3_\nu$ given in coordinates by*

$$\boldsymbol{\alpha}^t(s) = \boldsymbol{z}^t(u(s), v(s)).$$

Then:

(i) $\boldsymbol{\alpha}^t$ is an asymptotic line for $\boldsymbol{z}^t$ if and only if $\boldsymbol{\alpha}^{t+\pi/2}$ is a curvature line for $\boldsymbol{z}^{t+\pi/2}$.

(ii) $\boldsymbol{\alpha}^t$ is a curvature line for $\boldsymbol{z}^t$ if and only if $\boldsymbol{\alpha}^{t+\pi/2}$ is an asymptotic line for $\boldsymbol{z}^{t+\pi/2}$.

Proof: We will use Propositions 3.5.6 (p. 210) and 3.5.9 (p. 212):

(i) It suffices to note that, avoiding points of evaluation, we have:

$$\begin{vmatrix} v'^2 & -u'v' & u'^2 \\ \lambda^2 & 0 & \lambda^2 \\ e^{t+\pi/2} & f^{t+\pi/2} & -e^{t+\pi/2} \end{vmatrix} = \lambda^2(e^t u'^2 + 2f^t u'v' - e^t v'^2).$$

(ii) As in the previous item, the conclusion follows from the following identity:

$$\begin{vmatrix} v'^2 & -u'v' & u'^2 \\ \lambda^2 & 0 & \lambda^2 \\ e^t & f^t & -e^t \end{vmatrix} = \lambda^2(-e^{t+\pi/2}u'^2 - 2f^{t+\pi/2}u'v' + e^{t+\pi/2}v'^2).$$

□

The corresponding result for geodesics follows from the fact that all the surfaces obtained via a Bonnet rotation have the same coefficients for their First Fundamental Forms. We register that:

Proposition 4.3.39. *Let $\boldsymbol{x}, \boldsymbol{y}\colon U \to \mathbb{R}^3_\nu$ be parametrized regular surfaces, conjugate or Lorentz-conjugate, and isothermal. For each stage t, the Bonnet rotations map the geodesics in $\boldsymbol{x}$ into geodesics of $\boldsymbol{z}^t$ and $\boldsymbol{w}^t$.*

4.3.3 Enneper-Weierstrass representation formulas

Given a regular spacelike surface $M \subseteq \mathbb{R}^3_\nu$ and a parametrization $(U, \boldsymbol{x})$, the idea here is to make the usual identification of $\mathbb{R}^2$ with $\mathbb{C}$ and use $z = u + \mathrm{i}v$ as the surface parameter. Recall that

$$u = \frac{z + \overline{z}}{2} \quad \text{and} \quad v = \frac{z - \overline{z}}{2\mathrm{i}}.$$

We also have the *Wirtinger operators*

$$\frac{\partial}{\partial z} = \frac{1}{2}\left(\frac{\partial}{\partial u} - \mathrm{i}\frac{\partial}{\partial v}\right) \quad \text{and} \quad \frac{\partial}{\partial \overline{z}} = \frac{1}{2}\left(\frac{\partial}{\partial u} + \mathrm{i}\frac{\partial}{\partial v}\right).$$

With this, the Laplacian operator can be written as

$$\triangle = \frac{\partial^2}{\partial u^2} + \frac{\partial^2}{\partial v^2} = \left(\frac{\partial}{\partial u} + \mathrm{i}\frac{\partial}{\partial v}\right)\left(\frac{\partial}{\partial u} - \mathrm{i}\frac{\partial}{\partial v}\right) = 4\frac{\partial}{\partial \overline{z}}\frac{\partial}{\partial z}.$$

Abusing notation, we may also write

$$\boldsymbol{x}(z, \overline{z}) = (x^1(z, \overline{z}), x^2(z, \overline{z}), x^3(z, \overline{z})).$$

It will also be convenient to consider the above parametrizations as the real part of curves in $\mathbb{C}^3$. For this end, we consider an extension of the inner product of $\mathbb{R}^3$ to $\mathbb{C}^3$, to be also denoted by $\langle \cdot, \cdot \rangle_E$, defined by

$$\langle (z_1, z_2, z_3), (w_1, w_2, w_3) \rangle_E = z_1\overline{w}_1 + z_2\overline{w}_2 + z_3\overline{w}_3.$$

In a similar fashion, to study timelike surfaces in $\mathbb{L}^3$, we will use a split-complex parameter, and then extend the scalar product in $\mathbb{L}^3$ to the *complex Lorentzian space* $\mathbb{C}^3_1$, i.e., the vector space $\mathbb{C}^3$ equipped with the product $\langle \cdot, \cdot \rangle_L$, defined by

$$\langle (z_1, z_2, z_3), (w_1, w_2, w_3) \rangle_L = z_1\overline{w}_1 + z_2\overline{w}_2 - z_3\overline{w}_3.$$

We will keep the usual terminology about causal types in this new setting.

Definition 4.3.40. Let U be an open subset of $\mathbb{C}$ or $\mathbb{C}'$, and $\boldsymbol{x}\colon U \to \mathbb{R}^3_\nu$ be a non-degenerate parametrized regular surface.

(i) The *complex derivative* of $\boldsymbol{x}$ is

$$\boldsymbol{\phi} \equiv \frac{\partial \boldsymbol{x}}{\partial z} \equiv \boldsymbol{x}_z \doteq \frac{1}{2}(\boldsymbol{x}_u - \mathrm{i}\,\boldsymbol{x}_v).$$

(ii) The *split-complex derivative* of $\boldsymbol{x}$ is

$$\boldsymbol{\psi} \equiv \frac{\partial \boldsymbol{x}}{\partial w} \equiv \boldsymbol{x}_w \doteq \frac{1}{2}(\boldsymbol{x}_u + \mathrm{h}\,\boldsymbol{x}_v).$$

Remark. Note that $\langle \boldsymbol{\phi}, \boldsymbol{\phi} \rangle_E = 0$ does not imply that $\boldsymbol{\phi} = \boldsymbol{0}$, since $\boldsymbol{\phi}(z, \overline{z}) \in \mathbb{C}^3$ for all z, and not necessarily in $\mathbb{R}^3$. This holds *a fortiori* for $\boldsymbol{\psi}$.

Proposition 4.3.41. *If $M \subseteq \mathbb{R}^3_\nu$ is a non-degenerate regular surface and $(U, \boldsymbol{x})$ is a parametrization for M, then $\boldsymbol{x}$ is isothermal if and only if:*

(i) $\langle \boldsymbol{\phi}, \boldsymbol{\phi} \rangle_E = 0$, *for* $M \subseteq \mathbb{R}^3$;

(ii) $\langle \boldsymbol{\phi}, \boldsymbol{\phi} \rangle_L = 0$, *for spacelike* $M \subseteq \mathbb{L}^3$;

(iii) $\langle \boldsymbol{\psi}, \boldsymbol{\psi} \rangle_L = 0$, *for timelike* $M \subseteq \mathbb{L}^3$.

Proof: We'll do here the cases where M is spacelike, and just state the equivalent expression in the remaining case. If E, F and G are the coefficients of the First Fundamental Form of M relative to the parametrization $\boldsymbol{x}$, we have that

$$(x_z^j)^2 = \left(\frac{1}{2}(x_u^j - \mathrm{i}\, x_v^j)\right)^2 = \frac{1}{4}((x_u^j)^2 - (x_v^j)^2 - 2\mathrm{i} x_u^j x_v^j),$$

and summing over j we obtain

$$\langle \boldsymbol{\phi}, \boldsymbol{\phi} \rangle = \frac{1}{4}(E - G - 2\mathrm{i}F),$$

so that the conclusion follows from the fact that a complex number is zero if and only if both its real and imaginary parts are also zero. When M is timelike, we'll have

$$\langle \boldsymbol{\psi}, \boldsymbol{\psi} \rangle_L = \frac{1}{4}(E + G + 2\mathrm{h}F)$$

instead. □

Lemma 4.3.42. *Let* $M \subseteq \mathbb{R}^3_\nu$ *be a non-degenerate regular surface and* $(U, \boldsymbol{x})$ *be an isothermal parametrization for* M. *Then:*

(i) $\langle \boldsymbol{\phi}, \overline{\boldsymbol{\phi}} \rangle_E = \lambda^2/2 \neq 0$, *for* $M \subseteq \mathbb{R}^3$;

(ii) $\langle \boldsymbol{\phi}, \overline{\boldsymbol{\phi}} \rangle_L = \lambda^2/2 \neq 0$, *for spacelike* $M \subseteq \mathbb{L}^3$;

(iii) $\langle \boldsymbol{\psi}, \overline{\boldsymbol{\psi}} \rangle_L = \epsilon_u \lambda^2/2 \neq 0$, *for timelike* $M \subseteq \mathbb{L}^3$.

Proof: Again we'll do the proof only in the case where M is spacelike:

$$x_z^j \overline{x_z^j} = \frac{1}{4}(x_u^j - \mathrm{i}\, x_v^j)(x_u^j + \mathrm{i}\, x_v^j) = \frac{1}{4}((x_u^j)^2 + (x_v^j)^2 - 2\mathrm{i} x_u^j x_v^j)$$

and, summing over j, it follows that

$$\langle \boldsymbol{\phi}, \overline{\boldsymbol{\phi}} \rangle = \frac{1}{4}(\lambda^2 + \lambda^2 - 2\mathrm{i} \cdot 0) = \frac{\lambda^2}{2}.$$

The verification of this result for timelike surfaces is Exercise 4.3.18. □

Proposition 4.3.43. *Let* $M \subseteq \mathbb{R}^3_\nu$ *be a non-degenerate regular surface and* $(U, \boldsymbol{x})$ *be an isothermal parametrization for* M. *Then* $\boldsymbol{x}$ *is critical if and only if* $\boldsymbol{\phi}$ *is holomorphic or* $\boldsymbol{\psi}$ *is split-holomorphic, according to whether* M *is spacelike or timelike, respectively.*

Proof: This follows directly from the expressions

$$\frac{\partial \boldsymbol{\phi}}{\partial \bar{z}} = \frac{\partial^2 \boldsymbol{x}}{\partial \bar{z} \partial z} = \frac{1}{4} \triangle \boldsymbol{x} \quad \text{and} \quad \frac{\partial \boldsymbol{\psi}}{\partial \overline{w}} = \frac{\partial^2 \boldsymbol{x}}{\partial \overline{w} \partial w} = \frac{1}{4} \Box \boldsymbol{x}.$$

□

In view of this result we may conclude that every non-degenerate critical surface may be, at least locally, represented by a triple:

- $\boldsymbol{\phi} = (\phi^1, \phi^2, \phi^3)$ of holomorphic functions satisfying
$$(\phi^1)^2 + (\phi^2)^2 + (\phi^3)^2 = 0,$$
if $M \subseteq \mathbb{R}^3$;
- $\boldsymbol{\phi} = (\phi^1, \phi^2, \phi^3)$ of holomorphic functions satisfying
$$(\phi^1)^2 + (\phi^2)^2 - (\phi^3)^2 = 0,$$
if $M \subseteq \mathbb{L}^3$ is spacelike;
- $\boldsymbol{\psi} = (\psi^1, \psi^2, \psi^3)$ of split-holomorphic functions satisfying
$$(\psi^1)^2 + (\psi^2)^2 - (\psi^3)^2 = 0,$$
if $M \subseteq \mathbb{L}^3$ if timelike.

These motivate the following:

Definition 4.3.44. A spacelike (resp. timelike) *critical curve* is a map $\boldsymbol{\zeta}\colon U \to \mathbb{C}^3_\nu$ with holomorphic (resp. split-holomorphic) components satisfying $\langle \boldsymbol{\zeta}', \overline{\boldsymbol{\zeta}'} \rangle = 0$. The curve is said to be *regular* if $\langle \boldsymbol{\zeta}', \boldsymbol{\zeta}' \rangle \neq 0$.

The Bonnet rotations seen previously may be translated in terms of critical curves by using complex and split-complex numbers: regular critical curves give rise to families of associated critical surfaces via

$$\boldsymbol{z}^t(u,v) = \mathrm{Re}(\mathrm{e}^{-\mathrm{i}t}\boldsymbol{\zeta}(u + \mathrm{i}v)) \quad \text{and} \quad \boldsymbol{w}^t(u,v) = \mathrm{Re}(\mathrm{e}^{-\mathrm{h}t}\boldsymbol{\zeta}(u + \mathrm{h}v)),$$

which are spacelike and timelike for all t, according to whether $\boldsymbol{\zeta}$ takes values in $\mathbb{C}^3$ or $\mathbb{C}^3_1$. We emphasize that for spacelike surfaces, the regularity condition on $\boldsymbol{\zeta}$ is relative to the Euclidean or Lorentzian product, according to the ambient where the surface lies. With this in place, we may define the Gaussian curvature and the normal direction of a critical curve, from $\boldsymbol{z}^t$ and $\boldsymbol{w}^t$. For more details in the Euclidean case, see [27].

Proposition 4.3.45. *Let $M \subseteq \mathbb{R}^3_\nu$ be a non-degenerate regular and critical surface, U be a simply connected domain, and $(U, \boldsymbol{x})$ be an isothermal parametrization for M. Then the components of $\boldsymbol{x}$ satisfy:*

(i) $x^j(z,\overline{z}) = c_j + 2\,\mathrm{Re}\displaystyle\int_{z_0}^{z} \phi^j(\xi)\,\mathrm{d}\xi$*, for some $z_0 \in U$, if M is spacelike, and*

(ii) $x^j(w,\overline{w}) = c_j + 2\,\mathrm{Re}\displaystyle\int_{w_0}^{w} \psi^j(\omega)\,\mathrm{d}\omega$*, for some $w_0 \in U$, if M is timelike,*

where the $c_j \in \mathbb{R}$ are suitable constants.

Proof: For a change, let's do the proof in the case where M is timelike. First observe that since U is simply connected, $\boldsymbol{x}$ is isothermal and M is critical, then $\boldsymbol{\psi}$ is split-holomorphic, so that the integrals in the statement of the result are all path-independent. With differentials, we have that:

$$\psi^j\,\mathrm{d}w = \frac{1}{2}(x^j_u + \mathrm{h}\,x^j_v)(\mathrm{d}u + \mathrm{h}\,\mathrm{d}v) = \frac{1}{2}(x^j_u\,\mathrm{d}u + x^j_v\,\mathrm{d}v + \mathrm{h}(x^j_v\,\mathrm{d}u + x^j_u\,\mathrm{d}v))$$
$$\overline{\psi^j}\,\mathrm{d}\overline{w} = \frac{1}{2}(x^j_u - \mathrm{h}\,x^j_v)(\mathrm{d}u - \mathrm{h}\,\mathrm{d}v) = \frac{1}{2}(x^j_u\,\mathrm{d}u + x^j_v\,\mathrm{d}v - \mathrm{h}(x^j_v\,\mathrm{d}u + x^j_u\,\mathrm{d}v))$$

Adding, we get that

$$\mathrm{d}x^j = x^j_u \,\mathrm{d}u + x^j_v \,\mathrm{d}v = \psi^j \,\mathrm{d}w + \overline{\psi^j}\,\mathrm{d}\overline{w} = 2\,\mathrm{Re}\,\psi^j \,\mathrm{d}w,$$

whence

$$x^j(w,\overline{w}) = c_j + 2\,\mathrm{Re}\int_{w_0}^{w} \psi^j(\omega)\,\mathrm{d}\omega,$$

for some $c_j \in \mathbb{R}$ and $w_0 \in U$. □

Theorem 4.3.46 (Enneper-Weierstrass I). *Let $U \subseteq \mathbb{C}$ be simply connected, $z_0 \in U$, and $f, g\colon U \to \mathbb{C}$ be functions such that f is holomorphic, g is meromorphic, but fg^2 is holomorphic. Then the map $\boldsymbol{x}\colon U \to \mathbb{R}^3_\nu$ defined by $\boldsymbol{x}(z,\overline{z}) = (x^1(z,\overline{z}), x^2(z,\overline{z}), x^3(z,\overline{z}))$, where*

(i)
$$x^1(z,\overline{z}) = \mathrm{Re}\int_{z_0}^{z} f(\xi)(1-g(\xi)^2)\,\mathrm{d}\xi,$$
$$x^2(z,\overline{z}) = \mathrm{Re}\int_{z_0}^{z} \mathrm{i}f(\xi)(1+g(\xi)^2)\,\mathrm{d}\xi \text{ and,}$$
$$x^3(z,\overline{z}) = 2\,\mathrm{Re}\int_{z_0}^{z} f(\xi)g(\xi)\,\mathrm{d}\xi, \text{ for } \boldsymbol{x} \text{ in } \mathbb{R}^3 \text{ or;}$$

(ii)
$$x^1(z,\overline{z}) = \mathrm{Re}\int_{z_0}^{z} f(\xi)(1+g(\xi)^2)\,\mathrm{d}\xi,$$
$$x^2(z,\overline{z}) = \mathrm{Re}\int_{z_0}^{z} \mathrm{i}f(\xi)(1-g(\xi)^2)\,\mathrm{d}\xi \text{ and,}$$
$$x^3(z,\overline{z}) = 2\,\mathrm{Re}\int_{z_0}^{z} -f(\xi)g(\xi)\,\mathrm{d}\xi, \text{ for } \boldsymbol{x} \text{ in } \mathbb{L}^3$$

is a parametrized surface, regular on the points where the zeros of f have precisely twice the order of the poles of g, and $|g| \neq 1$ (this last condition only in $\mathbb{L}^3$). Moreover, its image is a critical spacelike surface.

Proof: The conditions on U, f and g ensure that all the integrals are path-independent. Also, in $\mathbb{R}^3$, the complex derivative of $\boldsymbol{x}$ is precisely

$$\boldsymbol{\phi} = \left(\frac{1}{2}f(1-g^2), \frac{\mathrm{i}}{2}f(1+g^2), fg\right),$$

which satisfies

$$\langle \boldsymbol{\phi}, \boldsymbol{\phi}\rangle_E = \left(\frac{1}{2}f(1-g^2)\right)^2 + \left(\frac{\mathrm{i}}{2}f(1+g^2)\right)^2 + (fg)^2 = 0,$$

so that the expression for $\langle \boldsymbol{\phi}, \boldsymbol{\phi}\rangle_E$ given in Proposition 4.3.41 ensures that $E = G$ and $F = 0$. A similar computation gives the same conclusion in $\mathbb{L}^3$. So, $\boldsymbol{x}$ is regular precisely when $E = G \neq 0$, which is equivalent to the condition given in the statement regarding the orders of the zeros of f and poles of g in $\mathbb{R}^3$. The condition $|g| \leq 1$ in $\mathbb{L}^3$ follows from the fact that if

$$\boldsymbol{\phi} = \left(\frac{1}{2}f(1+g^2), \frac{\mathrm{i}}{2}f(1-g^2), -fg\right),$$

then

$$\frac{\lambda^2}{2} = \langle \boldsymbol{\phi}, \overline{\boldsymbol{\phi}}\rangle_L = \left|\frac{1}{2}f(1+g^2)\right|^2 + \left|\frac{\mathrm{i}}{2}f(1-g^2)\right|^2 - |-fg|^2 = \frac{|f|^2}{2}(1-|g|^2)^2.$$

This being the case, $\boldsymbol{x}$ is isothermal and spacelike. Furthermore, $\boldsymbol{\phi}$ is holomorphic, and thus the image of $\boldsymbol{x}$ is critical. □

The pair (f,g) is known as the *Weierstrass data* for the surface. It is possible to write expressions for the coefficients of the First Fundamental Form and the Gaussian curvature in terms of the Weierstrass data (f,g). When $M \subseteq \mathbb{R}^3$, the function g is closely related to the Gauss map of the surface (which takes values in the sphere $\mathbb{S}^2$), via the stereographic projection (mentioned in Exercise 3.1.2, p. 152). For more details, see [57]. In the expression for the spacelike parametrization in $\mathbb{L}^3$, the sign in the third component represents a reflection about the plane $z = 0$, seemingly irrelevant, but which allows us to look for a relation with the Gauss map also in this case. For details, see [36].

When the function g is holomorphic and invertible, we may use it as a parameter to obtain an alternative representation:

Theorem 4.3.47 (Enneper-Weierstrass II). *Let $U \subseteq \mathbb{C}$ be open and simply connected, $z_0 \in U$, and $F\colon U \to \mathbb{C}$ be a holomorphic function. Then the map $\boldsymbol{x}\colon U \to \mathbb{R}^3_\nu$ given by $\boldsymbol{x}(z,\bar{z}) = (x^1(z,\bar{z}), x^2(z,\bar{z}), x^3(z,\bar{z}))$, where*

(i)
$$x^1(z,\bar{z}) = \operatorname{Re}\int_{z_0}^{z} (1-\xi^2)F(\xi)\,\mathrm{d}\xi,$$
$$x^2(z,\bar{z}) = \operatorname{Re}\int_{z_0}^{z} \mathrm{i}(1+\xi^2)F(\xi)\,\mathrm{d}\xi \text{ and,}$$
$$x^3(z,\bar{z}) = 2\operatorname{Re}\int_{z_0}^{z} \xi F(\xi)\,\mathrm{d}\xi, \text{ for } \boldsymbol{x} \text{ in } \mathbb{R}^3, \text{ or;}$$

(ii)
$$x^1(z,\bar{z}) = \operatorname{Re}\int_{z_0}^{z} (1+\xi^2)F(\xi)\,\mathrm{d}\xi,$$
$$x^2(z,\bar{z}) = \operatorname{Re}\int_{z_0}^{z} \mathrm{i}(1-\xi^2)F(\xi)\,\mathrm{d}\xi, \text{ and}$$
$$x^3(z,\bar{z}) = -2\operatorname{Re}\int_{z_0}^{z} \xi F(\xi)\,\mathrm{d}\xi, \text{ for } \boldsymbol{x} \text{ in } \mathbb{L}^3,$$

is a parametrized surface, which is regular on the points where $F(z) \neq 0$ (in $\mathbb{R}^3$) or $F(z) \neq 0$ and $|z| \neq 1$ (in $\mathbb{L}^3$). Moreover, its image is a critical spacelike surface.

Example 4.3.48 (Critical spacelike surfaces in $\mathbb{R}^3_\nu$).

(1) Enneper surface in $\mathbb{R}^3$: consider the Weierstrass data $f(z) = 1$ and $g(z) = z$. We obtain the parametrization $\boldsymbol{x} : \mathbb{R}^2 \to \mathbb{R}^3$ given by

$$\boldsymbol{x}(u,v) = \left(u - \frac{u^3}{3} + uv^2, -v + \frac{v^3}{3} - u^2v, u^2 - v^2\right).$$

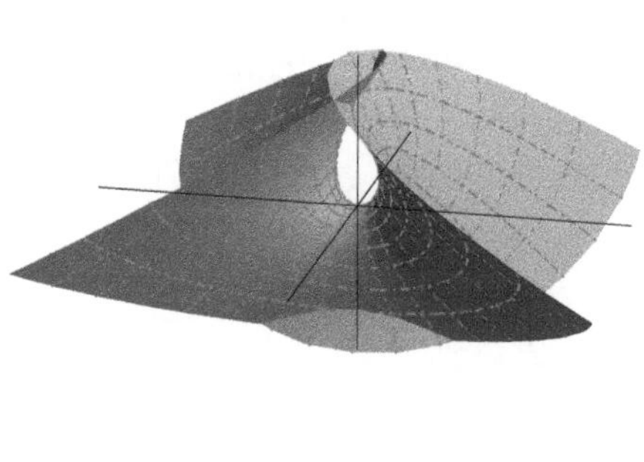

Figure 4.15: Enneper surface in $\mathbb{R}^3$.

The same surface could be obtained via the second representation with $F(z) = 1$.

(2) Spacelike Enneper surface in $\mathbb{L}^3$: consider the same Weierstrass data from the previous example, now in the Lorentzian setting. This time we obtain the parametrization $\boldsymbol{x} : \mathbb{R}^2 \to \mathbb{R}^3$ given by

$$\boldsymbol{x}(u,v) = \left(u + \frac{u^3}{3} - uv^2, -v - \frac{v^3}{3} + u^2v, v^2 - u^2\right).$$

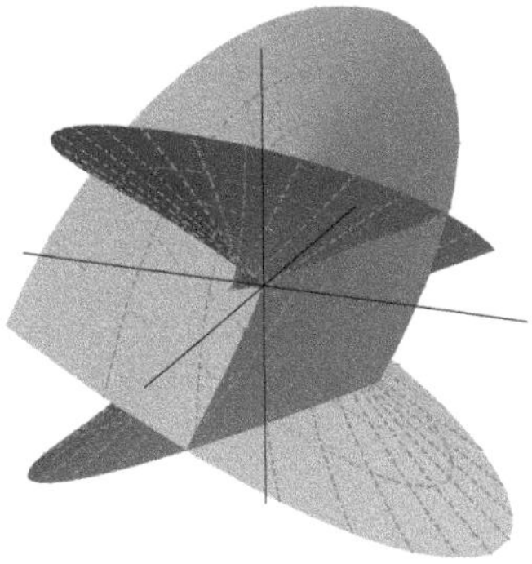

Figure 4.16: Spacelike Enneper surface in $\mathbb{L}^3$.

(3) Catalan surface in $\mathbb{R}^3$: using the single type II data $F(z) = \mathrm{i}\left(\frac{1}{z} - \frac{1}{z^3}\right)$, we produce the parametrization

$$\boldsymbol{x}(u,v) = \left(u - \sin u \cosh v, 1 - \cos u \cosh v, -4\sin\left(\frac{u}{2}\right)\sinh\left(\frac{v}{2}\right)\right)$$

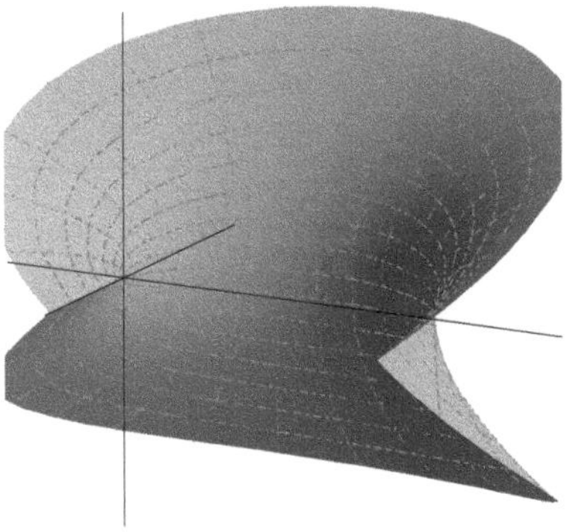

Figure 4.17: Catalan surface in $\mathbb{R}^3$.

(4) Spacelike catenoid in $\mathbb{L}^3$: this time we take the type II Weierstrass data $F(z) = 1/z^2$, which gives the parametrization

$$\boldsymbol{x}(u,v) = \left(u - \frac{u}{u^2+v^2}, v - \frac{v}{u^2+v^2}, -2\log(u^2+v^2)\right).$$

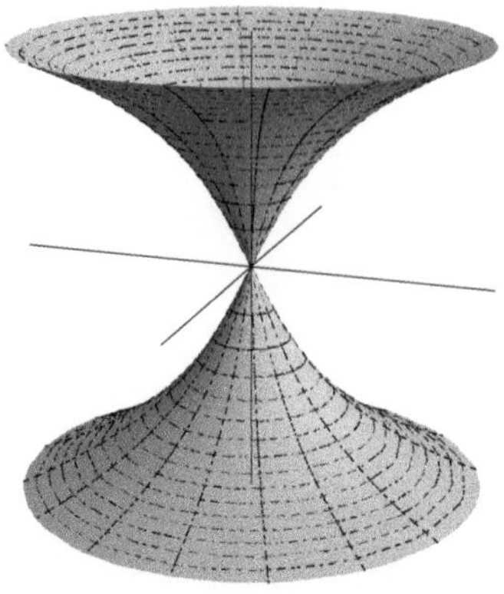

Figure 4.18: Spacelike Lorentzian catenoid.

(5) Henneberg surface in $\mathbb{R}^3$: it is given by $F(z) = 1 - \frac{1}{z^4}$, with the parametrization

$$\begin{aligned} x(u,v) = (&2\sinh u \cos v - (2/3)\sinh(3u)\cos(3v), \\ &2\sinh(u)\sin(v) + (2/3)\sinh(3u)\sin(3v), 2\cosh(2u)\cos(2v)). \end{aligned}$$

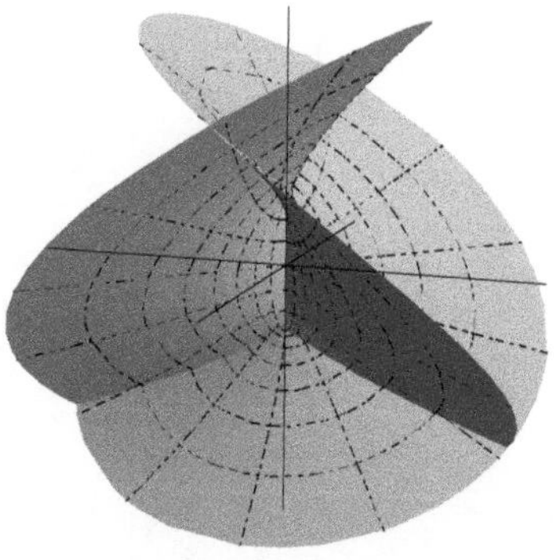

Figure 4.19: Henneberg surface.

Proceeding, we may express the Gaussian curvature in terms of the type II Weierstrass data. We register this in the:

Proposition 4.3.49. *Let $M \subseteq \mathbb{R}^3_\nu$ be a spacelike critical surface and (U, x) a type II Weierstrass parametrization induced by a holomorphic function F. Then its Gaussian curvature is given by*

$$K(x(u,v)) = (-1)^{\nu+1} \frac{4}{|F(u+\mathrm{i}v)|^2((-1)^\nu + u^2 + v^2)^4}.$$

Proof: To compute K, it suffices to know the value of $\lambda^2 = 2\langle \phi, \overline{\phi} \rangle$ and then apply the result from Exercise 3.7.3 (p. 256). In this case, we may study the situation in both ambient spaces at the same time. If $M \subseteq \mathbb{R}^3$, the complex derivative of $\boldsymbol{x}$ is

$$\boldsymbol{\phi}(z) = \left(\frac{1}{2}(1-z^2)F(z), \frac{\mathrm{i}}{2}(1+z^2)F(z), zF(z)\right),$$

while if $M \subseteq \mathbb{L}^3$ we have

$$\boldsymbol{\phi}(z) = \left(\frac{1}{2}(1+z^2)F(z), \frac{\mathrm{i}}{2}(1-z^2)F(z), -zF(z)\right),$$

but in any case we get that:

$$\begin{aligned}\lambda(z)^2 &= 2\left(\left|\frac{1}{2}(1-z^2)F(z)\right|^2 + \left|\frac{\mathrm{i}}{2}(1+z^2)F(z)\right|^2 + (-1)^\nu |zF(z)|^2\right) \\ &= \frac{1}{2}|F(z)|^2\left(|1-z^2|^2 + |1+z^2|^2 + 4(-1)^\nu |z|^2\right).\end{aligned}$$

Writing $z = u + \mathrm{i}v$, we have

$$\begin{cases} |1-z^2|^2 = (u^2-v^2-1)^2 + 4u^2v^2 \\ |1+z^2|^2 = (u^2-v^2+1)^2 + 4u^2v^2 \\ 4|z|^2 = 4(u^2+v^2). \end{cases}$$

Substituting that in the expression for λ^2, canceling and factoring terms, it follows that

$$\lambda(u,v)^2 = |F(u+\mathrm{i}v)|^2((-1)^\nu + u^2 + v^2)^2.$$

With this, we have

$$\log \lambda(u,v)^2 = \log F(z)\overline{F(z)} + 2\log((-1)^\nu + u^2 + v^2).$$

On one hand, a direct calculation shows that

$$\triangle \log((-1)^\nu + u^2 + v^2) = (-1)^\nu \frac{4}{((-1)^\nu + u^2 + v^2)^2}.$$

On the other hand, we claim that $\triangle \log F(z)\overline{F(z)} = 0$. To wit, this is trivial when F is constant. Else, F being holomorphic implies that $\partial F/\partial \overline{z} = 0$. And moreover, if F is holomorphic, $\overline{F}$ is not, so that $\partial \overline{F}/\partial z = 0$. Thus:

$$\begin{aligned}\triangle \log F(z)\overline{F(z)} &= \triangle(\log F(z) + \log \overline{F(z)}) \\ &= \triangle \log F(z) + \triangle \log \overline{F(z)} \\ &= 4\frac{\partial}{\partial z}\left(\frac{\partial}{\partial \overline{z}} \log F(z)\right) + 4\frac{\partial}{\partial \overline{z}}\left(\frac{\partial}{\partial z}\log \overline{F(z)}\right) \\ &= 4\frac{\partial}{\partial z}\left(\frac{1}{F(z)}\frac{\partial F}{\partial \overline{z}}(z)\right) + 4\frac{\partial}{\partial \overline{z}}\left(\frac{1}{\overline{F(z)}}\frac{\partial \overline{F}}{\partial z}(z)\right) \\ &= 0.\end{aligned}$$

Substituting everything, we obtain

$$K(\boldsymbol{x}(u,v)) = -\frac{\triangle \log \lambda(u,v)^2}{2\lambda(u,v)^2} = (-1)^{\nu+1}\frac{4}{|F(u+\mathrm{i}v)|^2((-1)^\nu + u^2 + v^2)^4},$$

as wanted. □

Now, we may establish the corresponding results for timelike surfaces in $\mathbb{L}^3$:

Theorem 4.3.50 (Enneper-Weierstrass I). *Let $U \subseteq \mathbb{C}'$ be open and simply connected, $w_0 \in U$, and $f, g\colon U \to \mathbb{C}$ be two functions such that f is split-holomorphic, g is split-meromorphic, and fg^2 is split-holomorphic. Then the parametrized surface $\boldsymbol{x}\colon U \to \mathbb{L}^3$ given by $\boldsymbol{x}(w,\overline{w}) = (x^1(w,\overline{w}), x^2(w,\overline{w}), x^3(w,\overline{w}))$, where*

$$\begin{aligned} x^1(w,\overline{w}) &= \operatorname{Re}\int_{w_0}^{w} f(\omega)(1-g(\omega)^2)\,d\omega \\ x^2(w,\overline{w}) &= 2\operatorname{Re}\int_{w_0}^{w} f(\omega)g(\omega)\,d\omega \\ x^3(w,\overline{w}) &= \operatorname{Re}\int_{w_0}^{w} f(\omega)(1+g(\omega)^2)\,d\omega, \end{aligned}$$

is regular on the points where the zeros of f have precisely twice the order of the poles of g, $f(w)$ is not a zero divisor and $g(w)$ is not real. Moreover, its image is a timelike critical surface.

Proof: The conditions on U, f and g again ensure that all the integrals are path-independent. In this case the split-complex derivative of $\boldsymbol{x}$ is

$$\boldsymbol{\psi} = \left(\frac{1}{2}f(1-g^2), fg, \frac{1}{2}f(1+g^2)\right).$$

We have that

$$\begin{aligned} \langle \boldsymbol{\psi}, \boldsymbol{\psi}\rangle_L &= \left(\frac{1}{2}f(1-g^2)\right)^2 + (fg)^2 - \left(\frac{1}{2}f(1+g^2)\right)^2 = 0 \text{ and} \\ \langle \boldsymbol{\psi}, \overline{\boldsymbol{\psi}}\rangle_L &= \frac{f\overline{f}}{4}\left((1-g^2)(1-\overline{g}^2) + 4g\overline{g} - (1+g^2)(1+\overline{g}^2)\right) = -\frac{f\overline{f}}{2}(g-\overline{g})^2, \end{aligned}$$

from where all conclusions follow. □

Theorem 4.3.51 (Enneper-Weierstrass II). *Let $U \subseteq \mathbb{C}'$ be open and simply connected with $U \cap \mathbb{R} = \varnothing$, $w_0 \in U$, and $F\colon U \to \mathbb{C}'$ be a split-holomorphic function. Then the parametrized surface $\boldsymbol{x}\colon U \to \mathbb{L}^3$ given by $\boldsymbol{x}(w,\overline{w}) = (x^1(w,\overline{w}), x^2(w,\overline{w}), x^3(w,\overline{w}))$, where*

$$\begin{aligned} x^1(w,\overline{w}) &= \operatorname{Re}\int_{w_0}^{w} (1-\omega^2)F(\omega)\,d\omega \\ x^2(w,\overline{w}) &= 2\operatorname{Re}\int_{w_0}^{w} \omega F(\omega)\,d\omega \\ x^3(w,\overline{w}) &= \operatorname{Re}\int_{w_0}^{w} (1+\omega^2)F(\omega)\,d\omega, \end{aligned}$$

is regular in the points where $F(w)$ is not a zero divisor. Moreover, its image is a timelike critical surface.

Example 4.3.52 (Timelike critical surfaces in $\mathbb{L}^3$).

(1) Timelike Enneper surface: for the data $F(w) = 1$ we obtain the parametrization

$$\boldsymbol{x}(u,v) = \left(v - u^2 v - \frac{v^3}{3}, 2uv, v + u^2 v + \frac{v^3}{3}\right).$$

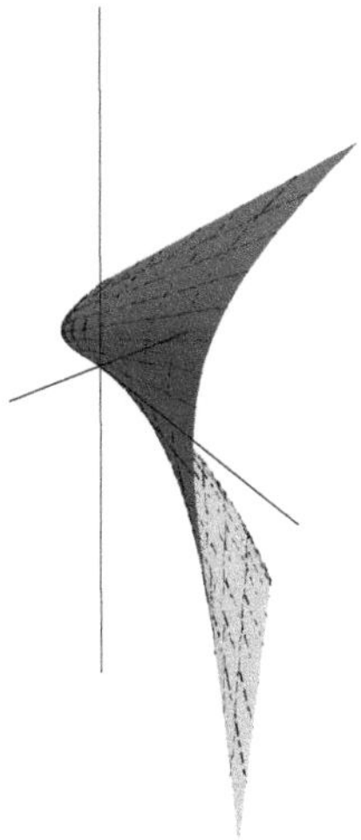

Figure 4.20: Timelike Enneper surface.

(2) Timelike catenoid: for the data $F(w) = 1/w^2$ we obtain the parametrization

$$x(u,v) = \left(-\frac{u}{u^2 - v^2} - u, \log\left((u^2 - v^2)^2 \right), -\frac{u}{u^2 - v^2} + u \right),$$

defined everywhere, except on the light rays ($u^2 = v^2$), and regular on its domain, except on the real axis ($v = 0$).

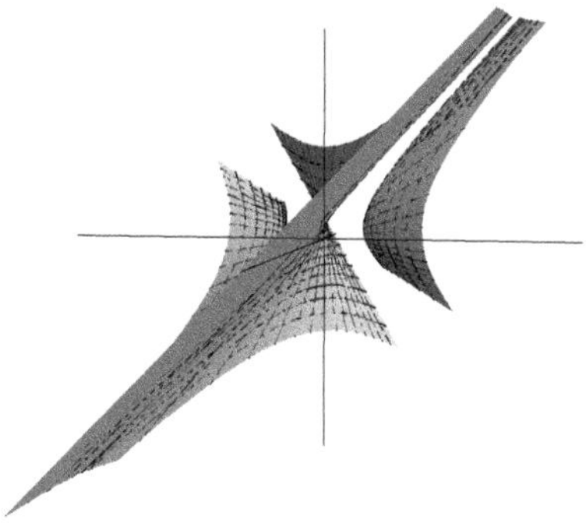

Figure 4.21: Timelike catenoid.

For more details about such representation formulas, you may consult [37] and [46]. Such techniques also have applications in the study of the so-called *Björling problems* — see, for example, [2], [12], [18], and [19].

Exercises

Exercise 4.3.15 (Mercator's Projection). We have seen that every surface locally admits isothermal parametrizations. However, it is not always easy to find them. Show that *Mercator's projection* $\boldsymbol{x}\colon \mathbb{R}\times\,]0,2\pi[\,\to\mathbb{R}^3$ given by

$$\boldsymbol{x}(u,v)=\left(\frac{\cos v}{\cosh u},\frac{\sin v}{\cosh u},\tanh u\right)$$

is a regular and isothermal parametrization of the sphere $\mathbb{S}^2$. Also verify that the meridians and parallels in $\mathbb{S}^2$ correspond via $\boldsymbol{x}$ to straight lines in the plane.

Exercise 4.3.16. Let $\boldsymbol{x},\boldsymbol{y}\colon U\to\mathbb{R}^3_\nu$ be two regular and conjugate (resp., Lorentz-conjugate) parametrized surfaces. Show that if $\boldsymbol{x}$ is isothermal, so is $\boldsymbol{y}$.

Exercise 4.3.17. Let $\theta\in\mathbb{R}$ be any angle. Show that the parametrized surface $\boldsymbol{x}\colon\,]0,2\pi[\,\times\mathbb{R}\to\mathbb{R}^3$ given by

$$\boldsymbol{x}(u,v)=(u\cos\theta\pm\sin u\cosh v,v\pm\cos\theta\cos u\sinh v,\pm\sin\theta\cos u\cosh v)$$

is isothermal and minimal.

Exercise 4.3.18. Prove Lemma 4.3.42 (p. 305) for spacelike surfaces.

Exercise 4.3.19. Let $(U,\boldsymbol{x})$ be a regular, injective, isothermal and minimal parametrized surface, and $\boldsymbol{N}$ be a Gauss map for $M\doteq\boldsymbol{x}(U)$, compatible with $\boldsymbol{x}$. Show that if M is not a plane, then $\boldsymbol{N}$ is a locally conformal map (in the sense of Exercise 3.2.28, p. 177).

Hint. Also compute the Gaussian curvature via the expression given in Proposition 3.3.10 (p. 182).

Exercise 4.3.20. There is a surprising "converse" to the result given in the previous exercise: let $M\subseteq\mathbb{R}^3_\nu$ be a non-degenerate and connected regular surface, and $\boldsymbol{N}$ be a Gauss map for M. Suppose that $\boldsymbol{N}$ is locally conformal. Show that M is a critical surface, or is a piece of a sphere, de Sitter space, or hyperbolic plane.

Hint. Consider an orthogonal parametrization $\boldsymbol{x}$ for M, compatible with $\boldsymbol{N}$. Use the conformality of $\boldsymbol{N}$ to show that $f(eG+Eg)=0$. If $eG+Eg=0$, then $\boldsymbol{x}$ is critical. If $f=0$, use again the conformality of $\boldsymbol{N}$ to show that $e/E=\pm g/G$. If the equality holds with the negative sign, then $\boldsymbol{x}$ is critical. Else, $\boldsymbol{x}$ is totally umbilical.

4.4 DIGRESSION: COMPLETENESS AND CAUSALITY

In Section 1.3 of Chapter 1, we discussed some interpretations of the definitions and results proven until then, in the context of Special Relativity, giving emphasis to Lorentz-Minkowski space $\mathbb{L}^n$. But now, with the language of manifolds, we can formalize such ideas. To motivate what we are going to do in this section, we first present the notions of intrinsic distance (in the Riemannian case) and geodesic completeness. We begin recalling a few definitions:

Definition 4.4.1. Let M be a smooth manifold. A *vector field* along M is a smooth mapping $\boldsymbol{V}\colon M \to \bigcup_{p\in M} T_pM$ such that $\boldsymbol{V}(p) \in T_pM$ for all $p \in M$.

Remark. The set $TM \doteq \bigcup_{p\in M} T_pM$ is called the *tangent bundle* of M, and it possesses a natural smooth manifold structure. This way, it makes sense to consider the smoothness of a map $M \to TM$.

Definition 4.4.2. Let M be a smooth manifold. A *pseudo-Riemannnian metric* in M is a choice of non-degenerate scalar products $\langle\cdot,\cdot\rangle_p$ in each tangent space T_pM such that

(i) the index of $\langle\cdot,\cdot\rangle_p$ is the same, for all $p \in M$;

(ii) the choice depends smoothly on p, that is: given any two smooth vector fields $\boldsymbol{V}$ and $\boldsymbol{W}$ along M, the map taking p to the number $\langle \boldsymbol{V}(p), \boldsymbol{W}(p)\rangle_p$ is smooth.

We will say that the pair $(M, \langle\cdot,\cdot\rangle)$ is a pseudo-Riemannian manifold.

Remark.

- If M is connected, condition (i) is automatically satisfied. We will assume from now on that this is the case.
- If the index of the pseudo-Riemannian metric is 0, the pair $(M, \langle\cdot,\cdot\rangle)$ is called a Riemannian manifold. If the index is 1, a Lorentz manifold.
- The causal type of a vector $\boldsymbol{v} \in T_pM$ is defined through the sign of $\langle \boldsymbol{v}, \boldsymbol{v}\rangle_p$ and the causal type of a curve is defined by the causal type of its velocity vectors, like always.

Definition 4.4.3. Let $(M, \langle\cdot,\cdot\rangle)$ be a pseudo-Riemannian manifold. The *energy functional* and the *arclength functional* associated to $\langle\cdot,\cdot\rangle$ are the functionals E and L defined by

$$E[\boldsymbol{\alpha}] \doteq \frac{1}{2}\int_I \langle \boldsymbol{\alpha}'(t), \boldsymbol{\alpha}'(t)\rangle_{\boldsymbol{\alpha}(t)}\,\mathrm{d}t \quad \text{and} \quad L[\boldsymbol{\alpha}] \doteq \int_I \sqrt{\left|\langle \boldsymbol{\alpha}'(t), \boldsymbol{\alpha}'(t)\rangle_{\boldsymbol{\alpha}(t)}\right|}\,\mathrm{d}t,$$

for all smooth curves $\boldsymbol{\alpha}\colon I \to M$. The critical points of E are called *geodesics*.

Remark.

- If $\boldsymbol{\alpha}\colon I \to M$ is timelike, we denote its *proper time* by

$$\mathfrak{t}[\boldsymbol{\alpha}] = \int_I \sqrt{-\langle \boldsymbol{\alpha}'(t), \boldsymbol{\alpha}'(t)\rangle_{\boldsymbol{\alpha}(t)}}\,\mathrm{d}t.$$

- The variational characterizations given in Section 3.6 of Chapter 3 still hold in this context. One can prove that a constant-speed curve is a critical point of one functional if and only if it is a critical point of the other one.

With these definitions in hand, we are ready to start the discussion.

A fundamental concept in geometry is that of distance between two points. When $(M, \langle\cdot,\cdot\rangle)$ is a Riemannian manifold, for each $p \in M$ we have that $\langle\cdot,\cdot\rangle_p$ is a positive-definite inner product. Since we are assuming that M is connected, we have that if

$$\Omega(p,q) \doteq \{\boldsymbol{\alpha}\colon [a,b] \to M \mid \boldsymbol{\alpha} \text{ is piecewise smooth, } \boldsymbol{\alpha}(a) = p \text{ and } \boldsymbol{\alpha}(b) = q\},$$

then $d\colon M \times M \to \mathbb{R}$ defined by

$$d(p,q) = \inf\{L[\boldsymbol{\alpha}] \mid \boldsymbol{\alpha} \in \Omega(p,q)\}$$

is a *distance function* on M, that is, it satisfies the following properties:

(i) $d(p,q) \geq 0$, and $d(p,q) = 0$ if and only if $p = q$;

(ii) $d(p,q) = d(q,p)$ and

(iii) $d(p,r) \leq d(p,q) + d(q,p)$.

In this context, d is called the *intrinsic distance* on M (see a concrete example of this in Exercise 4.4.1).

Of the above properties, the only one whose verification poses some difficulty is the first one, when it should be proven that if $p \neq q$, then $d(p,q) > 0$. Then, the pair (M,d) is a *metric space*. Bearing in mind the construction of the distance function d, it is reasonable to expect its properties to have a strong relation with the geometry and topology of M itself. This begins when we prove that d is, in fact, compatible with the original topology in M: every open ball according to d is an open set in M, and on the other hand, every open set in M is a union of open balls according to d.

The proof of these facts is made easier by the notion of *exponential map* (valid in any pseudo-Riemannian manifold): it can be shown that given $p \in M$, there is an open set $D_p \subseteq T_pM$ on which the map $\exp_p\colon D_p \to M$ given by $\exp_p(v) = \alpha(1)$, where α is the unique (maximal) geodesic such that $\alpha(0) = p$ and $\alpha'(0) = v$, is well-defined. That is, the exponential map at a point collects all the geodesics starting at p. It would be natural to wonder, now, how big can this domain D_p be. This leads us to the following general definition:

Definition 4.4.4. Let $(M, \langle \cdot, \cdot \rangle)$ be a pseudo-Riemannian manifold. We will call a geodesic $\alpha\colon I \to M$ (with I maximal) *complete* if $I = \mathbb{R}$. And we will say that M is *geodesically complete* if every geodesic in M is complete.

Besides the questioning about the size of D_p, we now have two notions of completeness for M (metric completeness and geodesic completeness), and it is natural to wonder if there is any relation between them. The following theorem answers all of our questions:

Theorem 4.4.5 (Hopf-Rinow). *Let $(M, \langle \cdot, \cdot \rangle)$ be a Riemannian manifold and $p_0 \in M$. The following are equivalent:*

(i) M is geodesically complete;

(ii) (M,d) is a complete metric space;

(iii) $\exp_{p_0}$ is defined on all of $T_{p_0}M$;

(iv) $\exp_p$ is defined on all of T_pM, for all $p \in M$;

(v) the closed and bounded subsets of M are compact;

(vi) there is a sequence of compact subsets $(K_n)_{n\geq 0}$ of M such that $K_n \subseteq \operatorname{int}(K_{n+1})$ for all $n \geq 0$, $M = \bigcup_{n\geq 0} K_n$, and $(K_n)_{n\geq 0}$ has the following property: for every sequence $(q_n)_{n\geq 0}$ of points in M such that $q_n \notin K_n$ for all $n \geq 0$, we have $d(p,q_n) \to +\infty$ for all $p \in M$.

Besides, each one of these properties implies that

(vii) given $p,q \in M$, there is a smooth geodesic $\alpha \in \Omega(p,q)$ such that $L[\alpha] = d(p,q)$.

For a proof, we suggest [16]. We have two important consequences:

Corollary 4.4.6. *Let $(M, \langle\cdot,\cdot\rangle)$ be a compact Riemannian manifold. Then M is geodesically complete.*

Definition 4.4.7. Let $(M, \langle\cdot,\cdot\rangle)$ be a Riemannian manifold. A (piecewise) smooth curve $\alpha\colon \mathbb{R}_{\geq 0} \to M$ is called *divergent* if for every $K \subseteq M$ compact there is $t_0 > 0$ such that $\alpha(t) \notin K$ for all $t > t_0$.

Remark. That is, α escapes from every compact set in M, without returning (so that its trace is, in a certain way, "divergent").

Corollary 4.4.8. *Let $(M, \langle\cdot,\cdot\rangle)$ be a Riemannian manifold. Then M is complete if and only if the arclength of every divergent curve in M is infinite.*

Remark. The arclength of a divergent curve $\alpha\colon \mathbb{R}_{\geq 0} \to M$ is $L[\alpha] = \lim_{b\to+\infty} L\left[\alpha\big|_{[0,b]}\right]$.

In general, this theorem allows us to use tools from analysis and topology to study the geometry of a Riemannian manifold $(M, \langle\cdot,\cdot\rangle)$. We would like very much to have an analogous result for pseudo-Riemannian manifolds. This is, in general, not possible: everything we have discussed so far fails, since we do not have a distance function anymore. Another difficulty that appears is that, for pseudo-Riemannian manifolds, we don't have only one notion of geodesic completeness, but one for each causal type, and it is not possible in general to control all three of them simultaneously.

Surprisingly, for certain Lorentz manifolds there is an analogous result, and the adequate tools for healing this deficiency in this case are the notions of *chronological precedence* ($\ll$) and *causal precedence* ($\preccurlyeq$), discussed briefly in Chapter 1. Let us see a few ideas:

Definition 4.4.9 (Spacetime)**.** Let $(M, \langle\cdot,\cdot\rangle)$ be a Lorentz manifold. We will say that M is *time-orientable* if there is a timelike vector field $\boldsymbol{V}$ defined globally on all of M (that is, $\langle \boldsymbol{V}(p), \boldsymbol{V}(p)\rangle_p < 0$ for all $p \in M$). When $\boldsymbol{V}$ is fixed, we will say that M is *time-oriented by $\boldsymbol{V}$*. A *spacetime* is a Lorentz manifold of dimension greater than or equal to 2, which is time-oriented. In this case, the points of M are called *events*.

To know if it is possible to time-orient a given manifold, we appeal to a result from algebraic topology (see [34] for a proof):

Theorem 4.4.10. *Let M be a smooth manifold. The following are equivalent:*

(i) There is a Lorentz metric on M.

(ii) There is a time-orientable Lorentz metric on M.

(iii) There is a non-vanishing vector field defined on all of M.

(iv) M is non-compact, or its Euler-Poincaré characteristic $\chi(M)$ is zero.

In particular, it follows that connected timelike surfaces in $\mathbb{L}^3$ are always time-orientable, perhaps seeing M as an abstract surface and changing its metric. The importance of all of this resides in the fact that a time orientation allows us to tell the direction of time in M, according to the:

Definition 4.4.11. Let $(M, \langle\cdot,\cdot\rangle)$ be a spacetime oriented by a timelike vector field $\boldsymbol{V}$. A timelike or lightlike vector $\boldsymbol{v} \in T_pM$ is called *future-directed* if $\langle \boldsymbol{v}, \boldsymbol{V}(p)\rangle_p \leq 0$, and *past-directed* if $\langle \boldsymbol{v}, \boldsymbol{V}(p)\rangle_p \geq 0$.

Remark.

- In Section 1.3, the standard time orientation for $\mathbb{L}^n$ is given by the field that associates to each point in $\mathbb{L}^n$ the vector $\boldsymbol{e}_n$.
- Note that in the above definition, if $\boldsymbol{v}$ is timelike, the inequalities must be strict (why?).
- As in Chapter 2, a timelike or lightlike curve is future-directed or past-directed according to whether one (hence all) of its velocity vectors is.

It is time to formalize Definition 1.3.3 (p. 20), given in Chapter 1. If $\boldsymbol{\alpha}\colon I \to M$ is a piecewise smooth timelike curve, and $t_0 \in I$, we will say that the causal type of $\boldsymbol{\alpha}$ in t_0 is the causal type of the one-sided derivatives of $\boldsymbol{\alpha}$ in t_0, when they coincide. This way, let

$$\Omega_c(p,q) \doteq \{\boldsymbol{\alpha} \in \Omega(p,q) \mid \boldsymbol{\alpha} \text{ is timelike or lightlike, and future-directed}\}.$$

We have:

Definition 4.4.12 (Future and Past)**.** Let $(M, \langle\cdot,\cdot\rangle)$ be a spacetime and $p \in M$. We define the *chronological future* and the *causal future* of p, respectively, as

$$I^+(p) \doteq \{q \in M \mid \text{there exists a timelike } \boldsymbol{\alpha} \in \Omega_c(p,q)\}$$
$$J^+(p) \doteq \{q \in M \mid \text{there exists } \boldsymbol{\alpha} \in \Omega_c(p,q)\}.$$

The definitions of *chronological past* and *causal past*, denoted respectively by $I^-(p)$ and $J^-(p)$, are dual to the given above, with "past-directed" instead of "future-directed". If $S \subseteq M$ is any subset, the *chronological future* of S is defined by

$$I^+(S) \doteq \bigcup_{p\in S} I^+(p),$$

and similarly for the other notions of future and past ($I^-(S)$, $J^+(S)$ and $J^-(S)$).

Remark. For some basic properties of I^+, see Exercise 4.4.7.

This allows us to define $\ll$ and $\preccurlyeq$ in M:

Definition 4.4.13 (Causality)**.** Let $(M, \langle\cdot,\cdot\rangle)$ be a spacetime, and $p, q \in M$. We will say that p *chronologically (resp. causally) precedes* q if $q \in I^+(p)$ (resp. $q \in J^+(p)$ or $q = p$). Such relations are denoted by $p \ll q$ and $p \preccurlyeq q$.

The futures and pasts of points in M end up being closely related to the topology in M. It is possible to prove, using the exponential map, the following result:

Proposition 4.4.14. *If $(M, \langle\cdot,\cdot\rangle)$ is a spacetime and we're given $p, q \in M$ such that $p \ll q$, then there are neighborhoods U and V of p and q in M such that $p' \ll q'$, for all $p' \in U$ and $q' \in V$. In other words, the binary relation $\ll$ is open in $M \times M$. In particular, for all $p \in M$, $I^+(p)$ and $I^-(p)$ are open sets in M.*

Assuming this result, we give an example of how the topology in M can influence its causality:

Proposition 4.4.15. *Let $(M, \langle\cdot,\cdot\rangle)$ be a compact spacetime. Then there is a (piecewise smooth) closed timelike curve in M.*

Proof: Consider the open cover $(I^+(p))_{p \in M}$ of M, consisting of the chronological futures of all points in M. By compactness of M, there are events $p_1, \ldots, p_k \in M$ such that $M = \bigcup_{i=1}^k I^+(p_i)$. If $1 \leq i \leq k$ is such that there is j distinct from i with $p_i \in I^+(p_j)$, we would have $I^+(p_i) \subseteq I^+(p_j)$, and so we could remove $I^+(p_i)$ from the open cover. With this in mind, we can assume that $p_i \notin I^+(p_j)$, for all distinct $1 \leq i, j \leq k$. So, $p_1 \in M$ and $p_1 \notin \bigcup_{i=2}^k I^+(p_i)$ means that $p_1 \in I^+(p_1)$, and the definition of chronological future gives us the desired curve. □

That is, in such a spacetime, causality breaks, since every point in the closed timelike curve is simultaneously in its own future and past. In other words, in this model, some observers could travel back in time. This indicates that this notion of "spacetime" is too broad. Bearing this in mind, we can impose several conditions on the causality of the spacetime to avoid pathological situations such as the one given by the above proposition. Let us register them:

Definition 4.4.16 (Causal hierarchy)**.** Let $(M, \langle\cdot,\cdot\rangle)$ be a spacetime. We will say that M is:

(i) *chronological*, if there is no closed timelike curve in M;

(ii) *causal*, if there is no closed timelike and no lightlike curve in M;

(iii) *distinguishing*, if given $p, q \in M$ with $I^+(p) = I^+(q)$ or $I^-(p) = I^-(q)$, then $p = q$;

(iv) *strongly causal*, if the collection of *chronological diamonds*

$$\mathscr{B}_{\Diamond,\ll} \doteq \{I^+(p) \cap I^-(q) \mid p, q \in M\}$$

form a basis for the original topology in M;

(v) *stably causal*, if M admits a *time function*, that is, a continuous real-valued function on M that is strictly increasing along every timelike or lightlike future-directed curve;

(vi) *causally simple* if it is causal, and the sets $J^+(p)$ and $J^-(p)$ are closed in M, for all $p \in M$;

(vii) *globally hyperbolic*, if it is causal, and each *causal diamond* $J^+(p) \cap J^-(q)$ is compact, for all $p, q \in M$.

Remark.

- This set of definitions receives the name of *causal hierarchy* (or *causal ladder*) because the following implications hold:

$$(vii) \implies (vi) \implies (v) \implies (iv) \implies (iii) \implies (ii) \implies (i).$$

However, *all* of the converses are false. We recommend [7] and [29] for more details.

- The topology generated by $\mathscr{B}_{\Diamond,\ll}$ is called the *Alexandrov topology* in M. In other words, a spacetime is strongly causal if the Alexandrov topology coincides with the original one.

- With this terminology, Proposition 4.4.15 above tells us that no compact spacetime is chronological. Despite that, there are still interesting examples of compact spacetimes, such as *Gödel's spacetime*: the first model of the universe where time travel is possible. This space has closed timelike curves, but no closed timelike geodesics. It's metric is a solution of the famous *Einstein's field equations*, and it was given to Einstein by Gödel as a birthday present in 1949. For more details, see [29].

These notions will allow us to state a result analogous to the Hopf-Rinow theorem for timelike surfaces. We start adapting the concept of distance:

Definition 4.4.17. Let $(M, \langle\cdot,\cdot\rangle)$ be a spacetime. The *time separation* (or *Lorentz distance*) in M is the map $\mathfrak{t}\colon M \times M \to [0,+\infty]$ defined by

$$\mathfrak{t}(p,q) \doteq \sup\{\mathfrak{t}[\alpha] \mid \alpha \in \Omega_c(p,q)\}$$

if $\Omega_c(p,q) \neq \varnothing$, and $\mathfrak{t}(p,q) = 0$ otherwise.

The wrong-way triangle inequality gives us some properties of $\mathfrak{t}$ analogous to some of the intrinsic distance d:

Proposition 4.4.18. *Let $(M, \langle\cdot,\cdot\rangle)$ be a spacetime, and $p,q,r \in M$. We have:*

(i) $\mathfrak{t}(p,q) > 0$ *if and only if* $p \in I^-(q)$*, if and only if* $q \in I^+(p)$.

(ii) $\mathfrak{t}(p,p) = +\infty$ *if there is a timelike* $\alpha \in \Omega_c(p,p)$*, and it is zero otherwise.*

(iii) If $0 \leq \mathfrak{t}(p,q) < +\infty$*, then* $\mathfrak{t}(q,p) = 0$*. In particular,* $\mathfrak{t}$ *is not, in general, symmetric.*

(iv) If $p \preccurlyeq q$ *and* $q \preccurlyeq r$*, then* $\mathfrak{t}(p,q) + \mathfrak{t}(q,r) \leq \mathfrak{t}(p,r)$.

Remark.

- There are spacetimes for which the time separation is constant and equal to $+\infty$. Such spaces are said to be *totally vicious*, and their Alexandrov topology is the chaotic one.

- The inequality in (iv) above still holds if $p = q$, $q = r$ or $p = r$.

Another important property is given by the:

Proposition 4.4.19. *Let $(M, \langle\cdot,\cdot\rangle)$ be a spacetime. Then $\mathfrak{t}$ is lower semi-continuous, that is, if $p,q \in M$ and $(p_n)_{n\geq 0}$ and $(q_n)_{n\geq 0}$ are two sequences in M such that $p_n \to p$ and $q_n \to q$, then*

$$\liminf_{n\to+\infty} \mathfrak{t}(p_n,q_n) \geq \mathfrak{t}(p,q).$$

In general, the time separation is not well-behaved, perhaps being discontinuous and assuming the value $+\infty$. However, $\mathfrak{t}$ has better properties according to how far we go in the causal ladder. This climb leads to the:

Theorem 4.4.20 (Avez-Seifert)**.** *Let $(M, \langle\cdot,\cdot\rangle)$ be a globally hyperbolic spacetime. Then:*

(i) if $p,q \in M$ *are such that* $p \preccurlyeq q$*, then there is* $\alpha \in \Omega_c(p,q)$ *such that* $\mathfrak{t}[\alpha] = \mathfrak{t}(p,q)$*;*

(ii) the time separation $\mathfrak{t}$ *is finite and continuous.*

For more technical details and deeper results about the topics discussed here, we refer the reader to [16], [54], [41] (for Riemannian Geometry) and [7], [29], [58], [34], [51] and [54] (for the interplay between Differential Geometry and General Relativity).

Exercises

Exercise† 4.4.1 (Introduction to Hyperbolic Geometry). We have seen that the hyperbolic plane $\mathbb{H}^2 \subseteq \mathbb{L}^3$ may be used as a model for hyperbolic geometry. Let's see a few basic facts about the lines in such geometry:

(a) Let $\boldsymbol{p}, \boldsymbol{q} \in \mathbb{H}^2$. Show that there is a unique geodesic in $\mathbb{H}^2$ joining $\boldsymbol{p}$ to $\boldsymbol{q}$.

Hint. Every geodesic in $\mathbb{H}^2$ is related to some spacelike direction in $\mathbb{L}^3$.

(b) Verify that the Parallel Postulate does not hold here: if $\boldsymbol{\alpha}$ is a geodesic in $\mathbb{H}^2$ and $\boldsymbol{p} \in \mathbb{H}^2$ is not in the trace of $\boldsymbol{\alpha}$, there are infinitely many geodesics containing $\boldsymbol{p}$ which do not intersect $\boldsymbol{\alpha}$.

Hint. Every geodesic is determined by a point and a tangent vector.

(c) Let $\boldsymbol{p}, \boldsymbol{q} \in \mathbb{H}^2$. Show that the length of the geodesic in $\mathbb{H}^2$ joining $\boldsymbol{p}$ to $\boldsymbol{q}$ is $\cosh^{-1}(-\langle \boldsymbol{p}, \boldsymbol{q}\rangle_L)$.

Hint. Parametrize the geodesic by $\boldsymbol{\alpha}(s) = (\cosh s)\boldsymbol{v} + (\sinh s)\boldsymbol{w}$, with $\boldsymbol{v}$ and $\boldsymbol{w}$ unit and orthogonal, $\boldsymbol{v}$ timelike, and $\boldsymbol{w}$ spacelike. Compute $\langle \boldsymbol{\alpha}(s_0), \boldsymbol{\alpha}(s_1)\rangle_L$, for any $s_0, s_1 \in \mathbb{R}$.

(d) Define[1] $d \colon \mathbb{H}^2 \times \mathbb{H}^2 \to \mathbb{R}$ by $d(\boldsymbol{p}, \boldsymbol{q}) = \cosh^{-1}(-\langle \boldsymbol{p}, \boldsymbol{q}\rangle_L)$. Show that d turns $\mathbb{H}^2$ into a *metric space*, that is, given any $\boldsymbol{p}, \boldsymbol{q}, \boldsymbol{r} \in \mathbb{H}^2$, the following properties hold:

(i) $d(\boldsymbol{p}, \boldsymbol{q}) \geq 0$, and $d(\boldsymbol{p}, \boldsymbol{q}) = 0 \iff \boldsymbol{p} = \boldsymbol{q}$.

(ii) $d(\boldsymbol{p}, \boldsymbol{q}) = d(\boldsymbol{q}, \boldsymbol{p})$.

(iii) $d(\boldsymbol{p}, \boldsymbol{q}) \leq d(\boldsymbol{p}, \boldsymbol{r}) + d(\boldsymbol{r}, \boldsymbol{q})$.

(e) A *hyperbolic isometry* is a function $\Lambda_0 \colon \mathbb{H}^2 \to \mathbb{H}^2$ such that

$$d(\Lambda_0(\boldsymbol{p}), \Lambda_0(\boldsymbol{q})) = d(\boldsymbol{p}, \boldsymbol{q}),$$

for any $\boldsymbol{p}, \boldsymbol{q} \in \mathbb{H}^2$. Prove that if Λ_0 is a hyperbolic isometry, then $\Lambda_0 = \Lambda\big|_{\mathbb{H}^2}$ for some orthochronous Lorentz transformation Λ in $\mathbb{L}^3$.

Remark. For more details about this model, see [62].

Exercise† 4.4.2. Show Corollary 4.4.8 (p. 317).

Exercise 4.4.3. Consider in $M = \mathbb{R} \times \mathbb{R}_{>0}$ the Riemannian metric

$$ds^2 = dx^2 + \frac{dy^2}{y}.$$

Show that the vertical segment $\{0\} \times \,]0,1[$ is divergent and that its length equals 2. Conclude that this metric is not complete.

[1]Recall that if $\boldsymbol{p}$ and $\boldsymbol{q}$ are timelike vectors in the same timecone, there is a unique *positive* φ such that $\langle \boldsymbol{p}, \boldsymbol{q}\rangle_L = -\|\boldsymbol{p}\|_L\|\boldsymbol{q}\|_L \cosh\varphi$.

Exercise 4.4.4. Let $(M, \langle\cdot,\cdot\rangle)$ be a connected Riemannian manifold. Show that any two points may be joined by a *geodesic polygonal*, that is, a piecewise smooth curve for which each smooth arc is a geodesic.

Hint. Note that if $p \in M$, each point in the image $\exp_p(D_p)$ may be joined to p by a geodesic there contained. Use this to show that a certain subset of M is both open and closed.

Exercise 4.4.5. Let $(M, \langle\cdot,\cdot\rangle)$ be a Riemannian manifold. Suppose that there is a function $f\colon M \to \mathbb{R}$ which is both:

- *Lipschitz-continuous*: there is $C > 0$ such that $|f(p) - f(q)| < Cd(p,q)$, for all $p, q \in M$, and
- *proper*: for all compact $K \subseteq \mathbb{R}$, $f^{-1}(K) \subseteq M$ is compact.

Show that $(M, \langle\cdot,\cdot\rangle)$ is geodesically complete.

Exercise 4.4.6. Show that the relation $\ll$ is transitive.

Remark. We have already seen this for $\preccurlyeq$ in Exercise 1.3.4 (p. 28, Chapter 1), when $M = \mathbb{L}^n$, but it is also true in this more general setting.

Exercise† 4.4.7. Let $(M, \langle\cdot,\cdot\rangle)$ be a spacetime.

(a) Show that $p \in I^+(q)$ if and only if $q \in I^-(p)$.

(b) Show that for all $S \subseteq M$, we have $I^+(S) = I^+(\overline{S})$.

(c) Show that if $(S_i)_{i\in I}$ is any collection of subsets of M, then $I^+\left(\bigcup_{i\in I} S_i\right) = \bigcup_{i\in I} I^+(S_i)$.

(d) Show that if $(S_i)_{i\in I}$ is any collection of subsets of M, then $I^+\left(\bigcap_{i\in I} S_i\right) \subseteq \bigcap_{i\in I} I^+(S_i)$, and look for an example showing that this inclusion is strict.

(e) Investigate if the relations given in items (c) and (d) also hold for causal futures instead of chronological futures.

Exercise 4.4.8. Consider the punctured plane $M = \mathbb{L}^2 \setminus \{(1,1)\}$ with the usual flat metric given by $\mathrm{d}s^2 = \mathrm{d}x^2 - \mathrm{d}y^2$. Show that $\overline{I^+(\mathbf{0})} \neq J^+(\mathbf{0})$ and that $J^+(\mathbf{0})$ is not closed. Deleting even a single event from a spacetime may have disastrous consequences for its causality.

Exercise 4.4.9. Let $(M, \langle\cdot,\cdot\rangle)$ be a spacetime and $F \subseteq M$ be any subset. We'll say that F is a *future-set* if $I^+(F) \subseteq F$.

(a) Show that for every $S \subseteq M$, $I^+(S)$ is a future-set (surprised?).

(b) Suppose that $F \subseteq M$ is open. Show that F is a future-set if and only if $F = I^+(S)$ for some $S \subseteq M$. Give a counter-example when F is not open.

Hint. If F is open, every $x \in F$ has an open neighborhood W contained in F such that every geodesic starting at x is contained in W, for small instants of time. In particular, past-directed timelike geodesics.

Exercise 4.4.10. Let $(M, \langle\cdot,\cdot\rangle)$ be a spacetime and $A \subseteq M$ be any subset. We'll say that A is *achronal* if any two events in A are not chronologically related or, equivalently, if $I^+(A) \cap A = \varnothing$. Show that if $F \subseteq M$ is a future-set, then its boundary ∂F is achronal.

Hint. Show the contrapositive statement by using the definition of ∂F twice, in a "good order":

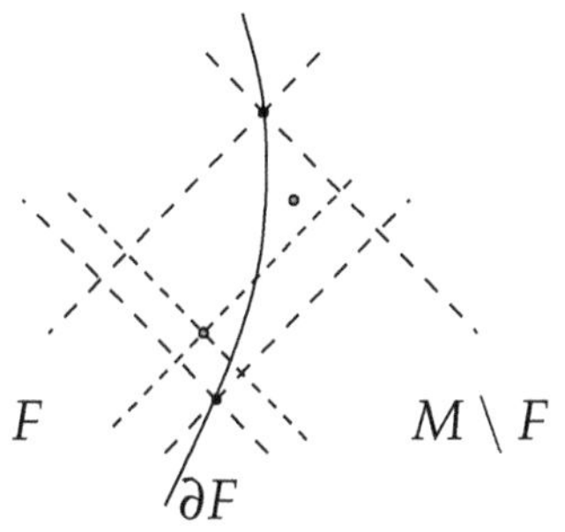

Figure 4.22: The indicated points separated by ∂F show that $I^+(F) \not\subseteq F$.

Exercise 4.4.11. Let $(M, \langle \cdot, \cdot \rangle)$ be a spacetime. Show that the chronological diamonds $\{I^+(p) \cap I^-(q) \mid p, q \in M\}$ indeed form a basis for a topology in M.

Hint. Since chronological futures and pasts are open in M, every event x in the intersection of two chronological diamonds has a neighborhood W still contained in such neighborhood and such that every event in W may be reached by a geodesic starting at x, completely inside W. Consider timelike geodesics, one going to the future and another one going to the past.

Exercise 4.4.12. Let $(M_1, \langle \cdot, \cdot \rangle_1)$ and $(M_2, \langle \cdot, \cdot \rangle_2)$ be two spacetimes. A bijection $\phi \colon M_1 \to M_2$ is called a *chronological isomorphism* if given $p, q \in M_1$, it holds that $p \ll q$ if and only if $\phi(p) \ll \phi(q)$. Show that ϕ is a homeomorphism, when M_1 and M_2 are equipped with their Alexandrov topologies.

Hint. Check that $\phi^{-1}(I^+(\phi(p)) \cap I^-(\phi(q))) = I^+(p) \cap I^-(q)$.

Exercise 4.4.13. Suppose that $(M, \langle \cdot, \cdot \rangle)$ is a Riemannian manifold, and that there is a non-vanishing smooth vector field V on all of M. Show that $\langle\!\langle \cdot, \cdot \rangle\!\rangle$ defined by

$$\langle\!\langle \boldsymbol{v}, \boldsymbol{w} \rangle\!\rangle_p \doteq \langle \boldsymbol{v}, \boldsymbol{w} \rangle_p - 2\frac{\langle \boldsymbol{v}, \boldsymbol{V}(p) \rangle_p \langle \boldsymbol{w}, \boldsymbol{V}(p) \rangle_p}{\langle \boldsymbol{V}(p), \boldsymbol{V}(p) \rangle_p}$$

is a Lorentz metric for which $\boldsymbol{V}(p)$ is timelike, for each $p \in M$.

Remark. Applying this construction to $\mathbb{R}^3$ equipped with the usual metric and taking the field $\boldsymbol{V}(x, y, z) = (0, 0, 1)$ produces $\mathbb{L}^3$, as $\langle \boldsymbol{V}, \cdot \rangle = \mathrm{d}z$.

Exercise 4.4.14. Give a counter-example for the Avez-Seifert Theorem when the spacetime is not globally hyperbolic.

Exercise 4.4.15 (Zeeman Topology)**.** The usual topology in Lorentz-Minkowski space $\mathbb{L}^n$ does not reflect in an adequate way its physical aspects, as it does not make any reference to causality. In this exercise we will see a more adequate topology that heals this deficiency.

We'll say that a subset $Z \subseteq \mathbb{L}^n$ is *Zeeman-open* if for every affine spacelike hyperplane or timelike line Σ in $\mathbb{L}^n$, we have that $Z \cap \Sigma$ is open in Σ.

(a) Show that the collection of Zeeman-open sets indeed forms a topology in $\mathbb{L}^n$, called the *Zeeman topology* if $\mathbb{L}^n$.

(b) Show that every open set is Zeeman-open, but that the converse is not in general true. In particular, the Zeeman topology is Hausdorff.

Hint. Given $p \in \mathbb{L}^n$ and $r > 0$, consider the natural candidate to "open ball": $Z_r(p) \doteq (B_r(p) \setminus C_L(p)) \cup \{p\}$.

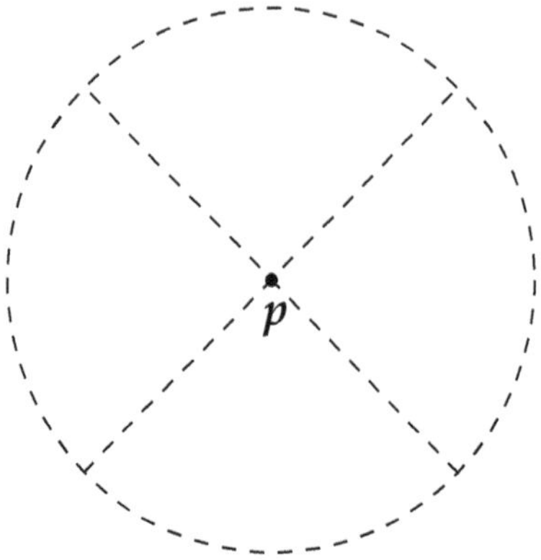

Figure 4.23: Illustrating $Z_r(p)$.

(c) Show that the Zeeman topology induced in lightlike rays is the discrete topology. In particular, conclude that every sequence of points in $\mathbb{L}^n$ which converges along a lightlike ray is eventually constant. This is motivated by the experimental evidence that photons may only be observed during discrete events of emission and absorption.

(d) Show that $(Z_r(p))_{r>0}$ is not a local basis at p of Zeeman-open sets.

Remark.

- Besides those basic facts, it is possible to show that the Zeeman topology is connected and locally connected, while it is not normal, compact, or locally compact. Non-normality, for instance, is an application of the Baire Category Theorem. In particular, the Zeeman topology is not metrizable. See [51] and also Zeeman's original paper, [72].

- The *Alexandrov-Zeeman Theorem* stated in Chapter 1 (Theorem 1.3.16, p. 26) may also be stated in this setting: all the homeomorphisms of $\mathbb{L}^n$ with the Zeeman topology are compositions of translations, positive homotheties and orthochronous Lorentz transformations.

Some Results from Differential Calculus

In this appendix, for completeness purposes, we will list a few results from Differential Calculus, convenient for a better understanding of the text. For proof of the results and more details, we recommend, for example, [43], [50] and [66]. Here, A will always denote an open subset of $\mathbb{R}^n$, $\mathrm{can}_n = (e_1, \ldots, e_n)$ the standard basis of $\mathbb{R}^n$, and $\mathrm{Lin}(\mathbb{R}^n, \mathbb{R}^k)$ the set of linear maps from $\mathbb{R}^n$ to $\mathbb{R}^k$.

Definition A.1. Let $f\colon A \subseteq \mathbb{R}^n \to \mathbb{R}^k$ and $p \in A$. The function f is said to be *differentiable* at p if there is $T \in \mathrm{Lin}(\mathbb{R}^n, \mathbb{R}^k)$ such that

$$\lim_{h \to 0} \frac{f(p+h) - f(p) - Th}{\|h\|} = 0.$$

If such transformation T exists, it is unique. In this case, T is called the *total derivative* of f at p, and it is denoted by $Df(p)$.

We will just say that f is differentiable, if it is differentiable at all points. And if f is bijective and differentiable, with differentiable inverse, f is called a *diffeomorphism.*

Remark. It immediately follows from the definition that

$$Df(p)(e_i) = \frac{\partial f}{\partial x_i}(p).$$

As particular cases of this, if I denotes an open interval in the real line, we have that:

- if $\alpha\colon I \subseteq \mathbb{R} \to \mathbb{R}^n$ is differentiable, then $D\alpha(t)(1) = \alpha'(t)$;
- if $f\colon I \to \mathbb{R}$ is differentiable, $Df(x)$ is a linear map from $\mathbb{R}$ to $\mathbb{R}$, and so it is a scalar multiple of the identity. This scalar is precisely the number $f'(x)$, seen in a first Calculus course;
- if $f\colon A \subseteq \mathbb{R}^n \to \mathbb{R}$ is differentiable, $Df(p)$ is a linear map from $\mathbb{R}^n$ to $\mathbb{R}$ and $Df(p)(v) = \langle \nabla f(p), v \rangle$, where

$$\nabla f(p) \doteq \left(\frac{\partial f}{\partial x_1}(p), \ldots, \frac{\partial f}{\partial x_n}(p) \right)$$

is the *gradient vector* of f at p, seen in a second Calculus course.

Definition A.2. Let $f\colon A \subseteq \mathbb{R}^n \to \mathbb{R}^k$ be differentiable at $\boldsymbol{p} \in A$. The matrix

$$Jf(\boldsymbol{p}) \doteq [Df(\boldsymbol{p})]_{\mathrm{can}_n,\mathrm{can}_k} = \left(\frac{\partial f_i}{\partial x_j}(\boldsymbol{p})\right)_{\substack{1\leq i\leq k\\ 1\leq j\leq n}}$$

is called the *Jacobian matrix of* f *at* $\boldsymbol{p}$, where $f = (f_1, \ldots, f_k)$.

Remark. In general, we'll identify $Jf(\boldsymbol{p})$ with $Df(\boldsymbol{p})$. In the case $n = k = 1$, we have $Jf(\boldsymbol{p}) = (f'(\boldsymbol{p}))$.

Theorem A.3. *If* $f\colon A \subseteq \mathbb{R}^n \to \mathbb{R}^k$ *is differentiable at a point* $\boldsymbol{p} \in A$*, then* f *is continuous at* $\boldsymbol{p}$.

Proposition A.4. *Let* $f\colon A \subseteq \mathbb{R}^n \to \mathbb{R}^k$ *and* $\boldsymbol{p} \in A$*. Then, writing* $f = (f_1, \ldots, f_k)$*, it holds that* f *is differentiable at* $\boldsymbol{p}$ *if and only if each component function* f_i *is. In this case, we have that* $Df(\boldsymbol{p}) = (Df_1(\boldsymbol{p}), \ldots, Df_k(\boldsymbol{p}))$ *or, in matrix terms,*

$$Jf(\boldsymbol{p}) = \begin{pmatrix} \nabla f_1(\boldsymbol{p}) \\ \vdots \\ \nabla f_k(\boldsymbol{p}) \end{pmatrix}.$$

Theorem A.5 (Chain Rule). *Let* $f\colon A_1 \subseteq \mathbb{R}^n \to \mathbb{R}^k$ *and* $g\colon A_2 \subseteq \mathbb{R}^k \to \mathbb{R}^p$ *be two functions, differentiable at* $\boldsymbol{p} \in A_1$ *and at* $f(\boldsymbol{p}) \in A_2$*, respectively, where* A_1 *and* A_2 *are open subsets with* $A_2 \subseteq f(A_1)$*, then* $g \circ f$ *is differentiable at* $\boldsymbol{p}$ *and, moreover, the formula*

$$D(g \circ f)(\boldsymbol{p}) = Dg(f(\boldsymbol{p})) \circ Df(\boldsymbol{p})$$

holds.

Example A.6.

(1) If $f\colon \mathbb{R}^n \to \mathbb{R}^k$ is constant, then f is differentiable and $Df(\boldsymbol{p})$ is the zero map, for every $\boldsymbol{p} \in \mathbb{R}^n$.

(2) Translations. Given $\boldsymbol{a} \in \mathbb{R}^n$, $T_{\boldsymbol{a}}\colon \mathbb{R}^n \to \mathbb{R}^n$ given by $T_{\boldsymbol{a}}(\boldsymbol{x}) = \boldsymbol{x} + \boldsymbol{a}$ is differentiable and $DT_{\boldsymbol{a}}(\boldsymbol{p}) = \mathrm{id}_{\mathbb{R}^n}$, for every $\boldsymbol{p} \in \mathbb{R}^n$.

(3) If $T\colon \mathbb{R}^n \to \mathbb{R}^k$ is linear, then T is differentiable and $DT(\boldsymbol{p}) = T$ for every $\boldsymbol{p} \in \mathbb{R}^n$. In particular, the sum $s\colon \mathbb{R}^n \times \mathbb{R}^n \to \mathbb{R}^n$ is linear when seen as a map $s\colon \mathbb{R}^{2n} \to \mathbb{R}^n$, and thus it is differentiable.

(4) If $f, g\colon A \subseteq \mathbb{R}^n \to \mathbb{R}^k$ are differentiable at $\boldsymbol{p} \in A$, then $f + g$ is also differentiable, and

$$D(f+g)(\boldsymbol{p}) = Df(\boldsymbol{p}) + Dg(\boldsymbol{p}).$$

To wit, $f + g = s \circ (f, g)$ is the composition of differentiable maps.

(5) If $B\colon \mathbb{R}^n \times \mathbb{R}^k \to \mathbb{R}^p$ is bilinear, then B is differentiable and

$$DB(\boldsymbol{x}, \boldsymbol{y})(\boldsymbol{h}, \boldsymbol{k}) = B(\boldsymbol{x}, \boldsymbol{k}) + B(\boldsymbol{h}, \boldsymbol{y}).$$

In particular, the multiplication $m\colon \mathbb{R} \times \mathbb{R} \to \mathbb{R}$ is bilinear, due to the distributive property in $\mathbb{R}$.

(6) If $f, g\colon A \subseteq \mathbb{R}^n \to \mathbb{R}$ are differentiable at $\boldsymbol{p} \in A$, then fg is also differentiable, and

$$D(fg)(\boldsymbol{p}) = g(\boldsymbol{p})Df(\boldsymbol{p}) + f(\boldsymbol{p})Dg(\boldsymbol{p}).$$

Indeed, $fg = m \circ (f, g)$ is the composition of differentiable maps. In particular, we have that if $\lambda \in \mathbb{R}$, then $D(\lambda f)(\boldsymbol{p}) = \lambda Df(\boldsymbol{p})$. Observe that if one of the functions takes values in $\mathbb{R}^k$, the result holds for each component function, and so the product is also differentiable.

(7) If $f\colon A \subseteq \mathbb{R}^n \to \mathbb{R}$ is differentiable at $\boldsymbol{p} \in A$ and $f(\boldsymbol{p}) \neq 0$, then $1/f$ is differentiable at $\boldsymbol{p}$ and we have that

$$D\left(\frac{1}{f}\right)(\boldsymbol{p}) = -\frac{1}{f(\boldsymbol{p})^2}Df(\boldsymbol{p}).$$

If $f\colon A \subseteq \mathbb{R}^n \to \mathbb{R}^k$ is differentiable, then $Df\colon A \to \mathrm{Lin}(\mathbb{R}^n, \mathbb{R}^k)$, and eventually identifying $\mathrm{Lin}(\mathbb{R}^n, \mathbb{R}^k)$ with $\mathbb{R}^{nk}$, it makes sense to wonder whether Df itself is continuous or differentiable. The latter being the case, we write

$$D^2f(\boldsymbol{p}) \doteq D(Df)(\boldsymbol{p})\colon \mathbb{R}^n \to \mathrm{Lin}(\mathbb{R}^n, \mathbb{R}^k).$$

If $\mathrm{Lin}_r(\mathbb{R}^n, \mathbb{R}^k)$ denotes the space of r-linear maps defined in $(\mathbb{R}^n)^r$ taking values in $\mathbb{R}^k$, we identify

$$\mathrm{Lin}_2(\mathbb{R}^n, \mathbb{R}^k) \cong \mathrm{Lin}(\mathbb{R}^n, \mathrm{Lin}(\mathbb{R}^n, \mathbb{R}^k))$$

and see $D^2f(\boldsymbol{p})$ as a bilinear map. A similar reasoning applies for derivaties of higher order, which are then identified with multilinear maps. For $k = 1$, this is used to study maxima and minima of functions.

Definition A.7. We'll say that a differentiable function $f\colon A \subseteq \mathbb{R}^n \to \mathbb{R}^k$ is of *class* $\mathscr{C}^1$ if $Df\colon \mathbb{R}^n \to \mathrm{Lin}(\mathbb{R}^n, \mathbb{R}^k)$ is continuous. In general, for $r > 1$, we'll say that f is of *class* $\mathscr{C}^r$ if Df is of class $\mathscr{C}^{r-1}$.

Theorem A.8 (Clairaut-Schwarz). *Let $f\colon A \subseteq \mathbb{R}^n \to \mathbb{R}$ be a function of class $\mathscr{C}^2$. Then*

$$\frac{\partial^2 f}{\partial x_i \partial x_j} = \frac{\partial^2 f}{\partial x_j \partial x_i}.$$

Remark. In the above result, it actually suffices for one of the mixed partial derivatives to be continuous and the other to exist, so they commute.

Theorem A.9. *Let $f\colon A \subseteq \mathbb{R}^n \to \mathbb{R}$ be a differentiable function. If $\boldsymbol{p} \in A$ is a local maximum or minimum of f, then $Df(\boldsymbol{p}) = 0$.*

Theorem A.10. *Let $f\colon A \subseteq \mathbb{R}^n \to \mathbb{R}$ be a function of class $\mathscr{C}^2$. If $\boldsymbol{p} \in A$ is a* critical point *of f, that is, with $Df(\boldsymbol{p}) = 0$, then we have that:*

(i) if $D^2f(\boldsymbol{p})$ is positive-definite, $\boldsymbol{p}$ is a local minimum for f;

(ii) if $D^2f(\boldsymbol{p})$ is negative-definite, $\boldsymbol{p}$ is a local maximum for f.

Remark. If f is of class $\mathscr{C}^2$, each $D^2f(\boldsymbol{p})$ is a symmetric bilinear form, so that results such as Sylvester's Criterion may be used to decide whether $D^2f(\boldsymbol{p})$ is positive-definite or not.

The above result is a consequence of the following one:

Theorem A.11 (Taylor Formula). *Let $f\colon A \subseteq \mathbb{R}^n \to \mathbb{R}^m$ be a function r times differentiable, and $\boldsymbol{p} \in A$ such that $D^{(r+1)}f(\boldsymbol{p})$ exists. So, if $\boldsymbol{h} \in \mathbb{R}^n$ is such that $\boldsymbol{p}+\boldsymbol{h} \in A$, we have that*

$$f(\boldsymbol{p}+\boldsymbol{h}) = \sum_{k=0}^{r} \frac{1}{k!} D^{(k)} f(\boldsymbol{p})(\boldsymbol{h}^{(k)}) + \boldsymbol{r}(\boldsymbol{h}),$$

where the remainder satisfies $\boldsymbol{r}(\boldsymbol{h})/\|\boldsymbol{h}\|^{r+1} \to \boldsymbol{0}$ if $\boldsymbol{h} \to \boldsymbol{0}$ and, for each k, $\boldsymbol{h}^{(k)}$ denotes the vector $(\boldsymbol{h},\ldots,\boldsymbol{h}) \in (\mathbb{R}^n)^k$.

Theorem A.12 (Inverse Function Theorem). *Let $f\colon A \subseteq \mathbb{R}^n \to \mathbb{R}^n$ be of class $\mathscr{C}^k$, with $k \geq 1$. If $\boldsymbol{p} \in A$ is such that $Df(\boldsymbol{p})$ is non-singular, there are open subsets $U \subseteq A$ and $V \subseteq \mathbb{R}^n$ containing $\boldsymbol{p}$ and $f(\boldsymbol{p})$, respectively, such that the restriction $f\colon U \to V$ is bijective, and the inverse $f^{-1}\colon V \to U$ is also of class $\mathscr{C}^k$. In this case, we have that*

$$D(f^{-1})(f(\boldsymbol{x})) = Df(\boldsymbol{x})^{-1}$$

for each $\boldsymbol{x} \in U$.

Every partition of the standard basis in $\mathbb{R}^m$ determines a decomposition of the form $\mathbb{R}^{n_1} \times \cdots \times \mathbb{R}^{n_r}$, with $m = n_1 + \cdots + n_r$, and each such decomposition induces "fat partial derivatives". For example, if $f\colon \mathbb{R}^n \times \mathbb{R}^k \to \mathbb{R}^p$ is differentiable at a point $(\boldsymbol{x}_0, \boldsymbol{y}_0)$, we have two linear maps $D_{\boldsymbol{x}}f(\boldsymbol{x}_0,\boldsymbol{y}_0)\colon \mathbb{R}^n \to \mathbb{R}^p$ and $D_{\boldsymbol{y}}f(\boldsymbol{x}_0,\boldsymbol{y}_0)\colon \mathbb{R}^k \to \mathbb{R}^p$ given by

$$D_{\boldsymbol{x}}f(\boldsymbol{x}_0,\boldsymbol{y}_0)(\boldsymbol{h}) \doteq Df(\boldsymbol{x}_0,\boldsymbol{y}_0)(\boldsymbol{h},\boldsymbol{0}) \quad \text{and} \quad D_{\boldsymbol{y}}f(\boldsymbol{x}_0,\boldsymbol{y}_0)(\boldsymbol{k}) \doteq Df(\boldsymbol{x}_0,\boldsymbol{y}_0)(\boldsymbol{0},\boldsymbol{k}).$$

Note that

$$Df(\boldsymbol{x}_0,\boldsymbol{y}_0)(\boldsymbol{h},\boldsymbol{k}) = D_{\boldsymbol{x}}f(\boldsymbol{x}_0,\boldsymbol{y}_0)(\boldsymbol{h}) + D_{\boldsymbol{y}}f(\boldsymbol{x}_0,\boldsymbol{y}_0)(\boldsymbol{k}).$$

With this, we may state the:

Theorem A.13 (Implicit Function Theorem). *Consider a function*

$$f\colon A \times B \subseteq \mathbb{R}^n \times \mathbb{R}^k \to \mathbb{R}^k$$

of class $\mathscr{C}^r$, with $r \geq 1$. Suppose that $(\boldsymbol{x}_0,\boldsymbol{y}_0) \in A \times B$ is such that $f(\boldsymbol{x}_0,\boldsymbol{y}_0) = \boldsymbol{c}$. If $D_{\boldsymbol{y}}f(\boldsymbol{x}_0,\boldsymbol{y}_0)$ is non-singular, there are open subsets $U \subseteq A$ and $V \subseteq B$ containing $\boldsymbol{x}_0$ and $\boldsymbol{y}_0$, and a function $\varphi\colon U \to V$ of class $\mathscr{C}^r$ such that

$$f(\boldsymbol{x},\varphi(\boldsymbol{x})) = \boldsymbol{c}$$

and $f(\boldsymbol{x},\boldsymbol{y}) = \boldsymbol{c}$ implies $\boldsymbol{y} = \varphi(\boldsymbol{x})$, for each $\boldsymbol{x} \in U$.

Remark. The equation $f(\boldsymbol{x},\varphi(\boldsymbol{x})) = \boldsymbol{c}$ allows us to effectively compute the derivative of the implicit function φ. We have that

$$D_{\boldsymbol{x}}f(\boldsymbol{x},\varphi(\boldsymbol{x})) + D_{\boldsymbol{y}}f(\boldsymbol{x},\varphi(\boldsymbol{x})) \circ D\varphi(\boldsymbol{x}) = 0,$$

and so $D\varphi(\boldsymbol{x}) = -D_{\boldsymbol{y}}f(\boldsymbol{x},\varphi(\boldsymbol{x}))^{-1} \circ D_{\boldsymbol{x}}f(\boldsymbol{x},\varphi(\boldsymbol{x}))$.

Exercises

Exercise A.1. Consider $f\colon \mathbb{R}^2 \to \mathbb{R}^2$ given by $f(x,y) = (e^x \cos y, e^x \sin y)$, and the linear map $T \doteq Df(3,\pi/6)$. Find the angle between $T^{2017}(1,0)$ and $T^{2018}(1,1)$.

Exercise A.2. Compute the total derivative of F in the following cases:

(a) $F(\boldsymbol{x},\boldsymbol{y}) = (\boldsymbol{x}, f(\boldsymbol{y}))$, where f is differentiable;

(b) $F(\boldsymbol{x}) = \langle \boldsymbol{x}, \boldsymbol{x}_0 \rangle$, where $\boldsymbol{x}_0$ is fixed;

(c) $F(\boldsymbol{x}) = \langle \boldsymbol{x}, T\boldsymbol{x} \rangle$, where T is a fixed linear operator;

(d) $F(\boldsymbol{x},\boldsymbol{y}) = f(\boldsymbol{x}) + g(\boldsymbol{y})$, where f and g are differentiable;

(e) $F(\boldsymbol{x},\boldsymbol{y}) = A(\boldsymbol{x})\boldsymbol{y}$, where $A\colon \mathbb{R}^n \to \mathrm{Lin}(\mathbb{R}^n, \mathbb{R}^k)$ is differentiable and $\boldsymbol{y} \in \mathbb{R}^k$.

Exercise A.3. Let $f\colon A \subseteq \mathbb{R}^n \to \mathbb{R}^k$ be a differentiable function and define two maps $\varphi\colon A \to \mathbb{R}^n \times \mathbb{R}^k$ and $F\colon A \times \mathbb{R}^k \to \mathbb{R}^k$, respectively, by $\varphi(\boldsymbol{x}) = (\boldsymbol{x}, f(\boldsymbol{x}))$ and $F(\boldsymbol{x},\boldsymbol{y}) = \boldsymbol{y} - f(\boldsymbol{x})$. Show that φ and F are both differentiable, and that for all $\boldsymbol{x}_0 \in A$ and $\boldsymbol{y}_0 \in \mathbb{R}^k$ we have that $\ker DF(\boldsymbol{x}_0, \boldsymbol{y}_0) = \mathrm{Im}\, D\varphi(\boldsymbol{x}_0)$.

Exercise A.4. Let $B\colon \mathbb{R}^n \times \mathbb{R}^k \to \mathbb{R}^p$ be a bilinear map. Show that there is $C > 0$ such that $\|B(\boldsymbol{x},\boldsymbol{y})\| \leq C\|\boldsymbol{x}\|\|\boldsymbol{y}\|$ for all $\boldsymbol{x} \in \mathbb{R}^n$ and $\boldsymbol{y} \in \mathbb{R}^k$. Use this to show that B is differentiable at all points, and the formula

$$DB(\boldsymbol{x},\boldsymbol{y})(\boldsymbol{h},\boldsymbol{k}) = B(\boldsymbol{x},\boldsymbol{k}) + B(\boldsymbol{h},\boldsymbol{y})$$

holds. State analogous results for multilinear maps $T\colon \mathbb{R}^{n_1} \times \cdots \times \mathbb{R}^{n_k} \to \mathbb{R}^p$.

Exercise A.5. Show that $\Phi\colon \mathrm{Lin}_2(\mathbb{R}^n, \mathbb{R}^k) \to \mathrm{Lin}(\mathbb{R}^n, \mathrm{Lin}(\mathbb{R}^n, \mathbb{R}^k))$ given by $\Phi(B)(\boldsymbol{x})(\boldsymbol{y}) = B(\boldsymbol{x},\boldsymbol{y})$ is an isomorphism of vector spaces.

Exercise A.6. Let $p_0(x) = a_0x^3 - b_0x^2 + c_0x - d_0$, with $a_0 \neq 0$, be a polynomial with real coefficients and three distinct real roots. Show that every polynomial of the form $p(x) = ax^3 - bx^2 + cx - d$ with coefficients (a,b,c,d) sufficiently close to (a_0,b_0,c_0,d_0) also has three distinct real roots, which depend smoothly on the coefficients of $p(x)$.

Hint. Vieta's Relations and Inverse Function Theorem.

Exercise A.7. Let $f\colon \mathbb{R}^n \to \mathbb{R}^n$ be differentiable, with $f(\boldsymbol{0}) = \boldsymbol{0}$. If $Df(\boldsymbol{0})$ does not have 1 as an eigenvalue, then $f(\boldsymbol{x}) \neq \boldsymbol{x}$ for $\boldsymbol{x}$ sufficiently close to $\boldsymbol{0}$, but not equal to $\boldsymbol{0}$.

Exercise A.8 (Local form of Immersions)**.** We'll say that a function $f\colon A \subseteq \mathbb{R}^n \to \mathbb{R}^{n+k}$ is an *immersion* if $Df(\boldsymbol{p})$ is injective for each $\boldsymbol{p} \in A$. Show that if f is an immersion of class $\mathscr{C}^1$, there are open subsets $U \times V$ and W of $\mathbb{R}^{n+k}$ containing $(\boldsymbol{p},\boldsymbol{0})$ and $f(\boldsymbol{p})$, and a diffeomorphism $\psi\colon W \to U \times V$ such that $\psi \circ f(\boldsymbol{x}) = (\boldsymbol{x},\boldsymbol{0})$, for each $\boldsymbol{x} \in U$.

Remark. That is, up to a diffeomorphism, every immersion is an inclusion.

Bibliography

[1] Alexandrov, D., A contribution to chronogeometry, *Canadian J. Math.* 19, pp. 1119–1128, 1967.

[2] Alias, L.; Chaves, R. M. B.; Mira, P.; Bjorling problem for maximal surfaces in Lorentz-Minkowski space, *Math. Proc. Camb. Phil. Soc.* (no. 134, pp. 289–316), 2003.

[3] Anciaux, H., Minimal Submanifolds in Pseudo-Riemannian Geometry, *World Scientific*, 2011.

[4] Antonuccio, F., Semi-Complex Analysis & Mathematical Physics (Corrected Version), *eprint arXiv:gr-qc/9311032*, 1993, `https://arxiv.org/pdf/gr-qc/9311032.pdf`.

[5] Araújo, P. V., Geometria Diferencial, *IMPA* (Coleção Matemática Universitária), 2012.

[6] Bär, C., Elementary Differential Geometry, *Cambridge University Press*, 2010.

[7] Beem, J. K.; Ehrlich, P. E.; Easley, K. L., Global Lorentzian Geometry, *CRC Press*, 1996.

[8] Calioli, C. A.; Domingues, H. H.; Costa, R. C. F., Álgebra Linear e Aplicações, *Atual*, 1995.

[9] Cannon, J. W.; Floyd W. J.; Kenyon, R.; Parry, W. R., Hyperbolic Geometry, *MSRI Publications, Flavors of Geometry* 31, 1997.

[10] Carroll, S., Spacetime and Geometry, An Introduction to General Relativity, *Pearson Education*, 2004.

[11] Catoni et al., Geometry of Minkowski Spacetime, *Springer-Verlag (Springer Briefs in Physics)*, 2011.

[12] Chaves, R. M. B.; Dussan, M. P.; Magid, M.; Bjorling Problem for timelike surfaces in the Lorentz-Minkowski space, *Journal of Mathematical Analysis and Applciations* (377, no. 2, pp; 481–494), 2011.

[13] Chen, B. Y.; Pseudo-Riemannian Geometry, δ-invariants and Applications, *World Scientific*, 2011.

[14] Chern, S. S., Curves and Surfaces in Euclidean Spaces, *Studies in Global Analysis, MAA Studies in Mathematics, The Mathematical Association of America*, 1967.

[15] Dieudonné, J., La Géométrie des Groupes Classiques, *Springer-Verlag*, 1971.

[16] do Carmo, M. P., Geometria Riemanniana, *IMPA (Coleção Projeto Euclides)*, 2005.

[17] do Carmo, M. P., Geometria Diferencial de Curvas e Superfícies, *SBM (Coleção Textos Universitários)*, 2014.

[18] Dussan, M. P.; Franco Filho, A. P.; Magid, M.; The Bjorling problem for timelike minimal surfaces in $\mathbb{R}^4_1$, *Annali di Matematica Pura ed Applicata* (pp. 1–19), 2016.

[19] Dussan, M. P.; Magid, M.; Bjorling problem for timelike surfaces in $\mathbb{R}^4_2$, *Journal of Geometry and Physics* (no. 73, pp. 187–199), 2013.

[20] Faber, R., Differential Geometry and Relativity Theory, *CRC Press*, 1983.

[21] Fang, Y., Lectures on Minimal Surfaces in $\mathbb{R}^3$, *Proceedings of the Centre for Mathematics and Its Applications, Australian National University* 35, 1996.

[22] Foldes, S., Hyperboloid preservation implies the Lorentz and Poincaré groups without dilations, *eprint arXiv:1009:3910*, `https://arxiv.org/pdf/1009.3910.pdf`, 2010.

[23] Fraleigh, J. B., A First Course in Abstract Algebra, *Addison-Wesley*, 1994.

[24] Gelfand, I. M.; Fomin, S. V., Calculus of Variations, *Prentice-Hall*, 1963.

[25] Goldman, R., Curvature formulas for implicit curves and surfaces, *Computer Aided Geometric Design - Special issue: Geometric modelling and differential geometry* 22, pp. 632–658, 2005.

[26] Goldman, W. M.; Margulis, G. A., Flat Lorentz 3-manifolds and cocompact Fuchsian groups. Crystallographic groups and their generalizations (Kortrijk, 1999) *Contemporary Mathematics* 262, pp. 135–145, 2000.

[27] Gray, A.; Abbena, E.; Salamon, S., Modern Differential Geometry of Curves and Surfaces with Mathematica, *Chapman and Hall*, 2016.

[28] Hall, B. C., An Elementary Introduction to Groups and Representations, *eprint arXiv:math-ph/0005032*, `https://arxiv.org/pdf/math-ph/0005032`, 2000.

[29] Hawking, S.; Ellis G., The Large Scale Structure of Spacetime, *Cambridge Monographs on Mathematical Physics*, 1973.

[30] Hitzer, E., Non-constant bounded holomorphic functions of hyperbolic numbers — Candidates for hyperbolic activation functions, *Proceedings of the First SICE Symposium on Computational Intelligence*, pp. 23–28, 2011.

[31] Hoffman, K.; Kunze, R., Linear Algebra, *Prentice-Hall*, 1971.

[32] Horn, R. A.; Johnson, C. R., Matrix Analysis, *Cambridge University Press*, 2013.

[33] Jaffe, A., Lorentz Transformations, Rotations and Boosts, *Lecture Notes*, 2013.

[34] Javaloyes, M. A.; Sánchez, M., An Introduction to Lorentzian Geometry and its Applications, *XVI Escola de Geometria Diferencial*, São Paulo, 2010.

[35] Khrennikov A.; Segre G., An Introduction to Hyperbolic Analysis, *eprint arXiv:math-ph/0507053*, `https://arxiv.org/pdf/math-ph/0507053.pdf`, 2005.

[36] Kobayashi, O., Maximal Surfaces in the 3-Dimensional Minkowski Space $\mathbb{L}^3$, *Tokyo J. of Math. Volume* 06, pp. 297–309, 1983.

[37] Konderak, J.; A Weierstrass Representation Theorem for Lorentz Surfaces, *Complex Variables* 50 (no. 5, pp. 319–332), 2005.

[38] Kosheleva, O.; Kreinovich, V., Observable Causality Implies Lorentz Group: Alexandrov-Zeeman-Type Theorem for Space-Time Regions, *Mathematical Structures and Modeling* 30, pp. 4–14, 2014.

[39] Kühnel, W., Differential Geometry: Curves – Surfaces – Manifolds, *AMS*, 2006.

[40] Lang, S., Complex Analysis, 4th. ed., *Springer-Verlag (Graduate Texts in Mathematics 103)*, 1999.

[41] Lee, J. M., Riemannian Manifolds: An Introduction to Curvature, *Springer-Verlag (Graduate Texts in Mathematics 176)*, 1997.

[42] Lima, E. L., Grupo Fundamental e Espaços de Recobrimento, *IMPA (Coleção Projeto Euclides)*, 2012.

[43] Lima, E. L., Análise no Espaço $\mathbb{R}^n$, *IMPA (Coleção Matemática Universitária)*, 2013.

[44] López, R., Differential Geometry of Curves and Surfaces in Lorentz-Minkowski space, *eprint arXiv:0810.3351*, `https://arxiv.org/pdf/0810.3351`, 2008.

[45] Lyusternik, L. A.; Fet, A. I., Variational problems on closed manifolds, *Dokl. Akad. Nauk. SSSR 81* (pp. 17–18), 1951.

[46] Magid, M.; Minimal Timelike Surfaces via the Split-Complex Numbers, *Proceedings of PADGE 2012, Shaker-Verlag, Aachen*, 2013.

[47] Matsuda, H., A note on an isometric imbedding of upper half-space into the anti de Sitter space, *Hokkaido Math. J. 13*, pp. 123–132, 1984.

[48] Mian, J., Fermi-Walker holonomy, second-order vector bundles, and EPR correlations. *Honours Thesis, National University of Singapore*, 2015.

[49] Müller, O.; Sánchez, M., Lorentzian Manifolds Isometrically Embeddable in $\mathbb{L}^N$, *Trans. Amer. Math. Soc.* 363, pp. 5367–5379, 2011.

[50] Munkres, J. R., Analysis on Manifolds, *Addison-Wesley*, 1991.

[51] Naber G. L., Spacetime and Singularities, An Introduction, *Cambridge University Press*, 1988.

[52] Naber, G. L., The Geometry of Minkowski Spacetime: An Introduction to the Mathematics of the Special Theory of Relativity, *Springer-Verlag (Applied Mathematical Sciences 92)*, 1992.

[53] Nomizu, K., The Lorentz–Poincaré metric on upper half-space and its extension, *Hokkaido Math. J. 11*, pp. 253–261, 1982.

[54] O'Neill, B., Semi–Riemannian Geometry with Applications to Relativity, *Academic Press*, 1983.

[55] O'Neill, B., Elementary Differential Geometry, *Academic Press*, 2006.

[56] Oprea, J., Differential Geometry and Its Applications, *Prentice-Hall*, 1997.

[57] Osserman, R., A Survey of Minimal Surfaces, *Dover*, 1986.

[58] Penrose, R., Techniques of Differential Topology in Relativity, *AMS Colloquium Publications (SIAM, Philadelphia)*, 1972.

[59] Petersen, P., Classical Differential Geometry, *eprint* **https://www.math.ucla.edu/~petersen/DGnotes.pdf**, 2012.

[60] Ratcliffe, J., Foundations of Hyperbolic Manifolds, *Springer-Verlag (Graduate Texts in Mathematics 149)*, 1994.

[61] Remmert, R., Theory of Complex Functions, *Springer-Verlag (Readings in Mathematics 122)*, 1991.

[62] Ryan, P. J., Euclidean and Non-Euclidean Geometry: An Analytic Approach, *Cambridge University Press*, 1986.

[63] Saloom, A., Curves in the Minkowski plane and Lorentzian surface, *Durham theses, Durham University*, 2012.

[64] Shilov, G. E., Linear Algebra, *Dover*, 1977.

[65] Simmonds, J. G., A Brief on Tensor Analysis, *Springer-Verlag (Undergraduate Texts in Mathematics)*, 1982.

[66] Spivak, M., Calculus on Manifolds, *Addison-Wesley*, 1965.

[67] Sternberg, S., Curvature in Mathematics and Physics, *Dover Publications, Inc.*, 2012.

[68] Stoker, J. J., Differential Geometry, *John Wiley & Sons, Inc.*, 1969.

[69] Tenenblat, K. Introdução à Geometria Diferencial, *Edgard Blücher*, 2008.

[70] Van Brunt, B., The Calculus of Variations, *Springer-Verlag (Universitext)*, 2004.

[71] Walrave, J., Curves and Surfaces in Minkowski Space, *Doctoral Thesis, K.U. Leuven, Fac. of Science, Leuven*, 1995.

[72] Zeeman, E. C.; The Topology of Minkowski Space, *Topology* (vol 6., pp. 161–170), *Pergamon Press*, 1967.

Index

Achronal set, 322
Admissible variation (of a parametrized surface), 200
Alexandrov topology, 319
Anti-de Sitter space, 278
Antipodal map, 155
Arc-photon, 71
Archimedes Spiral, 95
Arclength, 67, 160, 263, 315
Area
    functional, 200
    of a geometric surface, 263
    of a surface, 161
Artin space, 280
Asymptotic
    boundary, 275
    line, 211
    parametrization, 212
    vector, 207
Atlas, 260

Bernoulli equation, 203
Bilinear form
    indefinite, 5
    negative-definite, 5
    non-degenerate, 5
    positive-definite, 5
    symmetric, 5
    trace, 57
Binormal
    indicatrix, 104
    vector, 98, 118
Bonnet rotation, 299
Brioschi formula, 285
B-scroll (associated to a lightlike curve), 193

Cardioid, 93
Cartan
    curvature, *see also* Pseudo-torsion
    frame, 119
Catalan surface in $\mathbb{R}^3$, 309
Catenoid, 173, 301
    spacelike in $\mathbb{L}^3$, 309
    timelike in $\mathbb{L}^3$, 313
Cauchy-Riemann
    equations (revised), 177, 288
    equations, 177, 279
Causal
    automorphism, 25
    diamond, 319
    future, 20, 318
    hierarchy of a spacetime, 319
    type, 3, 262, 315
Cayley transform, 266
Center of curvature, 84
Central projection, 153
Chain rule, 146, 288, 326
Change of parameters, *see also* reparametrization
Christoffel Symbols, 228
Christoffel symbols, 267
Chronological
    diamond, 319
    future, 20, 264, 318
    isomorphism, 323
Clairaut
    parametrization, 240
    relation, 242
Clifford torus, 270
Codazzi-Mainardi equations, 251
Coindicator (of a curve), 98
Compatibility equations, *see also* Codazzi-Mainardi equations
Complex
    derivative, 304
    Lorentzian space, 304
Cone
    light, 4, 20
    time, 20
Conformal map (between surfaces), 177
Congruence of surfaces, 170
Coordinate curves, 132

Cotangent plane, 160
Covariant derivative, 216, 219
Critical
  curve, 306
  surface, 201
  value, 135
Cross product, 53
Curvature
  (of a space curve), 98
  Gaussian, 181, 269
  Mean, 181
  Mean (vector), 181
  principal, 179
Curve
  admissible, 98
  biregular, 98
  closed, 67
  congruence, 72
  divergent, 317
  energy of a, 235
  lightlike, 64
  Logarithmic spiral, 74
  Loxodromic, 74
  parametrized, 64
  regular, 66
  semi-lightlike, 118
  spacelike, 64
  timelike, 64

d'Alembertian, 248, 298
Darboux vector, 104
Darboux-Ribaucour frame, 222
de Sitter space, 80, 108
Diffeomorphism
  between open sets in $\mathbb{R}^n$, 325
  between surfaces, 142
  group (of a surface), 152
Differentiable manifold, 260
Differential, 145, 154
Dirichlet energy, 248
Distance function, 315
Dual numbers, 296
Dupin Indicatrix, 218

Elliptic
  helix, 110
  linear map, 44
  point, 197
Energy
  kinetic, 76, 234
  of a curve, 67, 160, 263
  potential, 76
Energy of a curve, 315
Energy-momentum map, 42
Enneper surface
  spacelike in $\mathbb{L}^3$, 309
  in $\mathbb{R}^3$, 308
  timelike in $\mathbb{L}^3$, 312
Enneper-Weierstrass representation formula, 307, 308, 312
Euler formulas, 208
Euler-Lagrange Equations, 232
Evolute of a curve, 85

Fermi chart, 281
Fermi-Walker Parallelism, 245
First Fundamental Form, 158
First variation (of a functional), 232
Flip operator, 92, 96, 166, 174, 177
Folium of Descartes, 92
Frenet-Serret frame, 78, 98
Future-directed, 20, 264, 317
Future-set, 322

Gauss
  equations, 251
  normal map, 178
Geodesic, 216, 220, 267, 281, 315
  completeness, 316
  curvature, 223
  polygonal, 322
  torsion, 223
Geometric surface, 262
Girard's Formula, 167
Gradient of a smooth function defined in a surface, 165
Gram matrix, 6
Gram-Schmidt process, 13
Group
  Euclidean, 34
  Holonomy, 244
  Klein, 42
  Lorentz, 33
  Lorentz special, 35
  orthogonal, 33
  Poincaré, 34
  pseudo-orthogonal, 33
  pseudo-orthogonal special, 35
  special orthochronous Lorentz, 38
Gödel's spacetime, 320

Hadamard product, 290
Hamilton's Principle, 235
Hamiltonian, 247
Harmonic function, 279
Helicoid, 133, 164, 301
  hyperbolic, 164
Hemisphere, 274
Henneberg surface (in $\mathbb{R}^3$), 310
Hessian, 92
Holomorphic function, 279
Horocycle, 153
Householder reflection, 41
Hyperbolic
  angle, 23, 160
  helix, 110
  isometry, 321
  linear map, 44
  numbers, *see also* Split-complex numbers
  plane, 108
  point, 197

Identity
  Jacobi, 59
  Lagrange, 56
  polarization, 34
Immersion, 329
Index (of an inner product), 20
Indicator
  of a curve, 64
  of a vector, 3, 262
Inequality
  reverse Cauchy-Schwarz, 23
  reverse triangular, 24
Inertial parametrization, 195, 257
Integral along a curve, 292
Intrinsic distance, 316
Isometry
  between surfaces, 169, 264
  Euclidean, 29
  group (of a surface), 174
  pseudo-Euclidean, 30
Isothermal parametrization, 256, 298

Klein
  bottle, 151
  disk, 273
Kronecker's Delta, 2

Lagrangian, 232
Laguerre transformation, 104
Lambert cylindrical projection, 143, 172
Lambert's Formula, 168
Laplace-Beltrami operator, 257
Laplacian, 248, 298
Lightlike
  subspace, 5
  vector, 3
Linear and affine parts, 34
Lines of curvature, 209
Lipschitz-continuous function, 322
Local canonical form
  of a curve in space, 107
  of a lightlike curve, 124
  of a plane curve, 84
Local coordinates, *see also* regular parametrization
Local form of immersions, 329
Lorentz
  boost, 48
  inner product, 2
  factor, 28
  manifold, 315
  surface, 262
Lorentz-harmonic function, 290
Lorentz-Minkowski space, 2
Lorentz-Poincaré half-plane, 277
Lorentzian
  manifold, 262
  metric, 158, 262

Margulis Invariant, 52
Maximal surface, 201
Mercator's projection, 314
Metric determinant (of a bilinear map), 57
Minimal surface, 201
Minkowski metric, *see also* Lorentz inner product
Monge parametrizations, 133
Möbius
  strip, 150
  transformation, 272

Neil's parabola, 76
Newton's Second Law, 235
Noether charge, 247
Normal
  curvature, 207
  plane, 107
  section of a surface, 207

vector, 82, 98, 118

Orientable Surface, 148
Oriented curvature (of a plane curve), 79
Orthochronous, 25, 35, 38
Orthogonal projection, 18
Osculating plane, 107
Outer semi-direct product for groups, 40

Parabolic
helix, 110
linear map, 44
point, 197
Parallel
field (along a curve), 219
surfaces, 189
translation, 244
Parametrization
compatible with a given orientation, 148
abstract, 260
Partial
differential, 155
indicator (of a parametrization), 156, 298
Past-directed, *see also* future-directed
Pauli Matrix, 51
Penrose basis, 17
Planar point, 197
Poincaré
disk, 265
half-plane, 266
Polar
coordinates, 92
parametrization, 218
Pole, 292
Positive basis (for a lightlike plane), 118
Pre-geodesic, 221
Principal
parametrization, 211
vector, 179
Proper
function, 322
time, 67, 315
Pseudo-
Euclidean norm, 3
Euclidean space of index ν, 2
orthogonal, 5
orthonormal, 5
Riemannian manifold, 315
Riemannian metric, 158, 262, 315
sphere, 272
torsion, 120

Radius of curvature, 84
Rectifying plane, 107
Regular
parametrization, 131
parametrized surface, 131
point, 135
surfaces, 130
value, 135
Reissner-Nordström space, 277
Relativistic addition of speeds, 52
Reparametrization, 69
Riemann
formula, 285
surface, 262
Riemannian
manifold, 262, 315
metric, 158, 262
Rigid motion, *see also* Isometry, Euclidean
Rindler
coordinates, 94
parametrization, 218
Rotation (pure), 47

Schwarzschild
half-plane, 276
horizon function, 276
space, 276
Second Fundamental Form, 180
Shape operator, *see also* Weingarten map
Sherlock hat, 205
Singular point, 66
Slant of a curve (in a surface), 243
Smooth function (defined on a surface), 141
Spacelike
subspace, 5
vector, 3
Spacetime, 317
Spatial and temporal parts, 36
Split-complex
derivative, 304
numbers, 95, 177, 287
Split-holomorphic function, 280, 288

Split-meromorphic function, 292
Stationary heat operator, *see also* Laplacian
Stationary wave operator, *see also* d'Alembertian
Stereographic Projection, 152
Structure constants, 295
Surface
  abstract, 260
  lightlike, 155
  of hyperbolic revolution, 166, 188, 249
  of revolution, 134
  spacelike, 155
  timelike, 155
Sylvester's Criterion, 8

Tangent
  bundle, 315
  indicatrix, 104, 112
  plane (to a surface), 261
  plane (to a parametrization), 140
  plane (to a surface), 139
  space, 261
  surface, 164, 176
  vector, 82, 98, 118
  vector (to a surface), 139
Tensor product (of curves), 270
Theorem
  Alexandrov-Zeeman, 26, 324
  Beltrami-Enneper, 218
  Bonnet, 253
  Calabi-Bernstein, 202
  Cartan–Dieudonné, 41
  Cauchy-Goursat (revised), 293
  Clairaut-Schwarz, 327
  Egregium, 186, 252
  Four Singularities, 90
  Four Vertex, 91
  Frobenius, 250
  Fundamental of Calculus (split-complex version), 292
  Fundamental of curves (lightlike/semi-lightlike case), 125
  Fundamental of Plane Curves, 88
  Fundamental para Curvas Admissíveis, 112
  Gauss-Bonnet, 168
  Green-Stokes, 201, 234
  Hopf-Rinow, 316
  Implicit Function (for products of surfaces), 155
  Implicit Function (in $\mathbb{R}^n$), 328
  Inverse Function (for products of surfaces), 154
  Inverse Function (for surfaces), 146
  Inverse Function (in $\mathbb{R}^n$), 328
  Korn-Lichtenstein, 298
  Lancret, 110
  Lancret (lightlike version), 125
  Liouville, 297
  Lyusternik-Fet, 237
  Meusnier, 208
  Noether, 247
  Riemann's classification (for surfaces with constant Gaussian curvature), 280
Third Fundamental Form, 187
Time
  orientability, 317
  orientation, 264
  separation, 320
Timelike
  subspace, 5
  vector, 3
Torsion, 99
Total derivative, 325
Totally umbilic surface, 190
Transformation
  isogonal, 42
  Lorentz, 25, 30
  Poincaré, *see also* Isometry, Lorentzian
  pseudo-orthogonal, 30
Translation surfaces, 205
Twins Paradox, 24

Umbilic point, 190
Unit normal field, 149

Variational derivative, *see also* First variation (of a functional)
Vector
  indicator, *see also* indicator of a vector
  field, 315
  field (along a curve), 219
  spatial velocity, 19
  velocity (of a curve), 64, 261

Viviani's window, 101

Weingarten Map, 179
Wirtinger operators
split-complex version, 289
complex version, 304

Zeeman Topology, 323

Milton Keynes UK
Ingram Content Group UK Ltd.
UKHW050441111024
449327UK00045B/2542